MICROWAVE and RF PRODUCT APPLICATIONS

Editor-in-Chief
MIKE GOLIO

CRC PRESS

Boca Raton London New York Washington, D.C.

Reprint, 2016

Library of Congress Cataloging-in-Publication Data

Catalog record is available from the Library of Congress.

The material included here first appeared in *The RF and Microwave Handbook* (CRC Press, 2001), Mike Golio, editor.

ISBN 978-1-4987-7207-5

Digitally Printed at Replika Press Pvt. Ltd

For sale in India, Pakistan, Nepal, Bhutan, Bangladesh and Sri Lanka only.

Preface

Microwave and RF Product Applications is a single-volume, comprehensive reference for high-frequency commercial, military, and medical applications. The introduction of the book defines the RF and microwave electromagnetic energy spectrum and examines key characteristics of that energy that are exploited for modern RF communications, sensor, and heating product applications. Individual chapters then examine cellular and mobile communications, broadband wireless access, wireless LANs and PANs, satellite communications, electronic navigation, avionics, radar, and therapeutic medical applications. Additional chapters describe RF and microwave analysis and simulation techniques and provide descriptions of the fundamental physical phenomena that govern electromagnetic applications. Written by leading researchers in the field, *Microwave and RF Product Applications* provides important information for engineers working with wireless RF or microwave applications. It also serves as an excellent source for those requiring information outside their area of expertise, such as managers, marketers, and technical support workers who need a better understanding of the fields driving their decisions.

Acknowledgments

Developing a book like this one is a big job — much bigger than I originally anticipated. This would simply never have been completed if it were not for the efforts of the managing editor, Janet Golio. Her focus, persistence, software expertise, and organizational skills were essential to the project. Her efforts can be measured in the nearly 10,000 pieces of e-mail correspondence she handled, the 100+ telephone conversations she initiated, or the tracking of approximately 80 articles from initial contact to commitment to receipt to review through final modification and submission. Yet all of these metrics combined do not completely capture the full magnitude of her contribution to this text. I cannot offer enough gratitude to compensate for the many long evenings she spent on this project.

I am also significantly indebted to the Handbook Editorial Board. This Board contributed to every phase of the handbook development. Their efforts are reflected in the organization and outline of the material, selection and recruitment of authors, article contributions, and review of the articles. Their labors were essential to the project and I am happy to acknowledge their help.

Special thanks is extended to Nora Konopka, Acquisitions Editor at CRC Press, who has worked most closely with the project during chapter development and has been more patient and encouraging than I deserve. Finally, Helena Redshaw, Production Manager, has taken the stacks of manuscripts, disks, and CDs, identified and added the missing bits and pieces, and turned them into a book. Thanks also to all the CRC staff that I have not had the pleasure to work closely with, but who have contributed to this effort.

Editor-in-Chief

Mike Golio received his B.S.E.E. degree from the University of Illinois in 1976 and completed his M.S.E.E. and Ph.D. degrees at North Carolina State University in 1980 and 1983, respectively. His research has resulted in 15 patents and more than 200 publications. Dr. Golio is the editor of four books, including *RF and Microwave Handbook* (CRC Press, 2000). He has served as the Distinguished Microwave Lecturer for the IEEE MTT Society and as co-editor of the *IEEE Microwave Magazine.* He was elected Fellow of the IEEE in 1996.

Managing Editor

Editorial Board

Contributors

Carl Andren
Intersil
Palm Bay, Florida

Saf Asghar
Advanced Micro Devices, Inc.
Austin, Texas

James L. Bartlett
Rockwell Collins
Cedar Rapids, Iowa

Melvin L. Belcher, Jr.
Georgia Institute of Technology
Smyrna, Georgia

Nicholas E. Buris
Motorola
Schaumburg, Illinois

W. R. Deal
Malibu Networks, Inc.
Calabasas, California

Stuart D. Edwards
Conway Stuart Medical Inc.
Sunnyvale, California

John Fakatselis
Intersil
Palm Bay, Florida

Patrick Fay
University of Notre Dame
Notre Dame, Indiana

Paul G. Flikkema
Northern Arizona University
Flagstaff, Arizona

Ian C. Gifford
M/A-COM, Inc.
Lowell, Massachusetts

Madhu S. Gupta
San Diego State University
San Diego, California

Ramesh K. Gupta
Comsat Laboratories
Clarksburg, Maryland

Robert D. Hayes
RDH Incorporated
Marietta, Georgia

T. Itoh
University of California
Los Angeles, California

Nils V. Jespersen
BAE Systems
Manassas, Virginia

Andy D. Kucar
4U Communications Research, Inc.
Ottawa, Ontario, Canada

Josh T. Nessmith
Georgia Institute of Technology
Smyrna, Georgia

Jim Paviol
PRISM Wireless Design Engineering
Palm Bay, Florida

Benjamin B. Peterson
U.S. Coast Guard Academy
New London, Connecticut

Brian Petry
3Com Corporation
San Diego, California

Y. Qian
Microsemi Integrated Products
Los Angeles, California

Vesna Radisic
Northrop Grumman
Redondo Beach, California

Alfy Riddle
Macallan Consulting
Milpitas, California

Arye Rosen
Drexel University
Philadelphia, Pennsylvania

Harel D. Rosen
UMDNJ/Robert Wood Johnson Medical School
New Brunswick, New Jersey

Matthew N.O. Sadiku
Prairie View A&M University
Prairie View, Texas

Thomas M. Siep
Texas Instruments
Dallas, Texas

Wayne E. Stark
University of Michigan
Ann Arbor, Michigan

Joseph Staudinger
Motorola
Tempe, Arizona

Manos M. Tentzeris
Georgia Institute of Technology
Atlanta, Georgia

James C. Wiltse
Georgia Institute of Technology
Atlanta, Georgia

Contents

Section I Introduction

Section II Applications

Section III System and Electromagnetic Simulation

Section IV Underlying Physics

I

Introduction

1 Introduction

Patrick Fay
University of Notre Dame

1.1 Overview of Microwave and Radio Frequency Engineering

Modern microwave and radio frequency (RF) engineering is an exciting and dynamic field, due in large part to the symbiosis between recent advances in modern electronic device technology and the current explosion in demand for voice, data, and video communication capacity. Prior to this revolution in communications, microwave technology was the nearly exclusive domain of the defense industry; the recent and dramatic increase in demand for communication systems with such applications as wireless paging, mobile telephony, broadcast video, and tethered as well as untethered computer networks is revolutionizing the industry. These systems are employed across a broad range of environments, including corporate offices, industrial and manufacturing facilities, infrastructure for municipalities, and private homes. The diversity of applications and operational environments has led, through the accompanying high production volumes, to tremendous advances in cost-efficient manufacturing capabilities of microwave and RF products. This, in turn, has lowered the implementation cost of a host of new and cost-effective wireless as well as wired RF and microwave services. Inexpensive handheld Global Positioning System (GPS) navigational aids, automotive collision-avoidance radar, and widely available broadband digital service access are among these. Microwave technology is naturally suited for these emerging applications in communications and sensing because the high operational frequencies permit both large numbers of independent channels for the wide variety of uses envisioned as well as significant available bandwidth per channel for high-speed communication. The interaction between microwave fields and biological tissues also enables exciting advances in medical diagnosis and treatment.

Loosely speaking, the fields of microwave and RF engineering together encompass the design and implementation of electronic systems utilizing frequencies in the electromagnetic spectrum from approximately 300 kHz to over 100 GHz. The term *RF engineering* is typically used to refer to circuits and systems having frequencies in the range from approximately 300 kHz at the low end to between 300 MHz and 1 GHz at the upper end. The term *microwave engineering*, meanwhile, is used rather loosely to refer to design and implementation of electronic systems with operating frequencies in the range from 300 MHz to 1 GHz on the low end to upward of 100 GHz. Figure 1.1 illustrates schematically the electromagnetic spectrum from audio frequencies through cosmic rays. The RF frequency spectrum covers the medium-frequency (MF), high-frequency (HF), and very high frequency (VHF) bands, while the microwave portion of the electromagnetic spectrum extends from the upper edge of the VHF frequency range to just below the THz radiation and far-infrared optical frequencies (approximately 0.3 THz and above). The wavelength of free-space radiation for frequencies in the RF frequency range is from approximately

0-8493-1732-0/03/$0.00+$1.50

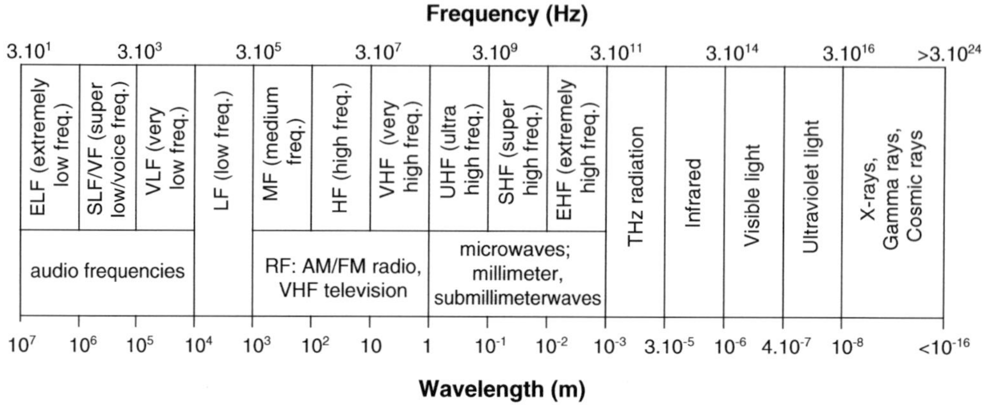

FIGURE 1.1 Electromagnetic frequency spectrum and associated wavelengths.

1 m (at 300 MHz) to 1 km (at 300 kHz), while those of the microwave range extend from 1 m to the vicinity of 1 mm (corresponding to 300 GHz) and below.

The boundary between RF and microwave design is both somewhat indistinct as well as one that is continually shifting as device technologies and design methodologies advance. This is due to implicit connotations that have come to be associated with the terms *RF* and *microwave* as the field has developed. In addition to the distinction based on the frequency ranges discussed previously, the fields of RF and microwave engineering are also often distinguished by other system features. For example, the particular active and passive devices used, the system applications pursued, and the design techniques and overall mindset employed all play a role in defining the fields of microwave and RF engineering. These connotations within the popular meaning of microwave and RF engineering arise fundamentally from the frequencies employed, but often not in a direct or absolute sense. For example, because advances in technology often considerably improve the high-frequency performance of electronic devices, the correlation between particular types of electronic devices and particular frequency ranges is a fluid one. Similarly, new system concepts and designs are reshaping the applications landscape, with mass-market designs utilizing ever higher frequencies rapidly breaking down conventional notions of microwave-frequency systems as serving "niche" markets.

The most fundamental characteristic that distinguishes RF engineering from microwave engineering is directly related to the frequency (and thus the wavelength) of the electronic signals processed. For low-frequency and RF circuits (with a few special exceptions such as antennas), the signal wavelength is much larger than the size of the electronic system and circuit components. In contrast, for a microwave system the sizes of typical electronic components are often comparable to (i.e., within approximately one order of magnitude) the signal wavelength. This gives rise to a reasonable working definition of the two areas based on the underlying approximations used in design. Because in conventional RF design the circuit components and interconnections are generally small compared to a wavelength, they can be modeled as lumped elements with parasitic inductances and capacitances incorporated to accurately model the frequency dependencies of devices and interconnects. For microwave frequencies, however, the finite propagation velocity of electromagnetic waves cannot be neglected because the time delay associated with signal propagation from one end of a component to the other is an appreciable fraction of the signal period. Consequently, lumped-element descriptions are no longer adequate to describe the electrical behavior; a distributed-element model is required to accurately capture the electrical behavior instead. The time delay associated with finite wave propagation velocity that gives rise to the distributed circuit effects is a distinguishing feature of the mindset of microwave engineering.

An alternative viewpoint is based on the observation that microwave engineering lies in a "middle ground" between traditional low-frequency electronics and optics, as shown in Fig. 1.1. As a consequence of RF, microwaves, and optics simply being different regimes within the same electromagnetic phenomena, there is a gradual transition between these regimes. The continuity of these regimes results in constant re-evaluation of the appropriate design strategies and trade-offs as device and circuit technology advances. For example, miniaturization of active and passive components often increases the frequencies at which lumped-element circuit models are sufficiently accurate because by reducing component dimensions, the time delay for propagation through a component is proportionally reduced. As a consequence, lumped-element components at traditionally microwave frequencies are becoming increasingly common in systems previously based on distributed elements due to significant advances in miniaturization, even though the operational frequencies remain unchanged. Component and circuit miniaturization also leads to tighter packing of interconnects and components, potentially introducing new parasitic coupling and distributed-element effects into circuits that could previously be treated using lumped-element RF models.

The comparable scales of components and signal wavelengths have other implications for the designer as well because neither the ray-tracing approach from optics nor the lumped-element approach from RF circuit design is valid in this middle ground. In this regard, microwave engineering can also be considered to be "applied electromagnetic engineering" because the design of guided-wave structures such as waveguides and transmission lines, transitions between different types of transmission lines, and antennas all require analysis and control of the underlying electromagnetic fields.

The distinction between RF and microwave engineering is further blurred by the trend of increasing commercialization and consumerization of systems using what have been traditionally considered to be microwave frequencies. Traditional microwave engineering, with its historical emphasis on military applications, has long been focused on delivering performance at any cost. As a consequence, special-purpose devices intended solely for use in high-performance microwave systems and often with somewhat narrow ranges of applicability were developed to achieve the required performance. With continuing advances in silicon microelectronics, including SiGe heterojunction bipolar transistors (HBTs) and conventional scaled CMOS, microwave frequency systems can now be reasonably implemented using the same devices as conventional low-frequency baseband electronics. In addition, the commercialization of low-cost III-V compound semiconductor electronics, including ion-implanted metal semiconductor field-effect transistors (MESFETs), pseudomorphic high electron mobility transistors (PHEMTs), and III-V HBTs, has dramatically decreased the cost of including these elements in high-volume consumer systems. This convergence, with silicon microelectronics moving ever higher in frequency into the microwave spectrum from the low-frequency side and compound semiconductors declining in price for the middle of the frequency range, blurs the distinction between microwave and RF engineering because microwave functions can now be realized with mainstream low-cost electronics. This is accompanied by a shift from physically large, low-integration-level hybrid implementations to highly integrated solutions based on monolithic microwave integrated circuits (MMICs). This shift has a dramatic effect not only on the design of systems and components but also on the manufacturing technology and economics of production and implementation.

Aside from these defining characteristics of RF and microwave systems, a number of physical effects that are negligible at lower frequencies become increasingly important at high frequencies. Two of these effects are the skin effect and radiation losses. The skin effect is caused by the finite penetration depth of an electromagnetic field into conducting material. This effect is a function of frequency; the depth of penetration is given by $\delta_s = \frac{1}{\sqrt{\pi f \mu \sigma}}$, where μ is the permeability, f is the frequency, and σ is the conductivity of the material. As the expression indicates, δ_s decreases with increasing frequency, and so the electromagnetic fields are confined to regions increasingly near the surface as the frequency increases. This results in the microwave currents flowing exclusively along the surface of the conductor, significantly increasing the effective resistance (and thus the loss) of metallic interconnects. Radiation losses also become increasingly important as the signal wavelengths approach the component and interconnect dimensions. For conductors and other components of comparable size to the signal wavelengths, standing waves caused by reflection of the electromagnetic waves from the boundaries of the component can greatly enhance the radiation of electromagnetic energy. These standing waves can be easily established either intentionally (in the case of antennas and resonant structures) or unintentionally (in the case of abrupt transitions, poor circuit layout, or other imperfections). Careful attention to transmission line geometry, placement relative to other components, transmission lines, and ground planes, as well as circuit packaging is essential for avoiding excessive signal attenuation and unintended coupling due to radiative effects.

A further distinction in the practice of RF and microwave engineering from conventional electronics is the methodology of testing. Due to the high frequencies involved, the capacitance and standing-wave effects associated with test cables and the parasitic capacitance of conventional test probes make the use of conventional low-frequency circuit characterization techniques impractical. Although advanced measurement techniques such as electro-optical sampling can sometimes be employed to circumvent these difficulties, in general, the loading effect of measurement equipment poses significant measurement challenges for debugging and analyzing circuit performance, especially for nodes at the interior of the circuit under test. In addition, for circuits employing dielectric or hollow guided-wave structures, voltage and current often cannot be uniquely defined. Even for structures in which voltage and current are well-defined, practical difficulties associated with accurately measuring such high-frequency signals make this difficult. Furthermore because a dc-coupled time-domain measurement of a microwave signal would have an extremely wide noise bandwidth, the sensitivity of the measurement would be inadequate for many purposes. For these reasons, components and low-level subsystems are characterized using specialized techniques, including s-parameter analysis, microwave transition analysis, and many others. A recent review of these techniques may be found in References 6 and 7.

1.2 Frequency Band Definitions

The field of microwave and RF engineering is driven by applications, originally for military purposes such as radar and more recently increasingly for commercial, scientific, and consumer applications. As a consequence of this diverse applications base, microwave terminology and frequency band designations are not entirely standardized, with various standards bodies, corporations, and other interested parties all contributing to the collective terminology of microwave engineering. Figure 1.2 shows graphically some of the most common frequency band designations, with their approximate upper and lower bounds.

As can be seen, some care must be exercised in the use of the "standard" letter designations; substantial differences in the definitions of these bands exist in the literature and in practice. Light shading at the ends of the frequency bands in Fig. 1.2 indicates variations in the definitions by different groups and authors; dark regions in the bars indicate frequencies for which there appears to be widespread agreement in the literature. The double-ended arrows appearing above some of the bands indicate the Institute of Electrical and Electronics Engineers (IEEE) definitions for these bands. Two distinct definitions of K-band are in use; the first of these defines the band as the range from 18 GHz to approximately 26.5 GHz, whereas the other definition extends from 10.9 to 36 GHz. Both of these definitions are illustrated in Fig. 1.2. Similarly, L-band has two overlapping frequency range definitions; this gives rise to the large "variation" regions shown in Fig. 1.2. In addition, some care must be taken with these letter designations because the IEEE and U.S. military specifications both define bands designated D, E, G, and L, but with very different frequencies. For example, the IEEE-defined L-band resides at the low end of the microwave spectrum, whereas the military definition of L-band is from 40 to 60 GHz. The IEEE designations (L-Y) are currently used widely in practice and the technical literature, with the newer U.S. military designations (A-N) having not yet gained widespread popularity outside the military community.

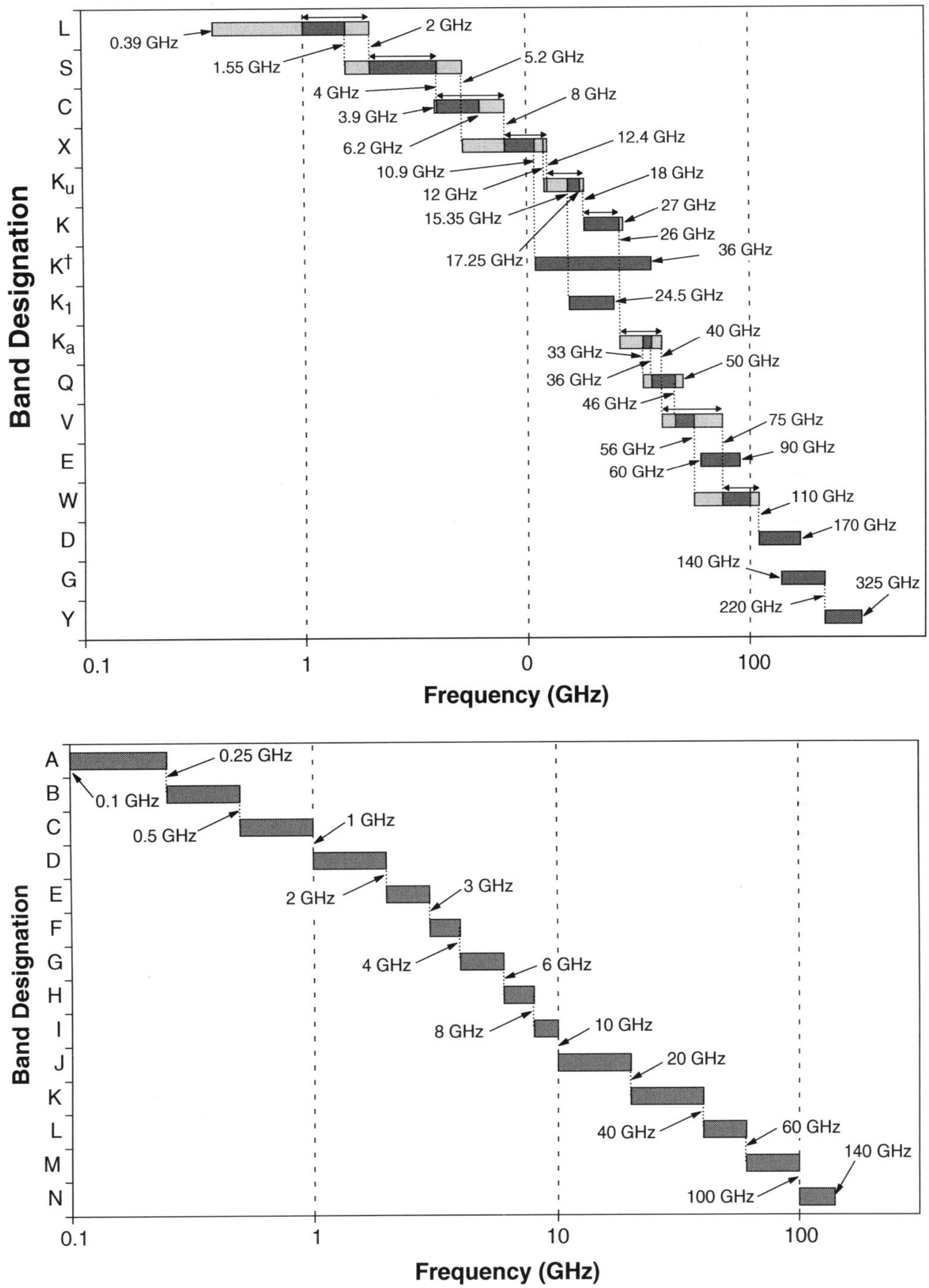

FIGURE 1.2 Microwave and RF frequency band designations.[1–5] (Top) Industrial and IEEE designations. Light shading indicates variation in the definitions found in literature; dark regions in the bars indicate frequencies for which there is widespread agreement. Double-ended arrows appearing above bands indicate the current IEEE definitions for these bands where they exist, and $K^{\dagger}$ denotes an alternative definition for K-band found in Reference 5. (Bottom) U.S. military frequency band designations.[1–3]

1.3 Applications

The field of microwave engineering is currently experiencing a radical transformation. Historically, the field has been driven by military applications, where performance at nearly any cost could be justified. Consequently, the primary emphasis was on achieving the highest performance, with less emphasis on cost or high-volume manufacturability. The current transformation of the field involves a dramatic shift from defense applications to those driven by the commercial and consumer sector, with an attendant shift in focus from design for performance to design for manufacturability. This transformation also entails a shift from small production volumes to mass production for the commercial market, and from a focus on performance without regard to cost to a focus on minimum cost while maintaining acceptable performance. For wireless applications, an additional shift from broadband systems to systems having very tightly regulated spectral characteristics also accompanies this transformation.

For many years the driving application of microwave technology was military radar. The small wavelength of microwaves permits the realization of narrowly focused beams to be achieved with antennas small enough to be practically steered, resulting in the ability to accurately detect and localize even small targets. Long-distance terrestrial communications for telephony as well as satellite uplink and downlink for voice and video were among the first commercially viable applications of microwave technology. These commercial communications applications were successful because microwave-frequency carriers (f_c) offer the possibility of very wide absolute signal bandwidths (Δf) while still maintaining relatively narrow fractional bandwidths (i.e., $\Delta f/f_c$). This allows many more voice and data channels to be accommodated than would be possible with lower frequency carriers or baseband transmission.

Among the current host of emerging applications, many are based largely on this same principle, namely, the need to transmit more and more data at high speed, and thus the need for many communication channels with wide bandwidths. Wireless communication of voice and data, both to and from individual users as well as from users and central offices in aggregate, and wired communication — including coaxial cable systems for video distribution and broadband digital access, fiber-optic communication systems for long- and short-haul telecommunication, and hybrid systems such as hybrid fiber-coaxial systems — are all poised to take advantage of the wide bandwidths and consequently high data carrying capacity of microwave-frequency electronic systems. In addition to the explosion in both diversity and capability of microwave-frequency communication systems, radar systems continue to be of importance with nonmilitary and nonnavigational applications including radar systems for automotive collision avoidance and weather and atmospheric sensing becoming increasingly widespread.

In addition to these traditional microwave applications, other fields of electronics are increasingly encroaching on the microwave-frequency range. Examples include wired data networks based on coaxial cable or twisted-pair transmission lines with bit rates of over 1 Gb/s, fiber-optic communication systems with data rates well in excess of 10 Gb/s, and inexpensive personal computers and other digital systems with clock rates into the gigahertz range. Medical advances in diagnosis and treatment are also emerging based on the interaction between high-frequency electromagnetic fields and biological molecules and tissues. In addition, continuing advances in the speed and capability of conventional microelectronics are pushing traditional circuit design ever further into the microwave frequency regime. These trends promise to both invigorate and reshape the field of microwave engineering in new and exciting ways.

References

1. Collin, R.E., *Foundations for Microwave Engineering*, McGraw-Hill, New York, 1992, 2.
2. Harsany, S.C., *Principles of Microwave Technology*, Prentice Hall, Upper Saddle River, NJ, 1997, 5.
3. Laverghetta, T.S., *Modern Microwave Measurements and Techniques*, Artech House, Norwood, MA, 1988, 479.
4. Rizzi, P.A., *Microwave Engineering*, Prentice-Hall, Englewood Cliffs, NJ, 1988, 1.
5. *Reference Data for Radio Engineers*, ITT Corp., New York, 1975.
6. Golio, M., Ed., *The RF and Microwave Handbook*, CRC Press, Boca Raton, FL, 2001.
7. Wartenberg, S.A., *RF Measurements of Die and Packages*, Artech House, Norwood, MA, 2002.

II

Applications

2

Cellular Mobile Telephony

Paul G. Flikkema
Northern Arizona University

The goal of modern cellular mobile telephone systems is to provide services to telephone users as efficiently as possible. In the past, this definition would have been restricted to mobile users. However, the cost of wireless infrastructure is less than wired infrastructure in new telephone service markets. Thus, wireless mobile telephony technology is adapted to provide in-home telephone service, the so-called wireless local loop (WLL). Indeed, it appears that wireless telephony can become dominant over traditional wired access worldwide.

The objective of this section is to familiarize the radio frequency (RF)/microwave engineer with the concepts and terminology of cellular mobile telephony (cellular), or mobile wireless networks. A capsule history and a summary form the two bookends of the section. In between, we start with the cellular concept and the basics of mobile wireless networks. Then we take a look at some of the standardization issues for cellular systems. Following that, we cover the focus of the standards battles: channel access methods. We then take a look at some of the basic aspects of cellular important to RF/microwave engineers: first, modulation, diversity, and spread spectrum; then coding, interleaving, and time diversity; and finally nonlinear channels. Before wrapping up, we take a glimpse at a topic of growing importance: antenna array technology.

2.1 A Brief History

Mobile telephone service was inaugurated in the U.S. in 1947 with six radio channels available per city. This evolved into the manual Mobile Telephone System (MTS) used in the 1950s and 1960s. The year

0-8493-1732-0/03/$0.00+$1.50

1964 brought the Improved MTS (IMTS) systems with eight channels per city with — finally — no telephone operator required. Later, the capacity was more than doubled to 18. Most importantly, the IMTS introduced narrowband frequency modulation (NBFM) technology. The first cellular service was introduced in 1983, called AMPS (Advanced Mobile Phone Service). Cities were covered by cells averaging about 1 km in radius, each serviced by a base station. This system used the 900 MHz frequency band still in use for mobile telephony. The cellular architecture allowed frequency reuse, dramatically increasing capacity to a maximum of 832 channels per cell.

The age of digital, or second-generation, cellular did not arrive until 1995 with the introduction of the IS-54 TDMA service and the competing IS-95 CDMA service. In 1996 and 1997, the U.S. Federal Communications Commission auctioned licenses for mobile telephony in most U.S. markets in the so-called PCS (Personal Communication System) bands at 1.9 GHz. These systems use a variety of standards, including TDMA, CDMA, and the GSM TDMA standard that originated in Europe. Outside the U.S., a similar evolution has occurred, with GSM deployed in Europe and the PDC (Personal Digital Cellular) system in Japan. In other countries there has been a pitched competition between all systems. While not succeeding in the U.S., so-called low-tier systems have been popular in Europe and Japan. These systems are less robust to channel variations and are therefore targeted to pedestrian use. The European system is called DECT (Digital European Cordless Telephony) and the Japanese system is called PHS (Personal Handyphone System).

Third-generation (or 3G) mobile telephone service will be rolled out in the 2001 to 2002 time frame. These services will be offered in the context of a long-lived standardization effort recently renamed IMT-2000 (International Mobile Telecommunications–2000) under the auspices of the Radio Communications Standardization Sector of the International Telecommunications Union (ITU-R; see http://www.itu.int). Key goals of IMT-2000 are:[13]

1. Use of a common frequency band over the globe
2. Worldwide roaming capability
3. Transmission rates higher than second-generation systems to handle new data-over-cellular applications

Another goal is to provide the capability to offer asymmetric rates, so that the subscriber can download data much faster than he can send it.

Finally, it is hoped that an architecture can be deployed that will allow hardware, software, and network commonality among services for a range of environments, such as those for vehicular, pedestrian, and fixed (nonmoving) subscribers. While also aiming for worldwide access and roaming, the main technical thrust of 3G systems will be to provide high-speed wireless data services, including 144 kb/s service to subscribers in moving vehicles, 384 kb/s to pedestrian users, 2 Mbps to indoor users, and service via satellites (where the other services do not reach) at up to 32 kb/s for mobile, handheld terminals.

2.2 The Cellular Concept

At first glance, a logical method to provide radio-based communication service to a metropolitan area is a single, centrally located antenna. However, radio-frequency spectrum is a limited commodity, and regulatory agencies, in order to meet the needs of a vast number of applications, have further limited the amount of RF spectrum for mobile telephony. The limited amount of allocated spectrum forced designers to adopt the cellular approach: using multiple antennas (base stations) to cover a geographic area, each base station covers a roughly circular area called a cell. Figure 2.1 shows how a large region can be split into seven smaller cells (approximated by hexagons). This allows different base stations to use the same frequencies for communication links as long as they are separated by a sufficient distance. This is known as frequency reuse, and allows thousands of mobile telephone users in a metropolitan area to share far fewer channels.

There is a second important aspect to the cellular concept. With each base station covering a smaller area, the mobile phones need less transmit power to reach any base station (and thus be connected with

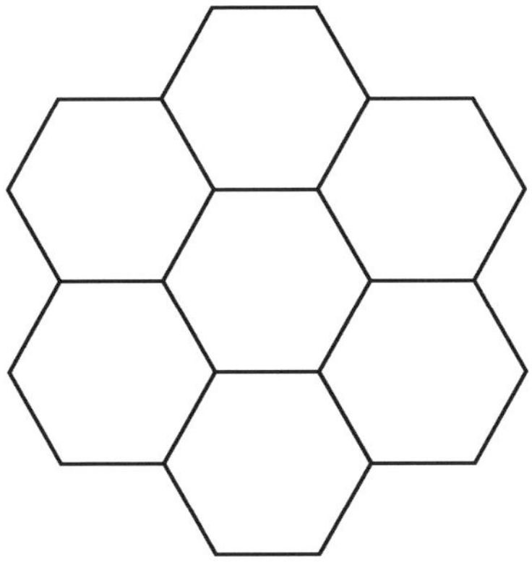

FIGURE 2.1 A region divided into cells. While normally the base stations are placed at the center of the cells, it is also possible to use edge-excited cells where base stations are placed at vertices.

the telephone network). This is a major advantage, since, with battery size and weight a major impediment to miniaturization, the importance of reducing power consumption of mobile phones is difficult to overestimate.

If two mobile units are connected to their respective base stations at the same frequency (or more generally, channel), interference between them, called *co-channel interference*, can result. Thus, there is a trade-off between frequency reuse and signal quality, and a great deal of effort has resulted in frequency assignment techniques that balance this trade-off. They are based on the idea of clustering: taking the available set of channels, allocating them in chunks to each cell, and arranging the cells into geographically local clusters. Figure 2.2 shows how clusters of seven cells (each allocated one of seven mutually exclusive channel subsets) are used to cover a large region; note that the arrangement of clusters maximizes the reuse distance — the distance between any two cells using the same frequency subset. Increasing the reuse distance has the effect of reducing co-channel interference.

Although it might seem attractive to make the cells smaller and smaller, there are diminishing returns. First, smaller cell sizes increase the need for management of mobile users as they move about. In other words, smaller cell sizes require more hand-offs, where the network must transfer users between base stations. Another constraint is antenna location, which is often limited by available space and aesthetics. Fortunately, both problems can be overcome by technology. Greater hand-off rates can be handled by increases in processing speed, and creative antenna placement techniques (such as on lamp posts or sides of buildings) are allowing higher base station densities.

Another issue is evolution: how can a cellular system grow with demand? Two methods have been successful. The first is cell splitting: by dividing a cell into several cells (and adjusting the reuse pattern), a cellular service provider can increase its capacity in high-demand areas. The second is sectoring: instead of a single omnidirectional antenna covering a cell, a typical approach is to sectorize the cell into N_S regions, each served by an antenna that covers an angular span of $2\pi/N_S$ ($N_S = 3$ is typical). Note that both approaches increase hand-off rates and thus require concurrent upgrading of network management. Later we will describe smart antennas, the logical extension to sectorization.

2.3 Networks for Mobile Telephony

A communication network that carries only voice — even a digital one — is relatively simple. Other than the usual digital communication system functions, such as channel coding, modulation, and synchronization, all that is required is call setup and takedown. However, current and future digital mobile telephony networks are expected to carry digital data traffic as well.

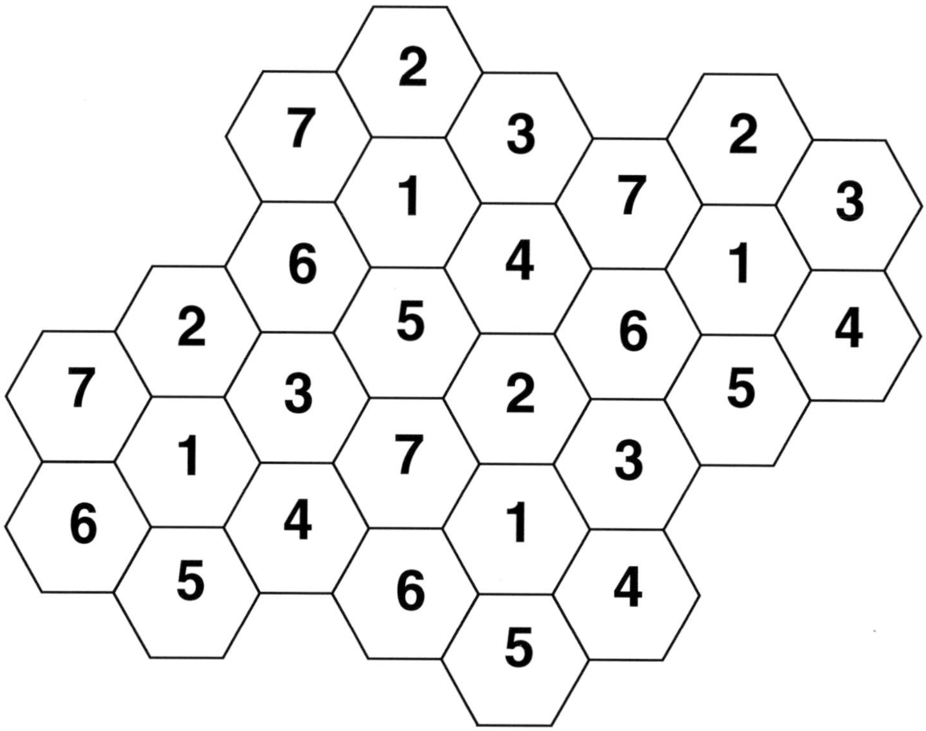

FIGURE 2.2 Cell planning with cluster size of 7. The number in each cell indexes the subset of channels allocated to the cell. Other cluster sizes, such as 4, 7, or 12 can be used.

Data traffic is by nature computer-to-computer, and requires that the network have an infrastructure that supports everything from the application (such as Web browsing) to the actual transfer of bits. The data are normally organized into chunks called packets (instead of streams as in voice), and requires a much higher level of reliability than digitized voice signals. These two properties imply that the network must also label the packets, and manage the detection and retransmission of packets that are received in error. It is important to note that packet retransmission, while required for data to guarantee fidelity, is not possible for voice because it would introduce delays that would be intolerable in a human conversation.

Other functions that a modern digital network must perform include encryption and decryption (for data security) and source coding and decoding. The latter functions minimize the amount of the channel resource (in essence, bandwidth) needed for transferring the information. For voice networks this involves the design of voice codecs (coder/decoders) that not only digitize voice signals, but strip out the redundant information in them. In addition to all the functions that wired networks provide, wireless networks with mobile users must also provide mobility management functions that keep track of calls as subscribers move from cell to cell.

The various network functions are organized into *layers* to rationalize the network design and to ease internetworking, or the transfer of data between networks.[11] RF/microwave engineering is part of the *physical layer* that is responsible for carrying the data over the wireless medium.

2.4 Standards and Standardization Efforts

The cellular industry is, if anything, dense with lingo and acronyms. Here we try to make sense of at least some of the important and hopefully longer lived terminology.

Worldwide, most of the cellular services are offered in two frequency bands: 900 and 1900 MHz. In each of the two bands, the exact spectrum allocated to terrestrial mobile services varies from country to country. In the U.S. cellular services are in the 800- to 900-MHz band, while similar services are in the 800- to 980-MHz band in Europe under the name GSM900. (GSM900 combines in one name a radio communication standard — GSM, or Global System for Mobile Communications — and the frequency band in which it is used. We will describe the radio communication, or *air interface*, standards later). In the mid-1990s, the U.S. allocated spectrum for PCS (Personal Communication Services) from 1850 to 2000 MHz; while many thought PCS would be different from cellular, they have converged and are interchangeable from the customer's perspective. Similarly, in Europe GSM1800 describes cellular services offered using the 1700- to 1880-MHz band.

The 1992 World Administrative Radio Conference (WARC '92) allocated spectrum for third-generation mobile radio in the 1885 to 1980 and 2110 to 2160 MHz bands. The ITU-Rs IMT-2000 standardization initiative adopted these bands for terrestrial mobile services. Note that the IMT-2000 effort is an umbrella that includes both terrestrial and satellite-based services — the latter for areas where terrestrial services are unavailable.

Please note that all figures here are approximate and subject to change in future WARCs; please consult References 2 and 13 for details.

The cellular *air interface* standards are designed to allow different manufacturers to develop both base station and subscriber (mobile user handset) equipment. The air interface standards are generally different for the downlink (base station to handset) and uplink (handset to base station). This reflects the asymmetry of resources available: the handsets are clearly constrained in terms of power consumption and antenna size, so that the downlink standards imply sophisticated transmitter design, while the uplink standards emphasize transmitter simplicity and advanced receive-side algorithms. The air interface standards address channel access protocols as well as traditional communication link design parameters such as modulation and coding. These issues are taken up in the following sections.

2.5 Channel Access

In a cellular system, a fixed amount of RF spectrum must somehow be shared among thousands of simultaneous phone conversations or data links. *Channel access* is about (1) dividing the allocated RF spectrum into pieces and (2) allocating the pieces to conversations/links in an efficient way.

The easiest channel access method to envision is FDMA (Frequency Division Multiple Access), where each link is allocated a sub-band (i.e., a specific carrier frequency; see Fig. 2.3). This is exactly the access method used by first-generation (analog) cellular systems. The second generation of cellular brought two newer channel access methods that were enabled by progress in digital process technology. One is TDMA (Time Division Multiple Access), wherein time is divided into *frames*, and links are given short *time slots* in each frame (Fig. 2.4). FDMA and TDMA can be seen as time/frequency duals, in that FDMA subdivides the band into narrow sub-bands in the frequency domain, while TDMA subdivides time into slots, during which a link (within a cell) uses the entire allocated bandwidth.

The second generation of cellular also brought CDMA (Code Division Multiple Access). In CDMA, all active links simultaneously use the entire allocated spectrum, but sophisticated codes are used that allow the signals to be separated in the receiver.[1] We will describe CDMA in more depth later.

It should be noted that both TDMA- and CDMA-based cellular systems also implicitly employ FDMA, although this is rarely mentioned. The reason is that the cellular bands are divided into smaller bands (a form of FDMA), and both TDMA and CDMA are used within these sub-bands.

[1]It is fashionable to depict CDMA graphically using a "code dimension" that is orthogonal to the time–frequency plane, but this is an unfortunate misrepresentation. Like any signals, CDMA signals exist (in fact, overlap) in the time–frequency plane, but have correlation-based properties that allow them to be distinguished.

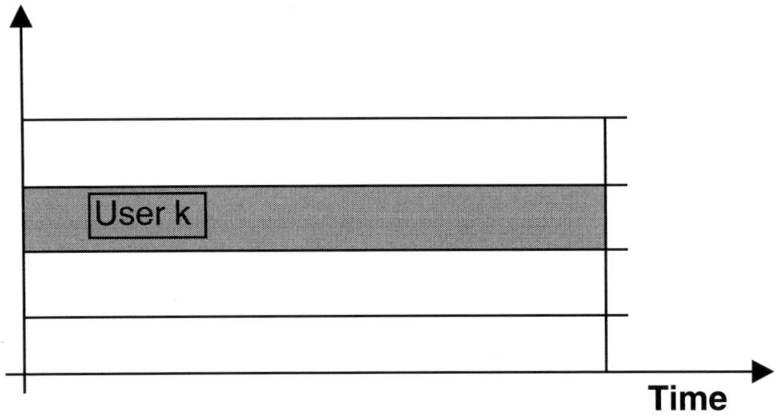

FIGURE 2.3 FDMA depicted on the time–frequency plane, with users assigned carrier frequencies, or channels. Not shown are guard bands between the channels to prevent interference between users' signals.

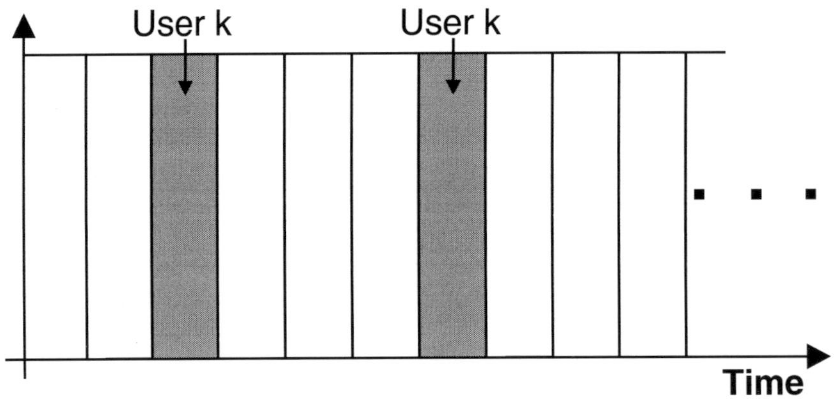

FIGURE 2.4 Depiction of TDMA on the time–frequency plane. Users are assigned time slots within a frame. Guard times (not shown) are needed between slots to compensate for timing inaccuracies.

In the U.S., the TDMA and CDMA standards are referred to by different acronyms. The TDMA standard originally was called IS-54, but with enhancements became IS-136. The CDMA standard was called IS-95, and has been re-christened as cdmaOne by its originator, Qualcomm. These standards were created under the auspices of the Telecommunications Industry Association (TIA) and the Electronic Industries Alliance (EIA).

In Europe, the second generation brought digital technology in the form of the GSM standard, which used TDMA. (The GSM acronym originally referred to Group Special Mobile, but was updated to capture its move to worldwide markets.) Japan also chose TDMA in its first digital offering, called PDC (Personal Digital Cellular).

The three multiple access approaches use different signal properties (frequency, time, or code) to allow the distinguishing of multiple signals. How do they compare? In the main, as we move from FDMA to TDMA to CDMA (in order of their technological development), complexity is transferred from the RF section to the digital section of the transmitters and receivers. The evolution of multiple access techniques has tracked the rapid evolution of digital processing technology as the latter has become cheaper and faster. For example, while FDMA requires a tunable RF section, both TDMA and CDMA need only a fixed-frequency front end. CDMA relieves one requirement of TDMA — strict synchronization among the various transmitters — but introduces a stronger requirement for synchronization of the receiver to the received signal. In addition, the properties of the CDMA signal provide a natural means to exploit the multipath nature of the digital signal for improved performance. However, these advantages come at the cost of massive increases in the capability of digital hardware. Luckily, Moore's law (i.e., that processing power roughly doubles every 18 months at similar cost) still remains in effect as of the turn of the century, and the amount of processing power that will be used in the digital phones in the 21st century will be unimaginable to the architects of the analog systems developed in the 1970s.

2.6 Modulation

The general purpose of modulation is to transform an information-bearing message signal into a related signal that is suitable for efficient transmission over a communication channel. In *analog modulation,* this is a relatively simple process: the information-bearing analog (or continuous-time) signal is used to alter a parameter (normally, the amplitude, frequency, or phase) of a sinusoidal signal (or carrier, the signal carrying the information). For example, in the NBFM modulation used in the AMPS system, the voice signal alters the frequency content of the modulated signal in a straightforward manner.

The purpose of *digital modulation* is to convert an information-bearing discrete-time symbol sequence into a continuous-time waveform. Digital modulation is easier to analyze than analog modulation, but more difficult to describe and implement.

2.6.1 Modulation in Digital Communication

Before digital modulation of the data in the transmitter, there are several processing steps that must be applied to the original message signal to obtain the discrete-time symbol sequence. A continuous-time message signal, such as the voice signal in telephony, is converted to digital form by sampling, quantization, and source coding. *Sampling* converts the original continuous-time waveform into discrete-time format, and *quantization* approximates each sample of the discrete-time signal using one of a finite number of levels. Then *source coding* jointly performs two functions: it strips redundancy out of the signal and converts it to a discrete-time sequence of symbols.

What if the original signal is already in discrete-time (sampled format), such as a computer file? In this case, no sampling or quantization is needed, but source coding is still used to remove redundancy.

Between source coding and modulation is a step critical to the efficiency of digital communications: channel coding. This is discussed later; it suffices for now to know that it converts the discrete-time sequence of symbols from the source coder into another (better) discrete-time symbol sequence for input to the modulator. Following modulation, the signal is upconverted, filtered (if required), and amplified in RF electronics before being sent to the antenna. All the steps described are shown in block-diagram form in Fig. 2.5. In the receiver, the signal from the antenna, following filtering (again, if required), is amplified and downconverted prior to demodulation, channel decoding, and source decoding (see Fig. 2.5).

What is the nature of the digital modulation process? The discrete-time symbol sequence from the channel coder is really a string of symbols (letters) from a *finite* alphabet. For example, in *binary* digital modulation, the input symbols are 0's and 1's. The modulator output converts those symbols into one of a *finite* set of waveforms that can be optimized for the channel.

While it is the finite set of waveforms that distinguishes digital modulation from analog modulation, that difference is only one manifestation of the entire paradigm of *digital communication.* In a good digital communication design, the source coder, channel coder, and modulator all work together to maximize the efficient use of the communication channel; even two of the three are not enough for good performance.

2.6.2 Selection of Digital Modulation Formats

There are several (often conflicting) criteria for selection of a modulation scheme. They are:

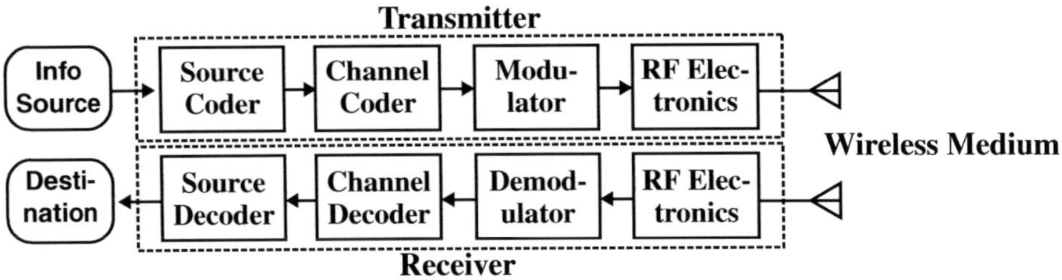

FIGURE 2.5 Communication system block diagram for wireless communication. In the wireless medium, multipath propagation and interference can be introduced. For system modeling purposes, the two blocks of RF electronics are combined with the wireless medium to form the wireless channel — a channel that distorts the signal, and adds noise and interference. The other blocks are designed to maximize the system performance for the channel.

- BER (bit error rate) performance
  - In wireless, particularly in cellular mobile channels, the scheme must operate under conditions of severe fading.
  - Cellular architectures imply co-channel interference.
  - Typically, a BER of 10^{-2} or better is required for voice telephony, and 10^{-5} or better is required for data.
- Spectral (or bandwidth) efficiency (measured in b/s/Hz)
- Power efficiency (especially for handheld/mobile terminals)
- Implementation complexity and cost

In the U.S. cellular market, complexity is of special importance: with the number of standards growing, many handsets are now dual and triple mode; for example, a phone might have both GSM and 3G capability. While some hardware can be shared, multimode handsets clearly place additional constraints on the allowable complexity for each mode.

2.6.3 Classification of Digital Modulation Schemes

Broadly, modulation techniques can be classified into two categories.

Linear methods include schemes that use combinations of amplitude and phase modulation of a pulse stream. They have higher spectral efficiencies than constant-envelope methods (see the following), but must use more expensive (or less efficient) linear amplifiers to maintain performance and to limit out-of-band emissions.

Examples of linear modulation schemes include PSK (phase-shift keying) and QAM (quadrature amplitude modulation). QAM can be viewed as a generalization of PSK in that both the amplitude and the phase of the modulated waveform are altered in response to the input symbols.

Constant-envelope methods are more complicated to describe, but usually are sophisticated methods based on frequency modulation. Their key characteristic is a constant envelope (resulting in a constant instantaneous signal power) regardless of the source symbol stream. They allow use of less expensive amplification and/or higher amplification efficiencies (e.g., running amplifiers in the nonlinear region), at the expense of out-of-band emissions. Historically, they are limited to spectral efficiencies of about 1 b/s/Hz.

Examples of constant envelope methods include FSK (frequency-shift keying) and more sophisticated methods such as minimum-shift keying (MSK) and Gaussian prefiltered minimum-shift keying (GMSK) (these will be described shortly). These methods can be thought of as digital (finite alphabet) FM in that the spectrum of the output signal is varied according to the input symbol stream.

The spectral occupancy of a modulated signal (per channel) is roughly

$$S_O = B + 2\Delta f$$

where B is the bandwidth occupied by signal power spectrum and Δf is the maximum one-way carrier frequency drift.[2] We can express the bandwidth

$$B = \frac{R_d}{\epsilon}$$

where R_d is the channel data rate (in b/s) and ϵ is the spectral efficiency (in b/s/Hz). Combining, we obtain

$$S_O = \frac{R_d}{\epsilon} + 2\Delta f$$

Thus, to minimize spectral occupancy (thus maximizing capacity in number of users) we can:

1. Reduce R_d by lowering the source coding rate (implying more complexity or lower fidelity)
2. Improve the spectral efficiency of the modulation (implying higher complexity)
3. Improve the transmitter/receiver oscillators (at greater cost)

2.6.4 Modulation, Up/Downconversion, and Demodulation

To transmit a string of binary information symbols (or bits — zeros and ones), $\{b_0, b_1, b_2, \ldots\}$, we can represent a 1 by a positive-valued pulse of amplitude one, and a 0 by a negative pulse of the sample amplitude. This mapping from the bit value at time n, b_n, to amplitude a_n can be accomplished using

$$a_n = 2b_n - 1$$

To complete the definition, we define a pulse of unit amplitude with start time of zero and stop time of T as $p_T(t)$. Then the modulated signal can be efficiently written as

$$u(t) = \sum_n a_n p_T(t - nT)$$

This signal is at baseband — centered at zero frequency — and is therefore unsuitable for wireless communication media. However, this signal can be upconverted to a desired RF by mixing with a sinusoid to get the passband signal

$$x(t) = u(t)\cos(2\pi f_c t) = \cos(2\pi f_c t)\sum_n a_n p_T(t - nT)$$

where f_c is the carrier frequency.

Multiplying a sinusoid by ± 1 is identical to changing its phase between 0 and π radians, so we have

$$x(t) = \cos\left(2\pi f_c t + \sum_n d_n p_T(t - nT)\right)$$

where we assign $d_n = 0$ when $a_n = -1$ and $d_n = \pi$ when $a_n = 1$. This equation shows that we are simply shifting the phase of the carrier between two different values: this is BPSK (binary phase-shift keying).

[2]This drift can be caused by oscillator instability or Doppler due to channel time variations.

Why not use more than two phase values? In fact, four are ordinarily used for better efficiency: pairs of bits are mapped to four different phase values, 0, $\pm\pi/2$, and π. For example, the CDMA standards employ this scheme, known as quaternary PSK (QPSK).

In general, the baseband signal will be complex-valued, which leads to the general form of upconversion from baseband to passband:

$$x(t) = \sqrt{2}\Re\left\{u(t)e^{j2\pi f_c t}\right\}$$

where the $\sqrt{2}$ factor is simply to maintain a consistency in measurement of signal power between passband and baseband. The motivation of using the baseband representation of a signal is twofold: first, it retains the amplitude and phase of the passband signal, and is thus independent of any particular carrier frequency; second, it provides the basis for modern baseband receiver implementations that use high-speed digital signal processing. The baseband representation is also known as the *complex envelope* representation.

BPSK and QPSK are linear modulation methods; in contrast, FSK is a constant-envelope modulation scheme. For binary FSK (BFSK), there are two possible signal pulses, given at baseband by

$$u_0(t) = Ae^{-j\pi\Delta ft} p_T(t), \quad u_1(t) = Ae^{j\pi\Delta ft} p_T(t)$$

where A is the amplitude. Notice that we have two (complex) tones separated by Δf. MSK and GMSK are special forms of FSK that provide greater spectral efficiency at the cost of higher implementation efficiency. The GSM standard and its next-generation version, currently known as EDGE (for Enhanced Data Rates for Global Evolution), use GMSK.

At the receiver, the RF signal is amplified and downconverted with appropriate filtering to remove interference and noise. The downconverted signal is then passed to the demodulator, whose function is to detect (guess in an optimum way) what symbol stream was transmitted. Following demodulation (also referred to as detection), the symbol stream is sent to subsequent processing steps (channel decoding and source decoding) before delivery to the destination.

At this point it is typical to consider the BERs and spectral efficiencies of various digital modulation formats, modulator and demodulator designs, and the performance of different detection strategies for mobile cellular channels. This is beyond the scope of this section, and we direct the reader to a good book on digital communications (e.g., References 1, 4, 6–8) for more information.

2.7 Diversity, Spread Spectrum, and CDMA

A mobile wireless channel causes the transmitted signal to arrive at the receiver via a number of paths due to reflections from objects in the environment. If the channel is linear (including transmit and receive amplifiers), a simple modeling approach for this multipath channel is to assume that it is specular (i.e., each path results in a specific amplitude, time delay, and phase change). If the channel is also at least approximately time-invariant, its impulse response under these conditions can be expressed as[3]

$$h(t) = \sum_{\lambda=0}^{\Lambda} \alpha_\lambda e^{j\theta_\lambda} \delta(t - \tau_\lambda)$$

where α_λ, τ_λ, and θ_λ are, respectively, the amplitude, time delay, and phase for the λ-th path.

Let the transmitted signal be

[3]Here $\delta(t)$ denotes the Dirac delta function.

$$s(t) = \sum_n a_n f_n(t)$$

a sequence of pulses $f_n(t)$ each modulated by a transmitted symbol a_n at a symbol rate of $1/T$. When transmitted via a specular multipath channel with Λ paths, the received signal — found by the convolution of the transmitted signal and the channel impulse response — is

$$y(t) = \sum_{\lambda=0}^{\Lambda} \alpha_\lambda e^{j\theta_\lambda} s(t - \tau_\lambda).$$

For simplicity, consider sending only three symbols a_{-1}, a_0, a_1. Then the received signal becomes

$$y(t) = \sum_{n=-1}^{1} a_n \sum_{\lambda=0}^{\Lambda} \alpha_\lambda e^{j\theta_\lambda} f_n(t - \tau_\lambda)$$

Two effects may result: fading and intersymbol interference. *Fading* occurs when superimposed replicas of the same symbol pulse nullify each other due to phase differences. *Intersymbol interference (ISI)* is caused by the convolutive mixing of the adjacent symbols in the channels. Fading and ISI may occur individually or together depending on the channel parameters and the symbol rate T^{-1} of the transmitted signal.

Let us consider in more detail the case where the channel *delay spread* is a significant fraction of T (i.e., τ_Λ is close to, but smaller than T). In this case, we can have both fading and ISI, which, if left untreated, can severely compromise the reliability of the communication link. Direct-sequence spread-spectrum (DS/SS) signaling is a technique that mitigates these problems by using clever designs for the pulses $f_n(t)$. These pulse designs are wide bandwidth (hence "spread spectrum"), and the extra bandwidth is used to endow them with properties that allow the receiver to separate the symbol replicas.

Suppose we have a two-path channel, and consider the received signal for symbol a_0. Then the DS/SS receiver separates the two replicas

$$\alpha_0 e^{j\theta_0} a_p f_0(t - \tau_0), \quad \alpha_1 e^{j\theta_1} a_0 f_0(t - \tau_1)$$

Then each replica is adjusted in phase by multiplying it by $e^{-j\theta\lambda}$, $\lambda = 0, 1$ yielding (since $zz^* = |z|^2$)

$$\alpha_0 a_0 f(t - \tau_0), \quad \alpha_1 a_0 f(t - \tau_1)$$

Now all that remains is to delay the first replica by $\tau_1 - \tau_0$ so they line up in time, and sum them, which gives

$$(\alpha_0 + \alpha_1) a_0 f(t - \tau_1)$$

Thus DS/SS can turn the multipath channel to advantage — instead of interfering with each other, the two replicas are now added constructively. This *multipath combining* exploits the received signal's inherent *multipath diversity,* and is the basic idea behind the technology of RAKE reception[4] used in the CDMA digital cellular telephony standards.

[4]The RAKE nomenclature can be traced to the block diagram representation of such a receiver — it is reminiscent of a garden rake.

It is important to note that this is the key idea behind all strategies for multipath fading channels: we somehow exploit the redundancy, or *diversity* of the channel (recall the multiple paths). In this case, we used the properties of DS/SS signaling to effectively split the problematic two-path channel into two benign one-path channels. Multipath diversity can also be viewed in the frequency domain, and is in effect a form of *frequency diversity*. As we will see later, frequency diversity can be used in conjunction with other forms of diversity afforded by wireless channels, including time diversity and antenna diversity.

CDMA takes the spread spectrum idea and extends it to the separation of signals from multiple *transmitters*. To see this, suppose M transmitters are sending signals simultaneously, and assume for simplicity that we have a single-path channel. Let the complex (magnitude/phase) gain for channel m be denoted by $\beta^{(m)}$. Finally, the transmitters use different spread-spectrum pulses, denoted by $f^{(m)}(t)$. If we just consider the zeroth transmitted symbols from each transmitter, we have the received signal

$$y(t) = \sum_{m=1}^{M} \beta^{(m)} a_0^{(m)} f^{(m)}\left(t - t_m\right)$$

where the time offset t_m indicates that the pulses do not necessarily arrive at the same time.

The above equation represents a complicated mixture of the signals from multiple transmitters. If narrowband pulses are used, they would be extremely difficult — probably impossible — to separate. However, if the pulses are spread spectrum, then the receiver can use algorithms to separate them from each other, and successfully demodulate the transmitted symbols. Of course, these ideas can be extended to many transmitters sending long strings of symbols over multipath channels.

Why is it called CDMA? It turns out that the special properties of the signal pulses $f^{(m)}(t)$ for each user (transmitter) m derive from high-speed *codes* consisting of periodic sequences of chips $c_k^{(m)}$ that modulate chip waveforms $\varphi(t)$. One way to envision it is to think of $\varphi(t)$ as a rectangular pulse of duration $T_c = T/N$. The pulse waveform for user m can then be written

$$f^{(m)}(t) = \sum_{k=0}^{N-1} c_k^{(m)} \varphi\left(t - kT_c\right)$$

The fact that we can separate the signals means that we are performing code-division multiple access — dividing up the channel resource by using codes. Recall that in FDMA this is done by allocating frequency bands, and in TDMA, time slots. The pulse waveforms in CDMA are designed so that many users' signals occupy the entire bandwidth simultaneously, yet can still be separated in the receiver. The signal-separating capability of CDMA is extremely important, and can extend beyond separating desired signals within a cell. For example, the IS-95 CDMA standard uses spread-spectrum pulse designs that enable the receiver to reject a substantial amount of co-channel interference (interference due to signals in other cells). This gives the IS-95 system (as well as its proposed 3G descendants) its well-known property of universal frequency reuse.

The advantages of DS/SS signals derive from what are called their *deterministic correlation* properties. For an arbitrary periodic sequence $\{c_k^{(m)}\}$, the deterministic *autocorrelation* is defined as

$$\phi^{(m)}(i) = \frac{1}{N} \sum_{k=0}^{N-1} c_k^{(m)} c_{k+i}^{(m)}$$

where i denotes the relative shift between two replicas of the sequence. If $\{c_k^{(m)}\}$ is a direct-sequence spreading code, then

$$\phi^{(m)}(i) \approx \begin{cases} 1, & i = 0 \\ 0, & 1 < |i| < N \end{cases}$$

This "thumbtack" autocorrelation implies that relative shifts of the sequence can be separated from each other. Noting that each chip is a fraction of a symbol duration, we see that multipath replicas of a symbol pulse can be separated even if their arrival times at the receiver differ by less than a symbol duration.

CDMA signal sets also exhibit special deterministic *cross-correlation* properties. Two spreading codes $\{c_k^{(l)}\}$, $\{c_k^{(m)}\}$ of a CDMA signal set have the cross-correlation property

$$\phi^{(l,m)}(i) = \frac{1}{N}\sum_{k=0}^{N-1} c_k^{(l)} c_{k+i}^{(m)} \approx \begin{cases} 1, & l = m, i = 0, \\ 0, & l = m, 0 < |i| < N \\ 0, & l \neq m. \end{cases}$$

Thus, we have a set of sequences with zero cross-correlations and "thumbtack" autocorrelations. (Note that this includes the earlier autocorrelation as a special case.) The basic idea of demodulation for CDMA is as follows: if the signal from user m is desired, the incoming received signal — a mixture of multiple transmitted signals — is correlated against $\{c_k^{(m)}\}$. Thus multiple replicas of a symbol from user m can be separated, delayed, and then combined, while all other users' signals (i.e., where $l \neq m$) are suppressed by the correlation.

Details of these properties, their consequences in demodulation, and descriptions of specific code designs can be found in References 3, 4, 7, and 10.

2.8 Channel Coding, Interleaving, and Time Diversity

As we have mentioned, channel coding is a transmitter function that is performed after source coding, but before modulation. The basic idea of channel coding is to introduce highly structured redundancy into the signal that will allow the receiver to easily detect or correct errors introduced in the transmission of the signal.

Channel coding is fundamental to the success of modern wireless communication. It can be considered the cornerstone of digital communication, since, without coding, it would not be possible to approach the fundamental limits established by Shannon's information theory.[9,12]

The easiest type of channel codes to understand are *block codes:* a sequence of input symbols of length k is transformed into a code sequence (code word) of length $n > k$. Codes are often identified by their rate R, where $R = k/n \leq 1$. Generally, codes with a lower rate are more powerful. Almost all block codes are *linear,* meaning that the sum of two code words is another code word. By enforcing this linear structure, coding theorists have found it easier to find codes that not only have good performance, but have reasonably simple decoding algorithms as well.

In wireless systems, *convolutional codes* are very popular. Instead of blocking the input stream into length-k sequences and encoding each one independently, convolutional coders are finite-state sequential machines. Therefore they have memory, so that a particular output symbol is determined by a contiguous sequence of input symbols. For example, a rate-1/2 convolutional coder outputs two code symbols for each information symbol that arrives at its input. Normally, these codes are also linear.

Error-correcting codes have enough power so that errors can actually be corrected in the receiver. Systems that use these codes are called *forward error-control (FEC)* systems. *Error-detecting codes* are simpler, but less effective: they can tell *whether* an error has occurred, but not where the error is located in the received sequence, so it cannot be corrected.

Error-detecting codes can be useful when it is possible for the receiver to request retransmission of a corrupted code word. Systems that employ this type of feedback are called *ARQ,* or Automatic Repeat-reQuest systems.

As we have seen, the fading in cellular systems is due to multipath. Of course, as the mobile unit and other objects in the environment move, the physical structure of the channel changes with time, causing the fading of the channel to vary with time. However, this fading process tends to be slow relative to the symbol rate, so a long string of coded symbols can be subjected to a deep channel fade. In other words, the fading from one symbol to the next will be highly correlated. Thus, the fades can cause a large string of demodulation (detection) errors, or an *error burst.* Thus, fading channels are often described from the point of view of coding as *burst-error channels.*

Most well-known block and convolutional codes are best suited to random errors, that is, errors that occur in an uncorrelated fashion and thus tend to occur as isolated single errors. While there have been a number of codes designed to correct burst errors, the theory of random error-correcting codes is so well developed that designers have often chosen to use these codes in concert with a method to "randomize" error bursts.

This randomization method, called *interleaving,* rearranges, or scrambles, the coded symbols in order to minimize this correlation so that errors are isolated and distributed across a number of code words. Thus, a modest random-error correcting code can be combined with interleaving that is inserted between the channel coder and the modulator to shuffle the symbols of the code words. Then, in the receiver, the de-interleaver is placed between the demodulator and the decoder is reassemble the code words for decoding.

We note that a well-designed coding/interleaving system does more than redistribute errors for easy correction: it also exploits *time diversity.* In our discussion of spread spectrum and CDMA, we saw how the DS/SS signal exploits the frequency diversity of the wireless channel via its multipath redundancy. Here, the redundancy added by channel coding/interleaving is designed so that, in addition to the usual performance increase due to just the code — the *coding gain* — there is also a benefit to distributing the redundancy in such a way that exploits the time variation of the channel, yielding a *time diversity gain.*

In this era of digital data over wireless, high link reliability is required. This is in spite of the fact that most wireless links have a raw BER on the order of 1 in 1000. Clearly, we would like to see an error rate of 1 in 10^{12} or better. How is this astounding improvement achieved? The following two-level approach has proved successful. The first level employs FEC to correct a large percentage of the errors. This code is used in tandem with a powerful error-detecting algorithm to find the rare errors that the FEC cannot find and correct. This combined FEC/ARQ approach limits the amount of feedback to an acceptable level while still achieving the necessary reliability.

2.9 Nonlinear Channels

Amplifiers are more power efficient if they are driven closer to saturation than if they are kept within their linear regions. Unfortunately, nonlinearities that occur as saturation is approached lead to *spectral spreading* of the signal. This can be illustrated by observing that an instantaneous (or memoryless) nonlinearity can be approximated by a polynomial. For example, a quadratic term effectively squares the signal; for a sinusoidal input this leads to double-frequency terms.

A more sophisticated perspective comes from noting that the nonlinear amplification can distort the symbol pulse shape, expanding the spectrum of the pulse. Nonlinearities of this type are said to cause AM/AM distortion. Amplifiers can also exhibit AM/PM conversion, where the output phase of a sinusoid is shifted by different amounts depending on its input power — a serious problem for PSK-type modulations.

A great deal of effort has gone into finding transmitter designs that allow more efficient amplifier operation. For example, constant-envelope modulation schemes are insensitive to nonlinearities, and signaling schemes that reduce the peak-to-average power ratio (PAPR) of the signal allow higher levels. Finally, methods to linearize amplifiers at higher efficiencies are receiving considerable attention.

Modeling and simulating nonlinear effects on system performance is a nontrivial task. AM/AM and AM/PM distortions are functions of frequency, so if wideband amplifier characterization is required, a family of curves is necessary. Even then the actual wideband response is only approximated, since these systems are limited in bandwidth and thus have memory. More accurate results in this case can be obtained using Volterra series expansions, or numerical solutions to nonlinear differential equations. Sophisticated approaches are becoming increasingly important in cellular as supported data rates move higher and higher. More information can be found in References 1 and 5 and the references therein.

2.10 Antenna Arrays

We have seen earlier how sectorized antennas can be used to increase system performance. They are one of the most economical forms of multielement antenna systems, and can be used to reduce interference or to increase user capacity. A second use of multielement systems is to exploit the *spatial diversity* of the wireless channel. Spatial diversity approaches assume that the received antenna elements are immersed in a signal field whose strength varies strongly with position due to a superposition of multipath signals arriving via various directions. The resulting element signal strengths are assumed to be at least somewhat statistically uncorrelated. This spatial uncorrelatedness is analogous to the uncorrelatedness over time or frequency that is exploited in mobile channels.

One of the simplest approaches is to use multiple (normally omnidirectional in azimuth) antenna elements at the receiver, and choose the one with the highest signal-to-noise ratio. More sophisticated schemes combine — rather than select just one of — the element signals to further improve the signal-to-noise ratio at the cost of higher receiver complexity. These approaches date from the 1950s, and do not take into account other interfering mobile units. These latter schemes are often grouped under the category of *antenna diversity* approaches.

More recently, a number of proposals for systems that combine error-control coding mechanisms with multiple elements have been made under the name of *space-time coding*. One of the main contributions of these efforts has been the recognition that multiple-element transmit antennas can, under certain conditions, dramatically increase the link capacity.

Another approach, beamforming or phased-array antennas, is also positioned to play a role in future systems under the new moniker *smart antennas*. Space-time coding and smart antenna methods can be seen as two approaches to exploiting the capabilities of multiple-input/multiple-output (MIMO) systems. However, in contrast to space-time coding approaches, strong interelement correlation based on the direction of arrival of plane waves is assumed in smart antennas. The basic idea of smart antennas is to employ an array of antenna elements connected to an amplitude- and phase-shifting network to adaptively tune (steer electronically) the antenna pattern based on the geographic placement of mobile units. Much of the groundwork for smart antenna systems was laid in the 1950s in military radar research. The ultimate goal of these approaches can be stated as follows: to track individual mobile units with optimized antenna patterns that maximize performance (by maximizing the ratio of the signal to the sum of interference and noise) minimize power consumption at the mobile unit, and optimize the capacity of the cellular system. One can conjecture that the ultimate solution to this problem will be a class of techniques that involve joint design of channel coding, modulation, and antenna array processing in an optimum fashion.

2.11 Summary

Almost all wireless networks are distinguished by the characteristic of a shared channel resource, and this is in fact the key difference between wireless and wired networks. Another important difference between wired and wireless channels is the presence of multipath in the latter, which makes diversity possible. What is it that distinguishes cellular from other wireless services and systems? First, it historically has been designed for mobile telephone users, and has been optimized for carrying human voice. This has led to the following key traits of cellular:

- Efficient use of spectrum via the cellular concept
- System designs, including channel access mechanisms, that efficiently handle large numbers of uniform (i.e., voice) links

- Difficult channels: user mobility causes fast variations in channel signal characteristics compared with other wireless applications such as wireless local area networks

We close by mentioning two apparent trends. First, as we mentioned at the outset of this article, wireless local loop services, where home telephone subscribers use wireless phones — and the "last mile" is wireless rather than copper — are a new application for mobile wireless technology. Second, at this time there is a great deal of effort to make next-generation cellular systems useful for data networking in addition to voice. Certainly, the amount of data traffic on these networks will grow. However, one of the largest questions for the next ten years is whether mobile wireless will win the growing data market, or if new data-oriented wireless networks will come to dominate.

References

1. Zeng, M., Annamalai, A., and Bhargava, V. K., Recent advances in cellular wireless communications, *IEEE Communications Magazine,* 37, 9, 128–138, September 1999.
2. Walrand, J., and Varaiya, P., *High-Performance Communication Networks,* Morgan Kaufman, San Francisco, CA, 1996.
3. Chaudhury, P., Mohr, W., and Onoe, S., The 3GPP proposal for IMT-2000, *IEEE Communications Magazine,* 37, 12, 72–81, December 1999.
4. Anderson, J. B., *Digital Transmission Engineering,* IEEE Press, Piscataway, NJ, 1999.
5. Lee, E. A., and Messerschmitt, D. G., *Digital Communication,* 2nd ed., Kluwer Academic, 1994.
6. Proakis, J. G., *Digital Communications,* 3rd ed., McGraw-Hill, New York, 1995.
7. Haykin, S., *Communication Systems,* Wiley, New York, 1994.
8. Proakis, J. G., and Salehi, M., *Communication Systems Engineering,* Prentice-Hall, Englewood Cliffs, NJ, 1994.
9. Flikkema, P., Introduction to spread spectrum for wireless communication: a signal processing perspective, *IEEE Spectrum Processing Magazine,* 14, 3, 26–36, May 1997.
10. Viterbi, A. J., *CDMA: Principles of Spread Spectrum Communication,* Addison-Wesley, Reading, MA, 1995.
11. Shannon, C. E., Communication in the presence of noise, *Proceedings of the IRE,* 37, 1, 10–21, January 1949.
12. Wyner, A. D., and Shamai (Shitz), S., Introduction to "Communication in the presence of noise" by C. E. Shannon, *Proceedings of the IEEE,* 86, 2, 442–446, February 1998. Reprinted in the *Proceedings of the IEEE,* 86, 2, February 1998, 447–457.
13. Jeruchim, M. C., Balaban, P., and Shanmugan, K. S., *Simulation of Communication Systems,* Plenum, New York, 1992.

3

Nomadic Communications

Andy D. Kucar
4U Communications Research, Inc.

Nomadic peoples of desert oases, tropical jungles, steppes, tundras, and polar regions have shown a limited interest in mobile radio communications, at the displeasure of some urbanite investors in mobile radio communications. The focus of this contribution with a delegated title *Nomadic Communications* is on terrestrial and satellite mobile radio communications used by urbanites while roaming urban canyons or golf courses, and by suburbanites who, every morning, assemble their sport utility vehicles and drive to urban jungles hunting for jobs. The habits and traffic patterns of these users are important parameters in the analysis and optimization of any mobile radio communications system. The mobile radio communications systems addressed in this contribution and illustrated in Fig. 3.1 include:

1. The first generation *analog cellular mobile radio systems* such as North American AMPS, Japanese MCS, Scandinavian NMT, and British TACS. These systems use analog voice data and frequency modulation (FM) for the transmission of voice, and coded digital data and a frequency-shift keying (FSK) modulation scheme for the transmission of control information. Conceived and designed in the 1970s, these systems were optimized for vehicle-based services such as police and ambulances operating at possibly high vehicle speeds. The first-generation *analog cordless telephones* include CT0 and CT1 cordless telephone systems, which were intended for use in the household environment;
2. The second-generation *digital cellular and personal mobile radio systems* such as *Global System for Mobile Communications* (GSM), *Digital* AMPS > IS–54/136, DCS 1800/1900, and *Personal Digital Cellular* (PDC), all Time Division Multiple Access (TDMA), and IS–95 *spread-spectrum Code* Division Multiple Access (CDMA) systems. All mentioned systems employ digital data for both voice and control purposes. The second-generation *digital cordless telephony systems* include CT2,

0-8493-1732-0/03/$0.00+$1.50

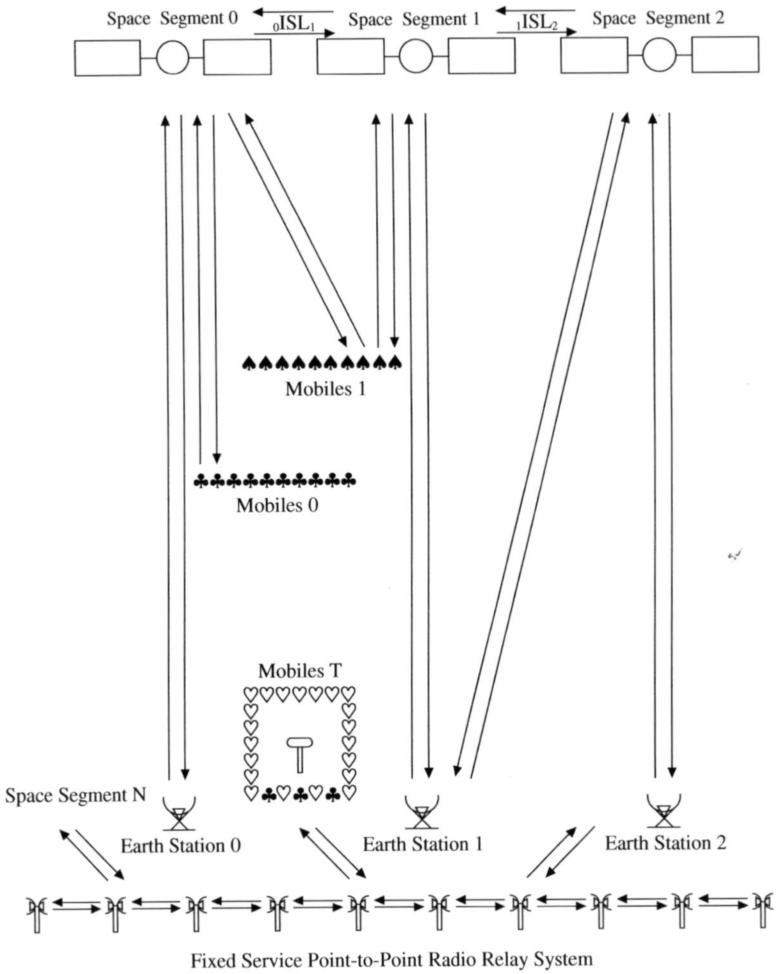

FIGURE 3.1 A model of fixed and mobile, satellite and terrestrial systems.

CT2Plus, CT3, *Digital Enhanced Cordless Telephone* (DECT), and *Personal Handyphone System* (PHS); *wireless data mobile radio systems* such as ARDIS, RAM, TETRA, *Cellular Digital Packet Data* (CDPD), IEEE 802.11 *Wireless Local Area Network* (WLAN), and recently announced Bluetooth; there are also projects known as *Personal Communication Network* (PCN), *Personal Communications Systems* (PCS) and FPLMTS > UMTS > IMT–2000 > 3G, where 3G stands for the third-generation systems. The second-generation systems also include *satellite mobile radio systems* such as INMARSAT, OmniTRACS, MSAT, AUSSAT, Iridium, Globalstar, and ORBCOMM.

After a brief prologue and historical overview, technical issues such as the repertoire of systems and services, the airwaves management, the operating environment, service quality, network issues and cell size, channel coding and modulation, speech coding, diversity, multiple broadcasting (FDMB, TDMB, CDMB), and multiple access (FDMA, TDMA, CDMA) are briefly discussed.

Many existing mobile radio communications systems collect some form of information on network behavior, users' positions, etc., with the purpose of enhancing the performance of communications, improving handover procedures and increasing the system capacity. Coarse positioning is usually achieved inherently, while more precise positioning and *navigation* can be achieved by employing LORAN-C and/

TABLE 3.1 Glossary of Terms

AMPS	Advanced Mobile Phone Service
ASIC	Application Specific Integrated Circuits
BER	Bit Error Rate
CAD	Computer-Aided Design
CB	Citizen Band (Mobile Radio)
CDMA	Spread-Spectrum Code Division Multiple Access
CEPT	Conference of European Postal and Telecommunications (Administrations)
CT	Cordless Telephony
DOC	Department of Communications (in Canada)
DSP	Digital Signal Processing
FCC	Federal Communications Commission (in U.S.)
FDMA	Frequency Division Multiple Access
FPLMTS	Future Public Land Mobile Telecommunications Systems
GDSS	Global Distress Safety System
GOES	Geostationary Operational Environmental Satellites
GPS	Global Positioning System
GSM	Groupe Spécial Mobile (now Global System for Mobile Communications)
ISDN	Integrated Service Digital Network
ITU	International Telecommunications Union
MOS	Mean Opinion Score
MMIC	Microwave Monolithic Integrated Circuits
NMC	Network Management Center
NMT	Nordic Mobile Telephone (System)
PCN	Personal Communications Networks
PCS	Personal Communications Systems
PSTN	Public Switched Telephone Network
SARSAT	Search and Rescue Satellite Aided Tracking System
SERES	SEarch and REscue Satellite
TACS	Total Access Communication System
TDMA	Time Division Multiple Access
WAAS	Wide Area Augmentation System
WARC	World Administrative Radio Conference
WRC	World Radiocommunications Conference

Source: 4U Communications Research Inc., 2000.06.10~00:09, Updated: 2000.05.03

or GPS, GLONASS, WAAS signals, or some other means, at an additional, usually modest, increase in cost and complexity.

3.1 Prologue

Mobile radio systems provide their users with opportunities to travel freely within the service area while being able to communicate with any telephone, fax, data modem, and electronic mail subscriber anywhere in the world; to determine their own positions; to track the precious cargo; to improve the management of fleets of vehicles and the distribution of goods; to improve traffic safety; to provide vital communication links during emergencies, search and rescue operations, to browse their favorites Websites, etc. These *tieless* (*wireless, cordless*) communications, the exchange of information, and the determination of position, course, and distance traveled are made possible by the unique property of the radio to employ an *aerial* (*antenna*) for radiating and receiving electromagnetic waves. When the user's radio antenna is stationary over a prolonged period of time, the term *fixed radio* is used. A radio transceiver capable of being carried or moved around, but stationary when in operation, is called a *portable radio.* A radio transceiver capable of being carried and used, by a vehicle or by a person on the move, is called *mobile radio, personal and/or handheld device.* Individual radio users may communicate directly, or via one or more intermediaries, which may be *passive radio repeater(s), base station(s),* or *switch(es).* When all intermediaries are located on Earth, the terms *terrestrial radio system* and *radio system* have been used. When at least one intermediary is satellite borne, the terms *satellite radio system* and *satellite system* have been used. According to the location of a user, the terms *land, maritime, aeronautical, space,* and *deep-space radio systems* have been used. The second unique property of all terrestrial and satellite radio systems is that they share the same natural resource — the *airways* (*frequency bands* and *space*).

Recent developments in microwave monolithic integrated circuit (MMIC), application specific integrated circuit (ASIC), analog/digital signal processing (A/DSP), and battery technology, supported by computer-aided design (CAD) and robotics manufacturing allow the viable implementation of miniature radio transceivers at radio frequencies as high as 6 GHz (i.e., at wavelengths as short as about 5 cm). Up to these frequencies additional spectra have been assigned to mobile services; corresponding shorter wavelengths allow a viable implementation of adaptive antennas necessary for improvement of the quality of transmission and spatial frequency spectrum efficiency. The continuous flux of market forces (excited by the possibilities of a myriad of new services and great profits), international and domestic standard forces (who manage a common natural resource — the airwaves), and technology forces (capable of creating viable products) acted harmoniously and created a broad choice of communications (voice and data), information, and navigation systems, which propelled the explosive growth of mobile radio services for travelers.

3.2 A Glimpse of History

Late in the 19th century, Heinrich Rudolf Hertz, Nikola Tesla, Alexander Popov, Edouard Branly, Oliver Lodge, Jagadis Chandra Bose, Guglielmo Marconi, Adolphus Slaby, and other engineers and scientists experimented with the transmission and reception of electromagnetic waves. In 1898 Tesla made a demonstration in Madison Square Garden of a radio remote controlled boat; later the same year Marconi established the first wireless ship-to-shore telegraph link with the royal yacht Osborne. These events are now accepted as the birth of the mobile radio. Since that time, mobile radio communications have provided safe navigation for ships and airplanes, saved many lives, dispatched diverse fleets of vehicles, won many battles, generated many new businesses, etc.

Satellite mobile radio systems launched in the seventies and early eighties use ultrahigh frequency (UHF) bands around 400 MHz and around 1.5 GHz for communications and navigation services.

In the fifties and sixties, numerous private mobile radio networks, citizen band (CB) mobile radio, ham operator mobile radio, and portable home radio telephones used diverse types and brands of radio equipment and chunks of airwaves located anywhere in the frequency band from near 30 MHz to 3 GHz. Then, in the seventies, Ericsson introduced the *Nordic Mobile Telephone* (NMT) system, and AT&T Bell Laboratories introduced *Advanced Mobile Phone Service* (AMPS). The impact of these two *public land mobile telecommunication systems* on the standardization and prospects of mobile radio communications may be compared with the impact of Apple and IBM on the personal computer industry. In Europe systems like AMPS competed with NMT systems; in the rest of the world, AMPS, backed by Bell Laboratories' reputation for technical excellence and the clout of AT&T, became *de facto* and *de jure* the technical standard (on which British TACS and Japanese MCS–L1 are based). In 1982, the Conference of European Postal and Telecommunications Administrations (CEPT) established Groupe Spécial Mobile (GSM) with the mandate to define future Pan-European cellular radio standards. On January 1, 1984, during the phase of explosive growth of AMPS and similar cellular mobile radio communications systems and services, came the divestiture (breakup) of AT&T.

3.3 Present and Future Trends

Based on the solid foundation established in 1970s the buildup of mobile radio systems and services at the end of the second millennium is continuing at an annual rate higher than 20%, worldwide. Terrestrial mobile radio systems offer analog and digital voice and low- to medium-rate data services compatible with existing public switching telephone networks in scope, but with poorer voice quality and lower data throughput. Wireless mobile radio data networks are expected to offer data rates as high as a few Mbit/s in the near future and even more in the portable environment.

Equipment miniaturization and price are important constraints on the systems providing these services. In the early 1950s, mobile radio equipment used a considerable amount of a car's trunk space and challenged the capacity of car's alternator/battery source while in transmit mode. Today, the pocket-size, ≈100-g handheld cellular radio telephone, manual and battery charger excluded, provides a few hours of talk capacity and dozens of hours in the standby mode. The average cost of the least expensive models of battery powered cellular mobile radio telephones has dropped proportionally and has broken the $100 U.S. barrier. However, one should take the price and growth numbers with a grain of salt, since some prices and growth itself might be subsidized. Many customers appear to have more than one telephone, at least during the promotion period, while they cannot use more than one telephone at the same time. These facts need to be taken into consideration while estimating growth, capacity, and efficiency of recent wireless mobile radio systems.

Mobile satellite systems are expanding in many directions: large and powerful single unit geostationary systems; medium-sized, low orbit multi-satellite systems; and small-sized, and low orbit multi-satellite systems, launched from a plane, see [Kucar, 1992], [Del Re, 1995]. Recently, some financial uncertainties experienced by a few technically advanced LEO satellite systems, operational and planned, slowed down explosive growth in this area. Satellite mobile radio systems currently offer analog and digital voice, low to medium rate data, radio determination, and global distress safety services for travelers.

During the last five years numerous new digital radio systems for mobile users have been deployed. Presently, users in many countries have been offered between 5 and 10 different mobile radio communications systems to choose from. There already exists radio units capable of operating on two or more different systems using the same frequency band or even using a few different frequency bands. Overviews of mobile radio communications systems and related technical issues can be found in [Davis, 1984], [Cox, 1987], [Mahmoud, 1989], [Kucar, 1991], [Rhee, 1991], [Steele, 1992], [Chuang, 1993], [Cox, 1995], [Kucar, 1991], [Cimini, March 1999], [Mitola, 1999], [Cimini, July 1999], [Ariyavisitakul, 1999], [Cimini, November 1999], [Oppermann, 1999] and [Oppermann, 2000].

3.4 Repertoire of Systems and Services

The variety of services offered to travelers essentially consists of information in analog and/or digital form. Although most of today's traffic consists of analog or digital voice transmitted by analog frequency modulation FM (or phase modulation PM), or digital quadrature amplitude modulation (QAM) schemes, digital signaling, and a combination of analog and digital traffic, might provide superior frequency reuse capacity, processing, and network interconnectivity. By using a powerful and affordable microprocessor and digital signal processing chips, a myriad of different services particularly well suited to the needs of people on the move could be realized economically. A brief description of a few elementary systems/services currently available to travelers follows. Some of these elementary services can be combined within the mobile radio units for a marginal increase in the cost and complexity with respect to the cost of a single service system; for example, a mobile radio communications system can include a positioning receiver, digital map, Web browser, etc.

Terrestrial systems. In a terrestrial mobile radio network labeled Mobiles T in Fig. 3.1, a repeater was usually located at the nearest summit offering maximum service area coverage. As the number of users increased, the available frequency spectrum became unable to handle the increase traffic, and a need for frequency reuse arose. The service area was split into many subareas called cells, and the term *cellular radio* was born. The frequency reuse offers an increased overall system capacity, while the smaller cell size can offer an increased service quality, but at the expense of increased complexity of the user's terminal and network infrastructure. The trade-offs between real estate complexity, and implementation dynamics dictate the shape and the size of the cellular network. Increase in the overall capacity calls for new frequency spectra, smaller cells, which requires reconfiguration of existing base station locations; this is usually not possible in many circumstances, which leads to suboptimal solutions and even less efficient use of the frequency spectrum.

The *satellite systems* shown in Fig. 3.1 employ one or more satellites to serve as base station(s) and/ or repeater(s) in a mobile radio network. The position of satellites relative to the service area is of crucial importance for the coverage, service quality, price, and complexity of the overall network. When a satellite encompasses Earth in 12-h, 24-h, etc. periods, the term *geosynchronous orbit* has been used. An orbit inclined with respect to the equatorial plane is called an *inclined orbit*; an orbit with an inclination of about 90° is called a *polar orbit*. A circular geosynchronous 24-h orbit in the equatorial plane (0° inclination) is known as the *geostationary orbit* (GSO), since from any point on the surface of Earth, the satellite appears to be stationary; this orbit is particularly suitable for land mobile services at low latitudes, and for maritime and aeronautical services at latitudes of $< |80|°$. Systems that use geostationary satellites include INMARSAT, MSAT, and AUSSAT. An elliptic geosynchronous orbit with the inclination angle of 63.4° is known as *Tundra orbit*. An elliptical 12-h orbit with the inclination angle of 63.4° is known as *Molniya orbit*. Both Tundra and Molniya orbits have been selected for coverage of the Northern latitudes and the area around the North Pole — for users at those latitudes the satellites appear to wander around the zenith for a prolonged period of time. The coverage of a particular region (*regional coverage*), and the whole globe (*global coverage*), can be provided by different constellations of satellites including ones in inclined and polar orbits. For example, inclined circular orbit constellations have been used by GPS (18 to 24 satellites, 55° to 63° inclination), Globalstar (48 satellites, 47° inclination), and Iridium (66 satellites, 90° inclination — polar orbits) system. All three systems provide global coverage. ORBCOM system employs Pegasus launchable low-orbit satellites to provide uninterrupted coverage of Earth below ±60° latitudes, and an intermittent, but frequent coverage over the polar regions.

Satellite antenna systems can have one (*single beam global system*) or more beams (*multibeam spot system*). The multibeam satellite system, similar to the terrestrial cellular system, employs antenna directivity to achieve better frequency reuse, at the expense of system complexity.

Radio paging is a non-speech, one-way (from base station toward travelers), personal selective calling system with alert, without message, or with defined messages such as numeric or alphanumeric. A person wishing to send a message contacts a system operator by public switched telephone network (PSTN), and delivers his message. After an acceptable time (queuing delay), a system operator forwards the message to the traveler, by radio repeater (FM broadcasting transmitter, VHF or UHF dedicated transmitter, satellite, or cellular radio system). After receiving the message, a traveler's small (roughly the size of a cigarette pack) receiver (pager) stores the message in its memory, and on demand either emits alerting tones or displays the message.

Global Distress Safety System (*GDSS*) geostationary and inclined orbit satellites transfer emergency calls sent by vehicles to the central earth station. Examples are: *COSPAS*, Search and Rescue Satellite-Aided Tracking system, *SARSAT*, Geostationary Operational Environmental Satellites *GOES*, and SEarch and REscue Satellite *SERES*). The recommended frequency for this transmission is 406.025 MHz.

Global Positioning System (*GPS*), [ION, 1980, 1984, 1986, 1993]. U.S. Department of Defense Navstar GPS 24–29 operating satellites in inclined orbits emit L-band (L1 = 1575.42 MHz, L2 = 1227.6 MHz) spread-spectrum signals from which an intelligent microprocessor-based receiver extracts extremely precise time and frequency information, and accurately determines its own three-dimensional position, velocity, and acceleration, worldwide. The coarse accuracy of <100 m available to commercial users has been demonstrated by using a handheld receiver. An accuracy of meters or centimeters is possible by using the precise (military) codes and/or differential GPS (additional reference) principals and kinematic phase tracking.

Glonass is Russia's counterpart of the U.S.'s GPS. It operates in an FDM mode and uses frequencies between 1602.56 and 1615.50 MHz to achieve goals similar to GPS.

Other systems have been studied by the European Space Agency (*Navsat*), and by West Germany (*Granas, Popsat,* and *Navcom*). In recent years many payloads carrying navigation transponders have been put on board GSO satellites; corresponding navigational signals enable an increase in overall availability and improved determination of user positions. The comprehensive project, which may include existing and new radionavigation payloads, has also been known as the *Wide Area Augmentation System* (*WAAS*).

LORAN-C is the 100-kHz frequency navigation system that provides a positional accuracy between 10 and 150 m. A user's receiver measures the time difference between the master station transmitter and secondary station signals, and defines his hyperbolic line of position. North American LORAN-C coverage includes the Great Lakes, Atlantic, and Pacific Coast, with decreasing signal strength and accuracy as the user approaches the Rocky Mountains from the east. Recently, new LORAN stations have been augmented worldwide. Similar radionavigation systems are the 100-kHz *Decca* and 10-kHz *Omega.*

Dispatch two-way radio land mobile or satellite systems, with or without connection to the PSTN, consist of an operating center controlling the operation of a fleet of vehicles such as aircraft, taxis, police cars, tracks, and rail cars.

A summary of some of the existing and planned terrestrial mobile radio systems, including MOBITEX RAM and ARDIS, is given in Table 3.2.

OmniTRACS dispatch system employs a Ku-band geostationary satellite located at 103° W to provide two-way digital message and position reporting (derived from incorporated satellite-aided LORAN-C receiver), throughout the contiguous U.S. (CONUS).

Cellular radio or public land mobile telephone systems offer a full range of services to the traveler similar to those provided by PSTN. The technical characteristics of some of the existing and planned systems are summarized in Table 3.3.

Vehicle Information System and Intelligent Highway Vehicle System are synonyms for the variety of systems and services aimed toward traffic safety and location. This includes: traffic management, vehicle identification, digitized map information and navigation, radio navigation, speed sensing and adaptive cruise control, collision warning and prevention, etc. Some of the vehicle information systems can easily be incorporated in mobile radio communications transceivers to enhance the service quality and capacity of respective communications systems.

TABLE 3.2 Comparison of Dispatch WAN/LAN Systems

Parameter	U.S.	Sweden	Japan	Australia	CDPD	IEEE 802.11
TX freq, MHz						
Base	935–941	76.0–77.5	850–860	865.00–870.00	869–894	2400–2483
	851–866			415.55–418.05		2470–2499
Mobile	896–902	81.0–82.5	905–915	820.00–825.00	824–849	2400–2483
	806–821			406.10–408.60		2470–2499
Duplexing method	sfFDD[a]	sFDD	sFDD	sfFDD	FDD	TDD
Channel spacing, kHz	12.5	25.0	12.5	25.0	30.0	1000
	25.00			12.5		
Channel rate, kb/s	≤9.6	1.2	1.2	≤	19.2	1000
# of Traffic channel	480	60	799	200	832	79
	600					
Modulation type:						
Voice	FM	FM	FM	FM		
Data	FSK	MSK-FM	MSK-FM	FSK	GMSK	DQPSK

[a] sfFDD stands for semi-duplex, full duplex, Frequency Division Duplex.
Similar systems are used in the Netherlands, U.K., USSR, and France.
ARDIS is a commercial system compatible with U.S. specs. 25-kHz spacing; 2 FSK, 4 FSK, ≤19.2 kb/s.
MOBITEX/RAM is a commercial system compatible with U.S. specs. 12.5-kHz spacing; GMSK, 8.0 kb/s.
Source: 4U Communications Research Inc., 2000.06.10~00:09, c:/tab/dispatch.sys

TABLE 3.3 Comparison of Cellular Mobile Radio Systems in Bands Below 1 GHz

Parameter	System Name									
	AMPS NAMPS	MCS L1 MCS L2	NMT 900	NMT 450	R.com 2000	C450	TACS UK	GSM	IS–54 IS–136	IS–95 USA	PDC Japan
TX freq, MHz											
Base	869–894	870–885	935–960	463–468	424.8–428	461–466	935–960	890–915	869–894	869–894	810–826
Mobile	824–849	925–940	890–915	453–458	414.8–418	451–456	890–915	935–960	824–849	824–849	890–915
Max b/m eirp, dBW	22/5	19/7	22/7	19/12	20.10	22/12	22/8	27/9	27/9	/–7	/5
Multiple access	F	F	F	F	F	F	F	F/T	F/T	F/C	F/T
Duplex method	FDD	FDD	FDD	FDD	FDD	FDD	FDD	FDD	FDD	FDD	FDD
Channel bw, kHz	30.0 10.0	25.0 12.5	12.5	25.0	12.5	20.0 10.0	25.0 12.5	200.0	30.0	1250	25
Channels/RF	1	1	1	1	1	1	1	8	3	42	3
Channels/band	832 2496	600 1200	1999	200	160	222	1000	125 × 8	832 × 3	n × 42	640 × 3
Voice/Traffic:	analog	analog	analog	analog	analog	analog	analog	RELP	VSELP	CELP	VSELP
comp. or kb/s	2:1	2:1	2:1	2:1	2:1	2:1	2:1	13.0	8.0	≤9.6	6.7
modulation	PM	PM	PM	PM	PM	PM	PM	GMSK π/4	B/OQ	π/4	
kHz and/or kb/s	±12	±5	±5	±5	±2.5	±4	±9.5	270.833	48.6	1228.8	42.0
Control:	digital	digital	digital	digital	digital	digital	digital	digital	digital	digital	digital
modulation	FSK	FSK	FFSK	FFSK	FFSK	FSK	FSK	GMSK	π/4	B/OQ	π/4
bb waveform	Manch. NRZ	Manch. NRZ	NRZ	NRZ	Manch.	NRZ	NRZ	NRZ	NRZ		
kHz and/or kb/s	±8.0/10	±4.5/0.3	±3.5/1.2	±3.5/1.2	±1.7/1.2	±2.5/5.3	±6.4/8.0	270.833	48.6	1228.8	42.0
Channel coding:	BCH	BCH	B1 Hag.	B1 Hag.	Hag.	BCH	BCH	RS	Conv.	Conv.	Conv.
base→mobile	(40,28)	(43,31)	burst	burst	(19,6)	(15,7)	(40,28)	(12,8)	1/2	6/11	9/17
mobile→base	(48,36)	a.(43,31) p.(11,07)	burst	burst	(19,6)	(15,7)	(48,36)	(12,8)	1/2	1/3	9/17

Note: Multiple Access: F = Frequency Division Multiple Access (FDMA), F/T = Hybrid Frequency/Time DMA, F/C = Hybrid Frequency/Code DMA.
π/4 corresponds to the π/4 shifted differentially encoded QPSK with $\alpha = 0.35$ square root raised–cosine filter for IS–136 and $\alpha = 0.5$ for PDC.
B/OQ corresponds to the BPSK outbound and OQPSK modulation scheme inbound.
comp. or kb/s stands for syllabic compandor or speech rate in kb/s; kHz and/or kb/s stands for peak deviation in kHz and/or channel rate kb/s.
IS–634 standard interface supports AMPS, NAMPS, TDMA, and CDMA capabilities.
IS–651 standard interface supports **A** GSM capabilities and **A**+ CDMA capabilities.
Source: 4U Communications Research Inc., 2000.06.10~00:09

3.5 Airwaves Management

The airwaves (frequency spectrum and the space surrounding us) are a limited natural resource shared by several different radio users (military, government, commercial, public, amateur, etc.). Its sharing of (among different users, services described in the previous section, TV and sound broadcasting, etc.), coordination, and administration is an ongoing process exercised on national, as well as on international levels. National administrations (Federal Communications Commission (FCC) in the U.S., Department of Communications (DOC), now Industry Canada, in Canada, etc.), in cooperation with users and industry, set the rules and procedures for planning and utilization of scarce frequency bands. These plans and utilizations have to be further coordinated internationally.

The International Telecommunications Union (ITU) is a specialized agency of the United Nations, stationed in Geneva, Switzerland, with more than 150 government and corporate members, responsible for all policies related to Radio, Telegraph, and Telephone. According to the ITU, the world is divided into three regions: Region 1 — Europe, including the former Soviet Union, Outer Mongolia, Africa, and the Middle East west of Iran; Region 2 — the Americas and Greenland; and Region 3 — Asia (excluding parts west of Iran and Outer Mongolia), Australia, and Oceania. Historically, these three regions have developed, more or less independently, their own frequency plans, which best suit local purposes. With the advent of satellite services and globalization trends, the coordination between different regions becomes more urgent. Frequency spectrum planning and coordination is performed through ITU's bodies such as: Comité Consultatif de International Radio (CCIR), now ITU-R, International Frequency Registration Board (IFRB), now ITU-R, World Administrative Radio Conference (WARC), and Regional Administrative Radio Conference (RARC).

ITU-R, through its study groups, deals with technical and operational aspects of radio communications. Results of these activities have been summarized in the form of reports and recommendations published every four years, or more often [ITU, 1990]. IFRB serves as a *custodian* of a common and scarce natural resource — the *airwaves*; in its capacity, the IFRB records radio frequencies, advises the members on technical issues, and contributes on other technical matters. Based on the work of ITU-R and the national administrations, ITU members convene at appropriate RARC and WARC meetings, where documents on frequency planning and utilization, the *Radio Regulations,* are updated. The actions on a national level follow, see [RadioRegs, 1986], [WARC, 1992], [WRC, 1997]. The far-reaching impact of mobile radio communications on economies and the well-being of the three main trading blocks, other developing and third-world countries, and potential manufacturers and users, makes the airways (frequency spectrum) even more important.

The International Telecommunications Union (ITU) recommends the composite *bandwidth-space-time* domain concept as a measure of spectrum utilization. The *spectrum utilization factor* $U = B \cdot S \cdot T$ is defined as a product of the frequency bandwidth B, spatial component S, and time component T. Since mobile radio communications systems employ single omnidirectional antennas, their S factor will be rather low; since they operate in a single channel arrangement, their B factor will be low; since new digital schemes tend to operate in a packet/block switching modes which are inherently loaded with a significant amount of overhead and idle traffic, their T factor will be low as well. Consequently, mobile radio communications systems will have poor spectrum utilization factors.

The model of a mobile radio environment, which may include different sharing scenarios with fixed service and other radio systems, can be described as follows. Objects of our concern are *events* (for example, conversation using a mobile radio, measurements of amplitude, phase and polarization at the receiver) occurring in *time* $\{u^0\}$, *space* $\{u^1, u^2, u^3\}$, *space time* $\{u^0, u^1, u^2, u^3\}$, *frequency* $\{u^4\}$, *polarization* $\{u^5, u^6\}$, and *airwaves* $\{u^0, u^1, u^2, u^3, u^4, u^5, u^6\}$, see Table 3.4. The coordinate $\{u^4\}$ represents frequency resource (i.e., bandwidth in the space time $\{u^0, u^1, u^2, u^3\}$). Our goal is to use a scarce natural resource — the airwaves in an environmentally friendly manner.

When users/events are divided (sorted, discriminated) along the time coordinate u^0, the term *time division* is employed for function $f(u^0)$. A division $f(u^4)$ along the frequency coordinate u^4 corresponds to the *frequency division.* A division $f(u^0, u^4)$ along the coordinates (u^0, u^4) is usually called a *code division*

TABLE 3.4 The Multidimensional Spaces Including the Airwaves

u^0	time		
u^1			
u^2	space	space time	
u^3			airwaves
u^4	frequency/bandwidth		
u^5	polarization		
u^6			
u^7			
u^8	Doppler		
u^9			
u^A	users: government/military, commercial/public, fixed/mobile, terrestrial/satellite …		
u^B			
⋮			
u^n			

Source: 4U Communications Research Inc., 2000.06.10~00:09, c:/tab/airwaves.1

or *frequency hopping*. A division $f(u^1, u^2, x^3)$ along the coordinates (u^1, u^2, u^3) is called the *space division*. Terrestrial cellular and multibeam satellite radio systems are vivid examples of the space division concepts. Coordinates $\{u^5, u^6\}$ may represent two orthogonal polarization components, horizontal and vertical or right-handed and left-handed circular; a division of users/events according to their polarization components may be called the *polarization division*. Any division $f(u^0, u^1, u^2, u^3, u^4, u^5, u^6)$ along the coordinates $(u^0, u^1, u^2, u^3, u^4, u^5, u^6)$ may be called the *airwaves division*. Coordinates $\{u^7, u^8, u^9\}$ may represent velocity (or Doppler frequency) components; a division of users/events according to their Doppler frequencies similar to the moving target indication (MTI) radars may be called the *Doppler frequency division*. We may also introduce coordinate $\{u^4\}$ to represent users, divide the airways along the coordinate $\{u^A\}$ (military, government, commercial, public, fixed, mobile, terrestrial, satellite, and others) and call it the *users division*. Generally, the segmentations of frequency spectra to different users lead to uneven use and uneven spectral efficiency factors among different segments.

In analogy with division, we may have time, space, frequency, code, airwaves, polarization, Doppler, users, $\{u^\alpha, \ldots, u^\omega\}$ *access* and *diversity*. Generally, the signal space may be described by m coordinates $\{u^0, \ldots, u^{m-1}\}$. Let each signal component has k degrees of freedom. At the transmitter site, each signal can be radiated via n_T antennas, and received at n_R receiver antennas. There is a total of $n = n_T + n_R$ antennas, two polarization components, and L multipath components, i.e., paths between transmitter and receiver antennas. Thus, the total number of degrees of freedom $m = k \times n \times 2 \times L$. For example, in a typical land mobile environment 16 multipath components can exist; if one wants to study a system with four antennas on the transmitter side and four antennas on the receiver side, and each antenna may employ both polarizations, then the number of degrees of freedom equals $512 \times k$. By selecting a proper coordinate system and using inherent system symmetries, one might be able to reduce the number of degrees of freedom to a manageable quantity.

3.6 Operating Environment

A general configuration of terrestrial FS radio systems, sharing the same space and frequency bands with FSS and/or MSS systems, is illustrated in Fig.3.1. The emphasis of this contribution is on mobile systems; however, it should be appreciated that mobile systems may be required to share the same frequency band with fixed systems. A *satellite system* usually consists of many earth stations, denoted Earth Station 0 … Earth Station 2 in Fig. 3.1, one or many space segments, denoted Space Segment 0 … Space Segment N, and in the case of a *mobile satellite system* different types of mobile segments denoted by ♣ and ♠ in the same figure. Links between different Space Segments and mobile users of MSS systems are called *service links*; links connecting Space Segments and corresponding Earth Stations are called *feeder links*. FSS systems employ Space Segments and fixed Earth Station segments only; corresponding connections are called *service links*. Thus, technically similar connections between Space Segments and fixed Earth Station segments perform different functions in MSS and FSS systems and are referred to by different names. Administratively, the feeder links of MSS systems are often referred to as FSS.

Let us briefly assess spectrum requirements of an MSS system. There exist many possibilities of how and where to communicate in the networks shown in Fig. 3.1. Each of these possibilities can use different spatial and frequency resources, which one needs to assess for sharing purposes. For example, a mobile user ♣ transmits at frequency f_0 using a small hemispherical antenna toward Space Segment 0. This space segment receives a signal at frequency f_0, transposes it to frequency F_{n+0}, amplifies it and transmits it toward Earth Station 0. This station processes the signal, makes decisions on the final destination, sends the signal back toward the same Space Segment 0, which receives the signal at frequency f_{m+0}. This signal is transposed further to frequency F_{k+0} and is emitted via *inter-satellite link* $_0ISL_1$ toward Space Segment 1, which receives this signal, processes it, transposes it, and emits toward Earth and mobile ♠ at frequency F_1. In this process a quintet of frequencies (f_0, F_{n+0}, f_{m+0}, F_{k+0}, F_1) is used in one direction. Once the mobile ♠ receives the signal, its set rings and sends back the signal in reverse directions at a different frequency quintet (or a different time slot, or a different code, or any combination of time, code, and frequency), thus establishing the two-way connection. Obviously, this type of MSS system uses significant parts of the frequency spectrum.

Mobile satellite systems consist of two directions with very distinct properties. A direction from an Earth Station, also called a *hub* or *base station,* which may include a *Network Management Center* (NMC), toward the satellite space segment and further toward a particular mobile user is known as the *forward* direction. In addition, we will call this direction the *dispatch* direction, *broadcast* direction, or *division* direction, since the NMC dispatches/broadcasts data to different users and data might be divided in frequency (F), time (T), code (C), or a hybrid (H) mode. The opposite direction from a mobile user toward the satellite space segment and further toward the NMC is known as the *return* direction. In addition, we will call this direction the *access* direction, since mobile users usually need to make attempts to access the mobile network before a connection with NMC can be established; in some networks the NMC may poll the mobile users, instead. A connection between NMC and a mobile user, or between two mobile users, may consist of two or more hops, including inter-satellite links, as shown in Fig. 3.1.

While traveling, a customer — a *user of a cellular mobile radio system* — may experience sudden changes in signal quality caused by his movements relative to the corresponding base station and surroundings, multipath propagation, and unintentional jamming such as man-made noise, adjacent channel interference, and co-channel interference inherent in cellular systems. Such an environment belongs to the class of nonstationary random fields, on which experimental data are difficult to obtain; their behavior hard to predict and model satisfactorily. When reflected signal components become comparable in level to the attenuated direct component, and their delays comparable to the inverse of the channel bandwidth, *frequency selective fading* occurs. The reception is further degraded due to movements of a user relative to reflection points and the relay station, causing Doppler frequency shifts. The simplified model of this environment is known as the *Doppler Multipath Rayleigh Channel.*

The existing and planned cellular mobile radio systems employ sophisticated narrowband and wideband filtering, interleaving, coding, modulation, equalization, decoding, carrier and timing recovery, and multiple access schemes. The cellular mobile radio channel involves a *dynamic interaction* of signals arriving via different paths, adjacent and co-channel interference, and noise. Most channels exhibit some degree of memory, the description of which requires higher order statistics of — *spatial and temporal* — multidimensional random vectors (amplitude, phase, multipath delay, Doppler frequency, etc.) to be employed.

A model of a multihop satellite system that incorporates interference and nonlinearities is illustrated and described in Fig. 3.2. The signal flow in the forward/broadcast direction, from Base to Mobile User, is shown on the left side of the figure; the right side of the figure corresponds to the reverse/access direction. For example, in the forward/broadcast direction, the transmitted signal at the Base, shown in the upper left of the figure, is distorted due to nonlinearities in the RF power amplifier. This signal distortion is expressed via differential phase and differential gain coefficients DP and DG, respectively.

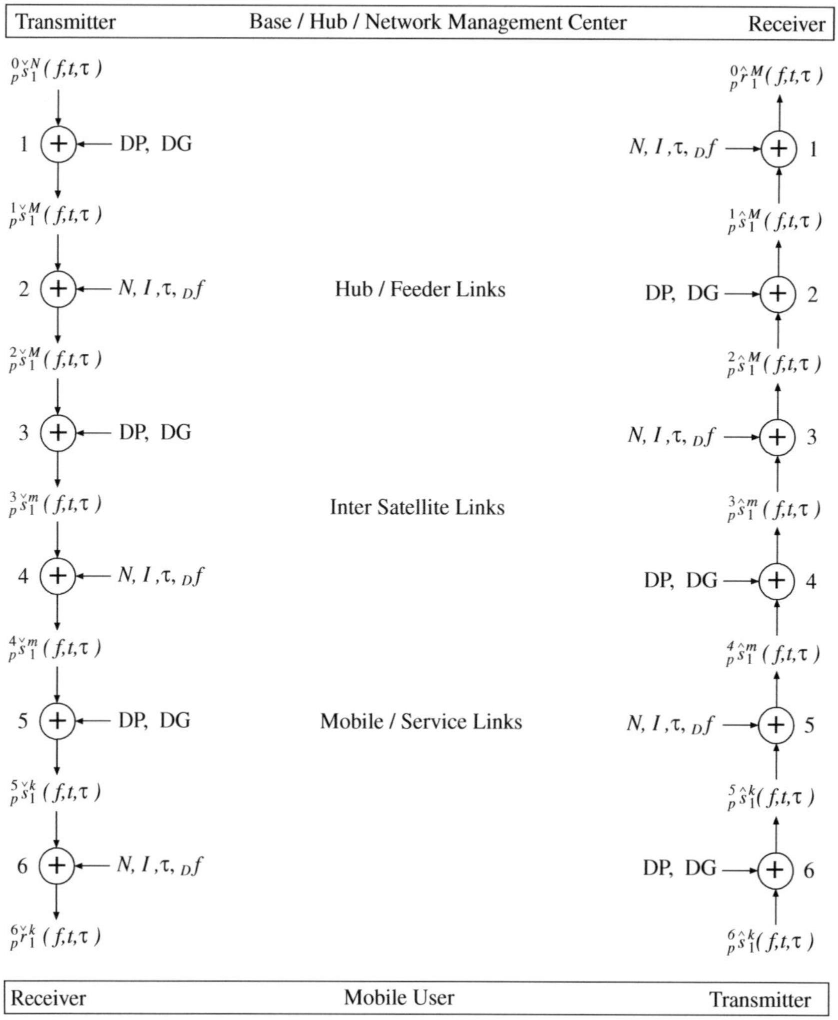

FIGURE 3.2 Signals and interference in a multihop satellite system. ${}^{q}_{p}x_1^M(f, t, \tau)$ represents signals $x = r, s$, where r is the received and s is the sent/transmitted signal, $\check{x}$ represents the dispatch/forward direction and $\hat{x}$ represents access/return direction; p is the polarization of the signal at location q and the number of signal components ranges from 1 to M; f, t, τ are frequency, time and delay of a signal at the location q, respectively. DP, DG are differential phase and differential gain (include AM/AM and AM/PM); N, I, τ, ${}_Df$ are the noise, interference, absolute delay, and Doppler frequency, respectively.

The same signal is emitted toward the satellite space segment receiver denoted as point 2; here, noise N, interference I, delay τ, and Doppler frequency ${}_Df$ symbolize the environment. The signals are further processed, amplified and distorted at stage 3, and radiated toward the receiver, 4. Here again, noise N, interference I, delay τ, and Doppler frequency ${}_Df$ symbolize the environment. The signals are translated and amplified at stage 5 and radiated toward the Mobile User at stage 6; here, additional noise N, interference I, delay τ, and Doppler frequency ${}_Df$ characterize the corresponding environment. This model is particularly suited for a detailed analysis of the link budget and for equipment design purposes. A system provider and cell designer may use a statistical description of a mobile channel environment, instead.

An FSS radio channel is described as the Gaussian; the mean value of the corresponding radio signal is practically constant and its value can be predicted with a standard deviation of a fraction of a dB. A terrestrial mobile radio channel could exhibit dynamics of about 80 dB and its mean signal could be predicted with a standard deviation of 5 to 10 dB. This may require the evaluation of usefulness of existing radio channel models and eventual development of more accurate ones.

Cell engineering and prediction of service area and service quality in an ever-changing mobile radio channel environment, is a very difficult task. The average path loss depends on terrain microstructure within a cell, with considerable variation between different types of cells (i.e., urban, suburban, and rural environments). A variety of models based on experimental and theoretic work have been developed to predict path radio propagation losses in a mobile channel. Unfortunately, none of them are universally applicable. In almost all cases, excessive transmitting power is necessary to provide adequate system performance.

The *first-generation* mobile satellite systems employ geostationary satellites (or payload piggy-backed on a host satellite) with small 18-dBi antennas covering the whole globe. When the satellite is positioned directly above the traveler (at zenith), a near constant signal environment, known as *Gaussian channel*, is experienced. The traveler's movement relative to the satellite is negligible (i.e., Doppler frequency is practically zero). As the traveler moves — north or south, east or west — the satellite appears lower on the horizon. In addition to the direct path, many significant strength-reflected components are present, resulting in a degraded performance. Frequencies of these components fluctuate due to movement of traveler relative to the reflection points and the satellite. This environment is known as the *Doppler Ricean Channel*. An inclined orbit satellite located for a prolonged period of time above 45° latitude north and 106° longitude west, could provide travelers all over the U.S. and Canada, including the far North, a service quality unsurpassed by either geostationary satellite or terrestrial cellular radio. Similarly, a satellite located at 45° latitude north and 15° longitude east, could provide travelers in Europe with improved service quality.

Inclined orbit satellite systems can offer a low start-up cost, a near Gaussian channel environment, and improved service quality. Low orbit satellites, positioned closer to the service area, can provide high signal levels and short (a few milliseconds long) delays, and offer compatibility with the cellular terrestrial systems. These advantages need to be weighted against network complexity, inter-satellite links, tracking facilities, etc.

Terrestrial mobile radio communications systems provide signal dynamics of about 80 dB and are able to penetrate walls and similar obstacles, thus providing inside building coverage. Satellite mobile radio communications systems are power limited and provide signal dynamics of less than 15 dB; the signal coverage is, in most cases, limited to the outdoors.

Let us compare the efficiency of a Mobile Satellite Service (MSS) with the Fixed Satellite Service (FSS); both services are assumed to be using the GSO space segments. A user at the equator can see the GSO arc reaching ±81°; if satellites are spaces 2° apart along the GSO, then the same user can reach 81 satellites simultaneously. An MSS user employs a hemispherical antenna having gain of about 3 dBi; consequently, he can effectively use only one satellite, but prevent all other satellite users from employing the same frequency. An FSS user employs a 43-dBi gain antenna that points toward a desired satellite. By using the same transmit power as an MSS user, but employing larger and more expensive antenna, this FSS user can effectively transmit about 40 dB (ten thousand times) wider bandwidth (i.e., 40 dB more information). The FSS user can, by adding 3 dB more power into an additional orthogonal polarization channel, reuse the same frequency band and double the capacity. Furthermore, the same FSS user can use additional antennas to reach each of 81 available satellites, thus increasing the GSO are capacity by 80 times. Consequently, the FSS is power-wise 10,000 times more efficient and spatially about 160 times more efficient than corresponding MSS. Similar comparisons can be made for terrestrial systems. The convenience and smallness of today's mobile systems user terminals are traded for low spatial and power efficiency, which may carry a substantial economic price penalty. The real cost of a mobile system seems to have been subsidized by some means beyond the cost of cellular telephone and traffic charges (both often $0).

3.7 Service Quality

The primary and the most important measure of service quality should be *customer satisfaction.* The customer's needs, both current and future, should provide guidance to a service offerer and an equipment manufacturer for both the system concept and product design stages. In the early stages of the product life, mobile radio was perceived as a necessary tool for performing important tasks; recently, mobile/personal/handheld radio devices are becoming more like status symbols and fashion. Acknowledging the importance of every single step of the complex service process and architecture, attention is limited here to a few technical merits of quality:

1. *Guaranteed quality level* is usually related to a percentage of the service area coverage for an adequate percentage of time.
2. *Data service quality* can be described by the average bit error rate (e.g., BER $< 10^{-5}$), packet BER (PBER $< 10^{-2}$), signal processing delay (1 to 10 ms), multiple access collision probability (< 20%), the probability of a false call (false alarm), the probability of a missed call (miss), the probability of a lost call (synchronization loss), etc.
3. *Voice quality* is usually expressed in terms of the mean opinion score (MOS) of subjective evaluations by service users. MOS marks are: bad = 0, poor = 1, fair = 2, good = 3, and excellent = 4. MOS for PSTN voice service, pooled by leading service providers, relates the poor MOS mark to a signal-to-noise ratio (S/N) in a voice channel of S/N ≈ 35 dB, while an excellent score corresponds to S/N > 45 dB. Currently, users of mobile radio services are giving poor marks to the voice quality associated with a S/N ≈ 15 dB and an excellent mark for S/N > 25 dB. It is evident that there is significant difference (20 dB) between the PSTN and mobile services. If digital speech is employed, both the speech and the speaker recognition have to be assessed. For more objective evaluation of speech quality under *real conditions* (with no impairments, in the presence of burst errors during fading, in the presence of random bit errors at BER $= 10^{-2}$, in the presence of Doppler frequency offsets, in the presence of truck acoustic background noise, in the presence of ignition noise, etc.), additional tests such as the diagnostic acceptability measure DAM, diagnostic rhyme test DRT, Youden square rank ordering, Sino-Graeco-Latin square tests, etc., can be performed.

3.8 Network Issues and Cell Size

To understand ideas and technical solutions offered in existing schemes, and in future systems one needs also to analyze the reasons for their introduction and success. Cellular mobile services are flourishing at an annual rate higher than 20%, worldwide. The first-generation systems, (such as AMPS, NMT, TACS, MCS, etc.), use *frequency division multiple access* FDMA and digital modulation schemes for access, command and control purposes, and analog phase/frequency modulation schemes for the transmission of an analog voice. Most of the network intelligence is concentrated at fixed elements of the network including base stations, which seem to be well suited to the networks with a modest number of medium to large-sized cells. To satisfy the growing number of potential customers, more cells and base stations were created by the cell splitting and frequency reuse process. Technically, the shape and size of a particular cell is dictated by the base station antenna pattern and the topography of the service area. Current terrestrial cellular radio systems employ cells with 0.5- to 50-km radius. The maximum cell size is usually dictated by the link budget, in particular the grain of a mobile antenna and available output power. This situation arises in a rural environment, where the demand on capacity is very low and cell splitting is not economical. The minimum cell size is usually dictated by the need for an increase in capacity, in particular in downtown cores. Practical constraints such as real estate availability and price, and construction dynamics limit the minimum cell size to 0.5 to 2 km. However, in such types of networks, the complexity of the network and the cost of service grow exponentially with the number of base stations, while the efficiency of present handover procedures becomes inadequate. Consequently, the second generation of all-digital schemes, which handle this increasing idle traffic more efficiently, were introduced. However, handling of the information, predominantly voice, has not been improved significantly, if at all.

In the 1980s extensive studies of then existing AMPS- and NMT-based systems were performed, see Davis et al. [1984] and Mahmoud et al. [1989], and the references therein. Based on particular service quality requirements, particular radio systems and particular cell topologies, few empirical rules had been established. Antennas with an omnidirectional pattern in a horizontal direction, but with about 10 dBi gain in vertical direction provide the frequency reuse efficiency of $N_{FDMA} = 1/12$. It was anticipated that base station antennas with similar directivity in a vertical direction and 60° directivity in a horizontal direction (a cell is divided into six sectors) can provide the reuse efficiency $N_{FDMA} = 1/4$, which results in a threefold increase in the system capacity; if CDMA is employed instead of FDMA, an increase in reuse efficiency $N_{FDMA} = 1/4 \rightarrow N_{CDMA} = 2/3$ may be expected. However, this does not necessarily mean that a CDMA system is more efficient than a FDMA system. The overall efficiency very much depends on spatiotemporal dynamics of a particular cell and the overall network.

Recognizing some of limitations of existing schemes and anticipating the market requirements, the research in *time division multiple access* (TDMA) schemes aimed at cellular mobile and DCT services, and in *code division multiple access* (CDMA) schemes aimed toward mobile satellite systems and cellular and personal mobile applications, followed with introduction of nearly ten different systems. Although employing different access schemes, TDMA (CDMA) network concepts rely on a smart mobile/portable unit that scans time slots (codes) to gain information on network behavior, free slots (codes), etc., improving frequency reuse and handover efficiency while hopefully keeping the complexity and cost of the overall network at reasonable levels. Some of the proposed system concepts depend on low gain (0 dBi) base station antennas deployed in a license-free, uncoordinated fashion; small size cells (10 to 1000 m in radius) and an emitted isotropic radiated power of about 10 mW (+10 dBm) per 100 kHz are anticipated. A frequency reuse efficiency of $N = 1/9$ to N to 1/36 has been projected for DCT systems. $N = 1/9$ corresponds to the highest user capacity with the lowest transmission quality, while $N = 1/36$ has the lowest user capacity with the highest transmission quality. This significantly reduced frequency reuse capability of proposed system concepts, will result in significantly reduced system capacity, which needs to be compensated for by other means including new spectra.

In practical networks, the need for a capacity (and frequency spectrum) is distributed unevenly in space and time. In such an environment, the capacity and frequency reuse efficiency of the network may be improved by *dynamic channel allocation,* where an increase in the capacity at a particular hot spot may be traded for a decrease in the capacity in cells surrounding the hot spot, the quality of the transmission, and network instability. The first-generation mobile radio communications systems used omnidirectional antennas at base stations. Today, three-sector 120°-wide cells are typical in a heavy traffic urban environment, while entry-level rural systems employ omnidirectional antennas; the most demanding environments with changing traffic patterns employ adaptive antenna solutions, instead.

To cover the same area (space) with smaller and smaller cells, one needs to employ more and more base stations. A linear increase in the number of base stations in a network usually requires an $(n(n-1)/2)$ increase in the number of connections between base stations, and increase in complexity of switches and network centers. These connections can be realized by fixed radio systems (providing more frequency spectra available for this purpose), or, more likely, by a cord (wire, cable, fiber, etc.).

An increase in overall capacity is attributed to

- Increase in available bandwidth, particularly above 1 GHz, but to the detriment of other services
- Increased use of adaptive antenna solutions which, through spatial filtering, increase capacity and quality of the service, but at a significant increase in cost

- Trade-offs between service quality, vehicular vs. pedestrian environments, analog vs. digital voice, etc.

The *first-generation* geostationary satellite system antenna beam covers the entire Earth (i.e., the cell radius equals ≈6500 km). The *second-generation* geostationary satellites use large multibeam antennas providing 10 to 20 beams (cells) with 800- to 1600-km radius. Low orbit satellites such as Iridium use up to 37 beams (cells) with 670 km radius. The *third-generation* geostationary satellite systems will be able to use very large reflector antennas (roughly the size of a baseball stadium), and provide 80 to 100 beams (cells) with a cell radius of ≈200 km. If such a satellite is tethered to a position 400 km above Earth, the cell size will decrease to ≈2 km in radius, which is comparable in size with today's small size cell in terrestrial systems. Yet, such a satellite system may have the potential to offer an improved service quality due to its near optimal location with respect to the service area. Similar to the terrestrial concepts, an increase in the number of satellites in a network will require an increase in the number of connections between satellites and/or Earth network management and satellite tracking centers, etc. Additional factors that need to be taken into consideration include price, availability, reliability, and timeliness of the launch procedures, a few large vs. many small satellites, tracking stations, etc.

3.9 Coding and Modulation

The conceptual transmitter and receiver of a mobile system may be described as follows. The transmitter signal processor accepts analog voice and/or data and transforms (by analog and/or digital means) these signals into a form suitable for a double-sided suppressed carrier amplitude modulator—also called quadrature amplitude modulator (QAM). Both analog and digital input signals may be supported, and either analog or digital modulation may result at the transmitter output. Coding and interleaving can also be included. Very often, the processes of coding and modulation are performed jointly; we will call this joint process *codulation.* A list of typical modulation schemes suitable for transmission of voice and/or data over Doppler-affected Ricean channel, which can be generated by this transmitter is given in Table 3.5. Details on modulation, coding, and system issues can be found in Kucar [2000], Proakis [1983], Sklar [1988], and Van Trees [1968–1971].

Existing cellular radio systems such as AMPS, TACS, MCS, and NMT employ hybrid (analog and digital) schemes. For example, in access mode, AMPS uses a digital codulation scheme (BCH coding and FSK modulation), while in information exchange mode, the frequency-modulated analog voice is merged with discrete *SAT* and/or *ST* signals and occasionally blanked to send a digital message. These hybrid codulation schemes exhibit a constant envelope and as such allow the use of power efficient radio frequency (RF) nonlinear amplifiers. On the receiver side, these schemes can be demodulated by an inexpensive, but efficient limiter/discriminator device. They require modest to high $C/N = 10 - 20$ dB, are very robust in adjacent (a spectrum is concentrated near the carrier) and co-channel interference (up to $C/I = 0$ dB, due to capture effect) cellular radio environment, and react quickly to the signal fade outages (no carrier, code, or frame synchronization). Frequency-selective and Doppler-affected mobile radio channels will cause modest to significant degradations known as the *random phase/frequency modulation.* By using modestly complex extended threshold devices C/N as low as 5 dB can provide satisfactory performance.

Tightly filtered codulation schemes, such as $\pi/4$ QPSK additionally filtered by a square root, raised-cosine filter, exhibit a nonconstant envelope that demands (quasi) linear, less D.C. power efficient amplifiers to be employed. On the receiver side, these schemes require complex demodulation receivers, a linear path for signal detection, and a nonlinear one for reference detection — differential detection or carrier recovery. When such a transceiver operates in a selective fading multipath channel environment, additional countermeasures (inherently sluggish equalizers, etc.) are necessary to improve the performance — reduce the *bit error rate floor.* These codulation schemes require modest $C/N = 8 - 16$ dB and perform modestly in adjacent and/or co-channel (up to $C/I = 8$ dB) interference environment.

TABLE 3.5 Modulation Schemes, Glossary of Terms

Abbreviation	Description	Remarks/Use
ACSSB	Amplitude Companded Single Sideband	Satellite transmission
AM	Amplitude Modulation	Broadcasting
APK	Amplitude Phase Keying Modulation	
BLQAM	Blackman Quadrature Amplitude Modulation	
BPSK	Binary Phase Shift Keying	Spread-spectrum systems
CPFSK	Continuous Phase Frequency-Shift Keying	
CPM	Continuous Phase Modulation	
DEPSK	Differentially Encoded PSK (with carrier recovery)	
DPM	Digital Phase Modulation	
DPSK	Differential Phase-Shift Keying (no carrier recovery)	
DSB-AM	Double Sideband Amplitude Modulation	
DSB-SC-AM	Double Sideband Suppressed Carrier AM	Includes digital schemes
FFSK	Fast Frequency-Shift Keying ≡ MSK	NMT data and control
FM	Frequency Modulation	Broadcasting, AMPS, voice
FSK	Frequency-Shift Keying	AMPS data and control
FSOQ	Frequency-Shift Offset Quadrature Modulation	
GMSK	Gaussian Minimum-Shift Keying	GSM voice, data, and control
GTFM	Generalized Tamed Frequency Modulation	
HMQAM	Hamming Quadrature Amplitude Modulation	
IJF	Intersymbol Jitter Free ≡ SQORC	
LPAM	L-ary Pulse Amplitude Modulation	
LRC	LT Symbols Long Raised Cosine Pulse Shape	
LREC	LT Symbols Long Rectangularly EnCoded Pulse Shape	
LSRC	LT Symbols Long Spectrally Raised Cosine Scheme	
MMSK	Modified Minimum Shift Keying ≡ FFSK	
MPSK	M-ary Phase-Shift Keying	
MQAM	M-ary Quadrature Amplitude Modulation	A subclass of DSB-SC-AM
MQPR	M-ary Quadrature Partial Response	Radio-relay transmission
MQPRS	M-ary Quadrature Partial Response System ≡ MQPR	
MSK	Minimum-Shift Keying	
m-*h*	Multi-*h* CPM	
OQPSK	Offset (staggered) Quadrature Phase-Shift Keying	
PM	Phase Modulation	Low capacity radio
PSK	Phase-Shift Keying	4PSK ≡ QPSK
QAM	Quadrature Amplitude Modulation	
QAPSK	Quadrature Amplitude Phase-Shift Keying	
QPSK	Quadrature Phase-Shift Keying ≡ 4 QAM	Low capacity radio
QORC	Quadrature Overlapped Raised Cosine	
SQAM	Staggered Quadrature Amplitude Modulation	
SQPSK	Staggered Quadrature Phase-Shift Keying	
SQORC	Staggered Quadrature Overlapped Raised Cosine	
SSB	Single Sideband	Low and high capacity radio
S3MQAM	Staggered Class 3 Quadrature Amplitude Modulation	
TFM	Tamed Frequency Modulation	
TSI QPSK	Two-Symbol-Interval QPSK	
VSB	Vestigial Sideband	TV
WQAM	Weighted Quadrature Amplitude Modulation	Includes most digital schemes
XPSK	Cross-correlated PSK	
$\pi/4$ DQPSK	$\pi/4$ Shift DQPSK with $\alpha = 0.35$ Raised Cosine Filtering	IS-54 TDMA voice and data
3MQAM	Class 3 Quadrature Amplitude Modulation	
4MQAM	Class 4 Quadrature Amplitude Modulation	
12PM3	12 State PM with 3 bit Correlation	

Source: 4U Communications Research Inc., 2000.06.10~00:09, c:/tab/modulat.tab

Codulation schemes employed in spread-spectrum systems use low-rate coding schemes and mildly filtered modulation schemes. When equipped with sophisticated amplitude gain control on the transmit and receive side, and robust rake receiver, these schemes can provide superior $C/N = 4 - 10$ dB and $C/I < 0$ dB performance. Unfortunately, a single transceiver has not been able to operate satisfactorily in a mobile channel environment. Consequently, a few additional signals have been employed to achieve the required quality of the transmission. These pilot signals significantly reduce the spectrum efficiency in the forward direction and many times in the reverse direction. Furthermore, two combined QPSK-like signals have up to (4×4) different baseband levels and may look like a 16-QAM signal, while three combined QPSK-like signals may look like a 64-QAM signal. These combined signals, one information and two pilot signals, at user's transmitter output, for example, exhibit high peak factors and total power that is by 3 to 5 dB higher than the C/N value necessary for a single information signal. Additionally, inherently power inefficient linear RF power amplifiers are needed; these three signal components of a CDMA scheme may have been optimized for minimal cross-correlation and ease of detection. As such, the same three signals may not necessarily have states in the QAM constellation that optimize the peak-to-average ratio, and vice versa.

3.10 Speech Coding

Human vocal tract and voice receptors, in conjunction with language redundancy (coding), are well suited for face-to-face conversation. As the channel changes (e.g., from telephone channel to mobile radio channel), different coding strategies are necessary to protect against the loss of information.

In (analog) companded PM/FM mobile radio systems, speech is limited to 4 kHz, compressed in amplitude (2:1), pre-emphasized, and phase/frequency modulated. At a receiver, inverse operations are performed. Degradation caused by these conversions and channel impairments results in lower voice quality. Finally, the human ear and brain have to perform the estimation and decision processes on the received signal.

In digital schemes for sampling and digitizing of an analog speech (source) are performed first. Then, by using knowledge of properties of the human vocal tract and the language itself, a spectrally efficient source coding is performed. A high rate 64 kb/s, 56 kb/s, and AD-PCM 32 kb/s digitized voice complies with ITU-T recommendations for toll quality, but may be less practical for the mobile environment. One is primarily interested in 8- to 16-kb/s rate speech coders, which might offer satisfactory quality, spectral efficiency, robustness, and acceptable processing delays in a mobile radio environment. A glossary of the major speech coding schemes is provided in Table 3.6.

At this point, a partial comparison between analog and digital voice should be made. The quality of 64 kb/s digital voice, transmitted over a telephone line, is essentially the same as the original analog voice (they receive nearly equal MOS). What does this *nearly equal MOS* mean in a radio environment? A mobile radio conversation consists of one (mobile to home) or a maximum of two (mobile to mobile) mobile radio paths, which dictate the quality of the overall connection. The results of a comparison between analog and digital voice schemes in different artificial mobile radio environments have been widely published. Generally, systems that employ digital voice and digital codulation schemes seem to perform well under modest conditions, while analog voice and analog codulation systems outperform their digital counterparts in fair and difficult (near threshold, in the presence of strong co-channel interference) conditions. Fortunately, present technology can offer a viable implementation of both analog and digital systems within the same mobile/portable radio telephone unit. This would give every individual a choice of either an analog or digital scheme, better service quality, and higher customer satisfaction. Trade-offs between the quality of digital speech, the complexity of speech and channel coding, as well as D.C. power consumption have to be assessed carefully, and compared with analog voice systems.

TABLE 3.6 Digitized Voice. Glossary of Terms

Abbreviation	Description	Remarks/Use
ADM	Adaptive Delta Modulation	
ADPCM	Adaptive Differential Pulse Code Modulation	Digital telephony, DECT
ACIT	Adaptive Code Sub-Band Exclted Transform	GTE
APC	Adaptive Predictive Coding	
APC–AB	APC with Adaptive Bit Allocation	
APC–HQ	APC with Hybrid Quantization	
APC–MQL	APC with Maximum Likelihood Quantization	
AQ	Adaptive Quantization	
ATC	Adaptive Transform Coding	
BAR	Backward Adaptive Reencoding	
CELP	Code Excited Linear Prediction	IS-95
CVSDM	Continuous Variable Slope Delta Modulation	
DAM	Diagnostic Acceptability Measure	
DM	Delta Modulation	A/D conversion
DPCM	Differential Pulse Code Modulation	
DRT	Diagnostic Rhyme Test	
DSI	Digital Speech Interpolation	TDMA FSS systems
DSP	Digital Signal Processing	
HCDM	Hybrid Companding Delta Modulation	
LDM	Linear Delta Modulation	
LPC	Linear Predictive Coding	
MPLPC	Multi Pulse LPC	
MSQ	Multipath Search Coding	
NIC	Nearly Instantaneous Companding	
PCM	Pulse Code Modulation	Digital Voice
PVXC	Pulse Vector EXcitation Coding	
PWA	Predicted Wordlength Assignment	
QMF	Quadrature Mirror Filter	
RELP	Residual Excited Linear Prediction	GSM
RPE	Regular Pulse Excitation	
SBC	Sub-Band Coding	
TASI	Time Assigned Speech Interpolation	TDMA FSS systems
TDHS	Time Domain Harmonic Scalling	
VAPC	Vector Adaptive Predictive Coding	
VCELP	Vector Code Excited Linear Prediction	
VEPC	Voice Excited Predictive Coding	
VQ	Vector Quantization	
VQL	Variable Quantum Level coding	
VSELP	Vector–Sum Excited Linear Prediction	IS–136, PDC
VXC	Vector EXcitation Coding	

Source: 4U Communications Research Inc., 2000.06.10~00:09, c:/tab/voice.tab

3.11 Macro and Micro Diversity

Macro diversity. Let us observe the typical evolution of a cellular system. In the beginning, the base station may be located in the barocenter of the service area (center of the cell). The base station antenna is omnidirectional in azimuth, but with about 6- to 10-dBi gain in elevation, and serves most of the cell area (e.g., > 95%). Some parts within the cell may experience a lower quality of service because the direct path signal may be attenuated due to obstruction losses caused by buildings, hills, trees, etc. The closest neighboring (the first tier) base stations serve corresponding neighboring area cells by using different sets of frequencies, eventually causing adjacent channel interference. The second closest neighboring (the second tier) base stations might use the same frequencies (frequency reuse) causing co-channel interference. When the need for additional capacity arises and/or a higher quality of service is required, the same nearly circular area may be divided into three 120°-wide sectors, six 60°-wide sectors, etc., all served from the same base

TABLE 3.7 Comparison of Cordless Telephone (CT) Systems

Parameter	System Name							
	CT0	CT1/+	JCT	CT2/+	CT3	DECT	CDMA	PHT
TX freq, MHz								
Base	22,26,30,31,46,48,45	914/885	254	864–8, 994–8	944–948	1880–1900		1895–1907
Mobile	48,41,39,40,69,74,48	960/932	380	864–8, 944–8	944–948	1880–1990		1895–1907
Multiple access band	FDMA	FDMA	FDMA	F/TDMA	TDMA	TDMA	CDMA	F/TDMA
Duplexing method	FDD	FDD	FDD	TDD	TDD	TDD	FDD	TDD
Ch. spacing, kHz	1.7,20,25,40	25	12.5	100	1000	1728	1250	300
Channel rate, kb/s				72	640	1152	1228.80	384
Channels/RF	1	1	1	1	8	12	32	4
Channels/band	10,12,15,20,25	40,80	89	20	2	5		20
Burst/frame length, ms				1/2	1/16	1/10	n/a	1/5
Modulation type	FM	FM	FM	GFSK	GMSK	GMSK	B/QPSK	π/4
Coding				Cyclic, RS	CRC 16	CRC 16	Conv 1/2, 1/3	
Transmit power, mW				≤10	≤80	≤100	≤10	≤80
Transmit power steps				2	1	1	many	many
TX power range, dB				16	0	0	≥80	
Vocoder type	analog	analog	analog	ADPCM	ADPCM	ADPCM	CELP	ADPCM
Vocoder rate, kb/s				fixed 32	fixed 32	fixed 32	≤9.6	fixed 32
Max data rate, kb/s				32	ISDN 144	ISDN 144	9.6	384
Processing delay, ms	0	0	0	2	16	16	80	
[3] Minimum				1/25	1/15	1/15	1/4	
Average				1.15	1/07	1/07	2/3	
Maximum				[1] 1/02	[1] 1/02	[1] 1/02	3/4	
[4]				100 × 1	10 × 8	6 × 12	4 × 32	
[5] Minimum				4	5–6	5–6	[2] 32 (08)	
Average				7	11–12	11–12	85 (21)	
Maximum				[1] 50	[1] 40	[1] 40	96 (24)	

Note: [1] The capacity (in the number of voice channels) for a single isolated cell. [2] The capacity in parentheses may correspond to a 32 kb/s vocoder. [3] Reuse efficiency. [4] Theoretical number of voice channels per cell and 10 MHz. [5] Practical number of voice channels per 10 MHz. Reuse efficiency and associate capacities reflect our own estimates.

Source: 4U Communications Research Inc., 2000.06.10~00:09 c:/tab/cordless.sys

station location; now, the same base station is located at the edge of respective sectors. Since the new sectorial antennas provide 5- and 8-dB larger gains than the old omnidirectional antenna, respectively, these systems with new antennas with higher gains have longer spatial reach and may cover areas belonging to neighboring cells of the old configuration. For example, if the same real estate (base stations) is used in conjunction with 120° directional (in azimuth) antennas, the new designated 120°-wide wedge area may be served by the previous base station and by two additional neighboring base stations now equipped with sectorial antennas with longer reach. Therefore, the same number of existing base stations equipped with new directional antennas and additional combining circuitry may be required to serve the same or different number of cells, yet in a different fashion. The mode of operation in which two or more base stations serve the same area is called the *macro diversity.* Statistically, three base stations are able to provide better coverage of an area similar in size to the system with a centrally located base station. The directivity of a base station antenna (120° or even 60°) provides additional discrimination against signals from neighboring cells, therefore, reducing adjacent and co-channel interference (i.e., improving reuse efficiency and capacity). Effective improvement depends on the terrain configuration, and the combining strategy and efficiency. However, it requires more complex antenna systems and combining devices.

Micro diversity is when two or more signals are received at one site (base or mobile):

1. *Space diversity* systems employ two or more antennas spaced a certain distance apart from one another. A separation of only $\lambda/2 = 15$ cm at $f = 1$ GHz, which is suitable for implementation on the mobile side, can provide a notable improvement in some mobile radio channel environments. Micro space diversity is routinely used on cellular base sites. Macro diversity, where in our example the base stations were located kilometers apart, is also a form of space diversity.
2. *Field-component diversity* systems employ different types of antennas receiving either the electric or the magnetic component of an electromagnetic signal.
3. *Frequency diversity* systems employ two or more different carrier frequencies to transmit the same information. Statistically, the same information signal may or may not fade at the same time at different carrier frequencies. Frequency hopping and very wide band signaling can be viewed as frequency diversity techniques.
4. *Time diversity* systems are primarily used for the transmission of data. The same data is sent through the channel as many times as necessary, until the required quality of transmission is achieved automatic repeat request (ARQ). *Would you please repeat your last sentence* is a form of time diversity used in a speech transmission.

The improvement of any diversity scheme is strongly dependent on the combining techniques employed (i.e., the selective—switched—combining), the maximal ratio combining, the equal gain combining, the feedforward combining, the feedback (Granlund) combining, majority vote, etc.); see [Jakes, 1974].

Continuous improvements in DSP and MMIC technologies and broader availability of ever-improving CAD electromagnetics tools are making adaptive antenna solutions more viable than ever before. This is particularly true for systems above 1 GHz, where the same necessary base station antenna gain can be achieved with smaller antenna dimensions. An adaptive antenna could follow spatially shifting traffic patterns, adjust its gain and pattern, and consequently improve the signal quality and capacity.

3.12 Multiple Broadcasting and Multiple Access

Communications networks for travelers have two distinct directions: the *forward link* — from the base station (via satellite) to all travelers within the footprint coverage area, and the *return link* — from a traveler (via satellite) to the base station. In the forward direction a base station distributes information to travelers according to the previously established protocol (i.e., no multiple access is involved); this way of operation is also called *broadcasting*. In the reverse direction many travelers make attempts to access one of the base stations; this way of operation is also called *access*. This occurs in so-called *control channels*, in a particular time slot, at a particular frequency, or by using a particular code. If collisions occur, customers have to wait in a queue and try again until success is achieved. If successful (i.e., no collision occurred), a particular customer will exchange (automatically) the necessary information for call setup. The network management center (NMC) will verify the customer's status, his credit rating, etc. Then, the NMC may assign a channel frequency, time slot, or code, on which the customer will be able to exchange information with his correspondent.

The optimization of the forward and reverse links may require different coding and modulation schemes and different bandwidths in each direction.

In *forward link*, there are three basic distribution (multiplex broadcasting) schemes: one that uses discrimination in frequency between different users and is called *frequency division multiplex broadcasting* (FDMB); another that discriminates in time and is called *time division multiplex broadcasting* (TDMB);

and the last having different codes based on spread-spectrum signaling, which is known as *code division multiplex broadcasting* (CDMB). It should be noted that hybrid schemes using a combination of basic schemes can also be developed. All existing mobile radio communications systems employ an FDM component; consequently, only FDMB schemes are pure, while the other two schemes are hybrid (i.e., TDMB/FDM and CDMB/FDM solutions are used); the two hybrid solutions inherit complexities of both parents (i.e., the need for an RF frequency synthesizer and a linear amplifier for *single channel per carrier* (SCPC) FDM solution, and the need for TDM and CDM overhead, respectively).

In *reverse link*, there are three basic access schemes: one that uses discrimination in frequency between different users and is called *frequency division multiple access* (FDMA); another that discriminates in time and is called *time division multiple access* (TDMA); and the last having different codes based on spread-spectrum signaling, which is known as *code division multiple access* (CDMA). It should be noted that hybrid schemes using combinations of basic schemes can also be developed.

A performance comparison of multiple access schemes is a very difficult task. The strengths of FDMA schemes seem to be fully exploited in narrowband channel environments. To avoid the use of equalizers, channel bandwidths as narrow as possible should be employed; yet in such narrowband channels, the quality of service is limited by the maximal expected Doppler frequency and practical stability of frequency sources. Current practical limits are about 5 kHz.

The strengths of both TDMA and CDMA schemes seem to be fully exploited in wideband channel environments. TDMA schemes need many slots (and bandwidth) to collect information on network behavior. Once the equalization is necessary (at bandwidths > 20 kHz), the data rate should be made as high as possible to increase frame efficiency and freeze the frame to ease equalization; yet, high data rates require high RF peak powers and a lot of signal processing power, which may be difficult to achieve in handheld units. Current practical bandwidths are about 0.1 to 1.0 MHz. All existing schemes that employ TDMA components are hybrid (i.e., the TDMA/FDM schemes in which the full strength of the TDMA scheme is not fully realized).

CDMA schemes need large spreading (processing) factors (and bandwidth) to realize spread-spectrum potentials; yet, high data rates require a lot of signal processing power, which may be difficult to achieve in handheld units. Current practical bandwidths are up to about 5 MHz. As mentioned before, a single transceiver has not been able to operate satisfactorily in a mobile channel environment. Consequently, a few CDMA elementary signals, information and pilot ones, may be necessary for successful transmission. This multisignal environment is equivalent to a MQAM signaling scheme with a not necessarily optimal state constellation. Significant increase in the equipment complexity is accompanied with a significant increase in the average and peak transmitter power. In addition, an RF synthesizer is needed to accommodate the CDMA/FDM mode of operation.

Narrow frequency bands seem to favor FDMA schemes, since both TDMA and CDMA schemes require more spectra to fully develop their potentials. However, once the adequate power spectrum is available, the later two schemes may be better suited for a complex (micro) cellular network environment. Multiple access schemes are also message sensitive. The length and type of message, and the kind of service will influence the choice of multiple access, ARQ, frame and coding, among others.

3.13 System Capacity

The recent surge in the popularity of cellular radio and mobile service, in general, has resulted in an overall increase in traffic and a shortage of available system capacity in large metropolitan areas. Current cellular systems exhibit a wide range of traffic densities, from low in rural areas to overloaded in downtown areas with large daily variations between peak hours and quiet night hours. It is a great system engineering challenge to design a system that will make optimal use of the available frequency spectrum, while offering a maximal traffic throughput (e.g., Erlangs/MHz/service area) at an acceptable service quality, constrained by the price and size of the mobile equipment. In a cellular environment, the overall system capacity in a given service area is a product of many (complexly interrelated) factors including the available frequency spectra, service quality, traffic statistics, type of traffic, type of protocol, shape and size of service area, selected antennas, diversity, frequency reuse capability, spectral efficiency of coding and modulation schemes, efficiency of multiple access, etc.

In the seventies, so-called analog cellular systems employed omnidirectional antennas and simple or no diversity schemes offering modest capacity, which satisfied a relatively low number of customers. Analog cellular systems of the nineties employ up to 60° sectorial antennas and improved diversity schemes. This combination results in a three- to fivefold increase in capacity. A further (twofold) increase in capacity can be expected from narrowband analog systems (25 $\rightarrow$ 12.5 kHz) and nearly threefold increase in capacity from the 5 kHz-wide narrowband AMPS; however, slight degradation in service quality might be expected. These improvements spurned the current growth in capacity, the overall success, and the prolonged life of analog cellular radio.

3.14 Conclusion

In this contribution, a broad repertoire of terrestrial and satellite systems and services for travelers is briefly described. The technical characteristics of the dispatch, cellular, and cordless telephony systems are tabulated for ease of comparison. Issues such as operating environment, service quality, network complexity, cell size, channel coding and modulation (codulation), speech coding, macro and micro diversity, multiplex and multiple access, and the mobile radio communications system capacity are discussed.

Presented data reveals significant differences between existing and planned terrestrial cellular mobile radio communications systems, and between terrestrial and satellite systems. These systems use different frequency bands, different bandwidths, different codulation schemes, different protocols, etc. (i.e., they are not compatible).

What are the technical reasons for this incompatibility? In this contribution, performance dependence on multipath delay (related to the cell size and terrain configuration), Doppler frequency (related to the carrier frequency, data rate, and the speed of vehicles), and message length (may dictate the choice of multiple access) are briefly discussed. A system optimized to serve the travelers in the Great Plains may not perform very well in mountainous Switzerland; a system optimized for downtown cores may not be well suited to a rural radio; a system employing geostationary (above equator) satellites may not be able to serve travelers at high latitudes very well; a system appropriate for slow moving vehicles may fail to function properly in a high Doppler shift environment; a system optimized for voice transmission may not be very good for data transmission, etc. A system designed to provide a broad range of services to everyone, everywhere, may not be as good as a system designed to provide a particular service in a particular local environment — as a decathlete world champion may not be as successful in competitions with specialists in particular disciplines.

However, there are plenty of opportunities where compatibility between systems, their integration, and frequency sharing may offer improvements in service quality, efficiency, cost and capacity (and therefore availability). Terrestrial systems offer a low start-up cost and a modest per user in densely populated areas. Satellite systems may offer a high quality of service and may be the most viable solution to serve travelers in scarcely populated areas, on oceans, and in the air. Terrestrial systems are confined to two dimensions and radio propagation occurs in the near horizontal sectors. Barostationary satellite systems use the narrow sectors in the user's zenith nearly perpendicular to Earth's surface having the potential for frequency reuse and an increase in the capacity in downtown areas during peak hours. A call setup in a forward direction (from the PSTN via base station to the traveler) may be a very cumbersome process in a terrestrial system when a traveler to whom a call is intended is roaming within an unknown cell. However, this may be realized earlier in a global beam satellite system.

References

Ariyavisitakul, S.L., Falconer, D.D., Adachi, F., and Sari, H. (Guest Editors), Special Issue on Broadband Wireless Techniques, *IEEE J. on Selected Areas in Commun.*, 17, 10, October 1999.

Chuang, J.C.-I., Anderson, J.B., Hattori, T., and Nettleton, R.W. (Guest Editors), Special Issue on Wireless Personal Communications: Part I, *IEEE J. on Selected Areas in Commun.*, 11, 6, August 1993, Part II, *IEEE J. on Selected Areas in Commun.*, 11, 7, September 1993.

Cimini, L.J. and Tranter W.H. (Guest Editors), Special Issue on Wireless Communication Series, *IEEE J. on Selected Areas in Commun.*, 17, 3, March 1999.

Cimini, L.J. and Tranter, W.H. (Guest Editors), Special Issue on Wireless Communications Series, *IEEE J. on Selected Areas in Commun.*, 17, 7, July 1999.

Cimini, L.J. and Tranter, W.H. (Guest Editors), Special Issue on Wireless Communications Series, *IEEE J. on Selected Areas in Commun.*, 17, 11, November 1999.

Cimini, L.J. and Tranter, W.H. (Guest Editors), Special Issue on Wireless Communications Series, *IEEE J. on Selected Areas in Commun.*, 18, 3, March 2000.

Cox, D.C., Hirade, K., and Mahmoud, S.A. (Guest Editors), Special Issue on Portable and Mobile Communications, *IEEE J. on Selected Areas in Commun.*, 5, 4, June 1987.

Cox, D.C. and Greenstein, L.J. (Guest Editors), Special Issue on Wireless Personal Communications, *IEEE Commun. Mag.*, 33, 1, January 1995.

Davis, J.H. (Guest Editor), Mikulski, J.J., and Porter, P.T. (Associated Guest Editors), King, B.L. (Guest Editorial Assistant), Special Issue on Mobile Radio Communications, *IEEE J. on Selected Areas in Commun.*, 2, 4, July 1984.

Del Re, E., Devieux Jr., C.L., Kato, S., Raghavan, S., Taylor, D., and Ziemer, R. (Guest Editors), Special Issue on Mobile Satellite Communications for Seamless PCS, *IEEE J. on Selected Areas in Commun.*, 13, 2, February 1995.

Graglia, R.D., Luebbers, R.J., and Wilton, D.R. (Guest Editors), Special Issue on Advanced Numerical Techniques in Electromagnetics. *IEEE Trans. on Antennas and Propagation,* 45, 3, March 1997.

Institute of Navigation (ION), *Global Positioning System,* Reprinted by The Institute of Navigation. Volume I. Washington, D.C., USA, 1980; Volume II. Alexandria, VA, USA, 1984; Volume III. Alexandria, VA, USA, 1986; Volume IV. Alexandria, VA, USA, 1993.

International Telecommunication Union (ITU), *Radio Regulations,* Edition of 1982, Revised in 1985 and 1986.

International Telecommunication Union (ITU), Recommendations of the CCIR, 1990 (also Resolutions and Opinions). *Mobile Radiodetermination, Amateur and Related Satellite Services,* Volume VIII, XVIIth Plenary Assembly, Düsseldorf, 1990. Reports of the CCIR, (also Decisions), *Land Mobile Service, Amateur Service, Amateur Satellite Service,* Annex 1 to Volume VIII, XVIIth Plenary Assembly, Düsseldorf, 1990. Reports of the CCIR, (also Decisions), *Maritime Mobile Service,* Annex 2 to Volume VIII, XVIIth Plenary Assembly, Düsseldorf, 1990.

Kucar, A.D. (Guest Editor), Special Issue on Satellite and Terrestrial Systems and Services for Travelers, *IEEE Commun. Mag.*, 29, 11, November 1991.

Kucar, A.D., Kato, S., Hirata, Y., and Lundberg, O. (Guest Editors), Special Issue on Satellite Systems and Services for Travelers, *IEEE J. on Selected Areas in Commun.*, 10, 8, October 1992.

Kucar, A.D. and Uddenfeldt, J. (Guest Editors), Special Issue on Mobile Radio Centennial, *Proceedings of the IEEE,* 86, 7, July 1998.

Mahmoud, S.A., Rappaport, S.S., and Öhrvik, S.O. (Guest Editors), Special Issue on Portable and Mobile Communications, *IEEE J. on Selected Areas in Commun.*, 7, 1, January 1989.

Mailloux, R.J. (Guest Editor), Special Issue on Phased Arrays, *IEEE Trans. on Antennas and Propagation,* 47, 3, March 1999.

Mitola, J. III, Bose, V., Leiner, B.M., Turletti, T., and Tennenhouse, D. (Guest Editors), Special Issue on Software Radios, *IEEE J. on Selected Areas in Commun.*, 17, 4, April 1999.

Oppermann, I., van Rooyen, P., and Kohno, R. (Guest Editors), Special Issue on Spread Spectrum for Global Communications I, *IEEE J. on Selected Areas in Commun.*, 17, 12, December 1999.

Oppermann, I., van Rooyen, P., and Kohno, R. (Guest Editors), Special Issue on Spread Spectrum for Global Communications II, *IEEE J. on Selected Areas in Commun.*, 18, 1, January 2000.

Rhee, S.B. (Editor) and Lee, W.C.Y. (Guest Editor), Special Issue on Digital Cellular Technologies, *IEEE Trans. on Vehicular Technol.*, 40, 2, May 1991.

Steele, R. (Guest Editor), Special Issue on PCS: The Second Generation, *IEEE Commun. Mag.*, 30, 12, December 1992.

World Administrative Radio Conference (WARC), *FINAL ACTS of the World Administrative Radio Conference for Dealing with Frequency Allocations in Certain Parts of the Spectrum* (WARC-92), Málaga-Torremolinos, 1992). ITU, Geneva, 1992.

World Radiocommunications Conference (WRC), *FINAL ACTS of the World Radiocommunications Conference* (WRC-97. ITU, Geneva, 1997.

Further Information

This trilogy, written by participants in AT&T Bell Labs projects on research and development in mobile radio, is the Holy Scripture of diverse cellular mobile radio topics:

Jakes, W.C. Jr. (Editor), *Microwave Mobile Communications,* John Wiley & Sons, Inc., New York, 1974.

AT&T Bell Labs Technical Personnel, Advanced Mobile Phone Service (AMPS), *Bell System Technical Journal,* 58, 1, January 1979.

Lee, W.C.Y., *Mobile Communications Engineering,* McGraw-Hill Book Company, New York, 1982.

An in-depth understanding of design, engineering, and use of cellular mobile radio networks, including PCS and PCN, requires knowledge of diverse subjects such as three-dimensional cartography, electromagnetic propagation and scattering, computerized analysis and design of microwave circuits, fixed and adaptive antennas, analog and digital communications, project engineering, etc. The following is a list of books relating to these topics:

Balanis, C.A., *Antenna Theory Analysis and Design,* Harper & Row Publishers, New York 1982; 2nd Edition, John Wiley & Sons, Inc., New York, 1997.

Bowman, J.J., Senior, T.B.A., and Uslenghi, P.L.E., *Electromagnetic and Acoustic Scattering by Simple Shapes,* Revised Printing. Hemisphere Publishing Corporation, 1987.

Hansen, R.C., *Phased Array Antennas,* John Wiley & Sons, Inc., New York, 1998.

James, J.R. and Hall, P.S. (Editors), *Handbook of Microstrip Antennas,* Volumes I and II. Peter Peregrinus Ltd., Great Britain, 1989.

Kucar, A.D., *Satellite and Terrestrial Wireless Radio Systems: Fixed, Mobile, PCS and PCN, Radio vs. Cable. A Practical Approach,* Stridon Press Inc., 2000.

Lo, Y.T. and Lee, S.W. (Editors), *The Antenna Handbook,* Volumes I–IV, Van Nostrand Reinhold, USA, 1993.

Mailloux, R.J., *Phased Array Antenna Handbook,* Artech House, Inc., Norwood, MA, 1994.

Proakis, John G., *Digital Communications,* McGraw-Hill Book Company, New York, 1983.

Silvester, P.P. and Ferrari, R.L., *Finite Elements for Electrical Engineers,* 3rd Edition, Cambridge University Press, Cambridge, 1996.

Sklar, B., *Digital Communications. Fundamentals and Applications,* Prentice-Hall Inc., Englewood Cliffs, NJ, 1988.

Snyder, J.P., *Map Projection — A Working Manual,* U.S. Geological Survey Professional Paper 1395, United States Government Printing Office, Washington: 1987 (Second Printing 1989).

Spilker, J.J., Jr., *Digital Communications by Satellite,* Prentice-Hall Inc., Englewood Cliffs, NJ, 1977.

Stutzman, W.L. and Thiele, G.A., *Antenna Theory and Design,* John Wiley & Sons, Inc., New York, 1981. 2nd Edition, John Wiley & Sons, Inc., New York, 1998.

Van Trees, H.L., *Detection, Estimation, and Modulation Theory,* Part I, 1968, Part II, 1971, Part III, John Wiley & Sons, Inc., New York, 1971.

Walker, J., *Advances in Mobile Information Systems,* Artech House, Inc., Norwood, MA, 1999.

4

Broadband Wireless Access: High Rate, Point to Multipoint, Fixed Antenna Systems

Brian Petry
3Com Corporation

Broadband Wireless Access (BWA) broadly applies to systems providing radio communications access to a core network. Access is the key word because a BWA system by itself does not form a complete network, but only the access part. As the "last mile" between core networks and customers, BWA provides access services for a wide range of customers (also called subscribers), from homes to large enterprises. For enterprises such as small to large businesses, BWA supports such core networks as the public Internet, Asynchronous Transfer Mode (ATM) networks, and the Public Switched Telephone Network (PSTN). Residential subscribers and small offices may not require access to such a broad set of core networks — Internet access is likely BWA's primary access function. BWA is meant to provide reliable, high throughput data services as an alternative to wired access technologies.

This chapter presents an overview of the requirements, functions, and protocols of BWA systems and describes some of today's efforts to standardize BWA interfaces.

4.1 Fundamental BWA Properties

Currently, the industry and standards committees are converging on a set of properties that BWA systems have, or should have, in common. A minimal BWA system consists of a single base station and a single subscriber station. The base station contains an interface, or interworking function (IWF), to a core network, and a radio "air" interface to the subscriber station. The subscriber station contains an interface to a customer premises network and of course, an air interface to the base station. Such a minimal system represents the point-to-point wireless transmission systems that have been in use for many years. Interesting BWA systems have more complex properties, the most central of which is point-to-multipoint (P-MP) capability. A single base station can service multiple subscriber stations using the same radio

0-8493-1732-0/03/$0.00+$1.50

channel. The P-MP property of BWA systems feature omnidirectional or shaped sector radio antennas at the base station that cover a geographic and spectral area that efficiently serves a set of customers given the allocation of radio spectrum. Multiple subscriber stations can receive the base station's downstream transmissions on the same radio channel. Depending on the density and data throughput requirements of subscribers in a given sector, multiple radio channels may be employed, thus overlaying sectors. The frequency bands used for BWA allow for conventional directional antennas. So, in the upstream transmission direction, a subscriber's radio antenna is usually highly directional, aimed at the base station. Such configuration of shaped sectors and directional antennas allow for flexible deployment of BWA systems and helps conserve radio spectrum by allowing frequency bands to be reused in nearby sectors.

With such P-MP functions and a sectorized approach, more BWA properties unfold and we find that BWA is similar to other well-known access systems. A BWA deployment is cellular in nature, and like a cellular telephone deployment, requires complicated rules and guidelines that impact power transmission limits, frequency reuse, channel assignment, cell placement, etc. Also, since subscriber stations can share spectrum in both the upstream and downstream directions, yet do not communicate with each other using the air interface, BWA systems have properties very similar to hybrid fiber coaxial (HFC) access networks that coexist with cable television service. HFC networks also employ a base station (called a head end) and subscriber stations (called cable modems). Subscriber stations share channels in both downstream and upstream directions. Such HFC networks are now popularized by both proprietary systems and the Data-over-Cable System Interface Specifications (DOCSIS) industry standards [1]. In the downstream direction, digital video broadcast systems have properties similar to BWA. They employ base stations on the ground or onboard satellites: multiple subscribers tune their receivers to the same channels. With properties similar to cellular, cable modems, and digital video broadcast, BWA systems borrow many technical features from them.

4.2 BWA Fills Technology Gaps

Since BWA is access technology, it naturally competes with other broadband, high data rate access technologies, such as high data rate digital cellular service, digital subscriber line (DSL) on copper telephone wires, cable modems on coaxial TV cables, satellite-based access systems, and even optical access technologies on fiber or free space. To some, the application of BWA overlaps with these access technologies and also appears to fill in the gaps left by them. Following are some examples of technology overlaps where BWA fills in gaps.

High data rate digital cellular data service is available. This service is built "on top of" digital cellular telephone service. The maximum achievable data rate for these new "third-generation" digital cellular systems is intended to be around 2.5 Mbps. At these maximum speeds, high data rate cellular competes with low-end BWA, but since BWA systems are not intended to be mobile, and utilize wider frequency bands, a BWA deployment should be able to offer higher data rates. Furthermore, a BWA service deployment does not require near ubiquitous service area coverage. Before service can be offered by mobile cellular services, service must be available throughout entire metropolitan areas. But for BWA, service can be offered where target customers are located before covering large areas. Thus, in addition to higher achievable data rates with BWA, the cost to reach the first subscribers should be much less.

Current DSL technology can reach out about 6 km from the telephone central office, but the achievable data rate degrades significantly as the maximum distance is reached. Currently, the maximum DSL data rate is around 8 Mb/s. Asymmetric DSL (ADSL) provides higher data rates downstream than upstream, which is ideal for residential Internet access, but can be limiting for some business applications. BWA can fill in by providing much higher data rates further from telephone central offices. BWA protocols and deployment strategies enable the flexibility necessary to offer both asymmetric and symmetric services.

HFC cable modem technology, which is also asymmetric in nature, is ideal for residential subscribers. But many subscribers — potentially thousands — often share the same downstream channels and contend heavily for access to a limited number of available upstream channels. A key advantage of HFC is consistent channel characteristics throughout the network. With few exceptions, the fiber and coaxial cables deliver a consistent signal to subscribers over very long distances. BWA fills in, giving a service provider the flexibility to locate base stations and configure sectors to best service customers who need consistent, high data rates dedicated to them.

Satellite access systems are usually unidirectional, whereas less available bidirectional satellite-based service is more expensive. Either type of satellite access is asymmetric in nature: unidirectional service requires some sort of terrestrial "upstream," and many subscribers contend for the "uplink" in bidirectional access systems. Satellites in geostationary Earth orbits (GEO) impose a minimum transit delay of 240 ms on transmissions between ground stations. Before a satellite access system can be profitable, it must overcome the notable initial expense of launching satellites or leasing bandwidth on a limited number of existing satellites by registering many subscribers. Yet, satellite access services offer extremely wide geographic coverage with no infrastructure planning, which is especially attractive for rural or remote service areas that DSL and cable modems do not reach. Perhaps high data rate, global service coverage by low Earth orbiting (LEO) satellites will someday overcome some of GEO's limitations. BWA fills in by allowing service providers to locate base stations and infrastructure near subscribers that should be more cost effective and impose less delay than satellite services.

Optical access technologies offer unbeatable performance in data rate, reliability, and range, where access to fiber-optic cable is available. But in most areas, only large businesses have access to fiber. New technology to overcome this limitation, and avoid digging trenches and pulling fiber into the customer premises is free space optical, which employs lasers to extend between a business and a point where fiber is more readily accessible. Since BWA base stations could also be employed at fiber access points to reach non-fiber-capable subscribers, both BWA and free space optical require less infrastructure planning such as digging, tunneling, and pulling cables under streets. Although optical can offer an order of magnitude increase in data rate over the comparatively lower frequency/higher wavelength BWA radio communications, BWA can have an advantage in some instances because BWA typically has a longer range and its sector-based coverage allows multiple subscribers to be serviced by a single base station.

Given these gaps left by other broadband access technologies, even with directly overlapping competition in many areas, the long-term success of BWA technology is virtually ensured.

4.3 BWA Frequency Bands and Market Factors

Globally, a wide range of frequency bands are available for use by BWA systems. To date, systems that implement BWA fall into roughly two categories: those that operate at high frequencies (roughly 10 to 60 GHz) and those that operate at low frequencies (2 to 11 GHz). Systems in the low frequency category may be further subdivided into those that operate in licensed vs. unlicensed bands. Unlicensed low frequency bands are sometimes considered separately because of the variations of emitted power restrictions imposed by regulatory agencies and larger potential for interference by other "unlicensed" technologies. The high frequencies have significantly different characteristics than the low frequencies that impact the expense of equipment, base station locations, range of coverage, and other factors. The key differing characteristics in turn impact the type of subscriber and types of services offered as will be seen later in this article.

Even though available spectrum varies, most nationalities and regulatory bodies recognize the vicinity of 30 GHz, with widebands typically available, for use by BWA. In the U.S., for instance, the FCC has allocated Local Multipoint Distribution Service (LMDS) bands for BWA. That, coupled with the availability of radio experience, borrowed from military purposes and satellite communications, influenced the BWA industry to focus their efforts in this area. BWA in the vicinity of 30 GHz is thus also a target area for standardization of interoperable BWA systems. Available spectrum for lower frequencies, 2 to 11 GHz, varies widely by geography and regulatory body. In the U.S., for instance, the FCC has allocated several bands called Multipoint/Multichannel Distribution Services (MDS) and licensed them for BWA use. The industry is also targeting the lower spectrum, both licensed and unlicensed, for standardization.

Radio communications around 30 GHz have some important implications for BWA. For subscriber stations, directional radio antennas are practical. For base stations, so are shaped sector antennas. But two key impairments limit how such BWA systems are deployed: line-of-site and rain. BWA at 30 GHz almost strictly requires a line-of-sight path to operate effectively. Even foliage can prohibitively absorb the radio energy. Some near line-of-sight scenarios, such as a radio beam that passes in close proximity to reflective surfaces like metal sheets or wet roofs, can also cause significant communications impairments. Rain can be penetrated, depending on the distance between subscriber and base station, the droplet size, and rate of precipitation. BWA service providers pay close attention to climate zones and historical weather data to plan deployments. In rainy areas where subscribers require high data rate services, small cell sizes can satisfy a particular guaranteed service availability. Also, to accommodate changing rain conditions, BWA systems offer adaptive transmit power control. As the rate of precipitation increases, transmit power is boosted as necessary. The base station and subscriber station coordinate with each other to boost or decrease transmit power.

Impairments aside, equipment cost is an important issue with 30-GHz BWA systems. As of today, of all the components in a BWA station, the radio power amplifier contributes most to system cost. Furthermore, since the subscriber antenna must be located outdoors (to overcome the aforementioned impairments), installation cost contributes to the equation. A subscriber installation consists of an indoor unit (IDU) that typically houses the digital equipment, modem, control functions, and interface to the subscriber network, and an outdoor unit (ODU), which typically houses the amplifier and antenna. Today these factors, combined with the aforementioned impairments, typically limit the use of 30-GHz BWA systems to businesses that both need the higher end of achievable data rates and can afford the equipment. BWA technology achieves data rates delivered to a subscriber in a wide range, 2 to 155 Mbps. The cost of 30-GHz BWA technology may render the lower end of the range impractical. However, many people project the cost of 30-GHz BWA equipment to drop as the years go by, to the point where residential service will be practical.

In the lower spectrum for BWA systems, in the range of approximately 2 to 11 GHz, line-of-sight and rain are not as critical impairments. Here, a key issue to contend with is interference due to reflections, also called multipath. A receiver, either base station or subscriber, may have to cope with multiple copies of the signal, received at different delays, due to reflections off buildings or other large objects in a sector. So, different modulation techniques may be employed in these lower frequency BWA systems, as opposed to high frequency systems, to compensate for multipath. Furthermore, if the additional expense can be justified, subscribers and/or base stations, could employ spatial processing to combine the main signal with its reflections and thus find a stronger signal that has more data capacity than the main signal by itself. Such spatial processing requires at least two antennas and radio receivers. In some cases, it may even be beneficial for a base station to employ induced multipath, using multiple transmit antennas; perhaps these are aimed at reflective objects to reach subscribers, even those hidden by obstructions, with a better combined signal than just one.

Unlike BWA near 30-GHz, BWA in the lower spectrum today has the advantage of less expensive equipment. Also, it may be feasible in some deployments for the subscriber antenna to be located indoors. Further, the achievable data rates are typically lower than at 30-GHz, with smaller channel bandwidths, in the range of about 2 to 15 Mb/s. Although some promise 30-GHz equipment costs will drop, these factors make lower frequency BWA more attractive to residences and small businesses today.

Due to the differing requirements of businesses and residences and the different capabilities of higher frequency BWA vs. lower, the types of service offered are naturally divided as well. Businesses will typically subscribe to BWA at the higher frequencies, around 30-GHz, and employ services that carry guaranteed quality of service for both data and voice communications. In the business category, multi-tenant office buildings and dwellings are also lumped in. At multi-tenant sites, multiple paying subscribers share one BWA radio and each subscriber may require different data or voice services. For data, Internet Protocol (IP) service is of prime importance, but large businesses also rely on wide area network technologies like asynchronous transfer mode (ATM) and frame relay that BWA must efficiently transport. To date, ATM's capabilities offer practical methods for dedicating, partitioning, and prioritizing data flows, generally called quality of service (QoS). But as time goes on, IP-based QoS capabilities will overtake ATM. So, for both residential and business purposes, IP service will be the data service of choice in the future. Besides data, businesses rely on traditional telephony links to local telephone service providers. Business telephony services, for medium-to-large enterprises, utilize time division multiplexed (TDM) telephone circuits on copper wires to aggregate voice calls. Some BWA systems have the means to efficiently transport such aggregated voice circuits. Due to the economic and performance differences between low frequency BWA and high frequency BWA, low frequency BWA generally carries residential- and small business-oriented services, whereas high frequency BWA carries small- to large-enterprise services.

Since BWA equipment for the lower frequencies may be less expensive and less sensitive to radio directionality, and therefore more practical to cover large areas such as residential environments, subscriber equipment can potentially be nomadic. Nomadic means that the equipment may be moved quickly and easily from one location to another, but is not expected to be usable while in transit. Whereas at the higher frequencies, with more expensive subscriber equipment, the decoupling of indoor and outdoor units, the highly directional nature of radio communications in that range, and subscriber-oriented services provisioned at the base station, subscriber stations are fixed. Once they are installed, they do not move unless the subscriber terminates service and re-subscribes somewhere else.

4.4 Standards Activities

Several standards activities are under way to enable interoperability between vendors of BWA equipment. The standardization efforts are primarily focused on an interoperable "air interface" that defines how compliant base stations interoperate with compliant subscriber stations. By this reading, some of the standards may have been completed — the reader is encouraged to check the status of BWA standardization. Some standards groups archive contributions by industry participants and those archives, along with the actual published standards, provide important insights into BWA technology. Currently, most activity is centered around the Institute for Electrical and Electronics Engineers (IEEE) Local Area Network/Metropolitan Area Network (LAN/MAN) Standards Committee (LMSC), which authors the IEEE 802 series of data network standards. Within LMSC, the 802.16 working group authors BWA standards. The other notable BWA standards effort, under the direction of the European Telecommunications Standards Institute (ETSI), is a project called Broadband Radio Access Networks/HyperAccess (BRAN/HA). The IEEE LMSC is an organization that has international membership and has the means to promote their standards to "international standard" status through the International Organization for Standardization (ISO) as does ETSI. But ETSI standards draw from a European base, whereas LMSC draws from a more international base of participation. Even so, the LMSC and BRAN/HA groups, although they strive to develop standards each with a different approach, have many common members who desire to promote a single, international standard. Hopefully, the reader will have discovered that the two groups have converged on one standard that enables internationally harmonized BWA interoperability.

To date, the IEEE 802.16 working group members have segmented their activities into three main areas: BWA interoperability at bands around 30 GHz (802.16.1), a recommended practice for the coexistence of BWA systems (802.16.2) and BWA interoperability for licensed bands between 2 and 11 GHz (802.16.3). By the time this book is published, more standards activities may have been added, such as interoperability for some unlicensed bands. The ETSI BRAN/HA group is focused on interoperability in bands around 30 GHz.

Past standards activities were efforts to agree on how to adapt existing technologies for BWA: cable modems and digital video broadcast. A BWA air interface, as similar to DOCSIS cable modems as possible, was standardized by the radio sector of the International Telecommunications Union (ITU) under the ITU-R Joint Rappateur's Group (JRG) 9B committee [2]. The Digital Audio-Video Council (DAVIC) has standardized audio and video transport using techniques similar to BWA [3]. Similarly, the Digital Video Broadcasting (DVB) industry consortium, noted for having published important standards for satellite digital video broadcast, has also published standards, through ETSI, for terrestrial-based digital television broadcast over both cable television networks and wireless. DVB has defined the means to broadcast digital video in both the "low" (<10 Gb/s) and "high" (>10 Gb/s) BWA spectra [4, 5]. These standards enabled interoperability of early BWA deployment by utilizing existing subsystems and components. Technology from them provided a basis for both the IEEE LMSC and ETSI BRAN/HA standardization processes. However, the current IEEE and ETSI efforts strive to define protocols with features and nuances more particular to efficient BWA communications.

4.5 Technical Issues: Interfaces and Protocols

A BWA access network is perhaps best described by its air interface: what goes on between the base station and subscriber stations. Other important interfaces exist in BWA systems, such as:

- Interfaces to core networks like ATM, Frame Relay, IP, and PSTN
- Interfaces to subscriber networks like ATM, Ethernet, Token Ring, and private branch exchange (PBX) telephone systems
- Interface between indoor unit (IDU) and outdoor unit (ODU)
- Interfaces to back-haul links, both wired and wireless, for remote base stations not co-located with core networks
- Air interface repeaters and reflectors

These other interfaces are outside the scope of this article. However, understanding their requirements is important to consider how a BWA air interface can best support external interfaces, particularly how the air interface supports their unique throughput, delay, and QoS requirements.

4.5.1 Protocols and Layering

Network subsystems following the IEEE LMSC reference model [6] focus on the lower two layers of the ISO Basic Reference Model for Open Systems Interconnection [7]. The air interface of a BWA system is also best described by these two layers. In LMSC standards, layers one and two, the physical and data link layers, are typically further subdivided. For BWA, the important subdivision of layer 2 is the medium access control (MAC) sublayer. This layer defines the protocols and procedures by which network nodes contend for access to a shared channel, or physical layer. In a BWA system, since frequency channels are shared among subscriber stations in both the downstream and upstream directions, MAC layer services are critical for efficient operation. The physical layer (PHY) of a BWA system is responsible for providing a raw communications channel, employing modulation and error correction technology appropriate for BWA.

Other critical functions, some of which may reside outside the MAC and PHY layers, must also be defined for an interoperable air interface: security and management. Security is divided two areas: a subscriber's authorized use of a base station and associated radio channels and privacy of transported data. Since the communications channel is wireless, it is subject to abuse by intruders, observers, and those seeking to deny service. BWA security protocols must be well defined to provide wire-like security and allow for interoperability. Since to a great extent, HFC cable TV access networks are very similar to BWA regarding security requirements, BWA borrows heavily from the security technology of such cable systems. Similarly, interoperable management mechanisms and protocols include the means to provision, control, and monitor subscribers stations and base stations.

4.5.1.1 The Physical Layer

The physical layer (PHY) is designed with several fundamental goals in mind: spectral efficiency, reliability, and performance. However, these are not independent goals. We cannot have the best of all three because each of those goals affects the others: too much of one means too little of the others. But reliability and performance levels are likely to be specified. Once they are specified, spectral efficiency can be somewhat optimized. One measure of reliability is the bit error ratio (BER), the ratio of the number of bit errors to the number of non-errored bits, delivered by a PHY receiver to the MAC layer. The physical layer must provide for better than 10^{-6} BER, and hopefully closer to 10^{-9}. The larger error ratio may only be suitable for some voice services, whereas a ratio closer to the lower end of the range is required for reliable data services that could offer equivalent error performance as local area networks (LANs). Reliability is related to availability. Business subscribers often require contracts that guarantee a certain level of availability. For instance, a service provider may promise that the air interface be available to offer guaranteed reliability and performance 99.99% (also called "four nines") of the time.

Performance goals specify minimum data rates. Since, in BWA systems, the spectrum is shared by subscribers, and allocation of capacity among them is up to the MAC layer, the PHY is more concerned with the aggregate capacity of a single radio channel in one sector of a base station than for capacity to a given subscriber. But if one subscriber would offer to purchase all the available capacity, the service provider would undoubtedly comply. For instance, a capacity goal currently set by the BRAN/HA committee is 25 Mb/s on a 28-MHz channel. Without considering deployment scenarios, however, PHY goals are meaningless. Obviously, higher capacity and reliability could be better achieved by shorter, narrower sectors (smaller cells) rather than wider, longer sectors (larger cells). The same sized sector in a rainy, or obstructed, terrain offers less guaranteed capacity than one in the flattest part of the desert. In any case, the industry seems to be converging on a goal to provide at least 1 bps/Hz capacity in an approximately 25-MHz wide channel with a BER of 10^{-8}. Many deployments should be able to offer much greater capacity.

In addition to such fundamental goals, other factors affect the choice of PHY protocols and procedures. One is duplex mode. The duplex mode can affect the cost of equipment, and some regulatory bodies may limit the choice of duplex mode in certain bands. Three duplex modes are considered for BWA: frequency division duplex (FDD), time division duplex (TDD), and half-duplex FDD (H-FDD). In FDD, a radio channel is designated for either upstream- or downstream-only use. Some bands are regulated such that a channel could only be upstream or downstream, thus requiring FDD if such bands are to be utilized. In TDD mode, one channel is used for both upstream and downstream communications. TDD-capable BWA equipment thus ping-pongs between transmit and receive mode within a single channel; all equipment in a sector is synchronized to divisions between transmit and receive. TDD is useful for bands in which the number of available, or licensed, channels is limited. TDD also allows for asymmetric service without reconfiguring the bandwidth of FDD channels. For instance, a service provider may determine that a residential deployment is more apt to utilize more downstream bandwidth than upstream. Then, rather than reallocating or sub-channeling FDD channels, the service provider can designate more time in a channel for downstream communications than upstream. Additionally, TDD equipment could potentially be less expensive than FDD equipment since components may be shared between the upstream and downstream paths and the cost of a duplexor may be eliminated. However, the third option, H-FDD, is a reasonable compromise between TDD and FDD. In H-FDD mode, a subscriber station decides when it can transmit and when it can receive, but cannot receive while transmitting. But the base station is usually full duplex, or FDD. For subscriber stations, H-FDD equipment can achieve the same cost savings as TDD, and offers the flexibility of asymmetric service. But H-FDD does not require all subscribers in a sector to synchronize on the same allocated time between transmit and receive.

Another important factor affecting spectral efficiency, upgradability, and flexible deployment scenarios, is adaptive modulation. In BWA, the channel characteristics vary much more widely than wired access systems. Rain, interference, and other factors can affect subscriber stations individually in a sector, whereas in wired networks, such as HFC cable TV, the channel characteristics are consistent. Thus, to make good use of available bandwidth in favorable channel conditions, subscribers that can take advantage of higher data rates should be allowed to do so. When it rains in one portion of a sector, or other impairments such as interference occur, subscriber stations can adapt to the channel conditions by reducing the data rate (although transmit power level adjustment is usually the first adaptive tool BWA stations use when it rains). Besides adapting to channel conditions, adaptive modulation facilitates future deployment of newer modulation techniques while retaining compatibility with currently installed subscriber stations. When the service provider upgrades a base station and offers better modulation to new customers, not all subscriber stations become obsolete. To achieve the most flexibility in adaptive modulation, BWA employs "per-subscriber" adaptive modulation to both downstream and upstream communications. Per-subscriber means that each subscriber station can communicate with the base station using a different modulation technique, within the same channel. Some BWA equipment offers per-subscriber adaptive modulation in both the downstream and upstream directions. But other equipment implements a compromise that allows for equipment or components, similar to cable modems or digital video broadcast systems, to require all subscribers to use the same modulation in the downstream direction at any one point in time. Most BWA equipment implements adaptive modulation in the upstream direction. The overriding factor for the PHY layer, with regard to adaptive modulation, is burst mode. Adaptive modulation generally requires burst mode communications at the PHY layer. Time is divided into small units in which stations transmit independent bursts of data. If the downstream employs per-subscriber adaptive modulation, the base station transmits independent bursts to the subscribers. Each burst contains enough information for the receiver to perform synchronization and equalization. However, if per-subscriber adaptive modulation is not employed in the downstream direction, the base station can transmit in continuous mode, in very large, continuous chunks, each chunk potentially containing data destined for multiple subscribers. In burst mode downstream communications, the base station informs subscriber stations, in advance, which burst is theirs. In this way, a subscriber station is not required to demodulate each burst to discover which bursts are for the station, but only to demodulate the "map." The base station encodes the map using the least common denominator modulation type so all subscriber stations can decode it. Conversely, continuous mode downstream, in which per-subscriber adaptive modulation is not used, requires all subscriber stations to demodulate prior to discovering which portions of data are destined for the station. So, per-subscriber adaptive modulation in the downstream affords more flexibility, but a continuous mode downstream may also be used. The standards efforts currently are attempting to work out how both downstream modes may be allowed and yet still have an interoperable standard.

Burst size and the choice of continuous downstream mode in turn affect the choice of error correction coding. Some coding schemes are more efficient with large block sizes, whereas others are more efficient with smaller block sizes.

The fundamental choice of modulation type for BWA varies between the upper BWA bands (~30 GHz) and lower bands (~2 to 11 GHz). In the upper bands, the industry seems to be converging on Quadrature Phase Shift Keying (QPSK) and various levels of Quadrature Amplitude Modulation (QAM). These techniques may also be used in the lower bands, but given the multipath effects that are much more prevalent in the lower bands, BWA equipment is likely to employ Orthogonal Frequency Division Multiplexing (OFDM) or Code Division Multiple Access (CDMA) technology that have inherent properties to mitigate the effects of multipath and spread transmit energy evenly throughout the channel spectrum.

4.5.1.2 The Medium Access Control Layer

The primary responsibility of the Medium Access Control Layer (MAC) is to allocate capacity among subscriber stations in a way that preserves quality-of-service (QoS) requirements of the services it transports. For instance, traditional telephony and video services could require a constant, dedicated capacity with fixed delay properties. But other data transport services could tolerate more bursty capacity allocations and a higher degree of delay variation. ATM service is notable for its QoS definitions [8]. Although not mature as of this writing, the Internet Protocol (IP) QoS definitions are also notable [9, 10]. Though QoS-based capacity allocation is a complex process, the BWA MAC protocol defines the mechanisms to preserve QoS as it transports data. Yet the MAC protocol does not fully define *how* MAC mechanisms are to be used. At first glance, this does not seem to make sense, but it allows the MAC protocol to be defined in as simple terms as possible and leave it up to implementations of base stations and subscriber stations how to best utilize the mechanism that the protocol defines. This approach also allows BWA vendors to differentiate their equipment and still retain interoperability. To simplify capacity allocation, the smarts of QoS implementation reside in the base station, since it is a central point in a BWA sector and is in constant communication with all of the subscriber stations in a sector. The base station is also administered by the service provider, and therefore can serve as the best point of control to keep subscribers from exceeding their contractual capacity limitations and priorities.

Capacity Allocation Mechanisms — An overview of the mechanisms employed by the MAC layer to allocate capacity follows. In the downstream direction, the MAC protocol informs subscriber stations what data belongs to what subscriber by means of per-subscriber addressing and within a subscriber, by per-data-flow addressing. All subscribers in a sector "listen" to the downstream data flow and pick off transmissions belonging to them. If the downstream channel employs per-subscriber adaptive modulation, some subscriber stations may not be able to decode the modulation destined to other subscribers. In this case, the base station informs subscribers what bursts it should observe, with a downstream "map." The downstream map indicates what offsets in a subsequent transmission may contain data for the specified subscriber. The MAC must communicate this information to the PHY layer to control its demodulation.

For upstream capacity allocation and reservation, the MAC employs slightly more complicated schemes. The upstream channel is the central point of contention: all subscriber stations in a channel are contending for access to transmit in the upstream channel. Some subscribers require constant periodic access, others require bursty access with minimum and maximum reservation limits. Still other data flows may not require any long-standing reservations but can request a chunk of capacity when needed and survive the inherent access delay until the base station satisfies the request. On top of these varying data flow requirements, which are specified by subscriber stations and granted by the base station, priorities increase complications. The base station administers both priorities and QoS parameters of each data flow in each subscriber station. How a base station keeps track of all the flows of subscribers and how it actually meets the reservation requirements are usually beyond the scope of the BWA air interface in standards documents. But base stations likely employ well-known queuing algorithms and reservation lists to ensure that they assign capacity fairly and meets subscribers' contractual obligations. Yet, as mentioned earlier, room is left for BWA base station vendors to employ proprietary "tricks" to differentiate their equipment from others. To communicate capacity allocation to subscribers, the base station divides time into multi-access frames (e.g., on the order of 1 to 5 ms) in which multiple subscribers are assigned capacity. To accomplish this, a fixed allocation unit, or time slot, is defined. So, the upstream channel is divided into small, fixed-length time slots (e.g., on the order of 10 μs) and the base station periodically transmits a "map" of slot assignments to all subscribers in a channel. The slot assignments inform the subscriber stations which slots are theirs for the upcoming multi-access frame.

Periodically, a set of upstream slots is reserved for "open contention." That is, any subscriber is authorized to transmit during an open contention period. A subscriber can utilize open contention for initial sign-on to the network (called "registration"), to transmit a request for upstream capacity, or even to transmit a small amount of data. Since a transmission may collide with that of another subscriber station, a collision avoidance scheme is used. A subscriber station initiates transmission in a randomly chosen open contention slot, but cannot immediately detect that its transmission collided with another. The only way a subscriber station can determine if its transmission collided is if it receives no acknowledgment from the base station. In this case, the subscriber backs off a random number of open contention slots before attempting another transmission. The process continues, with the random number range getting exponentially larger on each attempt, until the transmission succeeds. The random back-off interval is typically truncated at the sixteenth attempt, when the subscriber station starts over with its next attempt in the original random number range. This back-off scheme is called "truncated binary exponential back-off," and is employed by popular MAC protocols such as Ethernet [11].

To mitigate the effects of potentially excessive collisions during open contention, the MAC protocol defines a means to request bandwidth during assigned slots in which no collision would happen. For instance, active subscriber stations may receive from the base station a periodic slot for requesting capacity or requesting a change in a prior reservation. This form of allocation-for-a-reservation-request is called a "poll." Also, the MAC protocol provides a means to "piggy-back" a request for capacity with a normal upstream data transmission. Subscriber stations that have been inactive may receive less frequent polls from the base station so as to conserve bandwidth. So, with a means for contentionless bandwidth reservation, the only time subscriber stations need to use the open contention window is for initial registration.

Slot-based reservations require that the base stations and subscribers be synchronized. Of course, the base station provides a timing base for all subscriber stations. To achieve accurate timing, subscriber stations need to determine how far they are from the base station so their transmissions can be scheduled to reach the base station at the exact point in time, relative to each other. The procedure to determine this distance, which is not really a measured linear distance, but a measurement of time, is called "ranging." Each subscriber station, coordinating with the base station, performs ranging during its registration process.

To maintain efficient use of bandwidth and accommodate PHY requirements of transmit power control, and flexible duplex modes, the MAC protocol performs even more gyrations. If interested, the reader is encouraged to read BWA MAC protocol standards, or drafts in progress, to learn more.

4.5.1.3 Automatic Repeat Request (ARQ) Layer

Some BWA systems trade off the bandwidth normally consumed by the PHY's error correction coding for the potential delays of ARQ protocol. An ARQ protocol employs sequence numbering and retransmissions to provide a reliable air link between base station and subscriber. ARQ requires more buffering in both the base station and subscriber station than systems without ARQ. But even with a highly coded PHY, some subscriber stations may be located in high interference or burst-noise environments in which error correction falls apart. In such situations, ARQ can maintain performance, or ensure the service meets contractual availability and reliability requirements. Standards groups seem to be converging on allowing the use of ARQ, but not requiring it. The MAC protocol is then specified so that when ARQ is not used, no additional overhead is allocated just to allow the ARQ option.

4.6 Conclusion

This chapter has provided an overview of how BWA fits in with other broadband access technologies. It was also a short primer on BWA protocols and standards. To learn more about BWA, the reader is encouraged to read currently available standards documents and various radio communications technical journals, and consult with vendors of BWA equipment.

References

1. SCTE SP-RFI-105-981010: *Data-Over-Cable Service Interface Specifications: Radio Frequency Interface Specification,* The Society of Cable Telecommunications Engineers, Exton, PA, 1999.
2. Draft Recommendation F.9B/BWA. Radio transmission systems for fixed broadband wireless access (BWA) based on cable modem standard, International Telecommunications Union, Geneva, 1999.
3. DAVIC 1.4.1 *Specification Part 8: Lower Layer Protocols and Physical Interfaces,* Digital Audio-Visual Council, Geneva, 1999.
4. ETS 300 748, *Digital Video Broadcasting (DVB): Multipoint Video Distribution Systems (MVDS) at 10 GHz and above,* European Telecommunications Standards Institute, Geneva, 1997.
5. ETS 300 749, *Digital Video Broadcasting (DVB): Digital Video Broadcasting (DVB); Microwave Multipoint Distribution Systems (MMDS) below 10 GHz,* European Telecommunications Standards Institute, Geneva, 1997.
6. IEEE Std 802-1990, *IEEE Standards for Local and Metropolitan Area Networks: Overview and Architecture,* Institute for Electrical and Electronics Engineers, Piscataway, NJ, 1990.
7. ISO/IEC 7498-1:1994, *Information Technology — Open Systems Interconnection — Basic Reference Model: The Basic Model,* International Organization for Standardization, Geneva, 1994.

8. Bermejo, L. P. et al., Service characteristics and traffic models in broadband ISDN, *Electrical Commun.*, 64-2/3, 132–138, 1990.
9. Blake, S. et al., *RFC-2475 An Architecture for Differentiated Service*, Internet Engineering Task Force, 1998.
10. Braden, R. et al., *RFC-2205 Resource ReSerVation Protocol (RSVP) — Version 1 Functional Specification*, Internet Engineering Task Force, 1997.
11. IEEE Std 802.3, *Information Technology — Telecommunications and information exchange between systems — Local and metropolitan area networks — Specific requirements — Part 3: Carrier sense multiple access with collision detection (CSMA/CD) access method and physical layer specifications*, Institute for Electronics and Electrical Engineers, Piscataway, NJ, 1998.

5

Digital European Cordless Telephone

Saf Asghar
Advanced Micro Devices, Inc.

Cordless technology, in contrast to cellular radio, primarily offers access technology rather than fully specified networks. The digital European cordless telecommunications (DECT) standard, however, offers a proposed network architecture in addition to the air interface physical specification and protocols but without specifying all of the necessary procedures and facilities. During the early 1980s, a few proprietary digital cordless standards were designed in Europe purely as coexistence standards. The U.K. government in 1989 issued a few operator licenses to allow public-access cordless known as telepoint. Interoperability was a mandatory requirement leading to a common air interface (CAI) specification to allow roaming between systems. This particular standard (CT2/CAI), is described elsewhere in this book. The European Telecommunications Standards Institute (ETSI) in 1988 took over the responsibility for DECT. After formal approval of the specifications by the ETSI technical assembly in March 1992, DECT became a European telecommunications standard, ETS300-175, in August 1992. DECT has a guaranteed pan-European frequency allocation, supported and enforced by European Commission Directive 91/297. The CT2 specification has been adopted by ETSI alongside DECT as an interim standard I-ETSI 300 131 under review.

5.1 Application Areas

Initially, DECT was intended mainly to be a private system, to be connected to a private automatic branch exchange (PABX) to give users mobility, within PABX coverage, or to be used as a single cell at a small company or in a home. As the idea with telepoint was adopted and generalized to public access, DECT became part of the public network. DECT should not be regarded as a replacement of an existing network but as created to interface seamlessly to existing and future fixed networks such as public switched telephone network (PSTN), integrated services digital network (ISDN), global system for mobile communications (GSM), and PABX. Although telepoint is mainly associated with CT2, implying public access, the main drawback in CT2 is the ability to only make a call from a telepoint access point. Recently

0-8493-1732-0/03/$0.00+$1.50

there have been modifications made to the CT2 specification to provide a structure that enables users to make and receive calls. The DECT standard makes it possible for users to receive and make calls at various places, such as airport/railroad terminals, and shopping malls. Public access extends beyond telepoint to at least two other applications: replacement of the wired local loop, often called cordless local loop (CLL), (Fig. 5.1) and neighborhood access, Fig. 5.2. The CLL is a tool for the operator of the public network.

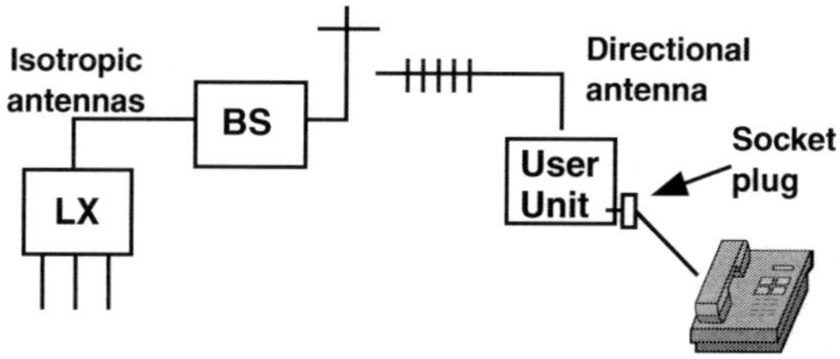

FIGURE 5.1

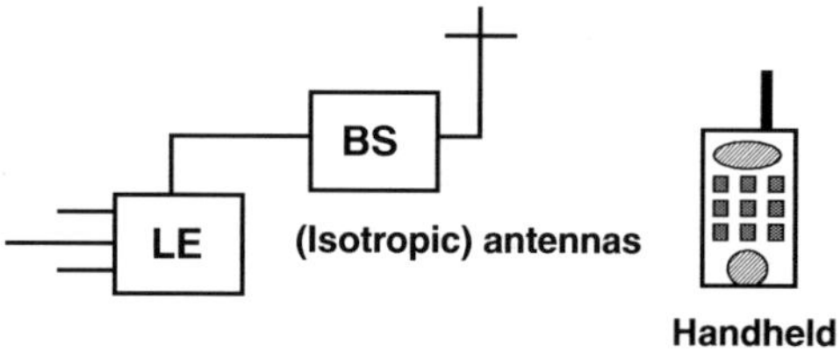

FIGURE 5.2

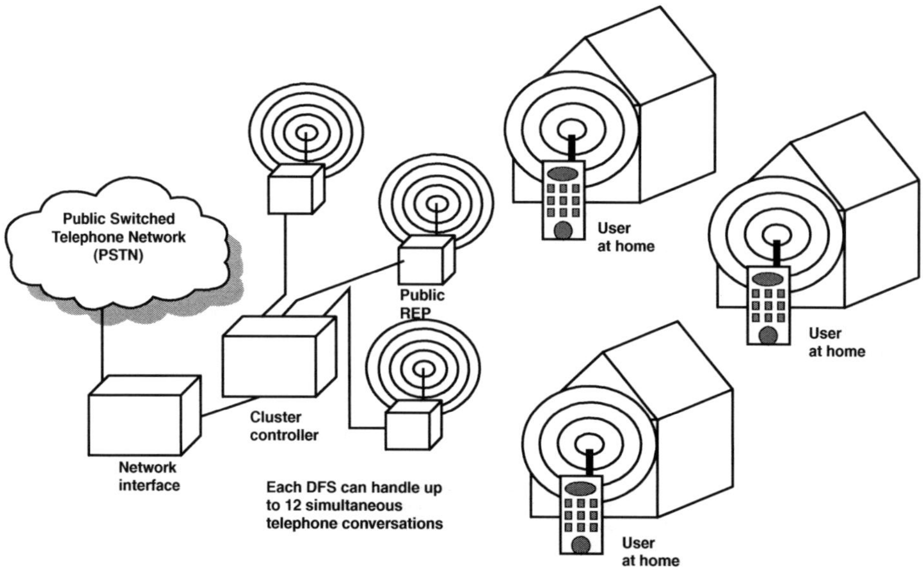

FIGURE 5.3

Essentially, the operator will install a multiuser base station in a suitable campus location for access to the public network at a subscriber's telephone hooked up to a unit coupled to a directional antenna. The advantages of CLL are high flexibility, fast installation, and possibly lower investments. CLL does not provide mobility. Neighborhoods access is quite different from CLL. First, it offers mobility to the users; and, second, the antennas are not generally directional, thus requiring higher field strength (higher output power or more densely packed base stations). It is not difficult to visualize that CLL systems could be merged with neighborhood access systems in the context of establishments, such as supermarkets, gas stations, and shops, where it might be desirable to set up a DECT system for their own use and at the same time also provide access to customers. The DECT standard already includes signaling for authentication, billing, etc. DECT opens possibilities for a new operator structure, with many diversified architectures connected to a global network operator (Fig. 5.3). DECT is designed to have extremely high capacity. A small size is used, which may seem an expensive approach for covering large areas. Repeaters placed at strategic locations overcome this problem.

5.2 DECT/ISDN Interworking

From the outset, a major objective of the DECT specification was to ensure that ISDN services were provided through the DECT network. Within the interworking profile two configurations have been defined: DECT end system and DECT intermediate system. In the end system the ISDN is terminated in the DECT fixed system (DFS). The DFS and the DECT portable system (DPS) may be seen as ISDN terminal equipment (TE1). The DFS can be connected to an S, S/T, or a P interface. The intermediate system is fully transparent to the ISDN. The S interface is regenerated even in the DPS. Both configurations have the following services specified: 3.1-kHz telephony (i.e., standard telephony); 7-kHz telephony (i.e., high-quality audio; video telephony; group III fax, modems, X.25 over the ISDN; and telematic services, such as group IV fax, telex, and videotax).

5.3 DECT/GSM Interworking

Groupe Speciale Mobile (GSM) is a pan-European standard for digital cellular radio operation throughout the European community. ETSI has the charter to define an interworking profile for GSM and DECT. The profile describes how DECT can be connected to the fixed network of GSM and the necessary air interface functions. The users obviously benefit from the mobility functions of GSM giving DECT a wide area mobility. The operators will gain access to another class of customer. The two systems when linked together will form the bridge between cordless and cellular technologies. Through the generic access profile, ETSI will specify a well-defined level of interoperability between DECT and GSM. The voice coding aspect in both of these standards is different; therefore, this subject will be revisited to provide a sensible compromise.

5.4 DECT Data Access

The DECT standard is specified for both voice and data applications. It is not surprising that ETSI confirmed a role for DECT to support cordless local area network (LAN) applications. A new technical committee, ETSI RES10, has been established to specify the high performance European radio LAN similar to IEEE 802.11 standard in the U.S. (Table 5.1).

5.5 How DECT Functions

DECT employs frequency division multiple access (FDMA), time division multiple access (TDMA), and time division duplex (TDD) technologies for transmission. Ten carrier frequencies in the 1.88- and 1.90-GHz band are employed in conjunction with 12 time slots per carrier TDMA and 10 carriers per 20 MHz of spectrum FDMA. Transmission is through TDD. Each channel has 24 times slots, 12 for

TABLE 5.1 DECT Characteristics

Parameters	DECT
Operating frequency, MHz	1880–1990 (Europe)
Radio carrier spacing, MHz	1.728
Transmitted data rate, Mb/s	1.152
Channel assignment method	DCA
Speech data rate, kb/s	32
Speech coding technique	ADPCM G.721
Control channels	In-call-embedded (various logical channels C, P, Q, N)
In-call control channel data rate, kb/s	4.8 (plus 1.6 CRC)
Total channel data rate, kb/s	41.6
Duplexing technique	TDD
Multiple access-TDMA	12 TDD time slots
Carrier usage-FDMA/MC	10 carriers
Bits per TDMA time slot, b	420 (424 including the 2 field)
Time slot duration (including guard time), μs	417
TDMA frame period, ms	10
Modulation technique	Gaussian-filtered FSK
Modulation index	0.45–0.55
Peak output power, mW	250
Mean output power, mW	10

transmission and 12 for receiving. A transmission channel is formed by the combination of a time slot and a frequency. DECT can, therefore, handle a maximum of 12 simultaneous conversations. TDMA allows the same frequency to use different time slots. Transmission takes place for ms, and during the rest of the time the telephone is free to perform other tasks, such as channel selection. By monitoring check bits in the signaling part of each burst, both ends of the link can tell if reception quality is satisfactory. The telephone is constantly searching for a channel for better signal quality, and this channel is accessed in parallel with the original channel to ensure a seamless changeover. Call handover is also seamless, each cell can handle up to 12 calls simultaneously, and users can roam around the infrastructure without the risk of losing a call. Dynamic channel assignment (DCA) allows the telephone and base station to automatically select a channel that will support a new traffic situation, particularly suited to a high-density office environment.

5.6 Architectural Overview

5.6.1 Baseband Architecture

A typical DECT portable or fixed unit consists of two sections: a baseband section and a radio frequency section. The baseband partitioning includes voice coding and protocol handling (Fig. 5.4).

5.6.2 Voice Coding and Telephony Requirements

This section addresses the audio aspects of the DECT specification. The CT2 system as described in the previous chapter requires adaptive differential pulse code modulation (ADPCM) for voice coding. The DECT standard also specifies 32-kb/s ADPCM as a requirement. In a mobile environment it is debatable whether the CCITT G.721 recommendation has to be mandatory. In the handset or the mobile it would be quite acceptable in most cases to implement a compatible or a less complex version of the recommendation. We are dealing with an air interface and communicating with a base station that in the residential situation terminates with the standard POTS line, hence compliance is not an issue. The situation changes in the PBX, however, where the termination is a digital line network. DECT is designed for this case, hence compliance with the voice coding recommendation becomes important. Adhering to this strategy for the base station and the handset has some marketing advantages.

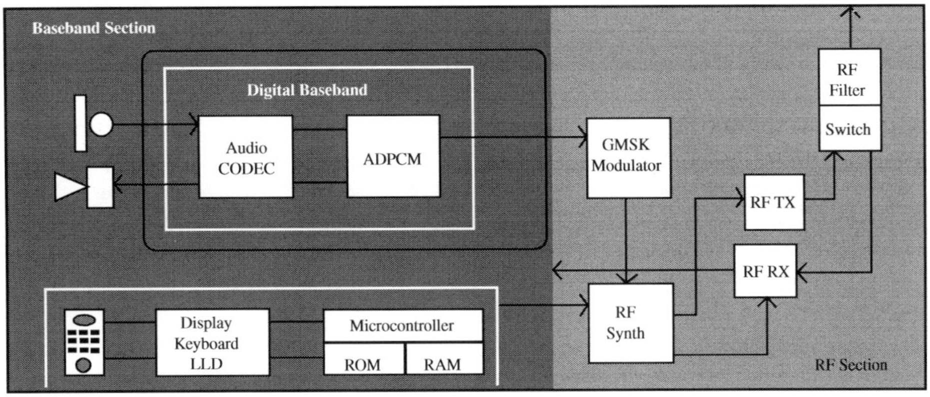

FIGURE 5.4

G.721 32-kb/s ADPCM from its inception was adopted to coexist with G.711 64-kb/s pulse code modulation (PCM) or work in tandem, the primary reason is an increase in channel capacity. For modem type signaling, the algorithm is suboptimal in handling medium-to-high data rates, which is probably one of the reasons why there really has not been a proliferation of this technology in the PSTN infrastructure. The theory of ADPCM transcoding is available in books on speech coding techniques (e.g., O'Shaughnessy [1987]).

The ADPCM transcoder consists of an encoder and a decoder. From Figs. 5.5 and 5.6 it is apparent that the decoder exists in the encoder structure. A benefit derived from this structure allows for efficient implementation of the transcoder.

The encoding process takes a linear speech input signal (the CCITT specification relates to a nonwireless medium such as a POTS infrastructure), and subtracts its estimate derived from earlier input signals to obtain a difference signal. This difference signal is 4-b code with a 16-level adaptive quantizer every 125 μs, resulting in a 32-kb/s bit stream. The signal estimate is constructed with the aid of the inverse adaptive quantizer that forms a quantized difference signal that added to the signal estimate is also used to update the adaptive predictor. The adaptive predictor is essentially a second-order recursive filter and a sixth-order nonrecursive filter,

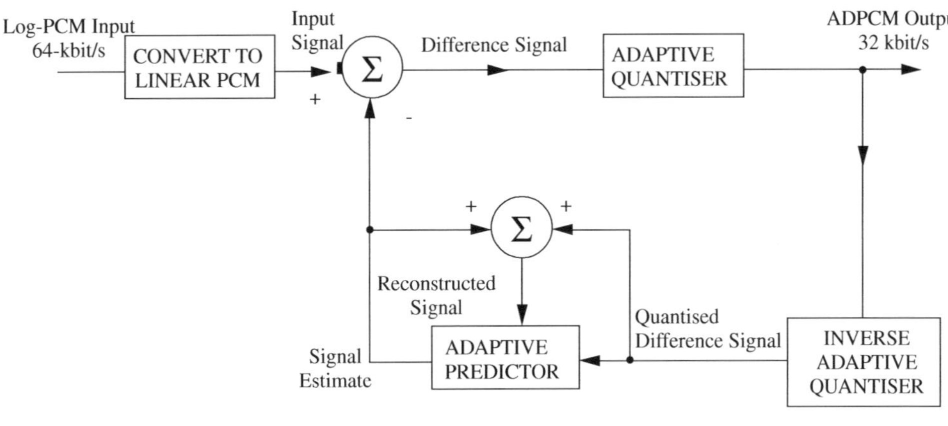

FIGURE 5.5 ADPCM encoder.

$$S_0(k)=\sum_{i=1}^{2}a_i(k-1)\varepsilon_r(k-i)+\sum_{i=1}^{6}b_i(k-1)d_q(k-i) \tag{5.1}$$

where coefficients a and b are updated using gradient algorithms.

As suggested, the decoder is really a part of the encoder; that is, the inverse adaptive quantizer reconstructs the quantized difference signal, and the adaptive predictor forms a signal estimate based on the quantized difference signal and earlier samples of the reconstructed signal, which is also the sum of the current estimate and the quantized difference signal as shown in Fig. 5.6. Synchronous coding adjustment tries to correct for errors accumulating in ADPCM from tandem connections of ADPCM transcoders.

ADPCM is basically developed from PCM. It has good speech reproduction quality, comparable to PSTN quality, which therefore led to its adoption in CT2 and DECT.

5.6.3 Telephony Requirements

A general cordless telephone system would include an acoustic interface (i.e., microphone and speaker at the handset coupled to a digitizing compressor/decompressor analog to uniform PCM to ADPCM at 32 kb/s enabling a 2:1 increase in channel capacity as a bonus). This digital stream is processed to be transmitted over the air interface to the base station where the reverse happens, resulting in a linear or a digital stream to be transported over the land-based network. The transmission plans for specific systems are described in detail in Tuttlebee [1995].

An important subject in telephony is the effect of network echoes [Weinstein, 1977]. Short delays are manageable even if an additional delay of, say, less than 15 μs is introduced by a cordless handset. Delays of a larger magnitude, in excess of 250 μs (such as satellite links [Madsen and Fague, 1993]), coupled to cordless systems can cause severe degradation in speech quality and transmission; a small delay introduced by the cordless link in the presence of strong network echoes is undesirable. The DECT standard actually specifies the requirement for network echo control. Additional material can be obtained from the relevant CCITT documents [CCITT, 1984–1985].

5.6.4 Modulation Method

The modulation method for DECT is Gaussian-filtered frequency-shift keying (GFSK) with a nominal deviation of 288 kHz [Madsen and Fague, 1993]. The *BT* (i.e., Gaussian filter bandwidth to bit ratio), is 0.5 and the bit rate is 1.152 Mb/s. Specification details can be obtained from the relevant ETSI documents listed in the reference section.

Digital transmission channels in the radio frequency bands, including the DECT systems, present serious problems of spectral schemes congestion and introduce severe adjacent/co-channel interference problems. There were several schemes employed to alleviate these problems: new allocations at high frequencies, use of frequency-reuse techniques, efficient source encoding, and spectrally efficient modulation techniques.

Any communication system is governed mainly by two criteria, transmitted power and channel bandwidth. These two variables have to be exploited in an optimum manner in order to achieve maximum bandwidth efficiency, defined as the ratio of data rate to channel bandwidth (units of bit/Hz/s) [Pasupathy, 1979]. GMSK/GFSK has the properties of constant envelope, relatively narrow bandwidth, and coherent detection capability. Minimum-shift keying (MSK) can be generated directly from FM (i.e., the output power spectrum of MSK can be created by using a premodulation low-pass filter). To ensure that the output power spectrum is constant, the low-pass filter should have a narrow bandwidth and sharp cutoff, a low overshoot, and the filter output should have a phase shift $\pi/2$, which is useful for coherent detection of MSK; see Fig. 5.7.

Properties of GMSK satisfy all of these characteristics. We replace the low-pass filter with a premodulation Gaussian low-pass filter [Murota and Hirade, 1981]. As shown in Fig. 5.8, it is relatively simple

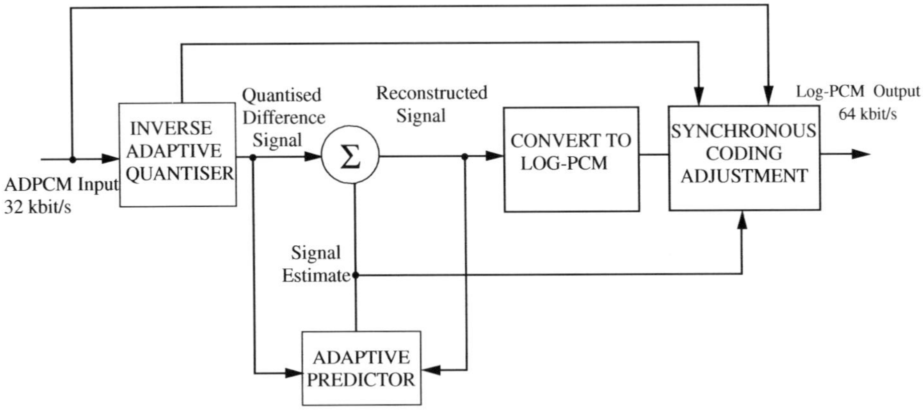

FIGURE 5.6 ADPCM decoder.

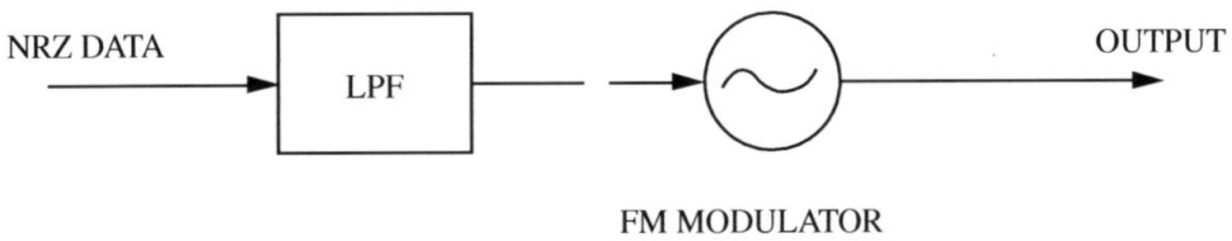

FIGURE 5.7 Premodulation baseband-filtered MSK.

to modulate the frequency of the VCO directly by the baseband Gaussian pulse stream; however, the difficulty lies in keeping the center frequency within the allowable value. This becomes more apparent when analog techniques are employed for generating such signals. A possible solution to this problem in the analog domain would be to use a phase-lock loop (PLL) modulator with a precise transfer function. It is desirable these days to employ digital techniques, which are far more robust in meeting the requirements talked about earlier. This would suggest an orthogonal modulator with digital waveform generators [de Jager and Dekker, 1978].

The demodulator structure in a GMSK/GFSK system is centered around orthogonal coherent detection, the main issue being recovery of the reference carrier and timing. A typical method, is described in de Buda [1972], where the reference carrier is recovered by dividing by four the sum of the two discrete frequencies contained in the frequency doubler output, and the timing is recovered directly from their difference. This method can also be considered to be equivalent to the Costas loop structure as shown in Fig. 5.9.

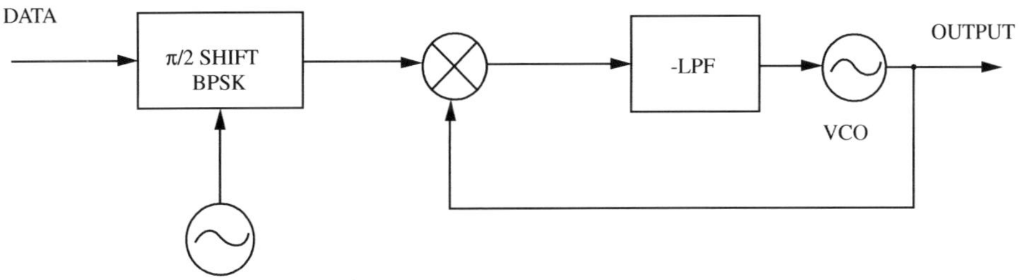

FIGURE 5.8 PLL-type GMSK modulator.

In the following are some theoretical and experimental representations of the modulation technique just described. Considerable literature is available on the subject of data and modulation schemes and the reader is advised to refer to Pasupathy [1979] and Murota and Hirade [1981] for further access to relevant study material.

5.6.5 Radio Frequency Architecture

We have discussed the need for low power consumption and low cost in designing cordless telephones. These days digital transmitter/single conversion receiver techniques are employed to provide highly accurate quadrature modulation formats and quadrature downconversion schemes that allow a great deal of flexibility to the baseband section. Generally, one would have used digital signal processors to perform most of the demodulation functions at the cost of high current consumption. With the advent of application-specific signal processing, solutions with these techniques have become more attractive.

From a system perspective, range, multipath, and voice quality influence the design of a DECT phone. A high bit rate coupled with multipath reflections in an indoor environment makes DECT design a challenging task. The delay spread (multipath) can be anywhere in the 100- to 200-ns range, and a DECT bit time is 880 ns. Therefore, a potential delay spread due to multipath reflections is 1 to 20% of a bit time. Typically, antenna diversity is used to overcome such effects.

DECT employs a TDMA/TDD method for transmission, which simplifies the complexity of the radio frequency end. The transmitter is on for 380 ms or so. The receiver is also only on for a similar length of time.

A single conversion radio architecture requires fast synthesizer switching speed in order to transmit and receive on as many as 24 time slots per frame. In this single conversion transmitter structure, the synthesizer has to make a large jump in frequency between transmitting and receiving, typically on the order of 110 MHz. For a DECT transceiver, the PLL synthesizer must have a wide tuning bandwidth at a high-frequency reference in addition to good noise performance and fast switching speed. The prescaler and PLL must consume as low a current as possible to preserve battery life.

In the receive mode the RF signal at the antenna is filtered with a low-loss antenna filter to reduce out-of-band interfering signals. This filter is also used on the transmit side to attenuate harmonics and reduce wideband noise. The signal is further filtered, shaped, and downconverted as shown in Fig. 5.10.

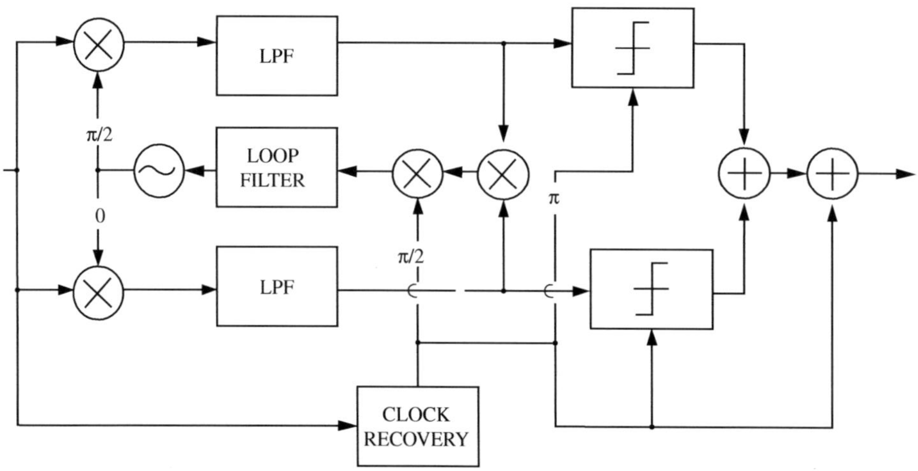

FIGURE 5.9 Costas loop.

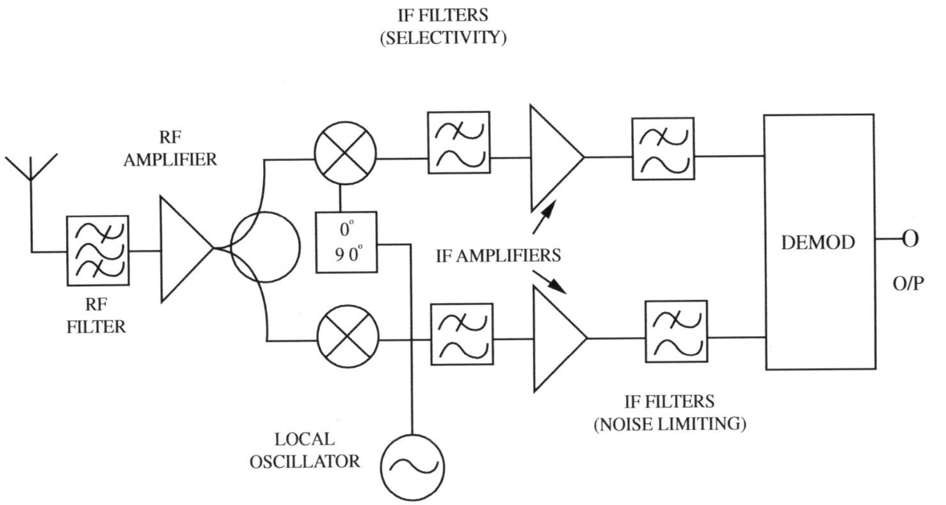

FIGURE 5.10 Direct conversion receiver architecture.

The signal path really is no different from most receiver structures. The challenges lie in the implementation, and this area has become quite a competitive segment, especially in the semiconductor world.

The direct conversion receiver usually has an intermediate frequency nominally at zero frequency, hence the term zero IF. The effect of this is to fold the spectrum about zero frequency, which results in the signal occupying only one-half the bandwidth. The zero IF architecture possesses several advantages over the normal superheterodyne approach. First, selectivity requirements for the RF filter are greatly reduced due to the fact that the IF is at zero frequency and the image response is coincident with the wanted signal frequency. Second, the choice of zero frequency means that the bandwidth for the IF paths is only half the wanted signal bandwidth. Third, channel selectivity can be performed simply by a pair of low-bandwidth low-pass filters.

For the twin IF chains of a direct conversion receiver, automatic gain control (AGC) is always required due the fact that each IF channel can vary between zero and the envelope peak at much lower rates than the highest signal bandwidth frequency. An additional requirement in newer systems is received signal strength indication (RSSI) to measure the signal or interference level on any given channel.

Moving on to the transmitter architecture (shown in Fig. 5.11 is a typical I-Q system), it is safe to say that the task of generating an RF signal is much simpler than receiving it. A transmitter consists of three

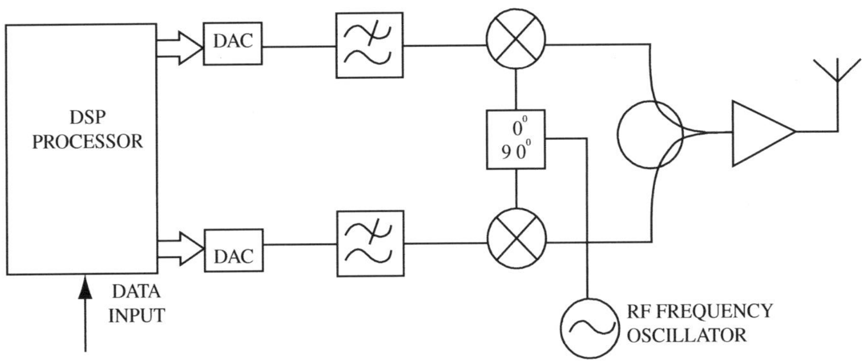

FIGURE 5.11 Transmit section.

main components: a final frequency generator, a modulator, and the power amplifier. These components can all be combined in common circuits (i.e., frequency synthesizer with inbuilt modulator). The problem of generating a carrier at a high frequency is largely one of frequency control. The main approach for accurately generating an output frequency from a crystal reference today is the PLL, and there is considerable literature available on the subject [Gardner, 1979]. In the modulation stage, depending upon the tightness of the phase accuracy specification of a cordless system, it may be necessary to apply tight control on the modulation index to ensure that the phase path of the signal jumps exactly in 90° increments.

Defining Terms

AGC: Automatic gain control.

ARQ: Automatic repeat request.

AWGN: Additive white Gaussian noise.

BABT: British approvals board for telecommunications.

Base Station: The fixed radio component of a cordless link. This may be single-channel (for domestic) or multichannel (for Telepoint and business).

BER: Bit error rate (or ratio).

CCITT: Comité Consultatif International des Télégraphes et Téléphones, part of the ITU.

CEPT: Conference of European Posts and Telecommunications Administrations.

CPFSK: Continuous phase frequency-shift keying.

CPP: Cordless portable part; the cordless telephone handset carried by the user.

CRC: Cyclic redundancy check.

CT2: Second generation cordless telephone-digital.

D Channel: Control and information data channel (16 kb/s in ISDN).

DCT: Digital cordless telephone.

DECT: Digital European cordless telecommunications.

DLC: Data link control layer, protocol layer in DECT.

DSP: Digital signal processing.

DTMF: Dual tone multiple frequency (audio tone signaling system).

ETSI: European Telecommunications Standards Institute.

FDMA: Frequency division multiple access.

FSK: Frequency-shift keying.

GMSK: Gaussian-filtered minimum-shift keying.

ISDN: Integrated services digital network.

ITU: International Telecommunications Union.

MPT 1375: U.K. standard for common air interface (CAI) digital cordless telephones.

MSK: Minimum-shift keying.

PSK: Phase-shift keying.

RES 3: Technical subcommittee, radio equipment and systems 3 of ETSI, responsible for the specification of DECT.

RSSI: Received signal strength indication.

SAW: Surface acoustic wave.

TDD: Time division duplex.

TDMA: Time division multiple access.

References

Cheer, A.P. 1985. Architectures for digitally implemented radios, IEE Colloquium on Digitally Implemented Radios, London.

Comité Consultatif International des Télégraphes et Téléphones. 1984. 32 kbits/sec Adaptive Differential Pulse Code Modulation (ADPCM), CCITT Red Book, Fascicle III.3, Rec. G721.

Comité Consultatif International des Télégraphes et Téléphones, 1984–1985. *General Characteristics of International Telephone Connections and Circuits,* CCITT Red Book, Vol. 3, Fascicle III.1, Rec. G101–G181.

de Buda, R. 1972. Coherent demodulation of frequency shifting with low deviation ratio. *IEEE Trans.* COM-20 (June):466–470.

de Jager, F. and Dekker, C.B. 1978. Tamed frequency modulation. A novel method to achieve spectrum economy in digital transmission, *IEEE Trans. in Comm.* COM-20 (May):534–542.

Dijkstra, S. and Owen, F. 1994. The case for DECT, *Mobile Comms. Int.* 60–65.

European Telecommunications Standards Inst. 1992. RES-3 DECT Ref. Doc. ETS 300 175-1 (Overview). Oct. ETSI Secretariat, Sophia Antipolis Cedex, France.

European Telecommunications Standards Inst. 1992. RES-3 DECT Ref. Doc. ETS 300 175-2 (Physical Layer) Oct. ETSI Secretariat, Sophia Antipolis Cedex, France.

European Telecommunications Standards Inst. 1992. RES-3 DECT Ref. Doc. ETS 300 175-3 (MAC Layer) Oct. ETSI Secretariat, Sophia Antipolis Cedex, France.

European Telecommunications Standards Inst. 1992. RES-3 DECT Ref. Doc. ETS 300 175-4 (Data Link Control Layer) Oct. ETSI Secretariat, Sophia Antipolis Cedex, France.

European Telecommunications Standards Inst. 1992. RES-3 DECT Ref. Doc. ETS 300 175-5 (Network Layer) Oct. ETSI Secretariat, Sophia Antipolis Cedex, France.

European Telecommunications Standards Inst. 1992. RES-3 DECT Ref. Doc. ETS 300 175-6 (Identities and Addressing) Oct. ETSI Secretariat, Sophia Antipolis Cedex, France.

European Telecommunications Standards Inst. 1992. RES-3 DECT Ref. Doc. ETS 300 175-7 (Security Features) Oct. ETSI Secretariat, Sophia Antipolis Cedex, France.

European Telecommunications Standards Inst. 1992. RES-3 DECT Ref. Doc. ETS 300 175-8 (Speech Coding & Transmission) Oct. ETSI Secretariat, Sophia Antipolis Cedex, France.

European Telecommunications Standards Inst. 1992. RES-3 DECT Ref. Doc. ETS 300 175-9 (Public Access Profile) Oct. ETSI Secretariat, Sophia Antipolis Cedex, France.

European Telecommunications Standards Inst. 1992. RES-3 DECT Ref. Doc. ETS 300 176 (Approval Test Spec) Oct. ETSI Secretariat, Sophia Antipolis Cedex, France.

Gardner, F.M. 1979. *Phase Lock Techniques,* Wiley-Interscience, New York.

Madsen, B. and Fague, D. 1993. Radios for the future: designing for DECT, *RF Design.* (April):48–54.

Murota, K. and Hirade, K. 1981. GMSK modulation for digital mobile telephony, *IEEE Trans.* COM-29(7):1044–1050.

Olander, P. 1994. DECT a powerful standard for multiple applications, *Mobile Comms. Int.* 14–16.

O'Shaughnessy, D. 1987. *Speech Communication,* Addison-Wesley, Reading, MA.

Pasupathy, S. 1979. Minimum shift keying: spectrally efficient modulation, *IEEE Comm. Soc. Mag.* 17(4):14–22.

Tuttlebee, W.H.W., ed., 1995. *Cordless Telecommunications Worldwide,* Springer-Verlag, Berlin.

Weinstein, S.B. 1977. Echo cancellation in the telephone network, *IEEE Comm. Soc. Mag.* 15(1):9–15.

6

Wireless Local Area Networks (WLAN)

Jim Paviol
PRISM Wireless Design Engineering

Carl Andren
Intersil

John Fakatselis
Intersil

Wireless Local Area Networks (WLANs) use radio transmissions to substitute for the traditional coaxial cables used with wired LANs like Ethernet. The first-generation WLAN products were targeted as wired-LAN extensions. They were originally intended to save money on relocation expenses and demonstrated the utility of wireless laptop operations. Wireless data technology and its application to local area networks introduced mobile computing. Centrally controlled wireless networks are most often used as part of a larger wired network. A radio base station or Access Point (AP) arbitrates access to the remote wireless stations by means of packetized data. In contrast, in a Peer-to-Peer wireless network, an ad hoc network can be formed at will by a group of wireless stations. New forms of network access protocols, such as Carrier Sense Multiple Access/Collision Avoidance (CSMA/CA) are needed for low error rate operation of wireless networks. Roaming is one of the main advantages of wireless networks, allowing users to freely move about while maintaining connectivity.

The WLAN is used in four major market segments, "Vertical" with factory, warehouse, and retail uses; "Enterprise" with corporate infrastructure mobile Internet uses; "SOHO" (Small Office/Home Office) with small rented space businesses; and "Consumer" with emerging uses. General WLAN trends are shown in Fig. 6.1.

0-8493-1732-0/03/$0.00+$1.50

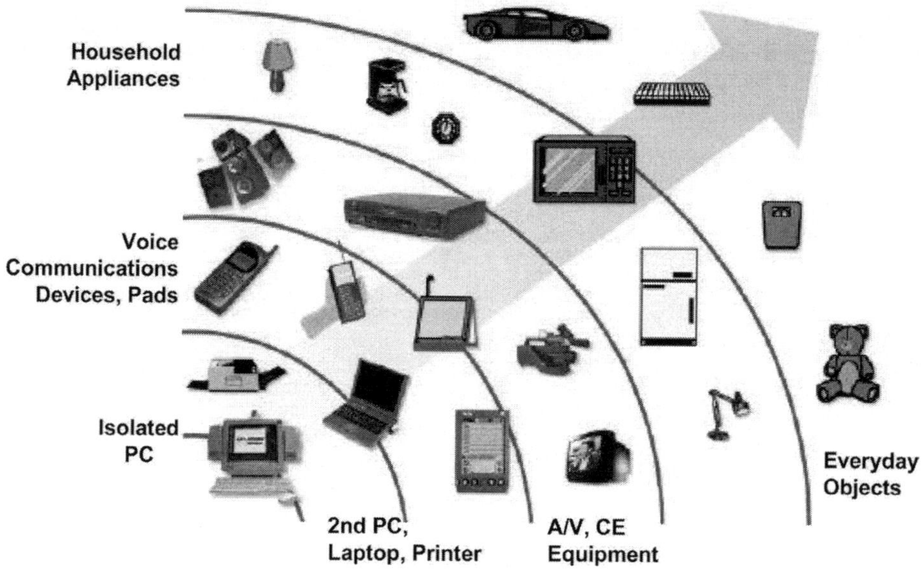

FIGURE 6.1 Commercial WLAN evolution.

6.1 WLAN RF ISM Bands

In 1985 the Federal Communications Commission (FCC) in the U.S. defined the ISM (Industrial, Scientific, and Medical) frequency bands allowing unlicensed spread-spectrum communications. Three of the ISM bands are illustrated in Fig. 6.2 with frequencies at 900-MHz, 2.4-GHz, and 5-GHz.

Most important to ISM band WLAN users is that no license for operation is required when the signal transmission is per the guidelines specified by the FCC or other regulatory agencies. Spread-spectrum technology in the ISM band is used to minimize interference and offers a degree of interference immunity from other jamming signals or noise. Other non-spread commercial applications have existed in the ISM bands for many years such as microwave ovens at 2.4-GHz. This is a major potential interference to WLAN in the 2.4-GHz ISM band and has been accounted for in the system design.

The IEEE 802.11 committee selected the 2.4-GHz ISM band for the first WLAN global standard. Unlike the 900-MHz band, 2.4-GHz is available worldwide; 2.4-GHz also has more available bandwidth than the 900-MHz band, and will support higher data rates and multiple adjacent channels in the band. In comparison with the 5.7-GHz band, it offers a good balance of equipment performance and cost. Increasing the transmit frequency impacts the power dissipation, availability of parts/processes and limits the indoor range. The 2.4-GHz band is ideal for a WLAN high-speed, unlicensed data link.

The 900-MHz band has been in use for some time, and component prices are very reasonable. Many cordless phones use this band. The 900 MHz band is quite crowded and it does not have global spectrum allocation. The 2.4-GHz band is less crowded, has global allocations, and the associated technology is very cost effective. This is the band in which IEEE 802.11b, Bluetooth, and HomeRF operate.

The 2.4-GHz band is most heavily used for WLAN. Operating channels are shown in Fig. 6.3.

The 5-GHz band has two 100-MHz segments for unlicensed use collectively known as the Unlicensed National Information Infrastructure (UNII) bands. There is a similar allocation in Europe, but it is currently reserved for devices that operate in compliance with the HIPERLAN standard. The 5-GHz components are more expensive, and radio propagation has higher losses and more severe multipath at these frequencies. These impairments can be overcome, and systems offering data rates in excess of 50-Mb/s in the 5-GHz band will become mainstream in the next several years.

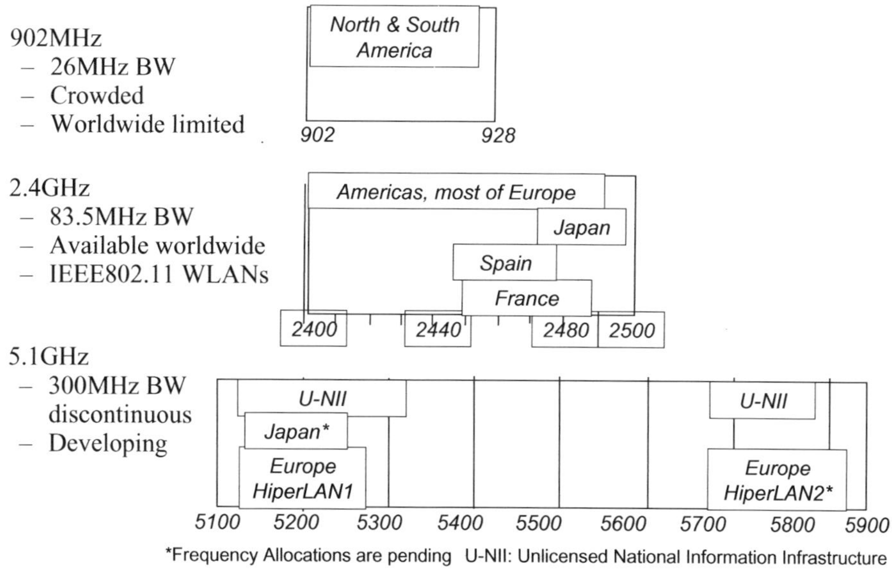

FIGURE 6.2 Unlicensed ISM RF band details.

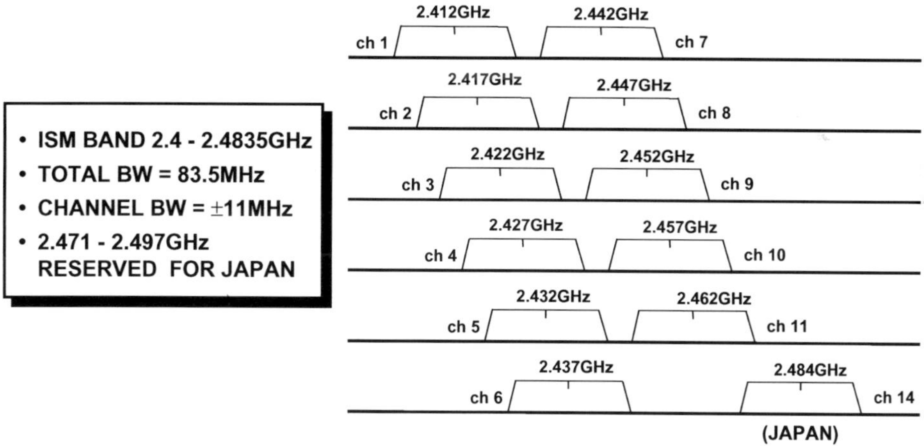

FIGURE 6.3 Operating channels for direct sequence.

6.2 WLAN Standardization at 2.4-GHz: IEEE 802.11b

Wired LAN standards were developed by the IEEE 802 committee such as IEEE 802.1 Systems Management/Networking, IEEE 802.3 Ethernet, IEEE 802.4 Token Ring, and IEEE 802.6 Metropolitan Area Networks. In 1990, the IEEE 802.11 Wireless LAN Working Group was formed and has in excess of 100 active voting members with global representation. It ratified the 802.11b high rate 2.4-GHz WLAN standard in 1999. The 802.11 standard fostered development of interoperable, inexpensive, and flexible equipment in the 2.4-GHz ISM band. Specified data rates for the IEEE 802.11, 2.4-GHz WLAN are 1, 2, 5.5, and 11-Mb/s. Spread-spectrum technology is specified in the 802.11 standard transceiver to provide a robust solution in a multi-user environment. One advantage to spread-spectrum techniques in the ISM bands are seen in the allowable transmit power levels. System transmitter power for IEEE 802.11 WLANs must conform to a regulatory agency's specified levels for unlicensed operation. As an example, the FCC states that non-spread-spectrum applications in this band are limited to a 50 mV/m at 3 m. This translates into 0.7 mW into a dipole antenna. Spread-spectrum applications in the U.S. are allowed up to 1 W of transmit power, clearly giving it a higher signal strength advantage over non-spread systems. The low spectral power density of a spread-spectrum system also limits interference to other in-band users.

Segmentation of a data communication system into layers allows different approaches from various vendors as long as the responsibilities of the individual layer are met. The IEEE 802.11 specification focuses on the Media Access Control (MAC) and Physical (PHY) layers for WLANs.

The MAC layer controls the protocol and physical layer management. The protocol used for IEEE 802.11 WLANs is the CSMA/CA (Carrier Sense Multiple Access/Collision Avoidance).

The Physical Layer controls the wireless transmission and reception of digital data from the MAC. It is the transceiver or radio for the WLAN. IEEE 802.11 specifies three different physical layer options, Direct Sequence Spread Spectrum (DSSS), Frequency Hopping Spread Spectrum (FHSS), and Diffused Infrared (DFIR). The DFIR method has the shortest range of operation and is limited to indoor operation due to interference from sunlight. DSSS and FHSS are RF technologies that must conform to the standards set by the regulatory agencies of various countries such as the FCC. These impact items such as the allowable bandwidth and transmit power levels.

Another feature of IEEE 802.11 is that the data are packetized. Packetized data are fixed numbers of data bytes sent in a single radio transmission of finite length. The data are grouped in frames up to 2304 bytes in length. A common data length is 1500 bytes. A header and preamble are attached in front of the data frame for control information. The preamble is the initial sequence at the start of the radio transmission that allows the demodulator to synchronize its timing and recognize key information concerning the data that follows. Short and long preambles exist. These are specified in greater detail within the standard. The packetization supports the CSMA/CA protocol.

6.3 Frequency Hopped (FH) vs. Direct Sequence Spread Spectrum (DSSS)

FH uses a form of FSK modulation called GFSK (Gaussian Frequency-Shift Keying). The baseline 1 Mb/s data rate for FH IEEE 802.11 has a 2-level GFSK modulation scheme. The symbol {1} is the center carrier frequency plus a peak deviation of (f+), whereas the symbol {o} is the center carrier frequency minus a peak deviation of (f–). The carrier frequency hops every 400 ms over the channel bandwidth per a prescribed periodic PN code. This channel is divided into 79 sub-bands. Each sub-band has a 1 MHz bandwidth. The minimum hop rate of 2.5 hops/s allows several complete data packets or frames to be sent at one carrier frequency before a hop.

DSSS in IEEE 802.11 specifies a DBPSK and DQPSK (D = Differential) modulation for 1 and 2 Mb/s data rates. Differential techniques use the received signal itself to demodulate the signal by delaying one symbol period to obtain clock information. In DBPSK, a logic 1-b input initiates a 180° phase change in the carrier and a ϕ-bit initiates no phase change in the carrier. DQPSK has a 0°-, 90°-, or 180°- phase transition on each symbol.

The carrier frequency in an IEEE 802.11 DSSS transmitter is spread by an 11-b Barker code. The chipping rate is 11 MHz for a 1-Mb data rate. This yields a processing gain of 11. The main lobe spacing is twice the chip rate and each side lobe is the chip rate as shown in Fig. 6.4. The DSSS receiver will filter the side lobes, downconvert the main lobe spectral component to baseband, and use a copy of the PN code in a correlator circuit to recover the transmitted signal. The FH scheme has been limited to 1 and 2 Mb by technical and regulatory issues. This is expected to limit the use of FH systems compared with the more versatile high-rate DSSS modulation in future applications. An illustration contrasting frequency hopping to spread spectrum is shown in Fig. 6.4.

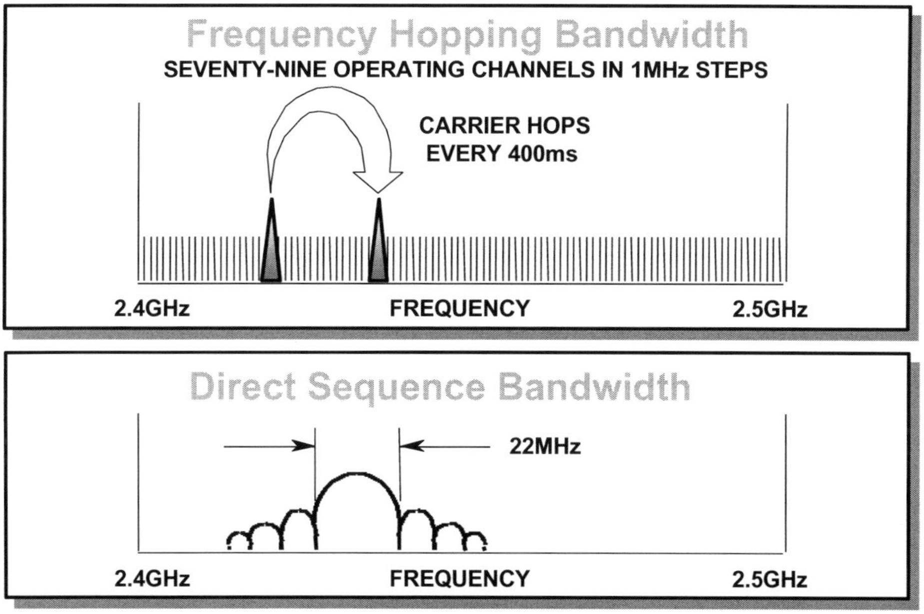

FIGURE 6.4 IEEE 802.11 frequency hopping vs. direct sequence spread spectrum.

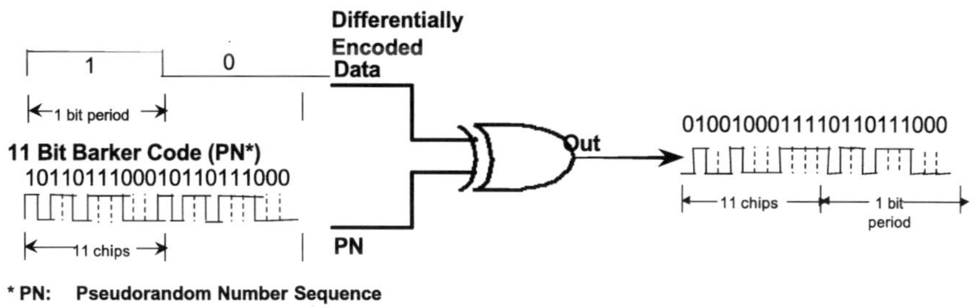

FIGURE 6.5 Direct sequence spectrum spreading. RF energy is spread by XOR of data with PN sequence.

6.4 Direct Sequence Spread-Spectrum Energy Spreading

In Direct Sequence Spread-Spectrum (DSSS) systems the spreading of the data is achieved by modulating the data with a Pseudorandom Number (PN) sequence of binary values called a PN code. If the PN code has a bandwidth much higher than the bandwidth of the data (approximately ×10 or greater), then the bandwidth of the modulated signal will have a spectrum that is nearly the same as a wideband PN signal. An 11-b Barker code is used with the IEEE 802.11 DSSS PN spreading. A Barker code was chosen for its unique short code properties.

By multiplying the information-bearing signal by the PN signal, each information bit is chopped up into a number of small time increments called "chips," as illustrated in the waveform diagram shown in Fig. 6.5. The rate at which the PN code is clocked to spread the data is called the chip rate. The PN sequence is a periodic binary sequence with a noise-like waveform. The acquisition of the data in the receiver is achieved by correlation of the received signal with the same PN code that was used to spread the signal at the transmitter. In DSSS systems the data is primarily PSK modulated before spreading. The spreading produces a "Processing Gain" dependent upon the PN spreading code to symbol rate ratio.

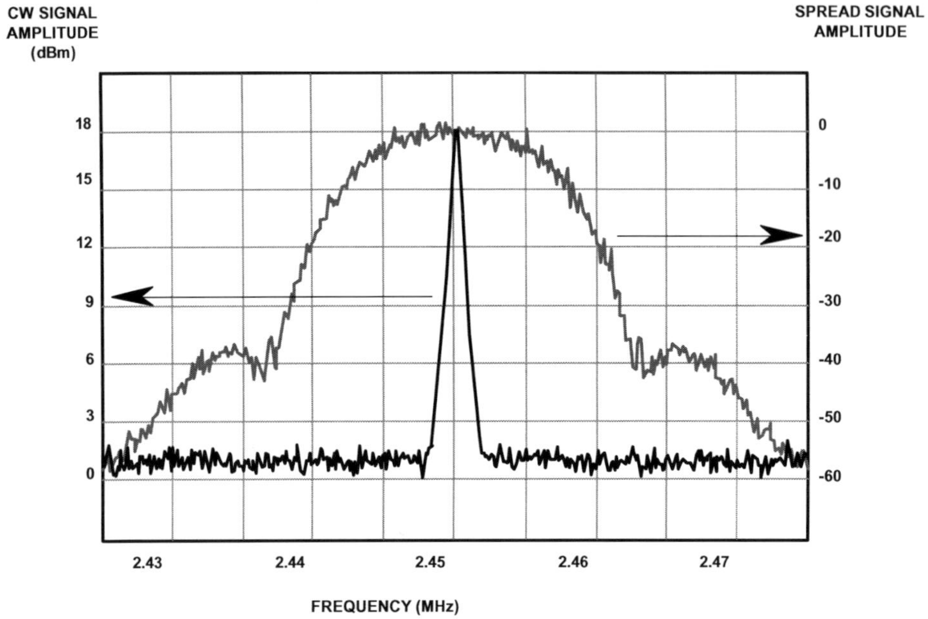

FIGURE 6.6 Direct sequence spread spectrum.

This value is a minimum of 10 dB for the IEEE 802.11 DSSS WLAN waveform. The spectrum after this PN code is used is wider and lower in signal level as illustrated in Fig. 6.6.

The primary advantage of a spread-spectrum system is its ability to reject interference whether it be the unintentional interference by another user transmitting on the same channel, or the intentional interference by a hostile transmitter attempting to jam the transmission. Due to its spread characteristic, a DSSS signal appears as noise to all receivers except the one meant to receive the signal. The intended receiver is able to recover the spread signal by means of correlation, which simultaneously recovers the signal of interest and suppresses the background noise by the amount of processing gain.

The reception of the DSSS signal in the presence of a narrowband interferer is accomplished by de-spreading the signal of interest while spreading the energy of the interfering signal by the amount equal to the processing gain. The result of the de-spreading process with an interference source is illustrated in Fig. 6.7 for narrowband interfering signals. Likewise, recovery of the signal of interest in presence of another spread signal having a different PN code is achieved by further spreading the interference while de-spreading the signal with the matching PN code. The worst-case interference to DSSS systems is a narrowband interference signal in the middle of the spectrum of the spread signal.

6.5 Modulation Techniques and Data Rates

The 1-Mb data rate is formed using Binary Phase-Shift Keying (BPSK) and the 2-Mb data rate uses Quadrature Phase-Shift Keying (QPSK). QPSK doubles the data rate by increasing the number of bits per symbol from one (BPSK) to two (QPSK) within the same bandwidth. Both rates use BPSK for the acquisition header known as the preamble.

The high rate modulation for 5.5 and 11 mb/s uses a form of M-Ary Orthogonal keying called Complementary Code Keying (CCK). CCK provides coding gain by the use of modified Walsh functions applied to the data stream. Two forms of this are used to provide multiple rates for stressed links. Altogether, four data rates are provided. To make 11-Mb CCK modulation, the input is formed into

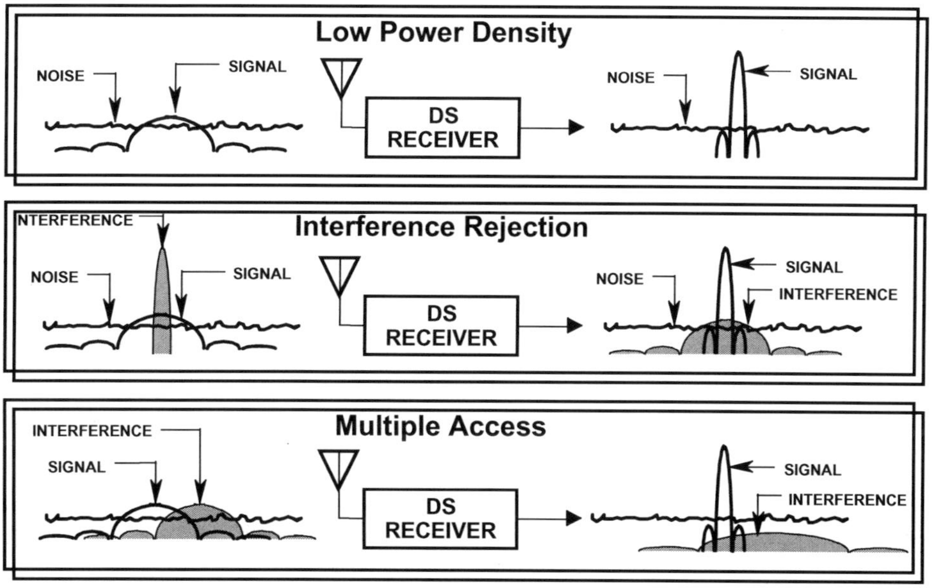

FIGURE 6.7 Direct sequence spread-spectrum properties.

bytes and then subgrouped into 2 and 6 b. The 6 b are used to pick one of 64 complex vectors of 8-chip length and the other 2-b DQPSK modulate the whole symbol vector. For 5.5-Mb CCK mode, the incoming data is grouped into 4-b nibbles where 2 of those bits select the spreading function out of a set of 4 (the 4 having the greatest distance of the 11-Mb set) while the remaining 2 b set the QPSK polarity of the symbol. The spreading sequence modulates the carrier by driving the I and Q modulators. Figure 6.8 illustrates modulation at the four data rates of 1, 2, 5.5, and 11 Mb.

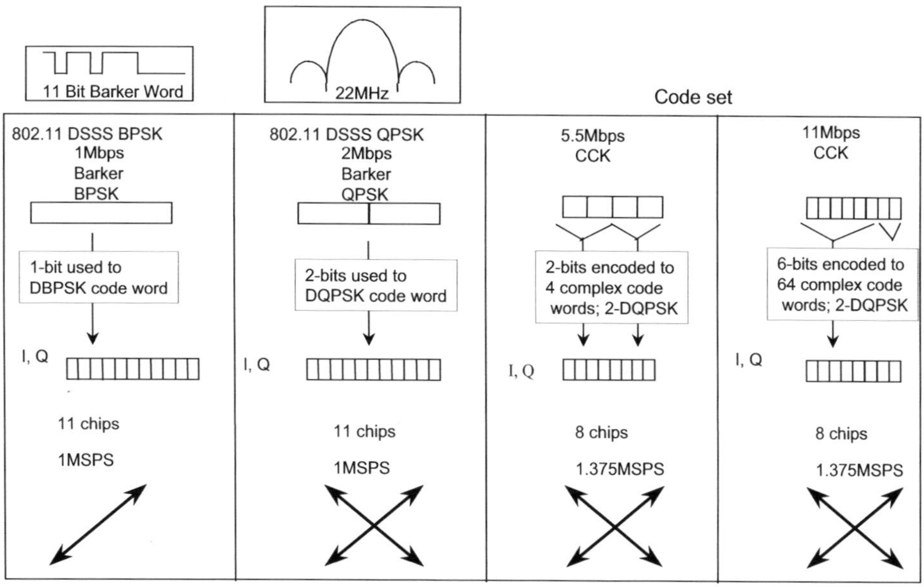

FIGURE 6.8 Modulation techniques and data rates.

6.6 Carrier Sense Multiple Access/Collision Avoidance (CSMA/CA)

The wireless environment offers a greater challenge to the WLAN designer when compared with the wired LAN environment. Wired LANs use a CSMA/CD (Carrier Sense Multiple Access/Collision Detect) protocol. Data collisions in a wired environment produce a unique voltage that can be monitored. This allows for random back-off time periods before the system initiates a data resend.

IEEE 802.11 WLANs use the CSMA/CA (Collision Avoidance) protocol. This protocol avoids data collisions by having the system listen to the channel and wait before sending a message. With CSMA/CA, only one node may talk at a time. This highlights the importance of performing a Clear Channel Assessment (CCA) to determine if the medium or channel is clear to transmit. Although the MAC controls the CSMA/CA protocol, the responsibility falls upon the physical layer to perform CCA.

IEEE 802.11 states that the physical layer must be able to provide at least one of three specified methods for CCA. CCA mode 1 simply detects energy above a programmable level. If no signal is present, the channel is clear to transmit. If the signal is present, the system will wait a set time period to check the channel again. CCA mode 2 provides the carrier sense function. Since this is a spread-spectrum system, correlating the received signal with the 11-b PN code performs carrier sense. No correlation indicates that the channel is clear to transmit. Correlation with a signal shows that the channel is busy and that the system will back off for a time period and try again. CCA mode 3 combines modes 1 and 2 by reporting a busy medium with both detection of energy and carrier sense. Figure 6.9 illustrates a four-station scenario where radio collisions are avoided using CSMA-CA.

6.7 Packet Data Frames in DSSS

The 802.11 WLAN standard specifies data packetization. This means that the data are segmented into frames with a preamble and header attached at the start of each frame. The preamble allows carrier and correlation lock as well as user identification. The header contains management and control information for the data transmission.

The sync field is made up of 128 one bits. Note that all bits are processed by a scrambler function, which is part of the IEEE 802.11 spread-spectrum physical layer so this original pattern of ones will be altered. The purpose of the sync field is to allow the receiver to lock on to the signal and to correlate to the PN code.

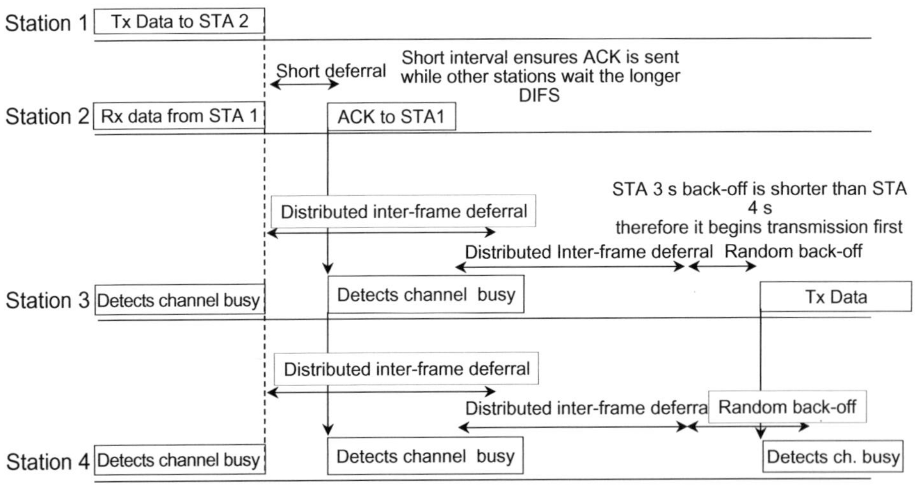

FIGURE 6.9 Carrier sense multiple access collision avoidance (CSMA/CA).

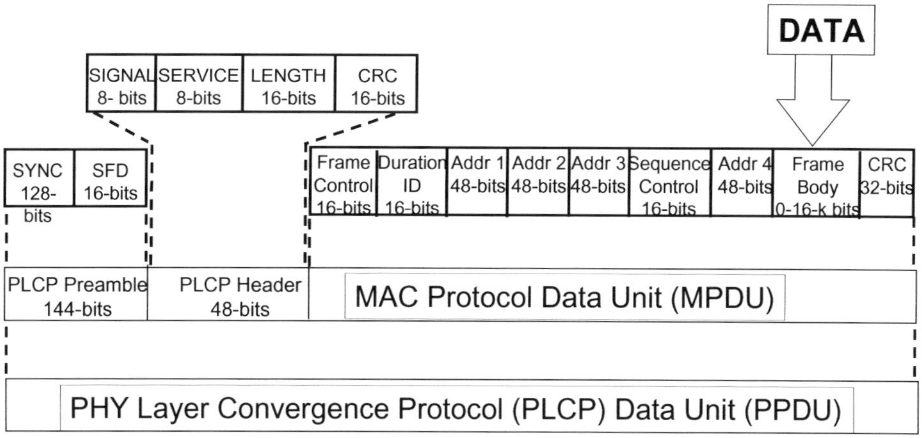

FIGURE 6.10 Frame format.

The start frame delimiter (SFD) field initiates the start of the data frame. The signal field indicates the data packet data wave. Note that the preamble and header are always transmitted as a DBPSK waveform. The length field defines the data packet size with a maximum length of 2304 bytes. Finally, one CRC (Cyclic Redundancy Check) protects the signal, service, and length fields with a frame check sequence and another protects the payload.

At the end of the data packet, the receiver will send an acknowledgment (ACK) indicating successfully transmitted data. If a data packet were lost either by multipath fading or interference, the sender would retransmit the packet. Figure 6.10 details frame format, preamble, header, and data.

6.8 IEEE 802.11 Network Modes

There are two network modes: Ad Hoc and Infrastructure. Ad Hoc mode permits users within range to set up a network among them without any infrastructure. Infrastructure mode uses an Access Point (AP) to coordinate the users and allow access to the wired network services.

6.8.1 Ad Hoc Mode

The Distributed Coordination Function (DCF) forms the basis to implement Ad Hoc networking. The provisions in the standard allow the creation and dissolution of an Ad Hoc network to be straightforward for users to set up. The CSMA/CA medium access method provides for fair access to the radio channel among all of the users. All data exchanges are directly between the individual stations.

When a single station is designated as the coordination function, the network is known as a Basic Service Set (BSS). This is illustrated in the peer-to-peer Ad Hoc network in Fig. 6.11.

6.8.2 Infrastructure Mode

An Access Point (AP) is a device that connects the wireless stations to the distribution system in a network. It typically will be configured to be the single Coordination Function in a BSS. The network planning for a large installation involves site surveys to do the cellular radio planning needed to determine the number and location of AP. The channel assignments for each AP can be optimized to reduce interference from adjacent cells depending on the physical layout. Many times, the installation will utilize a wired Ethernet network to connect a number of APs to the network server. Each AP will manage the traffic within its BSS. Stations in adjacent cells will recognize when the packet is not intended for its BSS. The same DCF mechanisms provided in the ad hoc mode help to manage radio interference from adjacent

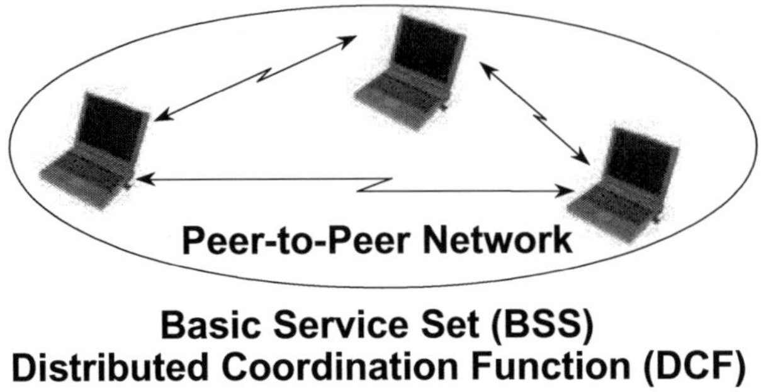

FIGURE 6.11 IEEE 802.11 ad hoc mode.

cells. Mechanisms such as Authentication, Association, and Wired Equivalency Privacy (WEP) provide security for the overall network. The authentication process is used to verify the identity of stations that are allowed access to the network. As users move between the various radios cells and they can no longer communicate with their AP, they can scan the channels to look for a new AP to associate with. With proper radio cell-planning, users can be assured of constant coverage as they roam through the facility. An infrastructure mode illustration is shown in Fig. 6.12.

6.8.3 "Hidden" Nodes

The planning for a WLAN ranges from none for an ad hoc IBSS, to careful radio surveys for AP-based infrastructure systems in large enterprise computing environments. It is impossible to plan for perfect radio coverage given the uncertain and time-varying conditions in the channel. Movement and location as well as indoor topology will affect radio coverage. In providing complete coverage of the facility there will be situations where there are two stations that can communicate with their common AP, but are out of range to hear each other. This is known as the Hidden Node problem. This could cause additional collisions to occur at the AP, since when one station is transmitting, the other will determine the channel is clear because it is out of range. After the normal deferral time it will transmit and will interfere with a packet from the other station. The standard provides an optional mechanism called Virtual Carrier Sense (VCS) to reduce the collision due to the hidden node problem. This mode is based on using a Request To Send (RTS) and Clear To Send (CTS) exchange between the station and the AP. When the station has determined it is clear to transmit, it will send a short request to send a message with a transaction duration included. The AP will respond with a CTS that also includes the transaction

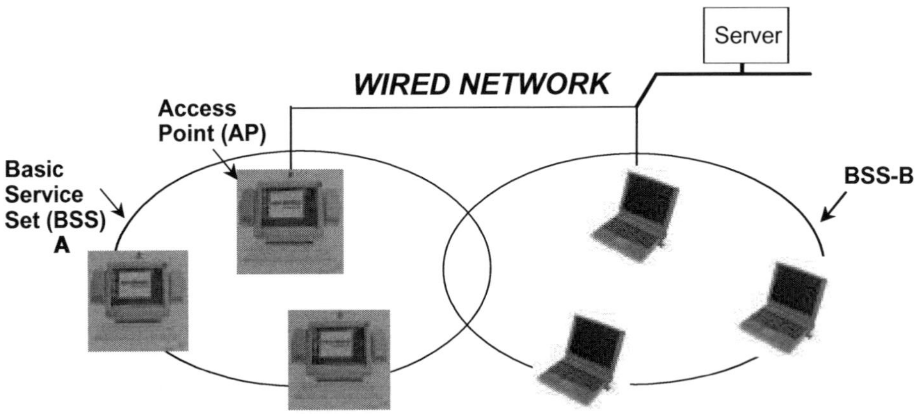

FIGURE 6.12 IEEE 802.11 infrastructure mode.

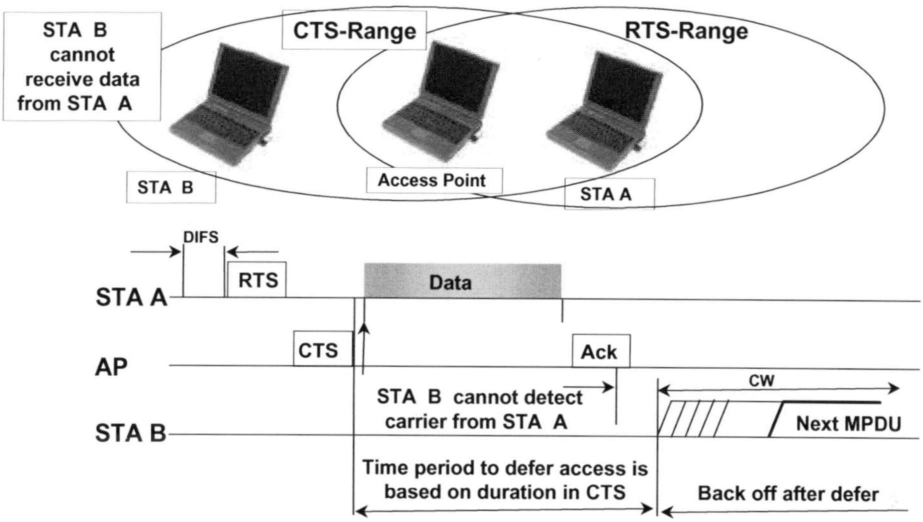

FIGURE 6.13 "Hidden node" provisions.

duration. The other station that is hidden cannot hear the RTS, but can hear the CTS. It will read the duration field and not begin looking for a clear channel until that time has elapsed. Hidden nodes are illustrated in Fig. 6.13 showing Stations A and B with an access point.

6.8.4 Point Coordination Function (PCF)

The Point Coordination Function (PCF) is an optional mode that can be selected in the installation of an 802.11 network. This mode is provided to optimize the network throughput. Rather than having each station contend for the channel using the CCA and the random back-off periods, the AP defines contention-free periods using a beacon frame sent at a regular interval. The Network Allocation Vector (NAV) is a variable that is transmitted in the control frames to tell all of the stations the duration to defer from accessing the channel. It is used to define the length of the contention-free period. The AP will then poll each station during the contention-free period. The poll will send data if there is some waiting for transmission to that station and request data from the station. As each station is polled, it will acknowledge reception and will include data if there is some pending transmission to the AP. One of the benefits of the PCF mode is that it allows the network planner to improve the probability of the delivery of data in a certain time bound. This is critical for real time data such as voice, audio, video, and multimedia. Minimizing the uncertainties of when a station can get the channel that exists in the DCF mode optimizes the system performance. If a packet of audio or video data has to be retransmitted due to lost packets and extended deferral times, the quality of the audio or video will suffer. The network can be optimized to permit a certain amount of retries within the contention-free period to improve the quality of service to the end application. The PCF timing is illustrated in Fig. 6.14.

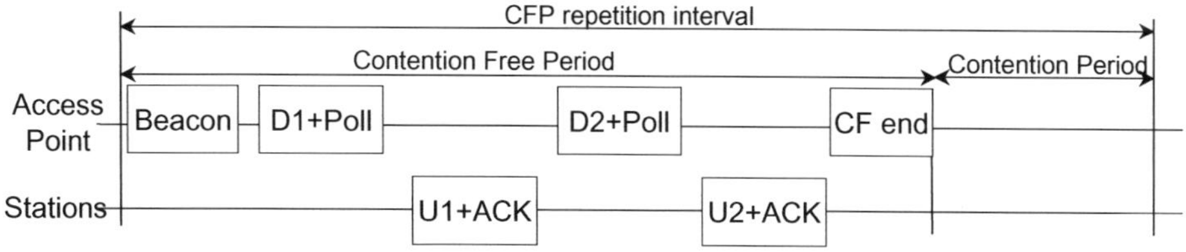

FIGURE 6.14 IEEE 802.11 point coordination function (PCF).

6.8.5 WLAN Security

Since they can be received outside of the controlled facility, wireless networks are more vulnerable to interference and theft than wired networks. The data security field known as cryptography is rapidly growing as more systems convert to wireless operation. Spread spectrum offers a little security by spreading the signal over a wide bandwidth.

There are a number of mechanisms provided in 802.11 to minimize the chances of someone either logging on the network without authorization or receiving and using data received off the air from authorized users. As an option the data can be encrypted to prevent someone who is receiving packets from the network to be able to interpret the data. The keys for the security are distributed to the users in the network by a secure key management procedure. Without the key the snooper will have to resort to complex code-breaking techniques to retrieve the original data. The level of encryption is defined as Wired Equivalency Privacy (WEP). It is strong enough to require effort equivalent to that required to get data from a wired LAN. This algorithm is licensed from RSA Data Security. The encryption mechanism is used in the authentication process as well.

6.8.6 Data Encryption

The WLAN Security: Authentication diagram illustrates a technique for authenticating the identity using the encryption features. In the "Challenge and Response Protocol" shown, the station transmits a "challenge" random message (r) to the AP. The AP receives the message, encrypts it by using a network algorithm (fK1), and transmits the encrypted information (y) "response" (fK1) back to A. System A has access to the network algorithm (Y) and compares it to the received response. If y = y′, the identification procedure criteria have been met and data transmission will follow. If it does not, then A will issue a new challenge with a different random message to B. Note that A and B share a common (private) key, k1. After the authentication process is successfully completed the station will then associate itself with an AP. The association tells the overall network which AP services any station. Once identification has occurred, both systems communicate using network encryption algorithms. In this case, the station encrypts data (x) and transmits a cipher (y) over the channel. The AP or another station has access to the encryption method and decodes the cipher to obtain the data. Because the stations in a BSS share a common key (k1), this type of encryption is called private-key cryptography.

Also included in the standard is an Integrity Check Vector (ICV). This is a variable added to an encrypted data packet as an additional error-detection mechanism. After the decryption process the transmitted ICV is compared with the ICV calculated from the plain text as an error check. Figure 6.15 illustrates data encryption with a block diagram.

6.9 5-GHz WLAN

With the FCC rule and order establishing the Unlicensed National Information Infrastructure (UNII), 300 more MHz of bandwidth were made available for WLAN users. The purpose was to expand the access of people to information without a lot of infrastructure build out. One of the frequently sited

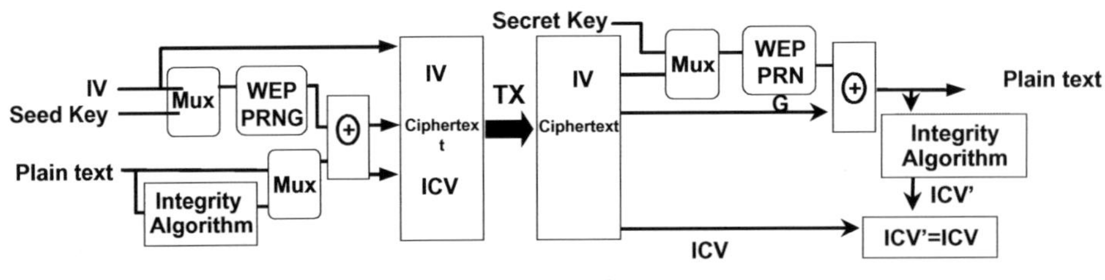

FIGURE 6.15 WLAN security: data encryption.

scenarios is the distribution of Internet access to schools without having to wire classrooms. The FCC chose to put in a minimum of specifications for the waveforms to be used in the band.

Spread spectrum is not a requirement. Transmit power and power spectral density are the primary specifications. No channelization or spectrum-sharing rules were included in the 15-part regulation. It remains to be seen how the various devices and standards will coexist in this new band. The band is split into three 100-MHz bands for defining maximum output power and spurious emissions levels. There are users of licensed bands on all sides of these new bands that argued for protection against interference from unlicensed devices. This drives the channel band edges for carriers and defines radio requirements for suppression of spurious emissions. The power levels allowed are 50 mW, 250 mW, and 1 W in the lowest, middle, and upper bands, respectively. A change was made in 1998 to permit higher antenna gain with a reduction in output power. This extends the range for point-to-point links.

In Europe 150 MHz of bandwidth is set aside for HIPERLAN1 devices. These are WLAN devices with the same functional requirements as an 802.11a WLAN. As opposed to the FCC, the ETSI regulation requires all devices to meet the HIPERLAN standard for the PHY and MAC layers. The maximum data rate is 54 Mb and the modulation is GMSK. Equalization is required for reliable operation. A small number of HIPERLAN products have been introduced since the standard was approved.

ETSI is defining a standard for the 5-GHz band that is oriented to wireless ATM traffic. This is suitable for Quality of Service applications and for wireless connectivity to an ATM system. The final spectral allocation has not been made for these devices.

Japan has started the development of standards for WLAN devices in the 5-GHz band. They have decided that devices will be required to use the same physical layer implementation that is defined in the IEEE 802.11a standard.

6.10 RF Link Considerations

The radio link performance can be characterized as consisting of radio design variables and link variables. When all the variables are understood, the link performance can be determined. The most important WLAN link performance measure is the Packet Error Rate (PER), which is usually expressed as a percentage. Most radio link parameters are given at a packet error rate of 0.1 or 10%. This is the highest practical acceptable PER and provides the design maximum. The IEEE 802.11 standard specifies an 8% PER for measurements.

All of the radio design variables combine to provide a given performance on a given link. The transmit power together with the receiver noise floor determines the ultimate range of the radio. Higher transmit power provides greater range but also higher battery drain. The antennas for wireless systems are generally omni-directional to allow mobility. The higher data rate requires a higher signal-to-noise ratio for the same error rate. The bit error rate and the length of the packet determine the packet error rate. Longer packets require lower bit error rates. Different modulation schemes require more or less power to achieve the same bit error rate. For instance FSK requires more power to achieve the same bit error rate as CCK. The link variables include the range, which is the distance from the transmitter to the receiver, the multipath environment, and the interference environment. Missed packets and corrupted (including recoverable) packet data are included in PER.

Radio signals radiated by an ideal isotropic antenna weaken with the square of the distance as they travel through free space (square power law). The attenuation also increases with frequency. At 2.4 GHz (Lambda = 0.125 m = 5") the path loss in free space is about 40 dB for 1 m. Propagation of RF signals in the presence of obstacles are governed by three phenomena: reflection, diffraction, and scattering. Reflection occurs when the dimension of obstacles is large compared to the wavelength of the radio wave. Diffraction occurs when obstacles are impenetrable by the radio wave. Based on Huygen's principle, secondary waves are formed behind the obstructing body even though there is no line of sight. Scattering occurs where the obstacles have dimensions that are on the order of the wavelength. The three propagation mechanisms all have impact on the instantaneous received signal in all different directions from the transmitting antenna.

TABLE 6.1 Indoor Propagation Path Loss

2.4 GHz signal attenuation through:	
Window in brick wall	2 dB
Metal frame, glass wall into building	6 dB
Office wall	6 dB
Metal door in office wall	6 dB
Cinder wall	4 dB
Metal door in brick wall	12.4 dB
Brick wall next to metal door	3 dB

Measurements have shown that propagation loss between floors does not increase linearly (in dB) with increasing separation of floors. Rather, the propagation loss between floors starts to diminish with increasing separation of floors. This phenomenon is thought to be caused by diffraction of radio waves along the side of a building as the radio waves penetrate the building's windows. Values for wall and door attenuation are shown in Table 6.1 and a plot of attenuation between building floors is shown in Fig. 6.16.

6.10.1 WLAN Power, Sensitivity, and Range

The range of a WLAN radio is influenced by data rate, bit error rate requirements, modulating waveform S/N ratio, power amplification, receiver sensitivity, and antenna gain. At higher data rates more power is required. The log of the data increase is the amount of power increase in dB to achieve the same range. Reliability is also a function of power. Higher power produces a lower bit error at the same range, or more range at the same bit error rate. The waveform is a significant contributor to performance. Phase-Shift Keying (PSK) type waveforms are the most power efficient. Frequency-Shift Keying (FSK) requires almost twice as much transmitted power to achieve the same range.

IEEE 802.11 sensitivity for 11 Mb CCK QPSK is specified as a minimum of –76 dBm. Typical radios attain 6 dB better than the minimum levels. Lower data rates such as 5.5-Mb CCK and 2-Mb QPSK have better sensitivities by approximately 5 dB than the 11-Mb Levels. The best sensitivity is obtained with the 1-Mb BPSK. Values are typically 3 dB better than the 2-Mb, QPSK; thus 1-Mb is used for the header/ preamble information.

In an ideal propagation environment such as free space (i.e., no reflectors), the transmitted power reaches the receiver some distance away attenuated as a function of the distance, r. As shown in Fig. 6.17, the Constant 40.2 is used for 2.4-GHz propagation and changes insignificantly across the ISM band. The

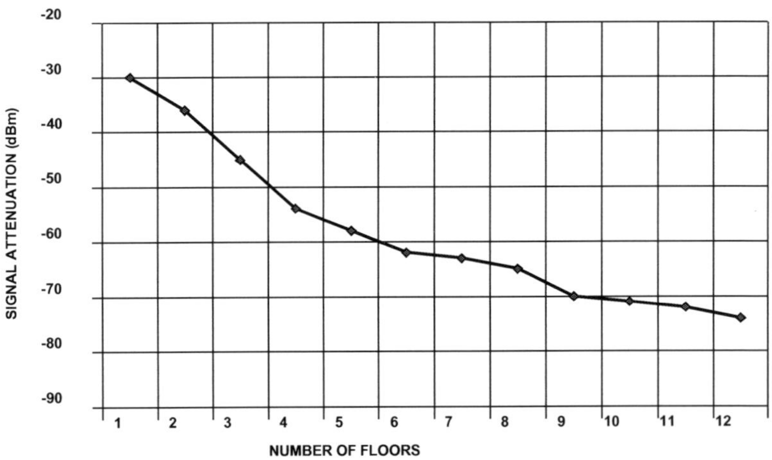

FIGURE 6.16 Path loss between building floors.

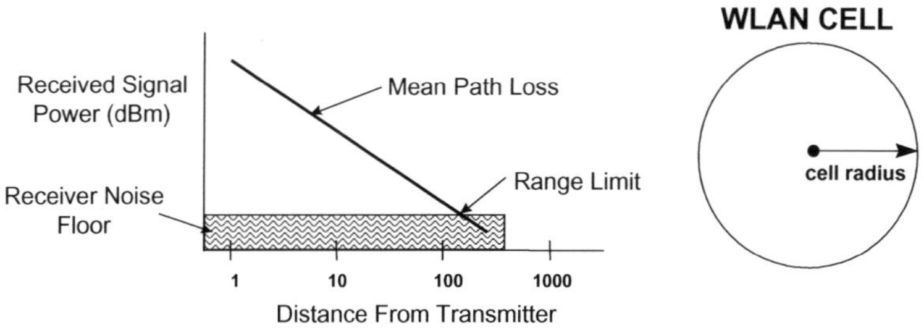

$$\text{Loss dB} = 40.2 + 10 \cdot \log(r^n)$$

Where Loss is transmit power/received power in dB

– r is the cell radius

– n is 2 for free space

FIGURE 6.17 Path loss — free space. Loss dB = 40.2 + 10 • log (r^n) where Loss is transmit power/received power in dB; r is the cell radius; and n is 2 for free space.

exponent of 2 is for free space and increases with multipath. As the receiver moves away from the transmitter, the received signal power reduces until it dips into the receiver noise floor, at which time the error rate becomes unacceptable. This is the first-order determination of the largest a cell can be. This model can be used for unobstructed line-of-sight propagation with highly directional antennas where the antenna gain allows a propagation path that is miles long.

6.10.2 Signal Fading and Multipath

As a transmitted radio wave undergoes reflection, diffraction, and scattering it reaches the receiving antenna via more than one path giving rise to a phenomenon called multipath. The multiple paths of the received signals cause them to have varying signal strengths as well as having different time delays (phase shifts), also known as delay spread. These signals are summed together (vector addition) by the receiving antenna according to their random instantaneous phase and strength giving rise to what is known as small-scale fading. Small-scale fading is a spatial phenomenon that manifests itself in the time domain having Rayleigh distribution; hence it is called Rayleigh fading. Small-scale fading produces instantaneous power levels that may vary as much as 30 or 40 dB while the local average signal level changes much more slowly with distance.

Just as the power law relationship between distance and received power is applied to path loss in free space, it may be used in the presence of obstacles. A general propagation loss model for local average received power uses a parameter, n, to denote the power law relationship where $n = 2$ for free space, and is generally higher for indoor wireless channels.

The "2.4-GHz Signal Path Loss" curve in Fig. 6.18 represents various path losses with different values of n. The first segment, $n = 2$, of the curve, loss is primarily free space loss. The second and last segments of the curve have values of 4 and 6 for n, respectively, representing more lossy channels. The instantaneous drop of the signal power as it transitions from –40 to –60 dB/dec is typical of a signal loss when a receiver loses line of sight to its respective transmitter.

In a multipath condition where the receiver also has a line-of-sight path to the transmitter, the statistical distribution of the local average signal level follows Rician distribution. Rician distribution is based on a factor, k, which specifies the ratio of direct path versus multipath power levels. Multipath is illustrated in Fig. 6.18.

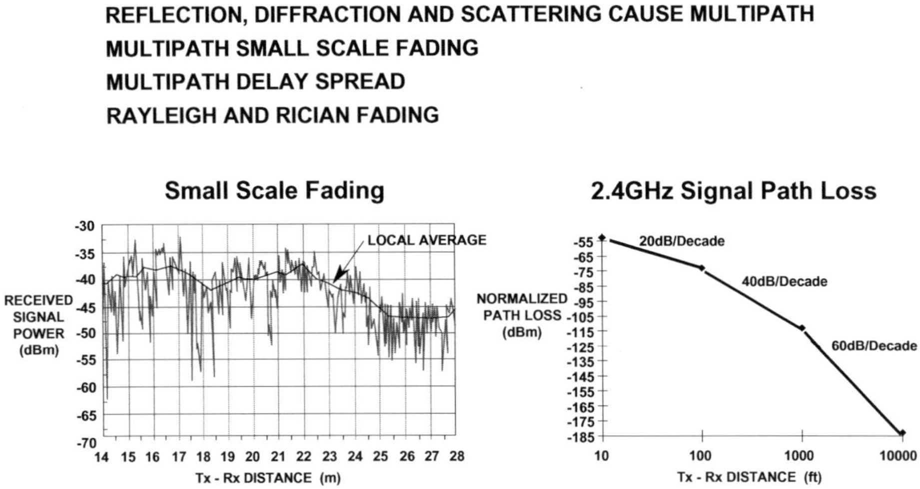

FIGURE 6.18 Signal fading and multipath. Reflection, diffraction, and scattering cause multipath; multipath small scale fading; multipath delay spread; Rayleigh and Rician fading.

6.10.2.1 Log Normal and Rayleigh Fading

The mechanism of the multipath fading can be viewed as being caused by two separate factors: the product of the reflection coefficients and the summation of the signal paths. These two mechanisms produce separate fading characteristics and can be described by their probability distribution functions. The first is characterized as having a log normal distribution and is called log normal fading. The second mechanism, the sum of the signal paths, produces a Rayleigh probability distribution function and is called Rayleigh fading. Figure 6.19 illustrates both multipath mechanisms.

Significant effort has gone into characterizing the multipath environment so that effective radio structures can be designed that operate in difficult high reflection environments.

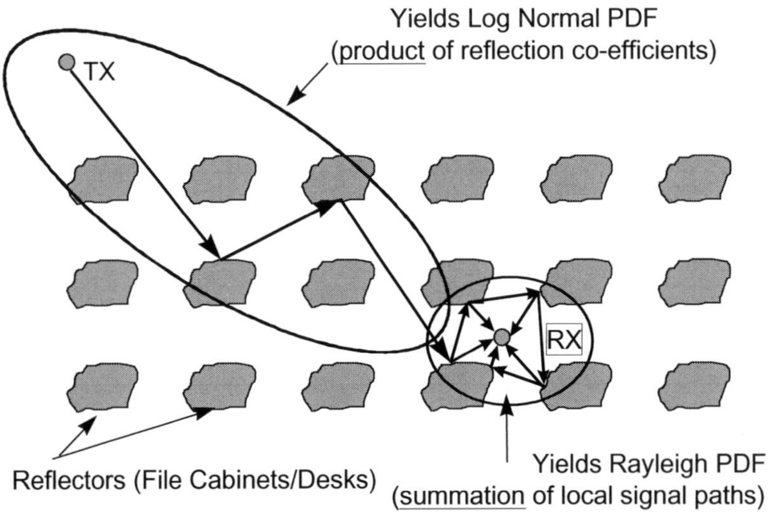

FIGURE 6.19 Log normal and Rayleigh fading.

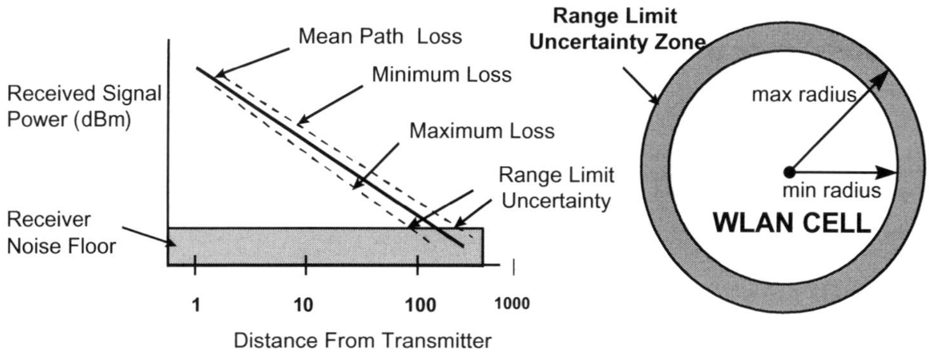

$$\text{Loss dB} = 40.2 + 10 \cdot \log(r^n) + K$$

FIGURE 6.20 Effects of multipath fading.

6.10.2.2 Effects of Multipath Fading

The value of (n) is 2 for free-space propagation, but in general (n) can take non-integer values greater or less than 2 depending on the environment. Random variable K is a log normal that is added to model the variability in the environment due to the different amounts and types of material through which the signal travels. The uncertainty is shown in Fig. 6.20.

Residential: $n = 1.4$–4.0, with $n = 2.8$ typical
Residential: Standard deviation of the lognormal distribution 7–12 dB with 8 dB typical
Office: $n = 1.74$–6.5, with $n = 3.7$ typical
Office: Standard deviation of the log normal distribution 6–16 dB with 10 dB typical
Light industrial: $n = 1.4$–4.0, with $n = 2.2$ typical (open plan),
Light industrial: Standard deviation of the log normal distribution: 4–12 dB with 10 dB typical

6.10.2.3 Delay Spread Craters

Multipath fading has been the chief performance criteria for selecting a new high-data-rate 802.11 DSSS standard waveform. Many independent multipath surveys published in the IEEE literature are in agreement in showing that high delay-spread holes exist anywhere within a cell. These holes can be a real difficulty for cell planners because stations may fail to operate even a short distance from the cell center. In commercial environments it is common to see the multipath spread reach 100 ns (rms). Multipath spread is one unit of measurement for the Rayleigh fading characteristic. A 100-nsEC multipath spread is commonly observed in cafeterias, atriums, and open Wal-Mart-like structures. Craters are illustrated in Fig. 6.21.

6.10.2.4 Multipath Mitigation

By itself, antenna diversity provides minimal relief from multipath craters. If a station is located within a crater, two antennas tend to see the same degree of multipath.

In a multipath crater, the SNR is often high. The average signal power in the crater is the same as described by the mean path loss shown before. If the crater is not near the cell boundary, the SNR is good. However, the multipath components can highly distort the signal so the conventional receiver still fails.

A simple parallel is the distortion caused by audio echoes. A speaker becomes unintelligible when severe echoes exist, even in the absence of other noise. The equivalent effect occurs in a receiver. The multipath echoes cause the code words and DSSS spreading chips to overlap in time. This self-interference causes receiver paralysis even in the absence of noise, unless the receiver has been designed to correct the echoes.

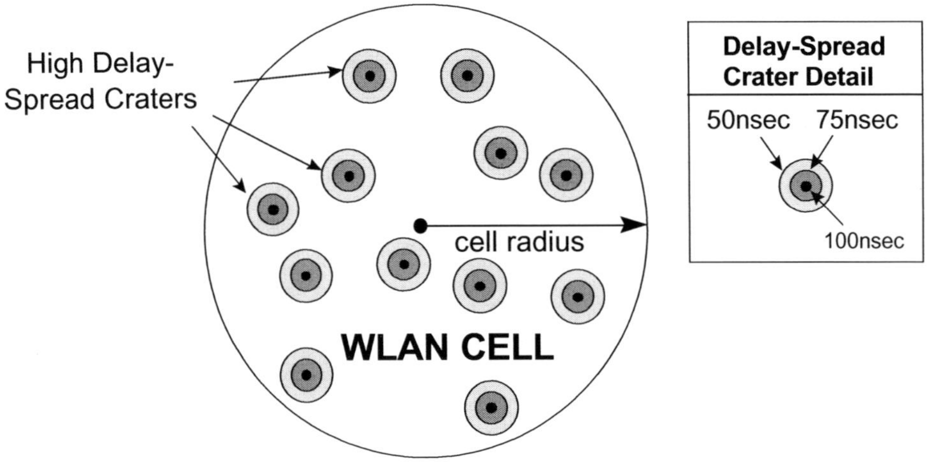

FIGURE 6.21 Delay spread craters.

6.10.2.5 Impulse Response Channel Models

In a room environment, two measurements of the same radio in the same place may not agree. This is due to the changing position of the people in the room and slight changes in the environment which, as we have seen, can produce significant changes in the signal power at the radio receiver. A consistent channel model is required to allow the comparison of different systems and to provide consistent results. Simulations may be run in software against models of the radio. But more valuable are hardware simulators that can be run against the radio itself. These simulators operate on the output of the radio and produce a simulated signal for the receiver from the transmitted signal. To be of value for comparison, a standard model should be used.

The IEEE 802.11 committee has been using quite a good simulation model that can be readily generalized to many different delay spreads.

Another model, which is more easily realized in hardware simulations, is the JTC '94 model. This model is a standard that provides three statistically based profiles for residential, office, and commercial environments. Two obscured line-of-sight profiles are provided for residential and commercial environments; these are illustrated in Fig. 6.22.

6.10.3 Interference Immunity and Processing Gain

Another significant parameter to be considered in the link is interference. In the ISM band the main source of interference is the microwave oven. Radios must be designed to operate in the presence of microwave ovens. The spread-spectrum nature of the waveform allows narrowband interference to be tolerated. Processing gain of 10 dB is available to provide protection from narrowband interferers of any type. Rate changes to lower data rates can be used to allow higher tolerance to microwave energy. The IEEE 802.11 protocol is designed to enable operation between microwave energy pulses. Within the 802.11 protocol is the ability to change frequencies to avoid a problem channel.

Other radios in the band can cause interference. Two types of interference must be considered. First is co-channel interference of our own system. This is the energy from a nearby cell of the same system that is on the same frequency. Frequency planning and good layout of access points can minimize this interference and keep it from being a system constraint. The second type of interference is from other systems. These might be Direct Sequence Spread Spectrum or Frequency Hop Spread Spectrum. Two mechanisms are available to mitigate this interference. The first is Clear Channel Assessment (CCA) provided in the IEEE 802.11 standard. MAC layer protocol provides collision avoidance using CSMA-

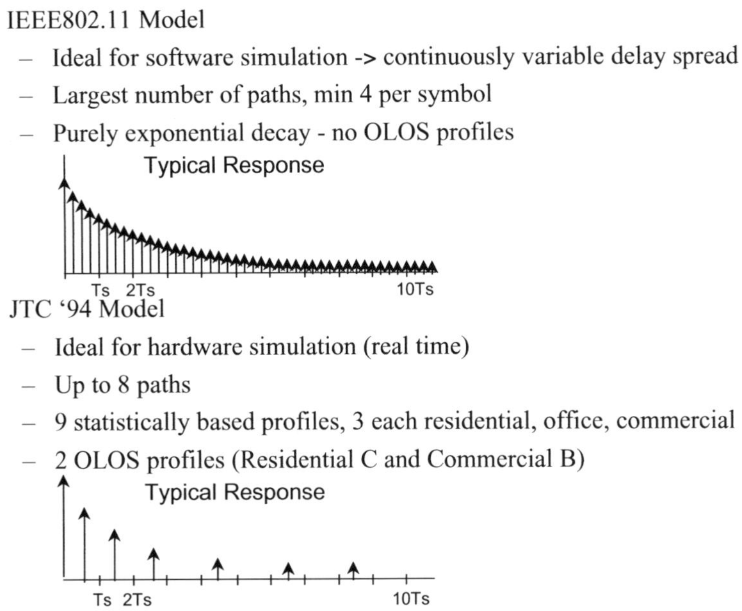

FIGURE 6.22 Impulse response channel models.

CA. The second is processing gain, which provides some protection from Frequency Hop Spread Spectrum radios, which appear as narrowband interferers. Some frequency planning is always required and not all systems will coexist without some performance degradation.

The radiated energy from a microwave oven can interfere with the wireless transmission of data. The two plots shown in Fig. 6.23 illustrate that the radiated energy from a microwave oven is centered in the 2.4-GHz ISM band.

The plots show the results of leakage tests conducted at 9.5 in. and 10 ft from a microwave oven. The test shown on the left plot was conducted at a distance of 9.5 in. from the oven. The leakage test shown on the right plot was conducted at a distance of 10 ft.

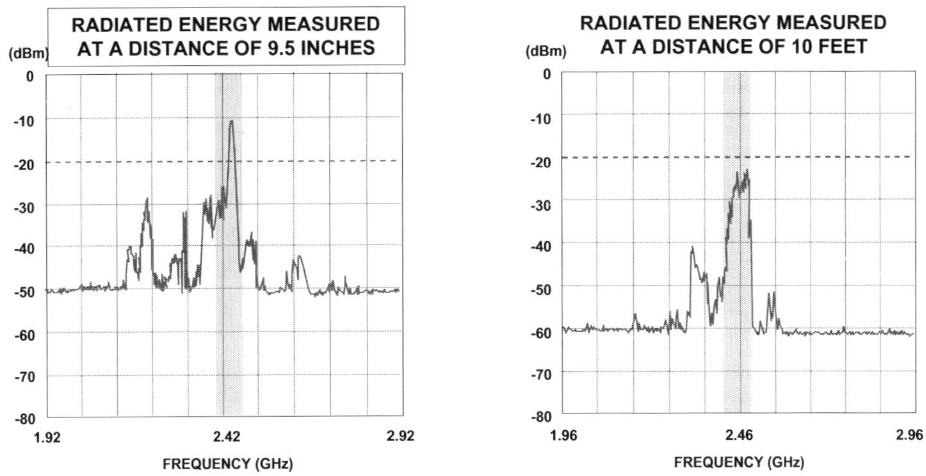

FIGURE 6.23 Microwave oven interference.

The two cases are typical plots selected from more that 20 leakage tests performed on different brands of ovens, at different distances and different antenna angles. The measurements show that the energy drops off dramatically when the distance between the measurement antenna and the oven exceeds 6 ft. At a distance of 10 ft, the energy level is typically below the level of –20 dBm as illustrated in the plot on the right in Fig. 6.23.

6.11 WLAN System Example: PRISM® II

The IEEE 802.11b high rate WLAN has been implemented with the Intersil PRISM® (Personal Radio using Industrial Scientific Medical bands) SiGe chipset. The block diagram is shown in Fig. 6.24. This implementation uses five chips and has a total component count of approximately 200. The major chips include an RF converter that operates using an RF frequency of 2.400 to 2.484 GHz and an IF frequency of 374 MHz. A synthesizer is included in the IC with ceramic filters and a VCO provided at the IC inputs. A PA boosts the signal to approximately +20 dBm prior to final T/R switching, filtering, and any diversity switching. An IF IC converts the signal to baseband or modulates the transmit signal following filtering with a single SAW filter at 374 MHz. The MODEM IC implements the IEEE 802.11 CCK modulation with special circuits for multipath corrections. Finally a digital IC implements the Media Access Controller (MAC) function for 11-Mb data and interfaces with the computer/controller.

A brief description of the signal paths begins on the upper left that the dual antennas which may be selected for diversity with a command from the Baseband Processor (BBP). The desired antenna is routed to a ceramic "roofing" filter to attenuate out-of-band signals and attenuate the 1.6-GHz image frequencies. Front-end losses with Diversity, T/R switches, and filters are typically 4 dB. The 3-dB NF LNA has a selectable high gain of +15 dB or low gain of –15 dB controlled by the BBP depending upon signal level. The range of signals from –90 to –30 dBm uses the LNA high gain mode and –29 to –4 dBm (IEEE 802.1 maximum) uses low gain on the LNA. The first active Gilbert Cell mixer has a gain of +8 dB and NF of 9 dB. A differential 374-MHz SAW filter with 8 dB loss follows the RF converter. An IF demodulator then processes the –80 to –20 dBm signal with a linearized AGC stage prior to the final baseband mixer conversion. The signal is low pass filtered and passed to the Baseband Processor (BBP). Digital BBP processing includes I/Q A/D, interpolating buffers with digital NCO phase locked carrier recovery and Rake receiver/Equalizer processing. Finally the received data are processed with the MAC CPU in accordance with the IEEE 802.11 protocol.

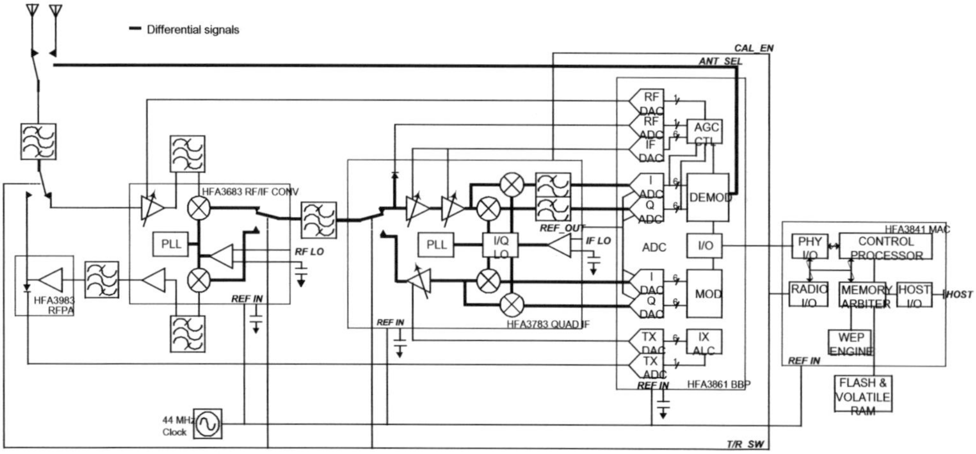

FIGURE 6.24 Intersil PRISM II® radio block diagram.

The transmit function begins with data to the MAC being processed for BBP modulation. The data modulator includes a self-synchronizing scrambler function that removes the periodicity of the short 11 chip Barker sequence. A BBP transmit Data Formatter and Spreading Table provides the PN spreading function. A digital filter then reduces the DQPSK –13 dB Side Lobe Level (SLL) to approximately –45 dBc. This digital baseband waveform is then sent to the BBP I and Q D/A converters. The IF chip's modulator upconverts these signals to the 374-MHz IF using an active Gilbert Cell mixer and combiner. This modulated signal is then Automatic Level Controlled (ALC) using a feedback algorithm with the HFA3983 PA's detector to maintain the IEEE 802.11, 30-dBc SLL. This IF signal is filtered by the same SAW as used in the receive path. A final RF upconversion from 374 MHz to 2.4 GHz is performed by the RF converter using the integrated on-chip synthesizer. The output level of approximately –5 dBm is filtered for image and harmonics and applied to the PA and finally the T/R and Diversity Switches. Schematics, application notes, and additional details for this example design may be obtained at www.Intersil.com.

7

Wireless Personal Area Network Communications: An Application Overview

Thomas M. Siep
Texas Instruments

Ian C. Gifford
M/A-Com, Inc.

Advancing technology is providing electronic aids that are increasingly becoming personal extensions of human beings. A prime example is the cellular telephone. It has replaced the pay phone as the means by which many of us maintain voice contact when away from our homes or businesses. The cell phone is a personal device and is customized to us. The pay phone is a generic, public device. The public device requires us to supply all necessary personal information, such as phone number and billing information, for its use. The personal device retains this slowly changing information for us. Each one of our personal information devices retains our stored information, ready for use. The number of these devices is increasing: Personal Digital Assistants (PDAs), lightweight laptop PCs, and pagers have joined the ranks of devices that keep our store of personal information.

If the stored personal information never changed, there would be no reason for anything but initial data entry. This is not the case. People move, appointments change, and access procedures are updated. This requires the modification of information in an information database retained in each of our devices.

The increasing number of electronic aids with their disparate and overlapping information databases fosters the need to interconnect them. There may be many devices within a relatively small area. This area is, however, more closely tied to an individual than any particular geographic location.

The obvious solution is to interconnect these increasingly intelligent devices and allow them to synchronize their information databases. Special-purpose wires and cables are the traditional method of interconnection. These are the wires that interconnect personal devices rather than connecting to public or local area networks. The problem is, nobody likes wires with personal, portable devices. They get

0-8493-1732-0/03/$0.00+$1.50

forgotten, lost, or broken. Plus, they are inconvenient to use. The solution to that problem is to use a wireless communications technique.

Three different wireless communications schemes could solve the problem: satellite, metropolitan, and short distance. The first two technologies use service-provider infrastructure and cost money each time they are used. The short distance infrastructure may be owned by the user and is "free" once the equipment is paid for.

Several short distance wireless solutions exist in the marketplace. These include the IEEE Standards for Wireless Local Area Networks (WLANs) (P802.11[1]) and Wireless Personal Area Networks (P802.15). Project 802.15 defines the WPAN™ topology. Whereas P802.11 is concerned with features such as Ethernet-matching speed and handoff support for devices in a localized area, WPAN communications are even more localized in their purview. They are concerned only with the immediate area about the person using the device. This concept has been dubbed a Personal Area Network. The untethered version of this is, of course, called a Wireless Personal Area Network™ or WPAN.

WPAN communications provide a Personal Operating Space (POS). This is the space about a person or object that typically extends up to 10 m in all directions and envelops the person whether stationary or in motion. WPAN communications are the wireless "last meter" for the so-called wearable computers or pervasive computing devices that began to emerge in the late 1990s.

The term WPAN was coined by the IEEE based on some of their members' work done in the IEEE Computer Society's IEEE 802 LAN/MAN Standards Committee. The IEEE has identified the Bluetooth™ technology as a solution for their applications for WPAN communications. The Bluetooth technology is a specification for small form factor, low-cost, wireless communication and networking between PCs, mobile phones, and other portable devices.[2]

The WPAN technology examined here is based on the work of the IEEE 802.15 Working Group. Other topologies and implementations exist for personal area networking, but generally they follow the characteristics of the protocol stack explained here.

7.1 Applications for WPAN Communications

WPAN communications are primarily for the user's personal convenience. Its purpose is to replace wires between objects that are more or less within easy reach. It may also hook to the larger network world when/if/while convenient. This wire replacement technology is intended as the primary method of connection between a large variety of devices, many with limited capabilities.

7.1.1 The Nature of WPAN Services

WPAN communications exist to tie together closely related objects, a function that is fundamentally different than the goal of the WLAN. The purpose of WLANs is to provide connectivity to the Ethernet plug in the wall at the workplace and Ethernet-like connectivity in ad hoc situations, such as conferences. Devices that attach to the WLANs are usually high-capability devices. Devices such as laptops and desktop computers are relatively expensive. Wireless connectivity has been justifiable for business entities as an infrastructure cost. The essential difference between the two technologies is topology: WLAN communications are outward-looking and WPAN are inward-looking. WLANs are oriented to connecting to the world as a whole and WPAN communications seek to unify communications among personal devices.

[1]The name IEEE P802.11 derives from the naming scheme used by the Institute of Electrical and Electronic Engineers for their standards bodies and resulting standards documents. Project 802 (P802) is the standards group that is concerned with Local and Metropolitan Area Networks (LAN/MAN). There are many aspects of communications that are covered by the charter of this organization, each of which has a "dot" number. A standard that many readers may be familiar with is P802.3 or Ethernet.

[2]See http:/www.bluetooth.com for more information.

FIGURE 7.1 WPAN™ application hardware.

In addition, WPAN technology is focused on supporting a certain kind of mobility. The backbone that WLANs attach to tends to stay in one place. WLAN devices may move about, but the infrastructure stays put (at least in the short term). People do not. They are their own infrastructure, and they expect the entire system to move with them wherever they go.

This kind of mobility has some interesting legal implications. It drives the need for a single standard that meets the worldwide regulatory requirements. WPAN technology is designed such that a single technology meets the spectrum power requirements of the world, since users don't want to break the law by crossing a border.

7.1.2 Application Example

Figure 7.1 shows the hardware involved in an example WPAN communications system. In this example there are three devices that are able to participate in the Wireless Personal Area Network:

- A headset consists of an in-the-ear speaker, a microphone, and an embedded WPAN radio transceiver.
- A cellular telephone has WPAN capabilities via an embedded radio transceiver as well as its normal macro-cellular radio.
- A laptop computer in the briefcase has a WPAN embedded radio transceiver or interface card with radio transceiver.

Each of these devices is capable of communicating with the other two.

Figure 7.2 demonstrates the WPAN hardware in action. It shows an example of how this kind of system can allow the assemblage of currently existing technologies into a new and powerful capability. The scene is a traveler hurrying in an airport who receives an urgent e-mail that requires an immediate answer before embarking on a plane. The following numbered list corresponds to the circled numbers in Fig. 7.2.

FIGURE 7.2 WPAN™ in action.

1. The cell phone receives the urgent e-mail. Since the cell phone is set up to handle e-mails via a laptop (if available), it does not ring the phone.
2. The cell phone, acting as a gateway, sends a message via a WPAN link, wakes up the laptop computer in the traveler's briefcase, and then relays the e-mail to it.
3. The laptop software inspects the incoming e-mail, determines that it is urgent, and translates the text to synthetic speech.
4. The laptop then uses the WPAN link with the headset that is normally used by the cell phone for voice communications and sends the translated e-mail to it.
5. The traveler then dictates a reply to the e-mail to the laptop via the WPAN link.
6. The laptop translates the voice to text and formats a reply e-mail.
7. The reply e-mail is then handed off to the cell phone for transmission.
8. The e-mail reply is sent within seconds of the arrival of the original e-mail, without the traveler having to stop running.

None of the actions described by Fig. 7.2 requires a WPAN connection. They all *could* be done with wires. The question is *would* they be done with wires? For the vast majority of mobile users the answer is no: this functionality requires wireless links to be feasible.

7.2 WPAN Architecture

As shown in the example above, a WPAN communications system can provide connectivity among a wide variety of personal devices, including personal mobile devices such as headsets, phones, PDAs, notebook computers, and the like. A WPAN can further provide connectivity between those devices and stationary electronic devices such as data access points to wire line LAN installations, or wireless modems connected to the PSTN.

WPAN systems typically use a short-range radio link that has been optimized for power-cautious, battery-operated, small size, lightweight personal devices. WPAN communications generally support both circuit switched channels for telephony grade voice communication and packet switched channels for data communications. As demonstrated in the example above, a cellular phone may use the circuit switched channels to carrying audio to and from a headset, while at the same time, use a packet switched channel to exchange data with a notebook computer.

7.2.1 WPAN Hardware

General-use WPANs operate in the worldwide unlicensed Industrial, Scientific, Medical (ISM) band at 2.4 GHz. A frequency-hopping technique is utilized to satisfy regulatory requirements and combat interference and fading.

The WPAN system consists of a radio unit, a link control unit, and a support unit for link management and host terminal interface functions. The discussion here is limited to the specifications of the WPAN Physical (PHY) layer and Media Access Control (MAC) sublayer that carries out the baseband protocols and other low-level link routines. This corresponds to the protocol layers covered by the IEEE P802.15 WPAN Standard.

The IEEE P802.15 architecture consists of several components that interact to provide a WPAN that supports station mobility transparently to upper layers. The following definitions are taken from the Standard:

Physical Layer (PHY): Protocol layer that directly controls radio transmissions.

Medium Access Control (MAC): Protocol layer that determines when and how to use the PHY.

Master Station: Any *device* that contains an IEEE P802.15 conformant MAC and PHY interface. It provides identity and clocking to WPAN peer devices.

Slave Station: Any *device* that contains an IEEE P802.15 conformant MAC and PHY interface. It receives and transmits to the master station only. Transmissions from the slave occur only when the master permits.

Piconet: Two or more WPAN Stations sharing the same RF channel form a *piconet*, controlled by one (and only one) master.

Scatternet: Multiple independent and non-synchronized piconets that may have some overlapping membership.

The discussions below will use these terms to describe the topology and behavior of WPAN systems.

7.2.2 Ad Hoc Networks

The WPAN networks are not created *a priori* and have a limited life span. They are created on an as-needed basis when necessary, then subsequently abandoned. These *ad hoc networks* are established whenever applications need to exchange data with matching applications in other devices. A WPAN will likely cease to exist when the applications involved have completed their task.

WPAN units that are within range of each other can set up ad hoc connections. WPANs™ support both point-to-point and point-to-multipoint connections. In principle, each unit is a peer with the same hardware capabilities. Two or more WPAN units that share a channel form a piconet with one unit being the master.

Each piconet is defined by a different frequency-hopping channel. This channel is a narrow band of radio transmission that is continually moving from frequency to frequency ("hopping") in a pseudorandom fashion. All units participating in the same piconet are synchronized to this channel.

7.2.3 A WPAN vs. a WLAN

As mentioned above, the IEEE also defines another type of short-distance wireless technology. It is IEEE P802.11 Wireless Local Area Networks (WLAN). On the surface the WLAN and WPAN seem quite similar.

WLAN technologies are "instant infrastructure" specifically designed for interconnecting peer devices in and around the office or home. A WPAN device is a cable replacement "communications bubble" that is designed to travel from country to country to be used in cars, airplanes, and boats. As a result, the topologies are different

A typical WLAN topology is shown in Fig. 7.3. It is essentially an extension of the wired LAN. Each of the communicating devices is an equal partner in the communications network. As such, any device can communicate directly with any other device, provided the radio waves can reach between them. If they cannot reach, there are provisions in the standard for relaying messages. There is no requirement for a hard-wired backbone, but it is typical for most installations. The topology is range-limited, with no upper ceiling on the number of participating units.

The WPAN topology is simpler. Figure 7.4 illustrates the WPAN topology. The devices interconnected are identical. How they are interconnected is different. Instead of a free-for-all competition for the airwaves, the WPAN has a master device that directs the traffic of its slaves through it. The slaves are "good children"; they never speak unless spoken to. Any device that cannot be directly reached through the master cannot be communicated with: there is no provision for message relay in the protocol.

The master/slave topology has both disadvantages and advantages. It is a disadvantage for two slaves that wish to converse to have to go through the master to do it. There are many advantages, however. Among them are:

- Spectral efficiency
- Simplicity
- Cost

The orderly use of the airwaves by WPAN technology yields a higher utilization of the spectrum. The WLAN uses a technique of Carrier Sense/Collision Avoidance (CS/CA) to assure fair sharing of the airwaves. It can use several schemes, the simplest of which is to Listen Before Talk (LBT). The WLAN

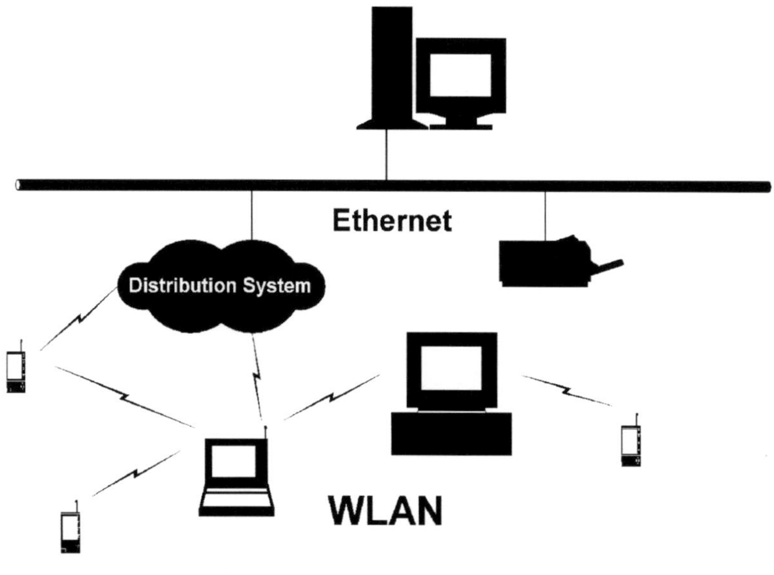

FIGURE 7.3 Wireless Local Area Network topology.

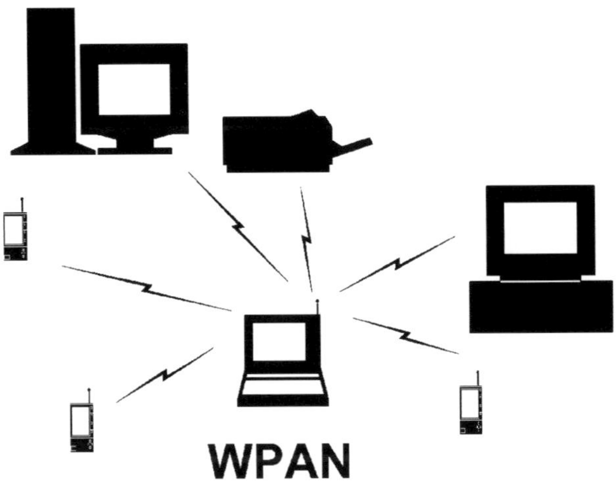

FIGURE 7.4 Wireless Personal Area Network topology.

device never knows when the medium will be occupied, so it must listen to the current environment before it transmits. In the event that the airwaves are busy, the device is required to do an exponential back off and retry.

The WPAN technology has a more controlled approach to using the airwaves. It establishes the master of the piconet, which dictates when each device can use the airwaves. There is no "lost time" in waiting for clear air or collisions due to two partners in the same net transmitting at the same time to the same device because they are out of range of each other.

Implementations of WPAN communications are inherently simpler. Without the need to support LAN technology, the WPAN implementation can forgo the extra buffers and logic that LANs require. The WPAN world is smaller and therefore simpler.

This simplicity of topology yields a less expensive system cost. Having simpler logic and less capability allows designers of WPAN systems to create communication subsystems that are a smaller percentage of total system cost. This, in turn, allows the technology to be embedded in more low cost devices. The effect of that will be to increase volume, which brings down the cost again.

7.2.4 WPAN Scatternet

Sometimes more complex communications schemes are required of the devices that support WPANs. In that case, the WPAN devices can form a *scatternet*: multiple independent and non-synchronized piconets. These are high aggregate-capacity systems. The scatternet structure also makes it possible to extend the radio range by simply adding additional WPAN units acting as bridges at strategic places.

To integrate the IEEE P802.15 architecture with a traditional wired LAN (or even Wireless LAN), an application is utilized to create an "*access point*." An access point is the logical point at which data to/from an integrated non-802.15 wired LAN leave/enter the IEEE 802.15 WPAN piconet. For example, an access point is shown in Fig. 7.3 connecting a WLAN to a wired 802 LAN: the same construct can be implemented with a WPAN, with the restrictions inherent in it.

7.3 WPAN Protocol Stack

The IEEE Project 802 LAN/MAN Standards Committee develops wired and wireless standards. IEEE 802 standards deal with the physical and data link layers as defined by the International Organization for Standardization (ISO) Open Systems Inter-connection (OSI) Basic Reference Model (ISO/IEC 7498-1: 1994).

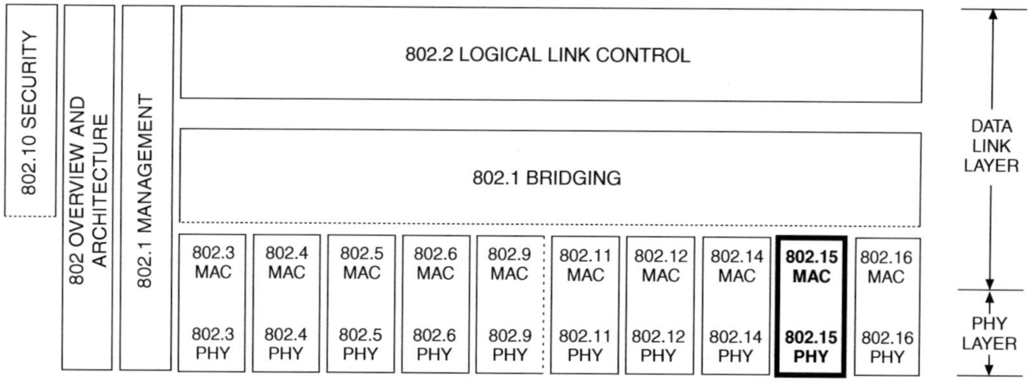

FIGURE 7.5 The IEEE P802 family of standards.

Currently, IEEE Project 802 standards define ten types of medium access technologies and associated physical media, each appropriate for particular applications or system objectives (see Fig. 7.5). The most popular of these standards is IEEE 802.3, the Ethernet standard.

7.3.1 WPAN Open Protocol Stack

All applications already developed by vendors according to the Bluetooth Foundation Specification can take immediate advantage of hardware and software systems that are compliant with the P802.15 Standard. The specification is open, which makes it possible for vendors to freely implement their own (proprietary) or commonly used application protocols on the top of the P802.15-specific protocols. Thus, the open specification permits the development of a large number of new applications that take full advantage of the capabilities of the WPAN technology.

7.3.2 The P802.15 Architecture

Project 802 LAN/MAN standards cover the two lowest layers of the ISO OSI protocol description.

Figure 7.6 shows the relationship of the ISO model to the purview of Project 802. Note that the lowest two levels of the ISO stack correspond to the three layers of the IEEE stack. The reason for that is that the interface that defines the Logical Link Control (LLC) is common for all 802 standards. P802.2 documents this interface and the other portions of the 802 family of standards just refer to it, rather than replicating it for each standard. Since the WPAN standard did not start out with the IEEE architecture in mind, the P802.15 group found it necessary to do a mapping from the Bluetooth structure to one that is compatible with the rest of the 802 family. Figure 7.7 shows this mapping. The radio definition of the Bluetooth specification directly corresponds to the PHY of a P802.15 standard. The MAC is made up of three explicit entities and one implicit entity. The Logical Link Control and Adaptation Protocol (L2CAP), Link Manager Protocol (LMP), and Baseband (BB) are specifically defined in the specification. The Link Manager (LM) is mentioned in several sections of the specification, but never explicitly defined. The P802.15 group has created a set of System Definition Language (SDL[3]) graphics that explicates this and other implied structures in the Bluetooth specification.

7.3.3 PHY Layer

The *physical layer* (PHY) enables the physical radio link between WPAN units. The P802.15 RF system is a Frequency-Hopping Spread-Spectrum system where packets are transmitted in defined time slots on

[3]ITU-T Recommendation Z.100.

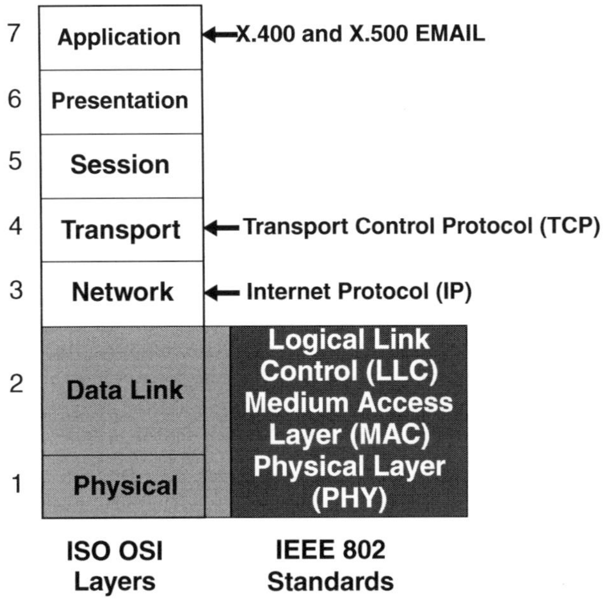

FIGURE 7.6 The relationship between ISO-OSI and IEEE P802.

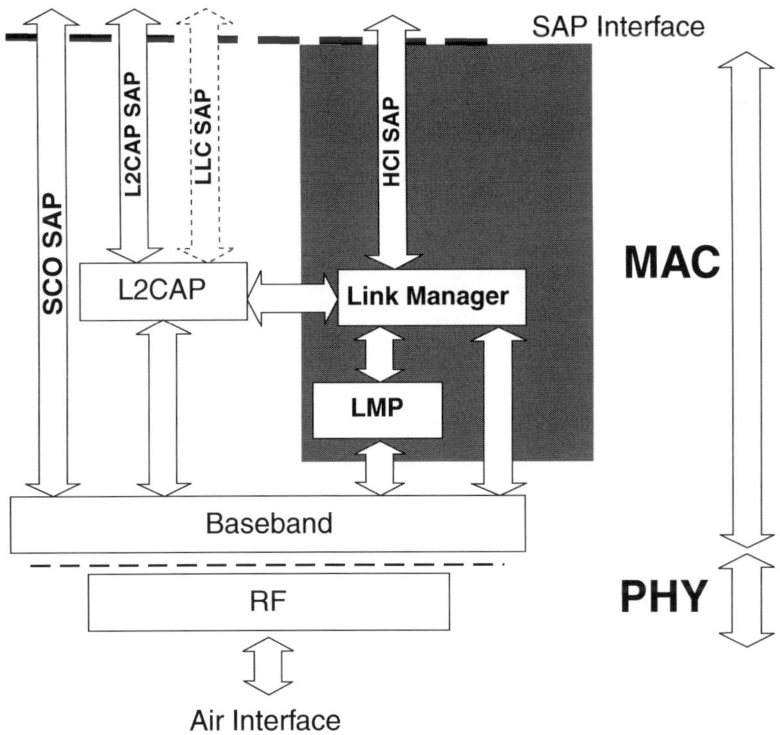

FIGURE 7.7 Bluetooth™ to P802.15 mapping.

defined frequencies. This hopping technique is used to achieve the highest possible robustness for noisy radio environments. The rate of hopping for this packet-based protocol is 1600 hops per second.

The P802.15 transceiver operates in the 2.4-GHz ISM band at a gross data rate of 1 Mbps. The required nominal range of a P802.15 radio is 10 m with 0-dBm output power.

A single unit can support a maximum data transfer rate of 723 kb/s or a maximum of 3 voice channels. A mixture of voice and data is possible to support multimedia applications. The modulation is Gaussian Frequency-Shift Keying (GFSK) with BT = 0.5.

The channel is represented by a pseudorandom hopping sequence. The channel is divided into time slots, where each slot corresponds to one RF hop. Time-division duplexing (TDD) is used for access to the channel.

The data is conveyed in packets. Each packet consists of an access code, header, and payload. Sixteen different packet types can be defined.

7.3.4 MAC Sublayer

The *Media Access Control* (MAC) portion of a P802 protocol decides when transmissions will occur and what will be transmitted. It marries the raw transmitter (the PHY) to the layer that establishes logical connections between devices, the Logical Link Controller (LLC). It sets the topology of the network.

The P802.15 MAC uses the Bluetooth ad hoc piconet and scatternet concepts. Both point-to-point and point-to-multipoint connections are supported.

In the P802.15 network all units have essentially the same capabilities, with identical hardware and software interfaces. Each device is distinguished by a unique IEEE 48-b address. At the start of a connection the initializing unit is temporarily assigned as master. This assignment is valid only during this connection. Every unit in the piconet uses the master identity and clock to track the hopping channel. Inter-piconet communication is achieved by selecting the proper master identity and clock offset to synchronize with the channel of the desired piconet. A corresponding set of identity and clock offsets is available for each piconet. A unit can act as a slave in several piconets. A master is also allowed to be a slave in another piconet.

7.3.4.1 Connection Types

The P802.15 standard provides two different kinds of logical connection types, Synchronous Connection-Oriented (SCO) and Asynchronous Connection-Less (ACL), which can be transmitted in a multiplexing manner on the same RF link. ACL packets are used for data only, while the SCO packet can contain audio only or a combination of audio and data.

The SCO link is a symmetric point-to-point link between the master and a specific slave. The SCO reserves two consecutive time slots (forward and return) at fixed intervals. It is considered a circuit-switched connection. Audio data can be transferred between one or more P802.15 devices, making various usage models possible. The typical model is relatively simple where two 802.15 devices send and receive audio data between each other by opening an audio link.

The ACL link supports symmetric and asymmetric packet-switched point-to-multipoint connections typically used for data. All data packets can be provided with differing levels of forward error control (FEC) or cyclic redundancy check (CRC) error correction and can be encrypted. Furthermore, different and multiple link types may apply between different master-slave pairs of the same piconet and the link type may change arbitrarily during a session.

7.3.4.2 Packets

Compared to wired physical media channels, the data packets defined by the MAC sublayer are smaller in size. The MAC sublayer segments large packet transmissions into multiple smaller packets. Similarly, on the receiving end it reassembles the multiple smaller packets back into the larger packet. Segmentation and reassembly (SAR) operations are used to improve efficiency by supporting a maximum transmission unit (MTU) size larger than the largest packet that can be transmitted.

A packet consists of three fields: access code, header, and variable length payload. The access code, described in the PHY layer, is used for synchronization, DC offset compensation, and identification. The packet header contains link-control information.

The channel is divided into time slots, each 625 μs in length. The packet must be completely transmitted in that time period. The slots are divided into two alternating groups, "even" and "odd." Even slots are master-to-slave, and odd slots are slave-to-master slots. Only the slave that was addressed in the preceding master-to-slave slot can transmit ACL data in the slave-to-master slot.

As stated above, the MAC sublayer provides connection-oriented and connectionless data services to upper layer protocols. Both of these services are supported with protocol multiplexing, segmentation and reassembly operation, and group abstraction capabilities. These features permit higher level protocols and applications to transmit and receive data packets up to 64 kilobytes in length.

7.3.4.3 In Case of Error

There are three error-correction schemes defined for 802.15: 1/3 rate FEC, 2/3 rate FEC, and Automatic Repeat-reQuest (ARQ) scheme for data. The purpose of the FEC scheme on the data payload is to reduce the number of retransmissions. The FEC of the packet payload can be turned on or off, while the packet header is always protected by a 1/3 rate FEC. In the case of a multi-slave operation, the ARQ protocol is carried out for each slave independently.

Two speech coding schemes, continuously variable slope delta (CVSD) modulation and logarithmic Pulse Coded Modulation (logPCM), are supported, both operating at 64 kb/s. The default is CVSD. Voice is never retransmitted, but CVSD is very resistant to bit errors, as errors are perceived as background noise, which intensifies as bit errors increase.

7.3.5 Security and Privacy

To provide user protection and information secrecy, the WPAN communications system provides security measures at the physical layer and allows for additional security at the applications layer. P802.15 specifies a base-level encryption, which is well suited for silicon implementation, and an authentication algorithm, which takes into consideration devices that do not have the processing capabilities. In addition, future cryptographic algorithms can be supported in a backwards-compatible way using version negotiation.

The main security features are:

- Challenge-response routine for authentication
- Session key generation, where session keys can be exchanged at any time during a connection
- Stream-cipher

Four different entities are used for maintaining security at the link layer: a public address that is unique for each unit (the IEEE 48-b address), two secret keys (authentication and encryption), and a random number that is different for each new connection. In a point-to-multipoint configuration the master may tell several slave units to use a common link key and broadcast the information encrypted. The packet payload can be encrypted for protection. The access code and packet header are never encrypted.

7.4 History of WPANs and P802.15

The chain of events leading to the formation of IEEE 802.15 began in June 1997 when the IEEE Ad Hoc "Wearables" Standards Committee was initiated during an IEEE Standards Board meeting. The purpose of the committee was to "encourage development of standards for wearable computing and solicit IEEE support to develop standards." The consensus recommendation was to encourage such standards development in the IEEE Portable Applications Standards Committee (PASC).

During the PASC Plenary Meeting in July 1997, an IEEE Ad Hoc Committee was assembled (17 attendees) to discuss "Wearables" Standards. The committee identified several areas that could be considered for standardization, including short range wireless networks or Personal Area Networks

(PANs), peripherals, nomadicity, wearable computers, and power management. Of these, the committee determined that the best area of focus was the wireless PAN because of its broad range of application. The IEEE Ad Hoc "Wearables" Standards Committee met twice more — once in December 1997 and again in January 1998. During the January 1998 meeting it was agreed that the 802 LAN/MAN Standards Committee (LMSC) was probably a more suitable home for the group's initial activities, especially with a WPAN focus. Two delegates were sent to the IEEE 802.11 interim meeting that same month to get reactions and to gain support for the proposal. At that meeting it was agreed to propose the formation of a study group under 802.11 at the March plenary of 802. A WPAN tutorial was organized to socialize the idea within 802. The result was that in March 1998, the "Wearables" Standards Ad Hoc Committee under PASC became the IEEE 802.11 Wireless Personal Area Network (WPAN) Study Group within LMSC with the goal of developing a Project Authorization Request (PAR) for the development of a WPAN standard.

At the time the study group was formed, there had been no other publicized initiatives in the WPAN space other than the Infrared Data Association's IrDA specification. IrDA's line-of-sight restrictions do not allow it to adequately address the problem. By the time the work of the study group concluded a year later, both the Bluetooth Special Interest Group and HomeRF™ were active in developing WPAN specifications. By March 1999, when the study group and 802.11 submitted the PAR to the 802 executive committee for approval, Bluetooth had over 600 adopter companies and HomeRF had over 60. Because of the significance of these groups in the WPAN market space, it was felt that the standards development process would be better served if a new working group were formed to address the problem rather than pursue it as a task group under 802.11. The PAR was approved and the new group was designated as IEEE 802.15 Working Group for Wireless Personal Area Networks.

7.5 Conclusions

Wireless Personal Area Networks will proliferate early in the next millennium and the IEEE 802.15 Working Group for Wireless Personal Area Networks (WPANs) is providing the leadership in the IEEE 802 standards committees to establish open standards for WPANs. WPANs, which are synonymous with the Bluetooth Foundation Specification v1.0, are relatively new emerging wireless network technologies and as such the applications are evolving.

The first standard derived by 802.15 from the Bluetooth Version 1.0 Specification Foundation Core, and Bluetooth Version 1.0 Specification Foundation Profiles is addressing the requirements for Wireless Personal Area Networking (WPAN) for a new class of computing devices. This class, collectively referred to as pervasive computing devices, includes PCs, PDAs, peripherals, cell phones, pagers, and consumer electronic devices to communicate and interoperate with one another. The authors anticipate that the IEEE Standards Board on or before March 2001 will approve this standard. The 802.15 working group is paving the way for Wireless Personal Area Network standards that will be — Networking the World™.

8

Satellite Communications Systems

Ramesh K. Gupta

Comsat Laboratories

The launch of commercial communications satellite services in the early 1960s ushered in a new era in international telecommunications that has affected every facet of human endeavor. Although communications satellite systems were first conceived to provide reliable telephone and facsimile services among nations around the globe, today satellites provide worldwide TV channels (24 h a day), global messaging services, positioning information, communications from ships and aircraft, communications to remote areas, disaster relief on land and sea, personal communications, and high-speed data services including Internet access. The percentage of voice traffic being carried over satellites — which stood at approximately 70% in the 1980s — is rapidly declining with the advent of undersea fiber-optic cables, and as new video and data services are added over existing satellite networks [1].

The demand for fixed satellite services and capacity continues to grow. Rapid deployment of private networks using very small aperture terminals (VSATs) — which interconnect widely dispersed corporate offices, manufacturing, supply, and distribution centers — has contributed to this growth. Approximately half a million VSATs [1] are in use around the world today employing a low-cost terminal with a relatively small aperture antenna (1 to 2 m). These terminals are inexpensive and easy to install, and are linked together with a large central hub station, through which all the communications take place. The majority of these VSAT applications use data rates of 64 or 128 kb/s. Demand for higher rate VSAT services with data rates up to 2 Mb/s is now growing due to emerging multimedia services, fueled by unprecedented growth in the Internet.

In the last two decades, wireless communications has emerged as one of the fastest growing services. Cellular wireless services, which have been leading the growth over the last decade, achieved cumulative annual growth rates in excess of 30% [2], with more than 200 million subscribers worldwide. To complement this rapidly emerging wireless telecommunications infrastructure around the world, a

0-8493-1732-0/03/$0.00+$1.50

number of satellite-based global personal communications systems (PCS) [3–5] have been deployed or are at advanced stages of implementation and deployment. Examples include low Earth orbit (LEO) systems such as Iridium and Globalstar, medium Earth orbit (MEO) systems such as ICO Global, and geostationary (GEO) systems such as Asia Cellular Satellite System (ACeS) and Thuraya. These systems are designed to provide narrowband voice and data services to small handheld terminals. Some of these systems have experienced problems with market penetration. These satellite systems will be discussed in detail in the next chapter.

In parallel with these developments, rapid growth in Internet traffic is creating an exponential increase in the demand for transmission bandwidth. For example, use of the Internet is projected to grow from its present estimate of 100 million households to more than 300 million households in the next few years. Telecommunications networks are currently being designed to support many new applications, including high-speed data, high-resolution imaging, and desktop videoconferencing, all of which require large transmission bandwidths. To serve these markets, the U.S. and other countries have been actively developing satellite-based broadband services for business and home users. One major advantage of satellite systems has been their ability to provide "instantaneous infrastructure," particularly in underserved areas. Currently, most satellite-based broadband services are being offered at Ku-band, using broadband VSAT-type terminals. Additional systems are being proposed for the recently available Ka- and V-Bands [6–7]. Deployment of these broadband systems could help overcome the "last mile" problem (also referred to as the "first mile" problem from the customer's perspective) encountered in the developed countries, in addition to offering a host of new services. In the U.S. alone since 1995, 14-Ka-band (30/20-GHz) and 16 V-band (50/40-GHz) satellite systems have been proposed to the Federal Communications Commission (FCC) in response to the Ka- and V-band frequency allotments. In addition, in 1996 Sky Station International proposed a V-band system employing a stabilized stratospheric platform at altitudes of 20 to 30 km. Deployment of the proposed systems will accelerate the implementation of broadband wireless infrastructure in regions of the world where terrestrial telecommunications infrastructure is inadequate for high-speed communications.

This chapter reviews the evolution of communications satellite systems, addresses various service offerings, and describes the characteristics of the newly proposed Ka- and V-band broadband satellite systems.

8.1 Evolution of Communications Satellites

Communications satellites have experienced an explosive growth in traffic over the past 25 years due to a rapid increase in global demand for voice, video, and data traffic. For example, the International Telecommunications Satellite Organization's INTELSAT I (Early Bird) satellite, launched in 1965, carried only one wideband transponder operating at C-band frequencies (6-GHz uplink and 4-GHz downlink) and could support 240 voice circuits. A large 30-m-diameter antenna was required. In contrast, the communications payload of the INTELSAT VI series of satellites, launched from 1989 onward, provides 50 distinct transponders operating over both C- and Ku-bands [8], providing the traffic-carrying capacity of approximately 33,000 telephone circuits. By using digital compression and multiplexing techniques, INTELSAT VI is capable of supporting 120,000 two-way telephone channels and three television channels [9]. The effective isotropically radiated power (EIRP) of these satellites is sufficient to allow use of much smaller Earth terminals at Ku-band (E1: 3.5 m) and C-bands (F1: 5 m), together with other larger earth stations. Several technological innovations have contributed to this increase in number of active transponders required to satisfy growing traffic demand. A range of voice, data, and facsimile services is available today through modern communications satellites to fixed and mobile users. The largest traffic growth areas in recent years have been video and data traffic. The satellite services may be classified into three broad categories: Fixed Satellite Services, Direct Broadcast Satellite Services, and Mobile Satellite Services.

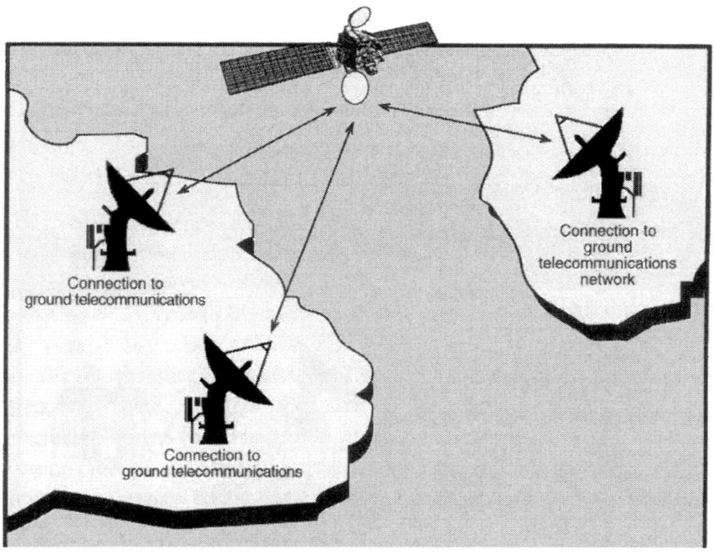

FIGURE 8.1 Fixed satellite communications services (FSS).

8.1.1 Fixed Satellite Services (FSS)

With FSS, signals are relayed between the satellite and relatively large fixed Earth stations. Terrestrial landlines are used to connect long-distance telephone voice, television, and data communications to these Earth stations. Figure 8.1 shows the FSS configuration. The FSS were the first services to be developed for global communications. The International Telecommunications Satellite Organization (INTELSAT), with headquarters in Washington, D.C., was formed to provide these services. INTELSAT has more than 130 signatories today, the largest being the U.S. signatory — Comsat Corporation. With satellites operating over the Atlantic, Indian, and Pacific Ocean regions, INTELSAT provides a truly global communications service to the world.

Over the past several years, with privatization of the communications industry, new global satellite operators have emerged, including PanAmSat, Skynet, and New Skies. Several regional satellite systems have been developed around the world for communications among nations within a continent or nations sharing common interests. Such systems include Eutelsat, Arabsat, and Asiasat. For large countries like the U.S., Canada, India, China, Australia, and Indonesia, FSS are being used to establish communications links for domestic use. The potential for global coverage, and the growth in communications traffic as world economies become more global, have attracted many private and regional satellite operators, with several new systems being planned beyond the year 2000. According to a recent International Telecommunication Union (ITU) report (March 1998), 75% of all communications traffic is now provided under competitive market conditions, as compared to just 35% in 1990. The FSS segment is expected to grow significantly as new satellite operators introduce services at higher frequency bands.

8.1.2 Direct Broadcast Satellite (DBS) Services

Direct broadcast satellites use relatively high-power satellites to distribute television programs directly to subscriber homes (Fig. 8.2), or to community antennas from which the signal is distributed to individual homes by cable. The use of medium-power satellites such as Asiasat has stimulated growth in the number of TV programs available in countries like India and China. Launch of DBS services such as DIRECTV (Hughes Networks) and Dish Network (Echostar) provides multiple pay channels via an 18-in. antenna. The growth of this market has been driven by the availability of low-cost receive equipment consisting of an antenna and a low-noise block downconverter (LNB). This market segment has seen

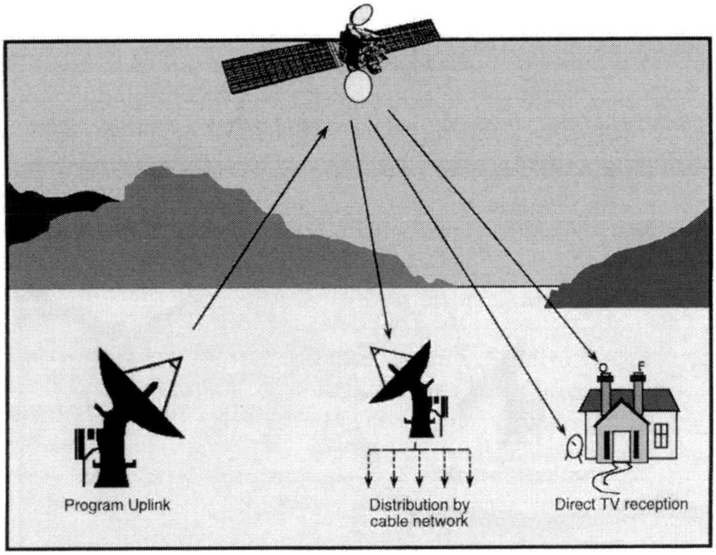

FIGURE 8.2 Direct broadcast satellite communications services (DBS). Direct-to-home (DTH) services are also offered with similar high-power Ku-band satellite systems.

impressive growth, with direct-to-home TV subscribers exceeding 10 million today. The future availability of two-way connection through an 18-in. satellite dish is expected to place the consumer subscribers in direct competition with cable for home-based interactive multimedia access.

8.1.3 Mobile Satellite Services (MSS)

With MSS, communication takes place between a large fixed Earth station and a number of smaller Earth terminals fitted on vehicles, ships, boats, or aircraft (Fig. 8.3). The International Maritime Satellite Organization (Inmarsat), headquartered in London [10], was created in 1976 to provide a space segment to improve maritime communications, improve distress communications, and enhance the safety of life at sea. Subsequently, Inmarsat's mandate was extended to include aeronautical services and land-mobile services to trucks and trains. Launch of the Inmarsat 2 series of satellites began in 1989 and improved communications capacity by providing a minimum of 150 simultaneous maritime telephone calls. This number was further increased by using digital technology. With the launch of Inmarsat 3 spacecraft, which cover Earth with seven spot beams, communications capacity increased tenfold.

The mobile services, which stimulated the growth of personal communications services (PCS) [4,5], have witnessed intense competition in the last decade. Failure of Iridium systems to provide low-cost services has created uncertainty in the market regarding the business viability of such systems. However, some of these systems are likely to modify their service offerings to data communications, rather than voice or low-data-rate messaging services. The mobile networks may evolve in the future to a universal mobile telecommunications system (UMTS) and IMT-2000 systems.

8.1.4 Frequency Allocations

The frequencies allocated for traditional commercial FSS in popular S-, C-, and Ku-bands [11] are listed in Table 8.1.

At the 1992 World Administrative Radio Conference (WARC) [11, 12], the L-band frequencies were made available for mobile communications (Fig. 8.4). The L-band allocations have been extended due to a rapid increase in mobile communications traffic. Recently, a number of new satellite systems such as Iridium, Globalstar, ICO-Global [5], and Ellipso were proposed to provide personal communications

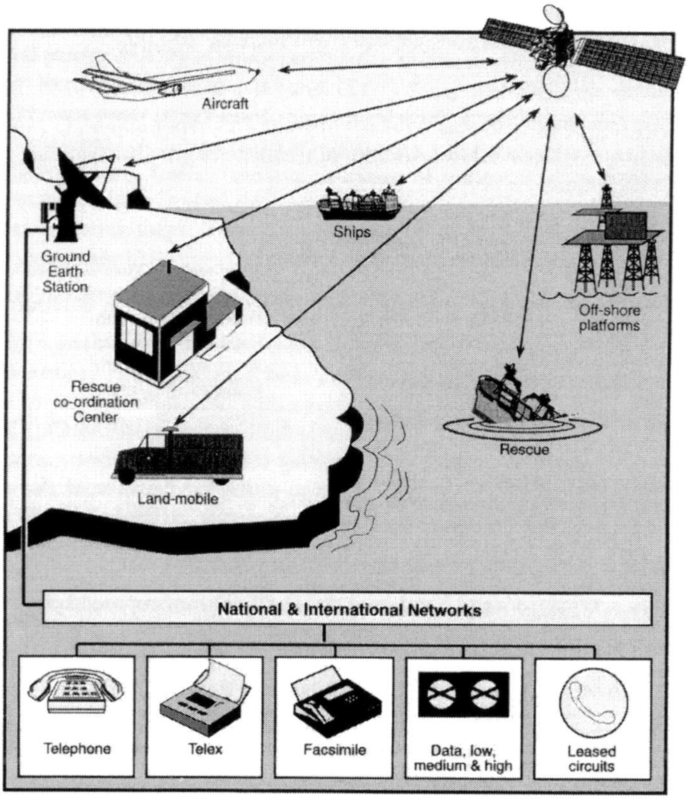

FIGURE 8.3 Mobile satellite communications services (MSS).

TABLE 8.1 S-, C-, and Ku- Band Frequency Allocations

Frequency Band (GHz)	Band Designation	Service
2.500–2.655	S	Fixed regional systems, space-to-Earth
2.655–2.690		Fixed regional systems, Earth-to-space
3.625–4.200	C	Fixed, space-to-Earth
5.925–6.245		Fixed, Earth-to-space
11.700–12.500	Ku	Fixed, space-to-Earth
14.000–14.500		Fixed, Earth-to-space

services to small handheld terminals. These systems used low earth orbits (LEO) between 600 and 800 km (Iridium, Ellipso, Globalstar), or an intermediate (10,000 km) altitude circular orbit (ICO). In addition, a number of geostationary (GEO) systems such as Asia Cellular Satellite System (ACeS) and Thuraya were launched or are in the process of being implemented. A detailed discussion of these systems is presented in the next section.

The 1997 World Radio Conference (WRC-97) adopted Ka-band frequency allocations for geostationary orbit (GSO) and non-geostationary orbit (NGSO) satellite services. Some of these bands require coordination with local multipoint distribution services (LMDS) and/or NGSO MSS feeder links (Fig. 8.5). In the United States, the Federal Communications Commission (FCC) has allocated V-band frequencies for satellite services. In November 1997, the ITU favored the use of 47.2- to 48.2-GHz spectrum for stratospheric platforms.

The Ka- and V-band frequency allocations are given in Table 8.2.

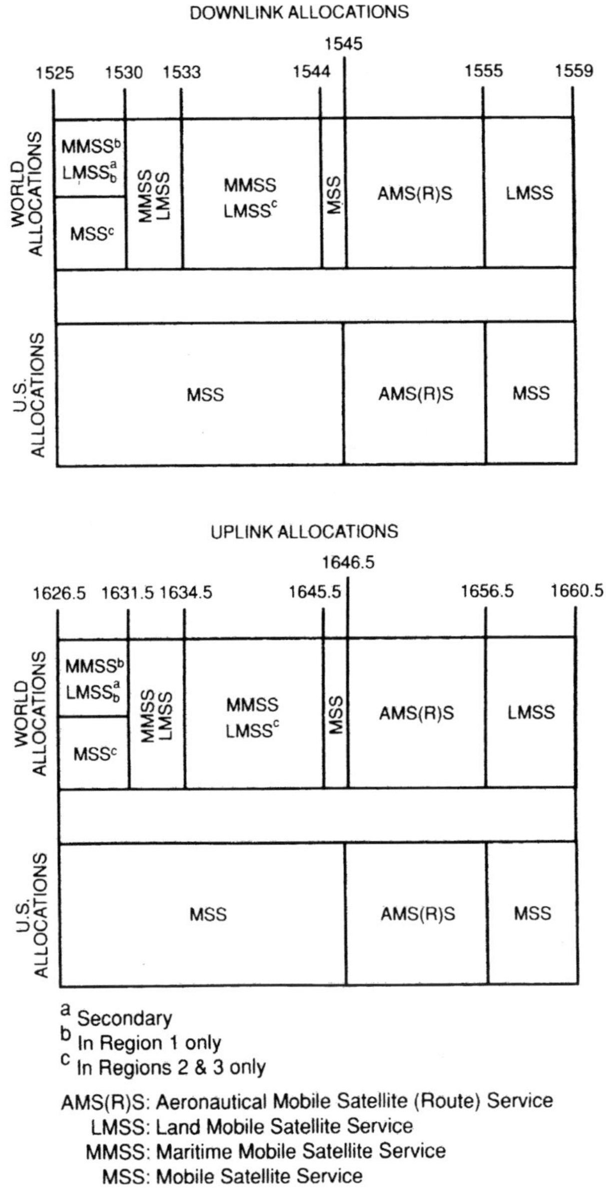

FIGURE 8.4 WARC-92 L-band mobile satellite service allocations.

8.1.5 Satellite Orbits

Commercial satellite systems were first implemented for operation in GEO orbit 35,700 m above Earth. With GEO satellite systems, a constellation of three satellites is sufficient to provide "approximately global" coverage. In the 1990s, satellite systems (lead by Iridium) were proposed using low Earth orbit (LEO) satellites (at 700- to 1800-km altitude) and medium Earth orbit (MEO) satellites (at 9,000- to 14,000-km altitude). Figure 8.6 shows the amount of Earth's coverage obtained from LEO, MEO, and GEO orbits. The higher the satellite altitude, the fewer satellites are required to provide global coverage. The minimum number of satellites required for a given orbital altitude and user elevation angle is shown in Fig. 8.7 and

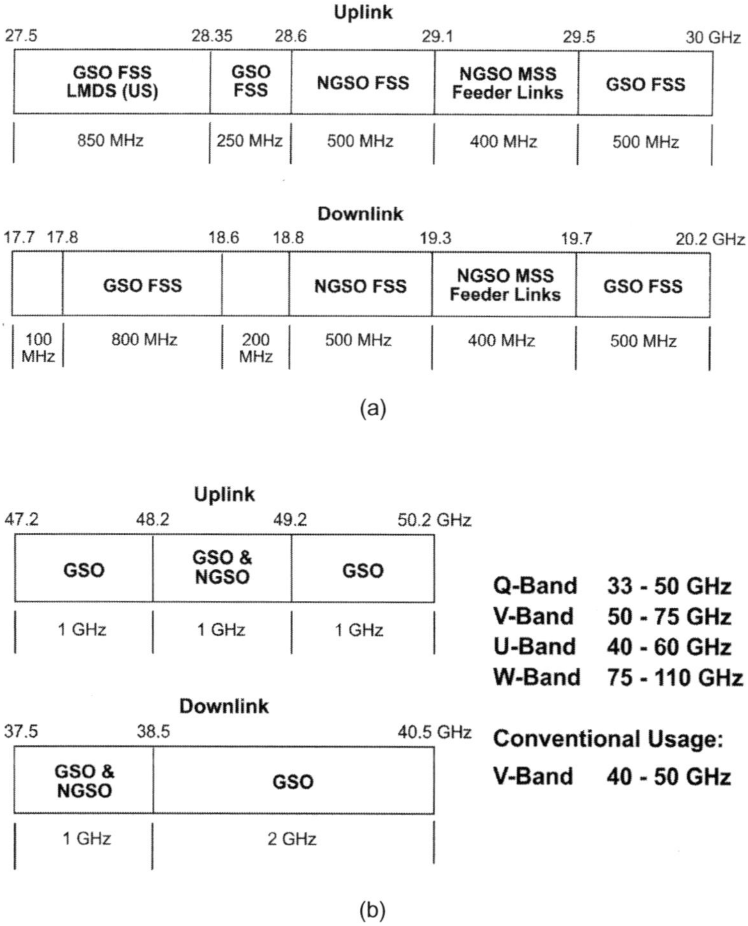

FIGURE 8.5 (a) WARC-97 Ka-band frequency allocations; (b) FCC V-band frequency allocations.

TABLE 8.2 Ka- and V-Band Frequency Allocations

Frequency Band (GHz)	Band Designation	Service
17.8–18.6	Ka	GSO FSS, space-to-Earth
19.7–20.2	Ka	GSO FSS, space-to-Earth
18.8–19.3	Ka	NGSO FSS, space-to-Earth
19.3–19.7	Ka	NGSO MSS feeder links, space-to-Earth
27.5–28.35	Ka	GSO FSS, Earth-to-space (coordination required with LMDS)
28.35–28.6	Ka	GSO FSS, Earth-to-space
29.5–30.0	Ka	GSO FSS, Earth-to-space
28.6–29.1	Ka	NGSO FSS, Earth-to-space
29.1–29.5	Ka	NGSO MSS feeder links, Earth-to-space
38.5–40.5	V	GSO, space-to-Earth
37.5–38.5	V	GSO and NGSO, space-to-Earth
47.2–48.2	V	GSO, Earth-to-space
49.2–50.2	V	GSO, Earth-to-space
48.2–49.2	V	GSO and NGSO, Earth-to-space

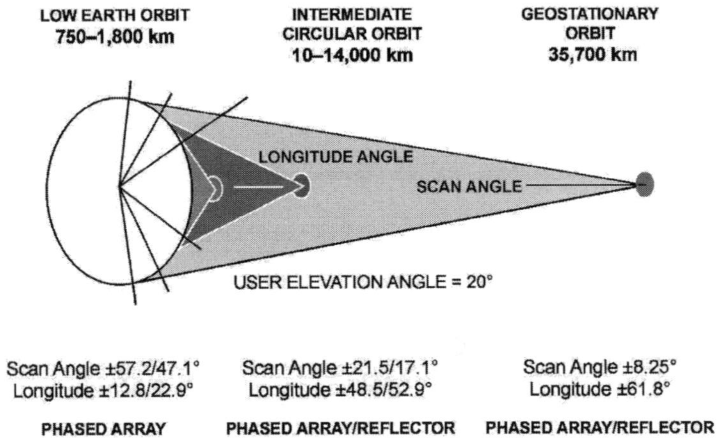

FIGURE 8.6 Relative Earth coverage by satellites in LEO, MEO, and GEO orbits [4].

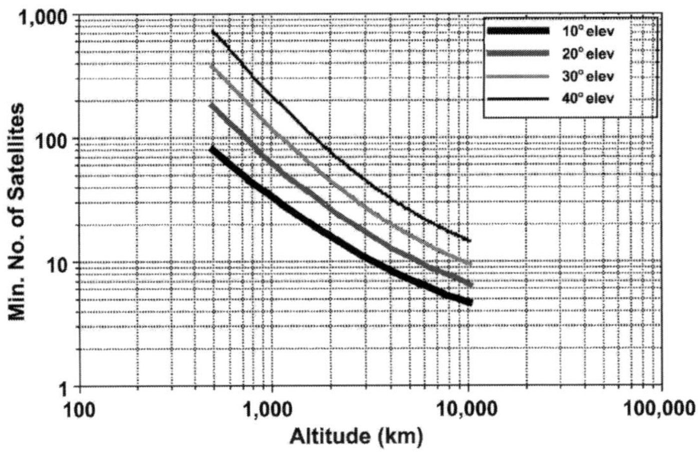

FIGURE 8.7 Number of orbiting satellites vs. orbital height [5].

the number of orbital planes is shown in Fig. 8.8. LEO satellites tend to be smaller and less expensive than MEOs, which are in turn likely to be less expensive than GEO satellites. In terms of space segment implementation, this represents an important system/cost trade-off. For example, the benefit of smaller and less expensive satellites for LEO, as compared to GEO, is offset by the greater number of satellites required.

Another key consideration is the link margin required for the terminal for different satellite orbits. The LEO system offers the advantage of lower satellite power and smaller satellite antennas; however, the user terminals must track the satellites, and an effective handover from satellite to satellite must be designed into the system. To preserve the link margins, the spot size of the beams must be kept small. This requires larger satellite antennas, the further out the satellite is placed. For example, required satellite antenna diameters range from 1 m for LEO systems, to 3 to 4 m for MEO systems, and to 10 to 12 m for GEO systems.

The important characteristics of the three orbital systems are summarized in Table 8.3. System costs, satellite lifetime, and system growth play a significant role in the orbit selection process. On the other hand, round-trip delay, availability, user terminal antenna scanning, and handover are critical to system utility and market acceptance.

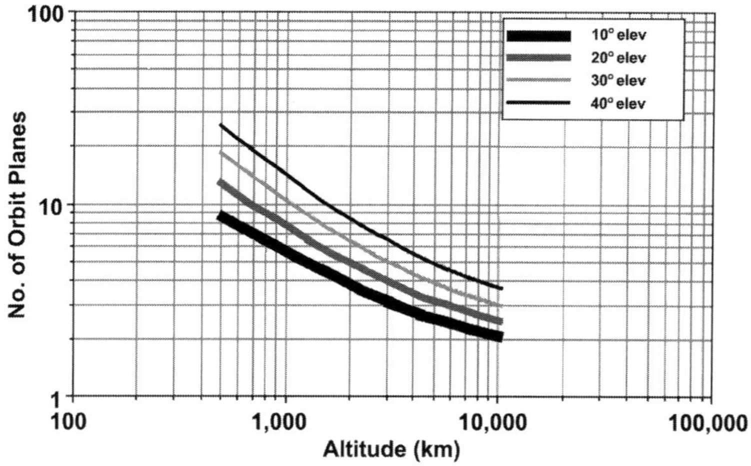

FIGURE 8.8 Number of orbit planes vs. orbital height [5].

TABLE 8.3 LEO, MEO, and GEO Satellite System Characteristics [5]

Characteristic	LEO	MEO	GEO
Space segment cost	Highest	Medium	Lowest
System cost	Highest	Medium	Lowest
Satellite lifetime, years	5–7	10–12	10–15
Terrestrial gateway cost	Highest	Medium	Lowest
Overall system capacity	Highest	Medium	Lowest
Round-trip time delay	Medium	Medium	Longest
Availability/elevation angles	Poor	Best	Restricted
Operational complexity	Complex	Medium	Simplest
Handover rate	Frequent	Infrequent	None
Building penetration	Limited	Limited	Very limited
Wide area connectivity	Intersatellite links	Good	Cable connectivity
Phased start-up	No	Yes	Yes
Development time	Longest	Medium	Shortest
Deployment time	Longest	Medium	Short
Satellite technology	Highest	Medium	Medium

8.2 INTELSAT System Example

The evolutionary trends in INTELSAT communications satellites are shown in Fig. 8.9. This illustration depicts an evolution from single-beam global coverage to multibeam coverages with frequency reuse. The number of transponders has increased from 2 on INTELSAT I to 50 on INTELSAT VI, with a corresponding increase in satellite EIRP from 11.5 to 30 dBW in the 4-GHz band, and in excess of 50 dBW at Ku-band. During the same time frame, Earth station size has decreased from 30 m (Standard A) to 1.2 m (Standard G) for VSAT data services (Fig. 8.10). Several technological innovations have contributed to the increase in the number of active transponders required to satisfy the growing traffic demand [13]. The development of lightweight elliptic function filters resulted in the channelization of allocated frequency spectrum into contiguous transponder channels of 40 and 80 MHz. This channelization provided useful bandwidth of 36 and 72 MHz, respectively, and reduced the number of carriers per transponder and the intermodulation interference generated by the nonlinearity of

The improved design of INTELSAT satellites has yielded increased capacity and reduced costs for service.

INTELSAT DESIGNATION	I	II	III	IV	IV-A	V	V-A	VI
Year of First Launch	1965	1967	1986	1971	1975	1980	1985	1989
Prime Contractor	Hughes	Hughes	TRW	Hughes	Hughes	Ford Aerospace	Ford Aerospace	Hughes
Width Dimensions, m.(Undeployed)	0.7	1.4	1.4	2.4	2.4	2.0	2.0	3.6
Height Dimensions, m.(Undeployed)	0.6	0.7	1.0	5.3	6.8	6.4	6.4	6.4
Launch Vehicles	Thor Delta	Thor Delta	Thor Delta	Atlas Centaur	Atlas Centaur	Atlas Centaur Ariane 1, 2	Atlas Centaur Ariane 1, 2	Ariane 4 or NASA STS (Shuttle)
Design Lifetime, Years	1.5	3	5	7	7	7	7	14
Bandwidth, MHz	50	130	300	500	800	2,144	2,250	3,300
Capacity								
Voice Circuits	240	240	1,500	4,000	6,000	12,000	15,000	120,000
Television Channels	-	-	-	2	2	2	2	3

FIGURE 8.9 INTELSAT Earth station size trend.

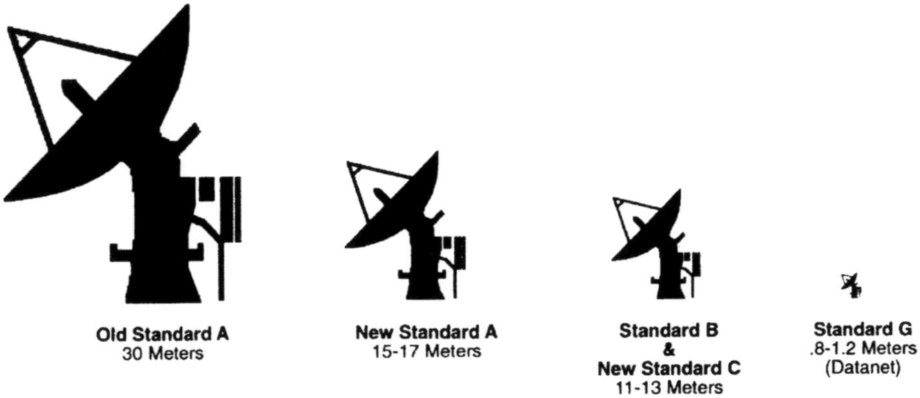

FIGURE 8.10 INTELSAT communications satellite trends. (Courtesy of INTELSAT.)

traveling wave tube amplifiers (TWTAs). For example, using filters and modifying the TWTA redundancy configuration resulted in the provision of twenty 40-MHz transponders in the 6/4-GHz frequency band for INTELSAT IVA satellites, as compared to twelve for INTELSAT IV.

Traffic capacity was further increased with the introduction of frequency reuse through spatial and polarization isolation. In INTELSAT V, for example, 14/11-GHz (Ku) band was introduced and fourfold frequency reuse was achieved at C-band. The use of spatially separated multiple beams also increased antenna gain due to beam shaping, hence increasing the EIRP and gain-to-noise temperature ratio (G/T) for these satellites [14]. This was made possible by significant advances in beamforming and reflector technologies. The increase in G/T on the uplink and the EIRP on the downlink, offer link budget advantages, enabling reductions in Earth terminal power amplifiers for a given antenna size. Consequently, Earth terminal costs could be reduced and terminals could be located closer to customer premises. Transition also occurred from the analog frequency-division multiple access (FDMA) techniques to time-division multiple access (TDMA) transmission using digitally modulated quadrature phase-shift keying (QPSK) signals. The multibeam satellite systems require onboard switch matrixes to provide connectivity among the isolated beams. In the INTELSAT V spacecraft, these interconnections were established by using electromechanical static switches that could be changed by ground command. One disadvantage of static interconnections is the inefficient use of satellite transponder capacity when there is not enough traffic to fill the capacity for the selected routing. In the INTELSAT VI spacecraft, satellite utilization efficiency was enhanced by providing cyclic and dynamic interconnection among six isolated beams using redundant microwave switch matrixes (MSMs) [15]. Satellite-switched TDMA (SS-TDMA) operation provided dynamic interconnections, allowing an Earth station in one beam to access Earth stations in all six beams in a cyclic manner in each TDMA frame [16]. Although INTELSAT VI MSMs were realized using hybrid MIC technology, use of GaAs monolithic microwave integrated circuit (MMIC) technology was demonstrated for such systems because it offered reproducibility, performance uniformity, and high reliability [17].

Improvements in quality of service and satellite operational flexibility can be achieved by using onboard regeneration and signal processing, which offer additional link budget advantages and improvements in bit error ratio performance through separation of additive uplink and downlink noise. Use of reconfigurable, narrow, high-gain pencil beams with phased-array antennas offers the additional flexibility of dynamic transfer of satellite resources (bandwidth and EIRP). Since these active phased-array antennas require several identical elements of high reliability, MMIC technology makes them feasible [18, 19]. These technological developments in microwave and antenna systems, which have helped improve satellite capacity per kilogram of dry mass and reduced cost (Fig. 8.11), have positioned satellite systems to launch even higher capacity broadband satellites. Advances in digital device technologies have lowered the cost of TDMA terminals and made them relatively easy to maintain.

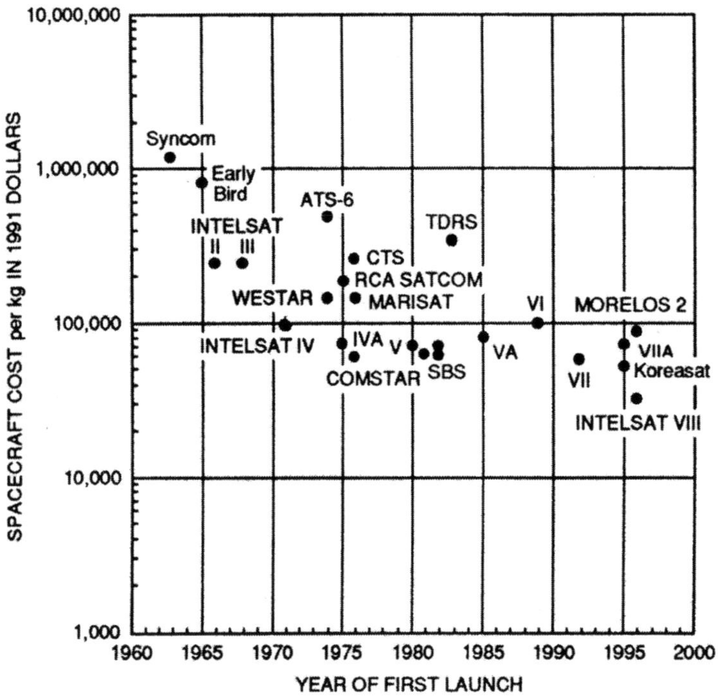

FIGURE 8.11 Communications satellite cost per kilogram. (Courtesy of Comsat.)

8.3 Broadband and Multimedia Satellite Systems

With the introduction of new services, the public switched telephone network (PSTN) has evolved toward Integrated Services Data Networks (ISDNs) and asynchronous transfer mode (ATM) [20]. ATM offers a suite of protocols that are well suited to handling a mix of voice, high-speed data, and video information, making it very attractive for multimedia applications. One of the unique virtues of satellite networks is that the satellite offers a shared bandwidth resource, which is available to many users spread over a large geographic area on Earth. This forms the basis for the concept of bandwidth-on-demand, in which terminals communicate with all other terminals (full-mesh connectivity), but use satellite capacity on an as-needed basis. By using multifrequency TDMA to achieve high-efficiency management and flexibility, commercial multiservice networks are now implemented using multi-carrier, multi-rate, TDMA, bandwidth-on-demand products such as Comsat Laboratories' LINKWAY™ 2000 or 2100 mesh networking platforms. Each of these platforms is capable of providing flexible data rates (up to 2 Mb/s) and can provide ATM, Frame Relay, ISDN, SS7, and IP interfaces [21].

NASA's Advanced Communications Technology Satellite (ACTS), which was launched in September 1993, demonstrated new system concepts, including the use of Ka-band spectrum, narrow spot beams (beamwidths between 0.25° and 0.75°), independently steerable hopping beams for up- and downlinks, a wide dynamic range MSM, and an onboard baseband processor. In addition to these technology developments, several experiments and demonstrations have been performed using ACTS system. These include demonstration of ISDN, ATM, and IP using Ka- band VSATs; high-definition video broadcasts, health care, and long-distance education; and several propagation experiments planned through September 2000. The success of the ACTS program, together with the development of key onboard technologies, a better understanding of high-frequency atmospheric effects, and the availability of higher frequency spectrum has resulted in proposals for a number of Ka- and V-band satellite systems [5, 6] for future multimedia services (Fig. 8.12).

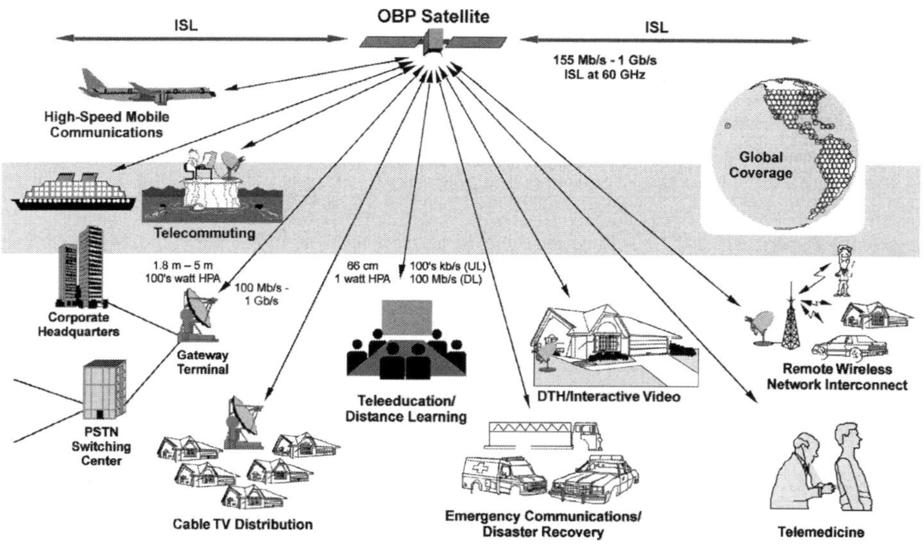

FIGURE 8.12 Multimedia system concept. (Courtesy of Comsat.)

8.3.1 Proposed Ka-Band Systems

The 14 proposed Ka-band systems in the U.S. are Astrolink, Cyberstar, Echostar, Galaxy/Spaceway, GE*Star, KaStar, Millennium, Morning Star, NetSat 28, Orion (F7, F8, and F9), PanAmSat (PAS-10 and PAS-11), Teledesic, VisionStar, and VoiceSpan. Table 8.4 gives an overview of these systems. Out of these original filings, some systems have been combined and others, for example, VoiceSpan, or have been withdrawn. The total number of satellites proposed for GEO systems is 69, in addition to 288 satellites proposed by Teledesic in an LEO constellation.

It should be noted, however, that development of these systems is contingent on overcoming significant challenges in terms of frequency coordination, technology readiness, and financing. Considering the current status and activities of the proponents, only a few of the proposed systems are expected to be deployed in the 2002 to 2005 time frame.

TABLE 8.4 Proposed Ka-band Satellite Communications Systems [6]

System	Orbit	Coverage	No. of Satellites	Capacity (Gb/s)	Intersatellite Link	Onboard Switching	Capital Investment ($B)
Astrolink	GEO	Global	9	9.6	1 Gb/s	FPS	4.00
Cyberstar	GEO	Limited Global	3	4.9	1 Gb/s	BBS	1.05
Echostar	GEO	U.S.	2	5.8	120 MHz	BBS	0.34
Galaxy/Spaceway	GEO	Global	20	4.4	1 Gb/s	BBS	5.10
GE*Star	GEO	Limited Global	9	4.7	None	None	4.00
KaStar	GEO	U.S.	2	7.5	155 Mb/s	FPS	0.65
Millennium[a]	GEO	U.S./Americas	4	5.2	1 Gb/s	FPS	2.30
Morning Star	GEO	Limited Global	4	0.5	None	None	0.82
NetSat 28	GEO	CONUS	1	772.0	None	Optical SW	0.25
Orion	GEO	U.S./IOR	3	2.9	TBD	FPS	0.73
PanAmSat	GEO	AOR	2	1.2	None	None	0.41
Teledesic/Celestri	LEO	Global	288[a]	13.3	1 Gb/s	FPS	9.00
VisionStar	GEO	CONUS	1	1.9	None	None	0.21
VoiceSpan[b]	GEO	Limited Global	12	5.9	0.5 Gb/s	FPS	N/A

Note: FPS: fast packet switch, BBS: baseband switch.

[a] Subject to revision as a result of Teledesic/Celestri combination.

[b] Withdrawn in 1997.

8.3.2 Proposed V-Band Systems

In September 1996, Motorola Satellite Systems, Inc. filed an application with the FCC requesting authorization to deploy a system of 72 LEO satellites (M-Star) which would operate in V-band and provide multimedia and other services. Motorola's application was followed by 15 other applications for V-band satellite systems, filed by 13 U.S. companies in 1997. Hughes Communications, Inc. alone filed for three of these systems (four including a PanAmSat filing). The 1997 filings were in response to a deadline of September 30, 1997, set by the FCC for such applications. A majority of the V-band applications were filed by the same companies that had filed for multimedia Ka-band systems in 1995. Generally, the V-band systems, filed by those who had already planned to operate at Ka-band are intended to supplement the capacity of the Ka-band systems, and especially to provide service at higher data rates to regions of high capacity demand.

The 16 proposed V-band systems are Aster (Spectrum Astro, Inc.), CAI (CAI Satellite Communications, Inc.), CyberPath (Loral Space and Communications), Expressway, SpaceCast, and StarLynx (Hughes Communications), GESN (TRW), GE*StarPlus (GE Americom), GS-40 (Globalstar), Leo One USA (Leo One USA Corp.), M-Star (Motorola), OrbLink (Orbital Sciences), Pentriad (Denali Telecom), Q/V-Band (Lockheed Martin), V-Band Supplement (VBS) (Teledesic Corp.), and V-Stream (PanAmSat). Table 8.5 lists some of the characteristics of these systems [6].

8.3.3 Key Technologies

The majority of the proposed systems employ either FDMA/TDMA or TDMA transmission. Uplink bit rates vary from 32 kb/s to 10 Mb/s for a typical user. Most systems propose to employ advanced technologies — in particular digital signal processing and onboard switching — owing to the successful technology demonstration provided by the ACTS program. Demonstrated ACTS technologies include hopping beams, onboard demodulation/remodulation, onboard FEC decoding/coding, baseband switching, adaptive fade control by FEC coding, and rate reduction. However, some of the proposed systems

TABLE 8.5 Proposed V-band Satellite Communications Systems [6]

System	Orbit	Coverage	No. of Satellites	Capacity (Gb/s)	ISL Capacity	Onboard Switching	Capital Investment ($B)
Aster	GEO	Global	25	6.2	Optical	SSTDMA & BBS	2.4
CAI	GEO	CONUS	1	1.86	None	Bent-pipe	0.3
CyberPath	GEO	Global	10	17.9	447.5 Mb/s	FPS	1.2[a]
Expressway	GEO	Global	14	65.0	Optical	SSTDMA (BB)	3.8
GESN	GEO & MEO	Global	4 & 15	50.0 & 75.0	Optical 2.5 Gb/s	BBS	3.4
GE[a]StarPlus	GEO	Global	11	65.0	Optical	Bent pipe	3.4
GS-40	LEO	Global	80	1.0	None	Bent pipe	Not avail.
Leo One USA	LEO	Global	48	0.007	None	BBS	0.03
M-Star	LEO	Global	72	3.7	830 Mb/s	Bent-pipe	6.4
OrbLink	MEO	Global	7	75.0	1.244 Gb/s	Bent-pipe	0.9
Pentriad	HEO	Limited global	9	30.2	None	Bent-pipe	1.9
Q/V-Band	GEO	Global	9	31.3	Optical & radio	FPS	4.8
SpaceCast	GEO	Limited global	6	60.0	Optical 3 Gb/s	SSTDMA (RF)	1.7
StarLynx	GEO & MEO	Global	4 & 20	5.9 & 6.3	Optical 3 Gb/s	BBS	2.9
VBS	LEO	Global	72	8.0	Optical 1 Gb/s	FPS	1.9
V-Stream	GEO	Global	12	32.0	1 GHz	Bent-pipe	3.5

Note: HEO: highly elliptical orbit, FPS: fast packet switching, BBS: baseband switching.

[a] For only four satellites.

are vastly more complex than ACTS (a system capacity of 220 Mb/s in ACTS vs. 10 Gb/s in the proposed systems, for example) and also employ new processing/switching technologies such as multicarrier demultiplexer/demodulators for several thousand carriers, fast packet switching (FPS), and intersatellite links (ISLs). In addition, network control functions traditionally performed by a ground control center will be partially implemented by an onboard network controller. Three of the key technology areas that influence payload and terminal design — onboard processing, multibeam antennas, and propagation effects — are discussed in the subsections that follow.

8.3.4 Onboard Processing

The majority of the proposed Ka- and V-band systems employ onboard baseband processing/switching [4], although some will use "bent-pipe" or SS-TDMA operation onboard the satellites. A majority of the processing systems will employ fast packet switching, which is also referred to as cell switching, packet switching, ATM switching, and packet-by-packet routing in the FCC filings. The remainder of the processing systems are currently undecided regarding their baseband switching (BBS) mechanisms and will probably use either FPS or circuit switching. Along with onboard baseband switching, most of the processing satellites will employ digital multicarrier demultiplexing and demodulation (MCDD). Onboard baseband switching allows optimized transmission link design based on user traffic volume, and flexible interconnection of all users in the network at the satellite. ISLs will provide user-to-user connection in many of these systems without assistance from ground stations.

Gallium arsenide (GaAs) monolithic microwave integrated circuit (MMIC) technology has been used successfully to develop microwave switch matrix (MSM) arrays for SS-TDMA operation and RF demodulator/remodulator hardware. Further development of low-power, application-specific integrated circuits (ASICs) with high integration densities and radiation tolerance is critical to the realization of relatively complex onboard processing and control functions.

8.3.5 Multibeam Antennas

The design of the satellite antennas depends on the beam definition, which in turn is a function of system capacity and projected traffic patterns. Several systems require a large number of fixed, narrow spot beams covering the whole service area and designed to deliver a high satellite EIRP of 50 to 60 dBW to user terminals. A single reflector with a large number of feeds may provide such coverage. However, if scanning loss is excessive due to the large number of beams scanned in one direction, multiple reflectors may be required. The coverage area is divided into a number of smaller areas, with each reflector boresight at the center of the corresponding area. Similarly, single or multiple phased arrays may be used. The phased array may have lower scan loss and, with the flexibility of digital beam formers, a single phased array can handle a large number of beams with no added complexity. If the system design calls for a small number of hopping beams instead of large number of fixed beams, the phased array solution becomes more attractive due to the flexibility and reliability of the beam-former vs. the switching arrangement in a focal-region-fed reflector antenna. For a small number of beams at a time, the microwave beam-former becomes a viable alternative and the choice between the microwave and the digital beam-formers becomes a payload system issue.

At Ka- and V-band, the manufacturing tolerance of the array feed elements and the surface tolerance of the reflectors play an important role in overall antenna performance. Waveguide-based elements are well developed at Ka-band, but lighter weight, lower profile printed circuit elements may need further development for space applications. For a large number of beams and a large number of frequency reuses, co- and cross-polarization interference becomes a major issue that imposes severe restrictions on the antenna sidelobe and cross-polarization isolations. Although the receive antenna may benefit from statistical averaging based on user distribution, the transmit antenna must satisfy a certain envelope in order to meet the required interference specifications.

8.3.6 Propagation Effects

Line-of-sight, rain attenuation, and atmospheric propagation effects are not significant at L-, S- and C-bands. At high elevation angles the communications between satellites and terminals at L- and S-bands is very reliable. In the mobile environment, however, multipath effects and signal blockages by buildings cause signal fades that require cooperative users willing to change their location. In addition, the links must be designed at worst-case user elevation and for operation at the beam edge. Because of these considerations, a 10- to 15-dB link margin is designed into the systems.

In comparison, the troposphere can produce significant signal impairments at the Ku-, Ka- and V-band frequencies, especially at lower elevation angles, thus limiting system availability and performance [7, 22]. Most systems are expected to operate at elevation angles above about 20°. Tropospheric radio wave propagation factors that influence satellite links include gaseous absorption, cloud attenuation, melting layer attenuation, rain attenuation, rain and ice depolarization, and tropospheric scintillation. Gaseous absorption and cloud attenuation determine the clear-sky performance of the system. Clouds are present for a large fraction of an average year, and gaseous absorption varies with the temperature and relative humidity. Rain attenuation — and to some extent melting layer attenuation — determine the availability of the system. Typical rain time is on the order of 5 to 10% of an average year. Depolarization produced by rain and ice particles must be factored into the co-channel interference budgets in frequency reuse systems. Signal amplitude fluctuations due to refractive index inhomogeneities (known as *tropospheric scintillation*) are most noticeable under hot, humid conditions at low elevation angles. Scintillation effects must be taken into account in establishing clear-sky margins and in designing uplink power control systems. Figure 8.13 shows the combined clear-sky attenuation distribution at Ka- and V-band frequencies for a site in the Washington, D.C. area at an elevation angle of 20°. Figure 8.14 shows the rain attenuation distribution for the same site. It can be seen that the required clear-sky margin can be several dBs at V-band, especially at 50 GHz, due to the elevated oxygen absorption. Figure 8.14 indicates that, to achieve reasonable availabilities, rain fade mitigation must be an integral part of the system design. Rain fade mitigation can be accomplished through power control, diversity, adaptive coding, and data rate reduction.

8.3.7 User Terminal Characteristics

A variety of user terminals have been proposed for the Ka- and V-band systems. A typical user terminal operates at an uplink bit rate between 128 kb/s and 1 Mb/s for Ka-band systems, and up to 10 Mb/s for V-band systems. These terminals employ small-aperture antennas with diameters of 50 to 66 cm (Ka-band), and as small as 20 cm for V-band, as well as a solid-state power amplifier (SSPA) of 1 to

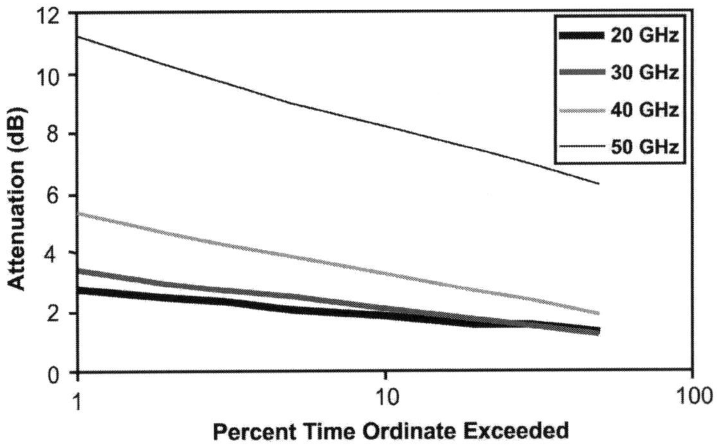

FIGURE 8.13 Probability distribution of clear-sky attenuation at 20° elevation, Washington, D.C. [22].

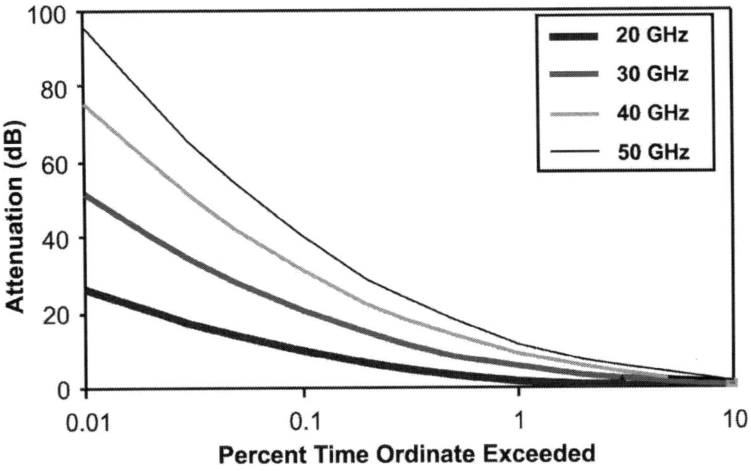

FIGURE 8.14 Probability distribution of rain attenuation at 20° elevation, Washington, D.C. [22].

10 W. Uplink power control is required for all of these terminals. Terminal antennas for LEO and MEO systems must provide tracking capability to perform handovers every few minutes. All RF components (SSPA, LNA, and up-downconverters) are integrated into a small outdoor unit (ODU) and antenna system. Development of low-cost antennas and low-cost, low-power consuming RF integrated circuits (RFICs) at Ka- and V-bands is critical to substantially reducing the cost of the terminals to targets of $1,000 to $3,000. Gateway terminals employ a larger antenna with a diameter of 2.4 to 7 m and an HPA of 50 to 200 W.

8.4 Summary

The overall telecommunications market is growing very rapidly, largely because of increased demand for video traffic and exponential growth in the Internet. Multimedia services using satellites are now beginning to emerge. Growth in international trade, reduced prices due to privatization of telecommunications services worldwide, access to the World Wide Web, and significant drops in prices of desktop and portable computers are all contributing to a heavy demand for such services. Satellites also provide "instant infrastructure" and are therefore viewed as a cost-effective solution for providing wide area coverage for developing countries. In the developed world, satellites could provide effective "last mile" connection to businesses and homes for broadband data.

A number of Ka- and V-band systems have been proposed, and many of Ka-band systems are at advanced stages of implementation. Many of these systems require capital investment of several billions of dollars to implement. As these systems become operational in the 2003 to 2010 time frame, several technological and market challenges must be overcome. Perhaps the most important challenge is to develop a low-cost ($1,000 to $3,000) customer premises terminal. Low-cost RF integrated circuits, multichip packaging techniques, and low-cost Ka-band antennas must be developed to meet these cost goals. While significant progress continues to be reported in Ka-band component technology, the use of V-band presents additional challenges.

Acknowledgments

The author gratefully acknowledges the contributions of many of his colleagues at Comsat Laboratories to the work described herein. In particular, the author thanks P. Chitre, A. Dissanayake, J. Evans, T. Inukai, and A. Zaghloul of Comsat Laboratories for many useful discussions.

References

1. Evans, J., Network interoperability meets multimedia, *Satellite Commun.*, 30–36, February 2000.
2. Pontano, B., Satellite communications: services, systems, and technologies. *1998 IEEE MTT-S International Microwave Symposium Digest*, June 1998, 1–4.
3. Evans, J., Satellite and personal communications, 15th AIAA International Communications Satellite Systems Conference, San Diego, CA, Feb./Mar. 1994, 1013–1024.
4. Evans, J., Satellites systems for personal communications, *IEEE Antenna Propagation Mag.*, 39, 7–10, June 1997.
5. Williams, A., Gupta, R., and Zaghloul, A., Evolution of personal handheld satellite communications, *Applied Microwave Wireless*, 72–83, Summer 1996.
6. Evans, J., The U.S. filings for multimedia satellites: a review, *Int. J. Satellite Commun.*, 18, 121–160, 2000.
7. Evans, J. and Dissanayake, A,, The prospectus for commercial satellite services at Q- and V-band, *Space Commun.*, 15, 1–19, 1998.
8. Bennett, S. and Braverman, D., INTELSAT VI — A continuing evolution, *Proc. IEEE,* 72, 11, 1457–1468, November 1984.
9. Wong, N., INTELSAT VI — A 50-channel communication satellite with SS-TDMA, *Satellite Communications Conference,* Ottawa, Canada, June 1983, 20.1.1–20.1.9.
10. Gallagher, B., ed., *Never Beyond Reach — The World of Mobile Satellite Communications,* International Maritime Satellite Organization, London, U.K., 1989.
11. Reinhart, E. and Taylor, R., Mobile communications and space communications, *IEEE Spectrum,* 27–29, February 1992.
12. Evans, J., Satellite and personal communications, *15th AIAA International Communications Satellite Systems Conference,* San Diego, CA, Feb.–Mar. 1994, 1013–1024.
13. Gupta, R. and Assal, F., Transponder RF Technologies Using MMICs for Communications Satellites, *14th AIAA International Communications Satellite Systems Conference,* Washington, D.C., March 1992, 446–454.
14. Sorbello, R. et al., A Ku-Band Multibeam Active Phased Array for Satellite Communications, *14th AIAA International Communications Satellite Systems Conference,* Washington, D.C., March 1992.
15. Assal, F., Gupta, R., Betaharon, K., Zaghloul, A., and Apple, J., A wideband satellite microwave switch matrix for SS-TDMA communications, *IEEE J. Selected Areas Comm.*, SAC-1(1), 223–231, Jan. 1983.
16. Gupta, R., Narayanan, J., Nakamura, A., Assal, F., and Gibson, B., INTELSAT VI On-board SS-TDMA subsystem design and performance, *Comsat Technical Review,* 21, 1, 191–225, Spring 1991.
17. Gupta, R., Assal, F., and Hampsch, T., A microwave switch matrix using MMICs for satellite applications, IEEE MTT-S International Microwave Symposium, Dallas, TX, May 1990, *Digest,* 885–888.
18. Gupta, R. et al., Beam-forming matrix design using MMICs for a multibeam phased-array antenna, IEEE GaAs IC Symposium, October 1991.
19. Gupta, R., MMIC insertion in communications satellite payloads, Workshop J: GaAs MMIC System Insertion and Multifunction Chip Design, IEEE MTT-S International Microwave Symposium, Boston, MA, June 1991.
20. Chitre, D., Gokhale, D., Henderson, T., Lunsford, J., and Matthews, N., Asynchronous Transfer Mode (ATM) Operation via Satellites: Issues, Challenges, and Resolutions, *Int. J. Satellite Commun.*, 12, 211–222, 1994.
21. Chitre, D., Interoperability of satellite and terrestrial networks, IEEE Sarnoff Symposium, April 2000.
22. Dissanayake, A., Allnutt, J., and Haidr, F., A prediction model that combines rain attenuation and other propagation impairments along the Earth-satellite path, *IEEE Trans. Antennas and Propagation*, 45, 10, October 1997.

9

Satellite-Based Cellular Communications

Nils V. Jespersen
BAE Systems

Communication satellite systems designed to serve the mobile user community have long held the promise of extending familiar handheld cellular communication to anywhere a traveler might find himself. One impetus to the fulfillment of this dream has been the success of the Inmarsat system of communication satellites. Founded in 1979 as an international consortium of signatories, Inmarsat provides worldwide communication services to portable and transportable terminals, thereby meeting one of its mandates by enhancing safety on the high seas and other remote areas. Although it does not support handheld, cellular-like operations (due to limitations in the satellite design), the current Inmarsat system is a successful business. Clearly, the next logical step would be to enhance the capabilities of the space segment to provide cell-phone utility with even greater ubiquity than available terrestrially. What we would then accomplish is to, essentially, raise the familiar cellular base station several hundreds of kilometers high and, thereby, extend the coverage many times over (Fig. 9.1).

Many elements comprise the optimum solution to a satellite cellular system, not the least of which is the business aspect. The best technical solution does not necessarily result in a successful overall solution. A significant market consolidation is in process at the time of this writing.

In this discussion, we will focus on the systems that are intended to provide voice and/or data services similar to terrestrial cellular communications. As such, we will not be discussing systems intended to provide low-rate messaging and asset tracking services (the so-called "little-LEO" systems such as Orbcomm and LEO-One). Of the many mobile satellite systems proposed in 1995 to provide cellular-like service, only a handful remain as viable. We will examine the salient characteristics of satellite cellular system design and, then consider some specific system examples.

9.1 Driving Factors

How a particular cellular satellite system is configured will depend on a variety of factors along with the importance assigned to each characteristic. In this section we will explore some of these top-level design

0-8493-1732-0/03/$0.00+$1.50

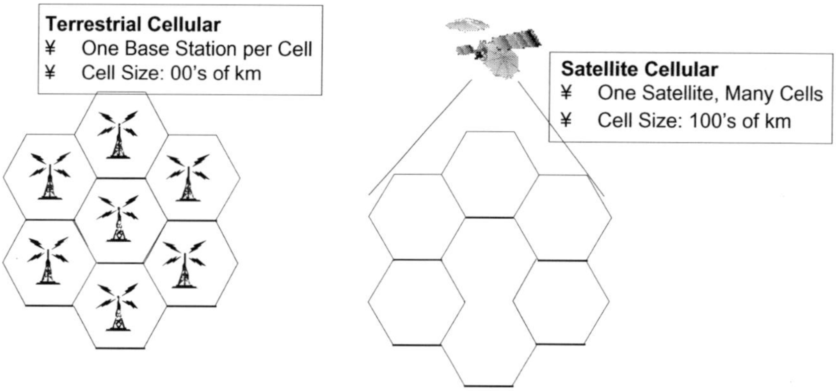

FIGURE 9.1 Comparison: terrestrial vs. satellite cellular.

drivers and consider their impact on the overall system design. While this listing is far from exhaustive, the parameters discussed do tend to be common considerations in every cellular satellite system design.

9.1.1 The User Terminal (UT)

The baseline assumption in this discussion is that cellular satellite system users interface with the system using "disadvantaged" terminals. By this terminology we mean that the UT is "disadvantaged" from an electronic design standpoint, implying that the terminal is small, lightweight, and convenient to carry on the user's person. This convenience in the UT is achieved at the cost of both antenna gain and transmit power (meaning low gain-to-system-temperature [G/T] and low effective isotropic radiated power [EIRP]). The UT G/T must be high enough to ensure acceptable, low-noise reception of the signal from the satellite. Similarly, the UT EIRP must be high enough to ensure that the satellite will be able to reliably process and translate the signal from the UT. We refer to this basic design consideration as "closing the link." Far from being a "nice to have" characteristic, in the commercial world of handheld wireless communications, UT convenience plays a *significant* role in customer acceptance; this fact is borne out within the Iridium cellular satellite system (discussed below). A small UT, in turn, makes the cellular satellite system engineer's job harder in that the performance burden now shifts to the space segment. The spacecraft, now, has to have both higher EIRP and G/T in order that the subscriber can have a small, sleek terminal that easily fits in his pocket. The impact on the system is, therefore, more complexity and cost.

9.2 Target Market

Arguably, the most important parameter to identify is the market that the cellular satellite system is targeted to serve. Military needs tend to be very different from those typically considered for a commercial system. For instance, guaranteed availability, reliable communication through multiple layers of jungle canopy, and nuclear event hardening and probability-of-intercept, which are often found in the requirements levied by the Department of Defense (DoD), are rarely, if ever, discussed in reference to commercial systems. Another consideration is whether the system is intended to serve a worldwide user group (including, perhaps, coverage of polar regions and open oceans), or the users are located in a general geographic region (e.g., Europe, Asia, specific landmasses). The state of terrestrial communication infrastructures in the target market — whether wired or wireless — must also be identified and evaluated.

These market considerations tend to drive several high-profile system design decisions. A DoD system, with a need to penetrate jungle canopy, would drive the selection of a particular operating frequency band to lower frequencies (e.g., UHF). Satellite orbit type, for instance, would need to be of either the low Earth or medium Earth orbiting type (LEO or MEO) if worldwide coverage is required. Regional coverage, on the other hand, could be met with a satellite at geosynchronous Earth orbit (GEO). The existence of adjacent terrestrial cellular coverage might influence the selection of a particular radio air interface structure (AIS). For instance, if the satellite cellular coverage area includes population centers operating on a particular AIS (Global System for Mobile communication, or GSM, for instance), then it might be prudent to base the satellite AIS on something similar in order to simplify UT design and, also facilitate roaming arrangements.

9.2.1 Expected Subscriber Population

An objective, well-researched estimate of the targeted subscriber population is an essential element in the viability of a cellular satellite system. Considerations include:

- Expected total number of subscribers
- Expected peak number of users in view of a given satellite compared to the expected average user loading
- Geographic distribution of the user population: whether they are evenly dispersed over the defined coverage area or concentrated in specific locations (population centers)

The total number of subscribers drives design elements such as the capacity of the overall system authentication registers: the so-called home location registers (HLR) and the visitor location registers (VLR). On the other hand, the number of active users in the satellite field-of-view has a direct impact on the satellite architecture). Generally, the satellite downlink tends to be the limiting constraint (i.e., the number of users that can be accommodated is based on the amount of available power in the downlink transmitter), since each downlink user signal receives a share of the total transmit power. In systems based on frequency division multiple access (FDMA), or a combination FDMA/TDMA (Time Division Multiple Access), it is important that the transmit system is operated in the linear region of the transmitter gain characteristic. As more carriers are added to the composite downlink signal (assuming equal power per carrier as necessary to close the link to the individual UT), the transmitter is driven closer to, and sometimes into, the nonlinear operating region. A transmitter driven with a multicarrier signal, at a nonlinear operating point, will generate intermodulation distortion (IMD) products due to the mixing of the individual carriers with one another (a classical heterodyning phenomenon). In a system transmitting many (e.g., hundreds) of carriers, particularly when the carriers have modulation, the IMD tends to take on a random characteristic and tends to appear as a uniform noise spectrum that is band limited by the output filter. The resulting effect is an increase in the transmitted noise of the system. This noise is characterized by the Noise Power Ratio (NPR) of the system and has a direct bearing on the downlink carrier-to-noise (C/N) ratio. System designs often include some type of automatic power control adjustment in order to maximize the number of downlink user signals that can be supported while maintaining the NPR at an acceptable level. Such a power control loop generally involves a measurement of received power level at the individual UT, which is then reported back to the Network Control Center (NCC) by way of a parallel control channel.

Assuming that the satellite transmitter can accommodate the expected peak traffic load, the next limitation becomes the switching capability within the satellite channelization equipment. The required switching capacity will generally dictate whether this channelization equipment is implemented in an analog or digital fashion. Lower capacity systems, such as the Inmarsat-3, are well served by conventional analog channelization and switching technology. By contrast, digital channelization and switching technology is required in modern cellular satellite systems that serve thousands of simultaneous users and have frequent associated call setups and teardowns.

Distribution of users in the satellite field of view drives the design of the antenna (cell coverage) pattern. In satellite cellular systems, the bulk of which operate in the crowded L-band or S-band spectra, efficiency of spectrum use is a critical concern. Operators of a proposed cellular satellite system must apply for an allocation of the spectrum pool, usually as a result of common meetings of the World Radio Conference (WRC) wherein many operators negotiate for spectrum rights. Since, inevitably, granted allocations are less than that requested, a major motivation exists to design the proposed system with a high level of frequency reuse. Frequency reuse implies that multiple cells (beams) in the satellite system can use and reuse the same frequency but with different user signals modulated onto them. Typical reuse values, for regional systems like ACeS and Thuraya, are on the order of 20 times. Thus, the same carrier frequency can be reused 20 or more times over the satellite field of view, thereby increasing the available capacity (the number of UTs that can be simultaneously served) by that same amount. By the same token, the link from the satellite to the Feeder, or Gateway, station must have an adequate spectrum allocation to accommodate the required terrestrial connectivity for the mobile UTs. Since Satellite-to-Feeder links are most often implemented in a higher frequency band (e.g., C-, Ku-, or Ka-band), wider operating bandwidths are generally more easily obtained and coordinated.

An operational scenario where many potentially active users are located in a concentrated area (population center), can put a strain on the amount of available spectrum. This concentration can lead to the need for smaller cell sizes (narrower beam patterns), which, in turn, drives the size of the satellite antenna (making it larger and more difficult to accommodate on the spacecraft). Given a specific coverage area, or satellite field of view, smaller cells mean more cells and, therefore, more switching hardware on the spacecraft. This increase in hardware means that the spacecraft becomes more complex and costly.

The users through busy signals, if reached, will negatively observe system capacity limits. How frequently a user encounters a busy signal is referred to as "call blocking rate" and is one measure of the system's quality of service (QoS). The Quality of Service Forum[1] defines QoS as "... consistent, predictable telecommunication service." Predictable service is what the subscriber demands and is, ultimately, the measure that will determine the success of the system. The elements that need to be assessed (estimated) for a planned cellular satellite system, in order to gain an estimate of required system capacity, include:

- The number of call attempts per second
- The average call holding time
- The expected distribution of call types
  - Percentage of Mobile Originated (MO) calls
  - Percentage of Mobile Terminated (MT) calls
  - Percentage of Mobile-to-Mobile (MM) calls

All of these factors impact system design decisions regarding:

- Satellite switching speed
- Satellite switching capacity
- Onboard buffering and memory capacity

9.2.2 Operating Environment

For a commercial system, putting aside political issues such as frequency coordination for the moment, we would want to select an operating frequency band that can penetrate at least *some* buildings. Additionally, the operating band should permit acceptable performance to be obtained with nondirectional antennas on the UT (i.e., a tolerably low space spreading loss with reasonably sized electrically small antennas). These constraints tend to drive toward lower frequency bands (e.g., L-band or S-band, which are roughly 1500 and 2500 MHz, respectively).

9.2.2.1 Link Margin Considerations

During the system design phase, the radio link between the satellite and the UT has to be given a great deal of detailed consideration. Both the typical and disadvantaged user conditions need to be handled. For instance, the typical case might find the user with a clear line of sight to the satellite. On the other hand, a disadvantaged condition might be where the user is in the middle of an office building, in a city (a "concrete jungle" where multipath propagation could be an issue) and, perhaps, at the edge of the satellite's coverage area (low elevation angle, greatest path distance to the satellite and, therefore, greatest signal loss). Further, meteorological conditions such as rain and, to a lesser extent, snow will degrade the link performance in proportion to both the rate of precipitation and the frequency of operation. Operation in a tropical area would clearly be more affected by rain attenuation than if the system were intended to serve, say, Northern Africa. In all of this, link margin is "king." If we take the traditional approach to assessing the amount of available link margin in the system, we can define link margin (non-rigorously, and in decibel terms) as

$Link\ Margin = (C/N)_{received} - (C/N)_{required}$, where:

- $(C/N)_{received}$ is the ratio of the power of the transmitted signal (at the receiver) to the total noise power (e.g., thermal noise, interference, and other degradations) impinging on the receiver. In the example of the link going to the UT, the satellite is the transmitter and the UT is the receiver. Mobile service providers will often refer to this transmission direction as the "Forward" direction. Transmission *from* the UT *to* the satellite is, likewise, called the "Return" transmission.
- $(C/N)_{required}$ is the minimum ratio of received signal to total noise required at the receiver in order to accomplish acceptable detection with the modulation method selected.

We can gain additional insight into what this equation means by breaking $(C/N)_{received}$ into its constituent components. Thus,

$Link\ Margin = \{EIRP - L_{impairments} + (G/T)\} - (C/N)_{required}$ where:

- *EIRP* is the Effective Isotropic Radiated Power of the transmitter (e.g., the satellite in the case of the link from the satellite to the UT).
- *(G/T)* is the figure of merit often applied to satellite receiving stations, in this case the UT. (G/T) is the ratio of the passive antenna gain (G), at the receiving station in the direction of the incoming signal, to the total system noise of the station expressed as an equivalent temperature (T). The primary component of the system noise is, usually, the noise added by the passive components (following the antenna) and the noise figure of the first amplifier, or low-noise amplifier (LNA), in the receiver. Additionally, the noise contributed by the subsequent components in the UT receiving chain is referred back to the antenna terminal (suitably scaled by the gains of the components ahead in the signal flow).
- $L_{impairments}$ is the combination of losses and effects on the transmission channel that tend to degrade the overall received signal quality.

In other words, link margin can be thought of as the excess desired signal power available at the receiver once we have accounted for the signal strength required by the demodulator in the receiver (commensurate with the chosen modulation format), the inevitable thermal noise, and the impairments suffered along the way. Some of the impairments that have the greatest detrimental impact on the link margin are:

- Transmitter intermodulation, as characterized by the NPR.
- Spreading loss: path loss, which is proportional to the distance between the UT and the satellite; very much driven by the slant range, or elevation angle at the UT.
- Atmospheric loss: signal absorption in the propagation path; proportional to the carrier frequency and the amount of moisture in the air.
- Polarization mismatch: orientation of the UT antenna relative to the satellite antenna.
- Body losses: absorption and blockage of the communication signal by the user's body.
- Multipath interference: in CDMA systems, rake receivers can actually take advantage of multipath to *enhance* the received signal. In most other multiple access techniques, multipath is detrimental.
- Co-channel interference: leakage from the other channels on the same frequency but in a different beam (or, for a Code Division Multiple Access, or CDMA-based system, signals in the same beam but with a different code).
- Adjacent channel interference: leakage from a channel on an adjacent frequency.

- Digital implementation loss: effects such as spectrum truncation due to the finite filtering bandwidth in the receiver's demodulator.

Each of these factors must be quantified and accounted for in the overall impairment budget in order to ensure the link margin needed to provide acceptable service.

Implementation Loss — It is of particular interest to consider link margin in light of the modern trend toward digital satellite systems where a substantial amount of onboard signal processing takes place. Many of these digitally processed systems will demultiplex, demodulate, process, switch, remodulate, and remultiplex the communication signals. These digital transponders stand in contrast to traditional "bent pipe" transponder approaches where the uplinked signal is merely filtered and translated in frequency before being downlinked. In systems that are all digital, the primary performance measure of interest is the demodulated E_b/N_0, which is the signal energy per bit (E_b) divided by the noise density (N_0). We can relate the classical signal- (or carrier-) to-noise ratio to E_b/N_0 by way of the raw transmission bit rate and the pre-detection bandwidth, as:

$(C/N)_{dB} = (E_b/N_0)_{dB} + (R/W)_{dB}$, where:

- R is the bit rate in bits per second.
- W is the pre-detection bandwidth in Hz.

Typical performance curves plot bit error rate as a function of E_b/N_0, the so-called "waterfall curves," as shown in Fig. 9.2.

Each impairment that we discussed above could be considered as a contributor to the overall digital implementation loss. In effect, for digital modulation schemes (e.g., m-ary PSK or n-ary QAM, where m and n are integers) there are three contributors to implementation loss that can be quantified and modeled as a characteristic of every block in the overall system. These contributors are:

- Additive White Gaussian Noise (AWGN): Thermal noise, interference, and other random noise effects that degrade the received signal-to-noise ratio (and, therefore, directly affect the E_b/N_0).

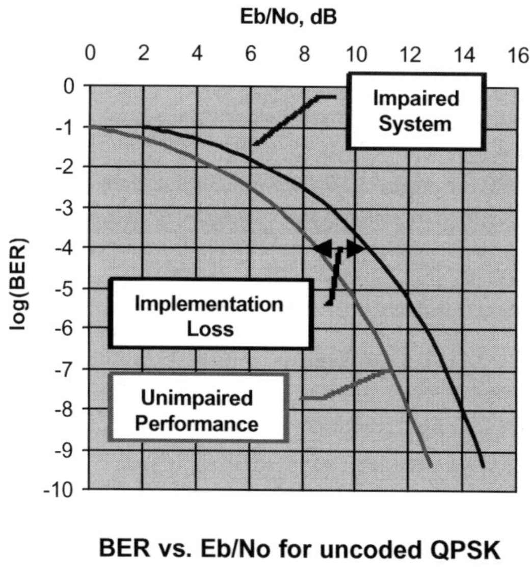

FIGURE 9.2 Typical "waterfall curve"; bit error rate vs. Eb/No for uncoded QPSK.

- Phase Distortion: Nonlinearities in the transmission of phase information.
- Amplitude Distortion: Nonlinearities in the transmission of amplitude information.

In fact, all of the impairments we have discussed could be couched within the above three terms. The amount of implementation loss suffered, at any given stage, will be a strong function of the type of modulation chosen. For instance, m-ary Phase-Shift Keyed (PSK), which has the intelligence coded in the phase of the signal only, will be much more affected by phase distortion than by amplitude distortion. Generally speaking, one can hard-limit a PSK signal and not lose its detectability. On the other hand, an n-ary Quadrature Amplitude Modulated (QAM) signal uses both amplitude and phase to code the intelligence. Such a signal would, clearly, be sensitive to both amplitude distortion as well as phase distortion. Each contributor to the overall implementation loss tends to shift the waterfall curve to the right (as shown in Fig. 9.2), implying a need for higher E_b/N_0 for a given (desired) bit error rate. As a consequence, for digital systems, there is justification for assessing link margin solely in terms of the various implementation losses encountered throughout the system. That is not to say that the conventional measures we previously discussed are obsolete. On the contrary, if we have limited control over the signals to be carried within the cellular satellite system we are designing, then the conventional measures offer the best common ground on which to specify the system performance. On the other hand, if we have complete control over both the ground and the space segments of the system, and it is an all-digital system (with a modulation format of our own choosing), we can streamline the entire analysis process, and drive more directly to the bottom line (i.e., E_b/N_0), if we model each segment in terms of its digital implementation loss to the end-to-end performance.[2]

The success of a given system hinges on customer satisfaction, and a large part of that satisfaction is derived from the ability to make successful calls "most of the time." The larger the link margin (i.e., the more degradation the link can suffer before communication is no longer possible), the more frequently a user will be able to complete successful calls, and the more often he is likely to use the system. The link design has to contain enough margin to cope with these "most of the time" situations, which is why cellular satellite system links are, as a rule, designed on a statistical basis.[3] No hard and fast rule exists here as the trade-off is subject to interrelated and, often, subjective criteria. Current cellular satellite systems, regardless of whether LEO, MEO, or GEO, tend to present the average user with about 10 dB of link margin after all impairments have been considered. This somewhat counter intuitive result (given that GEO constellation orbits are some 40 to 50 times higher than the orbits of LEO systems) is due to the fact that the lower complexity LEO satellites, as compared to GEO satellites, are smaller (lower overall transmit power capability) and must cover a broader angular field of view (lower antenna gain).

If sound engineering conservatism prevails against the sometimes exuberant optimism of the marketing department, the average user will be defined as one who is located closer to the edge of a typical cell beam (rather than at the peak), does not have his UT antenna ideally oriented with respect to the spacecraft, and will probably be located inside of a building (but not too far away from a window). Such a design approach will tend to meet the "most of the time" criterion.

Other methods exist to address the 5 to 10% of the conditions when the user is not ideally positioned. These methods generally include the assumption of a cooperative user. As an example, for a mobile terminated (MT) call (i.e., one in which the mobile user has an incoming call) the paging signal can be issued at a higher signal strength which, in turn, requests the user to move to a better line-of-sight position with respect to the satellite.

9.2.2.2 Latency

Latency refers to the amount of communication delay a cellular satellite system user experiences during a conversation or a data transaction. The effect of "substantial" latency can range from mild irritation (users at each end of the link talking over one another) to major transmission inefficiencies (multiple retransmissions under an IP environment), including dropped transactions, which could be disastrous at several levels ranging from economic to human safety considerations. The latency equation contains several contributing factors including:

- Propagation delay: the transit time (determined by the speed of light) for the signal to travel between the satellite and the UT. The total delay is, necessarily, twice the one-way delay since the signal must go from Earth to satellite and back down again. The approximate round trip (Earth-satellite-Earth) delay for the three main orbital configurations are:
  - GEO (orbital altitude of approximately 35,900 km): 239 ms
  - MEO (orbital altitude of approximately 10,300 km): 69 ms
  - LEO (orbital altitude of approximately 700 to 1400 km): 6 ms
- Coding delay: in digital systems the data to be communicated (be it digitized voice or computer files) is coded for various reasons (error correction and data compression are the most common reasons beside encryption and security concerns). The coding process implies that blocks of data must be stored before coding. Received blocks must, likewise, be buffered in blocks before decoding can take place. This buffering/coding process adds delay to the communication transmission. The amount of coding delay added to the transmission varies according to number of coding layers and the type of voice coder-decoder (codec or vocoder) employed. This coding delay can easily range between 50 to 150 ms.
- Relay delays: if a particular call needs to be routed through multiple nodes (for instance, satellite-to-satellite (ISLs), ground station-to-ground station, or relay station-to-relay station. Sometimes channels are demultiplexed and remultiplexed at these intermediate points (depending on the routing required and the multiplexing hierarchy required to effect the relay path). Again, the magnitude of the delay depends upon the path taken, but it will generally range between 10 and 40 ms.
- Other system delays: aggregated digital communication channels, in the terrestrial infrastructure, are frequently buffered for purposes such as synchronization. Also, echo cancellers and the relative placement of the channel carrier (e.g., if placed, particularly, at the band edge of the channel filters) contribute varying amounts of delay. These processes also contribute to the overall latency that the user experiences and can range from negligible to 120 ms (for the worst combinations of echo cancellers and channel filter group delay).
  - Emerging terrestrial systems are being designed to accommodate network protocol infrastructures such as H.323, which includes support for "Voice over IP" (VoIP). Considerations, which include variable grades of service, are an integral part of H.323 and imply additional latency contributions. These considerations will apply directly to cellular satellite systems that are configured to operate in compatible packetized modes.
  - Concerns about the impact of latency on Asynchronous Transmission Mode (ATM) and Transmission Control Protocol/Internet Protocol (TCP/IP) communications over the satellite channel have interested many workers in the field. Many different approaches have been analyzed and experimentally evaluated to deal with the transmission inefficiencies that could potentially arise.[4-7] It is evident that future cellular satellite systems will be required to have the capability of accommodating bursty packet modes such as these.

The relative impact of the various latency factors will, clearly, vary with the type of system. For a GEO system the latency is, generally, dominated by the propagation delay, which, in turn, is a direct function of the higher altitude of the satellite. LEO and MEO systems, with their lower orbital altitudes, have correspondingly lower values of propagation delay. This factor has been one of the main reasons that have caused several cellular satellite system designers to choose LEO or MEO constellations. On the basis of propagation delay alone, the decision would appear to be a "no-brainer" in favor of MEO or, especially LEO (particularly since the lower altitudes also provide a potential for lower spreading loss and, possibly, greater link margin). The subtleties of the other latency contributors, however, could easily conspire to erase the apparent advantage of the lower altitudes. For example, consider two users with LEO cellular satellite service located at opposite extremes of a GEO coverage area (separated by, say, 4000 km). If we assume that the two systems (LEO and GEO) have equal coding delays, the LEO path might have to traverse as many as four satellite planes and, perhaps, as many intermediate ground stations (if the system does not include ISLs).

Consequently, with each relay making its own contribution to the overall latency, it is easy to see how a point-to-point LEO communication could enter the realm of the GEO propagation delay. This discussion is not intended to favor one type of system over another with respect to latency, but merely to point out that one must consider the full complement of effects in carrying out an impartial trade study.

9.2.2.3 Orbit Altitude and the Van Allen Radiation Belts

Though considerations of latency, individual satellite coverage, and link margin all tend to enter into the selection of system orbit altitude, another physical constraint exists as to the placement of the specific orbit. This constraint is placed by the location of the Van Allen radiation belts. These radiation belts (first discovered and characterized by Van Allen in 1959) consist of two annular rings of radiation that encircle Earth and are centered on a plane defined by the equatorial latitude.[8-9] Sensitive spacecraft electronic equipment is readily damaged by high doses of radioactivity. Mitigation techniques usually involve component shielding of one form or another (e.g., thick metal boxes augmented with spot shielding with strips of tantalum), all of which imply additional spacecraft mass. Since mass is a premium item in spacecraft design (launch costs are proportional to launch mass), it is, clearly, desirable to make the satellite as light as possible. Flying a satellite within the Van Allen belts would entail massive amounts of shielding in order to achieve reasonable mission life. Consequently, orbits tend to be placed either to avoid the Van Allen belts altogether or to make only occasional (and very rapid) transitions through them. A schematic view of the radiation belt locations, from a polar perspective, is shown in Fig. 9.3. This figure also shows the relative locations of the three most common satellite orbits.

Other orbit types have been proposed for use in cellular satellite designs. Most notably, the elliptical orbit is planned for use in the Ellipso system.[10] In this system, a pair of elliptical orbits (perigee of 520 km and apogee of 7800 km) slips between the two Van Allen belts during the north–south transit.

9.2.2.4 Operating Environment Summary

Within the context of evaluating the operating environment of the system, we are primarily interested in assessing the amount of available link margin and determining the best way to maximize it. Higher link margin, therefore, primarily drives:

- The size of the satellite antenna (dramatically impacting cost, mass, and complexity)
- The optimum frequency band of operation (which, more often than not, has political ramifications that swamp the technical considerations)
- The power and linearity performance of the satellite transmitter (also a major cost element of the system)

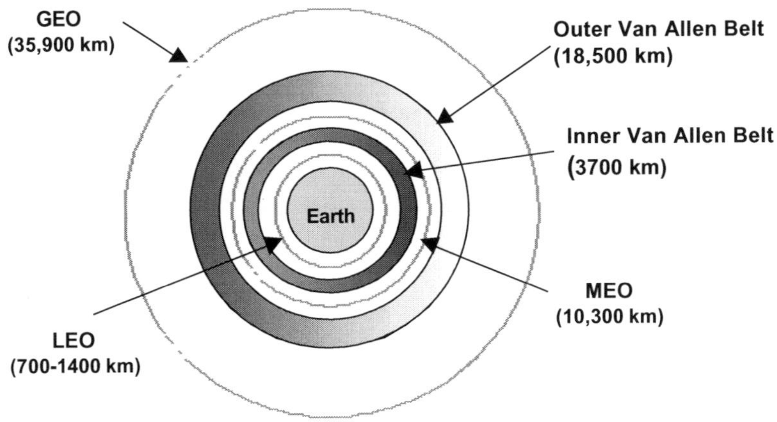

FIGURE 9.3 Locations of Van Allen radiation belts and typical cellular satellite orbits.

- The noise performance (sensitivity) of the satellite LNA and receive system

Where system latency is determined to be of high priority with respect to cellular satellite system performance, the main impacted design parameters tend to be:

- Satellite orbit altitude: impacts propagation delay, which tends to exhibit the greatest variability of all of the components of the latency equation
- System coding approach: while important with respect to the latency assessment, may be more strongly driven by requirements to provide error detection and correction (EDAC) as well as communication security

9.2.3 Service Offerings

Another major system design driver relates to the types of services that are to be offered to the user community. Many of the cellular satellite systems currently fielded were designed, primarily, to provide voice communication services with some facility to provide low-rate data (9.6 kb/s or less) and facsimile services. The fact that some systems were not "future-proofed" (i.e., physically incapable of supporting anything other than voice and low-rate data) has proven to be a considerable shortcoming and, in fact, has contributed to the demise of at least one early system. The explosive advent of the Internet in recent years has given rise to major advances in the wired communications infrastructure on a worldwide basis. The ubiquity of data communications at virtually every social stratum has fueled the tremendous growth of so-called e-commerce, a market that is forecasted to grow from \$233 billion to over \$1.44 trillion between the years 2000 and 2003.[11] Of this huge market, a very significant \$200 billion slice (in 2004) is predicted to be transacted over some sort of wireless infrastructure.[12] Consequently, there exists a large incentive for a satellite cellular system operator to ensure that the system is capable of supporting wireless data services at attractive data rates (on the order of 100 kb/s or more).

If the system is to support data services, then there is a decision to be made as to whether the services will be sold as circuit switched or packet switched. Circuit switched services represent the traditional approach to providing communication services: a channel is allocated, and dedicated, to the customer for the duration of the call. The customer is subsequently billed for the amount of time that the call was active, whether or not any data was passed during that time. Packet switching, on the other hand, is metered on the basis of the amount of data that is transmitted. The customer does not use any system resources while idle. It is a "service-on-demand" paradigm in a mode of being "always connected" (similar to having a computer on a local area network in the wired world). Compatibility with data transmission frameworks, such as the TCP/IP, is more streamlined and more efficient in a packet-based system. Infrastructures such as those developed for the General Packet Radio Service (GPRS) need to be incorporated into the design of the cellular satellite system if it is to support packet switching.

In summary, decisions on service offerings involve considerations as to whether the service backbone is to be voice-only or voice and data. If data is included, then the choice between circuit switching and packet switching needs to be considered. Besides the network equipment complement needed to execute these services, these decisions drive the basic system parameters of:

- Air interface: TDMA, FDMA, CDMA.
- Channel spacing: higher data rates require wider spectral bandwidths which, in turn, mean that the channels need to be spaced further apart. CDMA approaches, likewise, require wider spread bandwidths for higher data rates.
- Modulation type: bandwidth-efficient modulation methods should be employed in order to conserve system resources by transmitting the most amount of data for the lowest power and the narrowest bandwidth. Multilevel modulation types can be traded, such as various types of PSK or QAM.

9.2.4 Value Proposition

Will the typical user of the cellular satellite system be a business traveler, or will the business case assume a broader user population (penetrating lower economic strata)? Experience has shown that a target subscriber population based primarily on the affluent, worldwide business traveler places costly constraints on the system design and may not, necessarily, be attractive to the targeted consumer. For one thing, such a business case requires a worldwide coverage of satellites. This kind of coverage, in turn, dictates many satellites with narrow beams covering the entire surface of Earth. The most obvious approach to such worldwide coverage is to deploy a large fleet of satellites in a LEO configuration. The LEO constellation, if properly configured, provides the desired continuous global coverage, but it turns out to be extremely expensive to deploy such a system. Although the individual satellites, in a LEO system, tend to be small and inexpensive (in a relative sense considering the overall context of spacecraft technology), many satellites are required in order to achieve the desired coverage (tens to hundreds). The network infrastructure also demands a large number of gateways, with active features to track the rapidly moving satellites, in order to handle the connection of the mobile traffic to the public wired backbones. Clearly, the satellites could be designed to perform the call-by-call traffic switching, including relaying between satellites via intersatellite links (ISLs), which, in turn reduces the quantity of required gateways, but this approach adds complexity to the space segment and increases the overall system cost. Additionally, it is virtually required to have the entire constellation in place before a reasonable level of service can be offered for sale. A gradual ramp-up in service, based on a partially deployed constellation, is very difficult (if not impossible).

A system design based on the MEO configuration also has the potential of providing worldwide, or nearly worldwide, coverage with fewer satellites (on the order of tens or less) than a LEO version. MEO spacecraft tend to be more complex than LEO spacecraft, but less complex than their GEO counterparts. Fewer gateway stations are required to service a MEO constellation (relative to LEO) and, since the spacecraft are at a higher altitude, they move relatively slowly so that tracking is easier. Also because of the slow orbit, call handoffs between spacecraft tend to be less frequent. As a consequence, the total system cost for a MEO system tends to be somewhat lower than the cost of an equivalent LEO system. Also, depending on the actual orbits designed into the system and the quantity of spacecraft per orbital plane, MEO systems have the possibility of offering start-up service to selected areas on Earth. Thus, some revenue can be returned to the enterprise prior to the complete deployment of the satellite constellation.

Total system cost is a major issue with regard to service pricing. The price point is generally set so as to provide a profitable return within a set time frame consistent with the business plan of the enterprise. It is evident that the base system cost is a large part of the initial investment against which a threshold to profitability is set. Obviously, the sooner the system finances cross this profitability threshold the sooner the system can be declared a success, making for a happy investor community. In this regard, GEO systems tend to be more cost effective, with a greater probability of being profitable at a lower price point. While GEO spacecraft tend to be larger, more complex, and more costly than those destined for either LEO or MEO service, only one gateway is required to service the communication links. Connectivity is instantly available with the arrival of the spacecraft on station, and commercial service can generally be offered within a few months thereafter. Although GEO systems are, by nature, regional, careful selection of both service region and market mean that revenues can start flowing relatively quickly. Since the overall cost of deploying a regional GEO cellular satellite system is, by and large, lower than either a LEO or a MEO system, the threshold to profitability is potentially closer. The downside is, of course, that regional GEO systems, individually, cannot provide global coverage. Theoretically (except for the Polar Regions), three spacecraft placed at 120° longitude intervals can provide global coverage. The requisite narrow beams needed to cover the earth field of view would, however, require multiple, very large and possibly impractical antennas in order to close the link to UTs. Alternatively, one could take the approach of concentrating on the major potential revenue producing areas on Earth's surface (major landmasses), largely ignoring the open ocean areas. Under such a scenario "Earth coverage" could be accomplished with four to six GEO spacecraft, with two collocated spacecraft at longitudes over North–South America, and over Scandinavia–Europe–Africa.

Regardless of the approach taken, a successful cellular satellite system value proposition hinges on meeting the defined technical and service requirements with a system embodiment that minimizes system implementation cost and maximizes return on investment. The probability of obtaining a positive return on investment is inversely proportional to the risk taken in the design and deployment of the system. Several approaches can be taken to reduce the overall risk of the project. These risk reduction approaches include:[13]

- Minimizing development cost: reuse previously qualified designs to the maximum extent possible without compromising required system performance. Maximum reuse of qualified hardware also enhances (shortens) project schedules, and speeds system deployment and time to market.
- Maximizing system flexibility: anticipate future developments and plan to accommodate them. For instance, future data services will require wider channel bandwidths. A design that includes wider channel bandwidths will serve both current needs (e.g., voice transmission) in addition to allowing system migration to data services in the future.
- Well-researched business plan: diligently considered markets, along with a firm financing plan, helps to ensure steady progress in the deployment of the system.

9.3 Approaches

Having considered the fundamental driving factors of a cellular satellite system design, we now turn to specific implementations and the trade-off considerations that they imply. For convenience, a summary of the driving parameters already considered is shown in Table 9.1.

Regardless of the system embodiment selected, certain elements are common and will be found in any cellular satellite system. These main elements are depicted in Fig. 9.4.

Technology will, inevitably, continue to evolve. Unfortunately, once a satellite system is launched it is impractical, in most cases, to make modifications to the space segment. The typical 10 to 15 year on-orbit design lifetime of a spacecraft is close to an eternity when viewed against the backdrop of historic technology trends. These two incompatible facts make it incumbent upon system designers to anticipate the future and strive to make accommodations for upgrades to the extent practical. As we have previously discussed, system flexibility is key to future proofing the design. Flexibility allows the system to support evolving services with minimum modification. For instance, a space segment able to support wideband channels will be able to grow with the relentless trend toward higher data rates. Beam-forming flexibility (allowing the coverage area to evolve with the market) is another approach to future proofing, either by way of digitally programmed onboard beam forming or ground-based beam forming, or by way of an overdesigned conventional analog approach. In the next section we will look at some specific system designs, the embodiments of which are the results of decisions made by previous cellular satellite system designers after having grappled with the concepts we have discussed.

9.4 Example Architectures

9.4.1 LEO

In this section we will examine two of the so-called "Big LEO" cellular satellite systems. These systems are "big" since they have relatively large, complex spacecraft and are designed to handle large quantities of information. These characteristics stand in contrast to the antithetical "Little LEOs" whose main mission is to provide short message service, paging, and asset tracking, and are not considered cellular satellite systems for our purposes here. Common to all LEO cellular satellite systems is a large quantity of satellites in orbits that range from 700 km to about 1400 km (avoiding the inner Van Allen belt). LEO satellites also tend to be the simplest (or, at least, the lightest) of the major communication satellite types,

TABLE 9.1 Summary of Cellular Satellite System Driving Parameters

Driving Parameter	Trade Issues
User terminal convenience	Satellite EIRP and G/T
Target market and its location	Operating frequency
	Orbit type
	Air interface structure
Subscriber population and geographic distribution	System capacity
	Authentication register size
	Satellite aggregate EIRP
	Multiple access method
	Satellite linearity requirements
	Satellite switching speed
	Satellite switching capacity
	Satellite memory capacity
	Channelization approach
	Degree of frequency reuse
	Cell size
Operating environment	Multiple access method
	Satellite EIRP and G/T
	Modulation method
	Phase and amplitude linearity
	Available link margin
	Orbit type
	Latency
	Band of operation
	Coding method (digital system)
	High penetration alerting method
Service offerings	Air interface structure
	Channel spacing
	Channel bandwidth
	Modulation type
Value proposition (profitability)	Overall system cost
	Risk element (amount of new technology)
	Incremental revenue possibilities
	System flexibility (future proofing)
	Business plan
	Funding base

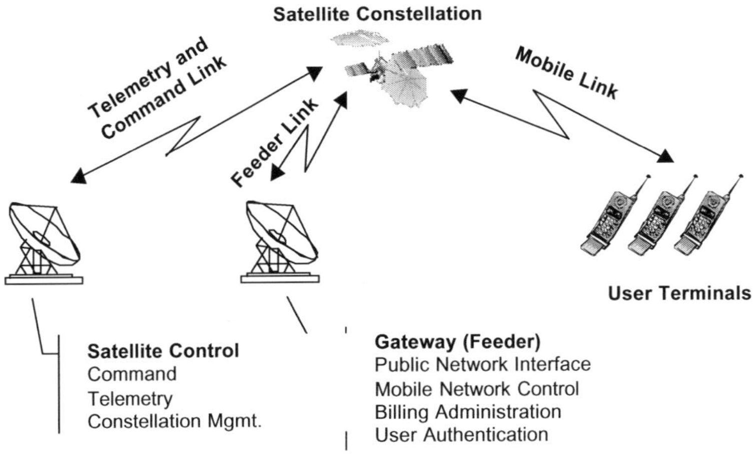

FIGURE 9.4 Common elements of a cellular satellite system.

and tend to have the shortest service life (5 to 7 years) due to limited onboard capacity fuel. Although somewhat counterintuitive, Big LEO systems tend to be the most costly to deploy (of the LEO, MEO, GEO varieties), mainly due to the large quantity of satellites required and the extensive, globally distributed ground infrastructure required to support the system. The main marketing point for LEOs is the low latency between the ground and the satellite due to the close proximity of the satellite to the user. The two best known of the Big LEOs are Iridium and Globalstar.

9.4.1.1 Iridium[14,15]

The Iridium cellular satellite system is owned by Iridium, LLC, of which Motorola is an 18% owner. The main contractors supplying system hardware are Raytheon (main mission antennas), Lockheed Martin (spacecraft bus), and Scientific Atlanta (Earth terminal equipment). In 6 polar orbital planes of 11 satellites each, 66 active satellites ring Earth at an altitude of 780 km. The Iridium system is targeted at providing, primarily, voice service to the globe-trotting business professional. But the system is also designed to support low-rate data communications as well as facsimile and paging.

Iridium uses a protocol stack that is partially built on GSM and partially Iridium unique. Therefore, the system is compatible with the GSM infrastructure at the service level, even though the physical layer (the radio link) is not in accordance with GSM standards. The main elements of the Iridium system are shown in Fig. 9.5.

The main mission links (to the UTs) are accomplished by a combined FDMA/TDMA time division duplex (TDD) method in the L-band (specifically, 1610 to 1626.5 MHz) and with QPSK modulation. Voice communication is digitized in a vocoder, and the individual voice channels operate at a data rate of 2.4 kb/s. Traffic is passed between the individual satellites for the purposes of traffic routing and call hand-off, as one satellite transits out of view and another is needed to pick up the connection. These cross-links are also operated at K-band (23 GHz), but at the higher data rate of 25 Mb/s. One of the motivations for including the complexities of cross-links in the spacecraft design was to enable efficient call routing (less dependent upon the physical location of the gateways) which, in turn, also allowed full service coverage to the oceanic regions.

Globally distributed gateways, each with a 3.3-m antenna, provide the interface between the Iridium system and the public wired networks. Links between the satellites and these stations are done via the spacecraft feeder antennas on a K-band carrier (19 GHz down and 28 GHz up on QPSK modulation at a coded data rate of 6.25 Mb/s). Call setup and teardown is controlled at these local gateways, and this is also where billing records are generated. Each gateway also includes the necessary user authentication equipment (HLR, VLR) as well as the necessary infrastructure to enable UT position location. Position

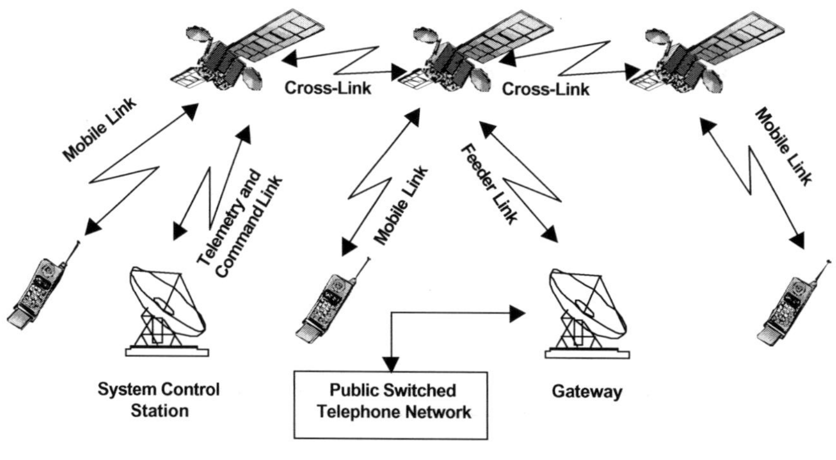

FIGURE 9.5 Elements of the Iridium communication system.

location is important for political reasons, among others, so that local jurisdictions can maintain control over telecommunication traffic in their respective regions. At least two gateway antennas are required at each site in order to properly track and smoothly maintain contact with spacecraft in the field of view.

Management of the whole system is done at the system control station, which is physically located in Lansdowne, Virginia. Here the network infrastructure is monitored and controlled. This station also takes care of satellite maintenance and control, monitors status and system health, and serves as the center for any troubleshooting required.

The 700-kg Iridium spacecraft is very sophisticated. Its design includes a complex digital signal processor that demodulates incoming signals, switches them, and remodulates them as needed. One great advantage of this method is that uplink implementation loss can be significantly isolated from the downlink, thereby improving the overall bit error rate performance. The downside to this sophistication is, naturally, a major increase in the complexity and cost of the spacecraft. Three main mission antennas form the 48 cellular beams per satellite by way of direct radiating phased array technology. These beams cover a footprint on Earth some 4700 km across. Each satellite has the switching capacity to handle 3840 simultaneous calls, but power considerations limit the practical number to around 1100.

The Iridium system has fallen on hard times as of this writing with the service having been discontinued and plans being made to actually deorbit the satellite constellation. This system is an object lesson in how even the most sophisticated, well-engineered system can become a failure if mishandled from a business perspective. Of the many problems that beset the Iridium system, some of the more significant ones include:

- Market share erosion: at the time of conception (1988–1990) the enormous build-out of inexpensive terrestrial cellular was not anticipated. This problem is the same one that contributed to the lackluster business performance of the American Mobile Satellite system.
- High system cost: estimates vary, but the cost of the Iridium system was some $5 to $7 billion. This high cost, in turn, implied the need for high service cost ($3 to $5 per minute) if returns were to be realized according to the business plan schedule. Additionally, the acquisition cost of the UT to the subscriber was high ($2000 to $3000). This combination of high service and terminal costs put the system out of reach to all but the most affluent consumers.
- Market overestimation: Iridium targeted the global business traveler who would often be outside of terrestrial cellular coverage. Most high-intensity business destinations today are well served by inexpensive terrestrial cellular coverage, thereby diminishing the need for Iridium's ubiquitous coverage.
- Ineffective marketing: obtaining Iridium service was not a streamlined process. Also, advertising was underwhelming and tended to depict users in polar ice fields or deserts (a very limited revenue population).
- Poor UT form factor: the Iridium UT has been variously described as a "brick" or a "club." The bottom line is that the UT was awkward and heavy in a user environment used to small pocket-sized cellular telephones. The UT was designed by engineers with very limited regard to user appeal and aesthetics.

9.4.1.2 Globalstar[15,16]

Globalstar is one of the other Big LEO systems in process of being fielded. This system is owned by a partnership of several well-known companies led by Loral and Qualcomm as the general partners. System hardware suppliers include Alcatel (gateway equipment) and Alenia (satellite integration).

The Globalstar system consists of 48 satellites, in 8 planes of 6 active satellites apiece, flying at an altitude of 1410 km. Thus, the Globalstar orbits are the highest of the Big LEOs fielded to date, which also means that the individual satellites move more slowly and are, consequently, easier to track at the gateway stations. Each orbit is inclined at 52°, thereby concentrating service resources in a band bordered by the 70° north and south latitudes. These latitude limits were carefully chosen in that the majority of Earth's population is located in this region. The system uses a CDMA air interface that is based on the popular IS-95 terrestrial standard, which includes such features as soft handoffs, dynamic power control (essential to preserve capacity in a CDMA system, and preserves handset battery life in the bargain), and soft capacity limits (2000 to 3000 simultaneous calls per satellite). The intelligence signal is spread to 1.25 MHz, and these CDMA channels are spaced, in an FDMA fashion, on 1.23-MHz centers. Globalstar is targeting, essentially the same customer base as Iridium (i.e., the global business traveler), and offers an equivalent suite of services. For instance, the Globalstar offerings include variable-rate voice (2.4, 4.8, and 9.6 kb/s), low-rate data (on the order of 7.2 kb/s), facsimile, paging, and position location.

Mobile links are realized at S-band down (2483.5 to 2500.0 MHz) and L-band up (1610.0 to 1626.5 MHz) with QPSK modulation. The satellites are simple, bent-pipe transponders (amplification and translation only) with gateway links (satellite–gateway) at C-band (5019 to 5250 MHz up, and 6875 to 7055 MHz down). Rake receivers are included in both the UTs and the gateways in order to use multipath signals to advantage in strengthening the links. As a result, typical forward E_b/N_0 (the weakest link due to satellite power limitations) is reported to be on the order of 4 dB, while the return link weighs in at around 6 dB.

The main elements of the Globalstar system are depicted in Fig. 9.6. The Gateway Operations Control Center (GOCC) and the Satellite Operations Control Center (SOCC) are collocated in San Jose, California. The individual, globally distributed gateways, with their 5.5-m tracking antennas, perform the local functions of the network administration. This administration includes user authentication (HLR, VLR functions), call control and local switching, local PSTN connections, and satellite TT&C processing.

Each satellite has a pair of direct radiating, fixed-beam phased array antennas to form the L-band cellular beam patterns. The transmit antenna consists of 91 elements, each having its own SSPA, phased with a fixed beam former into 16 transmit beams. The 16 receive beams (which are congruent with the transmit beams) are similarly formed with a fixed, low power beam-forming system, but the array consists of only 61 elements, each with its own LNA. The graceful degradation properties inherent in a phased array mean that redundant SSPAs and LNAs are not required. The satellites have a design lifetime of between 7 and 8 years. Figure 9.7 shows a generalized depiction of the Globalstar satellite architecture. Immediately obvious is the simplicity of the design. There is no onboard processing here; all of the technology is low risk. A hallmark of the Globalstar system is that the complexity, where needed, is relegated to the ground segment.

The Globalstar system is active at the time of this writing. Its commercial success has not been established at this stage. As a Big LEO system, one would think that Globalstar would approach service launch with great trepidation, given the negative example of the Iridium system. On the other hand, there are a number of factors about Globalstar that make its value proposition quite different, portending a brighter future. Some of these factors include:

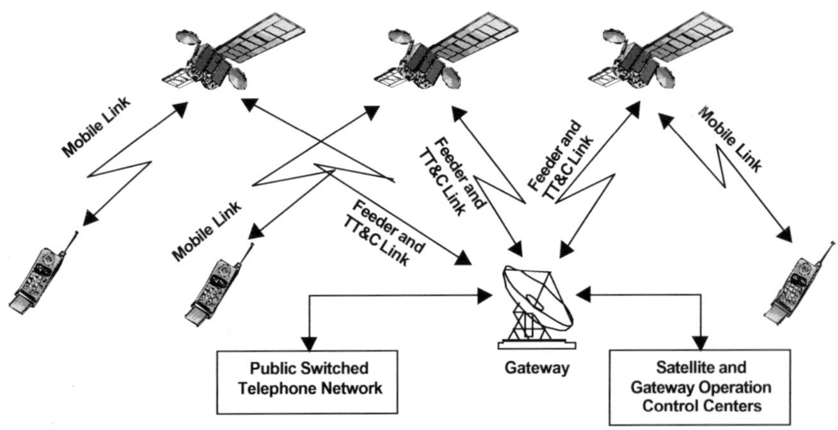

FIGURE 9.6 Elements of the Globalstar communication system.

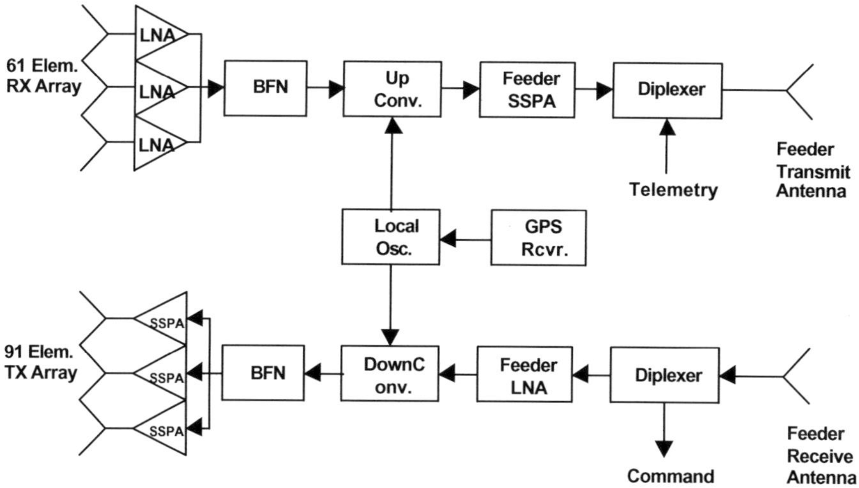

FIGURE 9.7 General architecture of the Globalstar satellite.

- Simple architecture: the space segment is constructed with low risk technology. The complexity of the system is kept on the ground where upgrades, as necessary, are readily accomplished. This approach keeps the overall cost of the system down, and permits lower service pricing to be offered. In fact, the start of service has been offered at $1.79 per minute, and a handset price of around $1500,[17] a far cry from Iridium's $7 per minute and a $3000 handset.
- Proven air interface: the CDMA-based IS-95 air interface has been well proven in terrestrial cellular systems. Additionally, this approach makes the job of constructing multimode UTs (terrestrial-satellite) easier.
- Phased service rollout: Globalstar observed the techniques employed by its Iridium forerunner and decided to gradually roll out service in well-researched market areas. This technique allowed the company to gain experience and make corrections as needed with a fault-tolerant, methodical process.
- Revenue area focus: Globalstar is not attempting to service low population areas like the oceans or the polar regions. The focus is on the parts of the globe where the people are. Precious resources are not expended in areas where return is marginal.
- More aesthetically pleasing UT: although not quite the size of a modern-day cellular handset, and the antenna is still quite a bit larger than desirable, the Globalstar UT is much smaller than its Iridium counterpart. Broad user acceptance has still to be proved, but the prospects are good.

9.4.2 MEO

MEO systems, in general, tend to be orbitally located in the region between the inner and outer Van Allen radiation belts (around 10,300 km). Consequently, they tend to take on a blend of characteristics of which some are LEO-like and others are GEO-like. For instance, MEO propagation delay works out to be around 40 to 50 ms, not as good as LEO, but clearly better than GEO. The satellites also tend to be fairly complex and approach the GEOs in terms of design life (on the order of 10 to 12 years). Because of their relatively high orbit, MEO satellites move fairly slowly across the sky, greatly simplifying tracking requirements and reducing the number of handoffs during a typical call holding period. Two of the better known MEO systems are ICO and Ellipso, and we will consider them in this section.

9.4.2.1 ICO[10,18]

The design of the ICO system began with a study program conducted by Inmarsat in the early part of the 1990s. Dubbed "Project 21," Inmarsat's objective was to move into the satellite cellular communications business. LEO, MEO, and GEO constellations were considered during the study, which eventually settled on a MEO configuration (an "intermediate circular orbit," or ICO), subsequent to which the project was spun off as a separate company. ICO was owned by ICO Global Communications Holdings, Ltd., with an additional 17 subsidiaries. That was until the company sought bankruptcy protection around the same time as Iridium faced a similar difficulty. Recently,[19] Craig McCaw and his Eagle River organization, along with Subash Chandra of ASC Enterprises (Ascel), both saw potential in the system and have, essentially, taken over the company. The new owner organization is known as New Satco Holdings, Inc. The major equipment contractors for the system include Hughes Space and Communications (satellites) and a team, led by NEC, which includes Ericsson and Hughes Network Systems (ground infrastructure).

The original focus of the ICO system was to complement terrestrial cellular communication by servicing customers who reside outside of normal cellular coverage, providing cellular extension service to those who often travel outside of terrestrial cellular, maritime customers, and also government users. In other words, ICO was already targeting a fairly broad market. Craig McCaw, on the other hand, saw opportunities to enhance the system to provide data services. It has been reported that he is driving modifications into the design so that packet-based medium data rate services are supported (GPRS-like rates of around 384 kb/s), with early support for Wireless Application Protocol (WAP) services. WAP enables thin clients, like cell phones, to access enterprise services like e-mail and Internet information.

The ICO constellation consists of 10 active satellites in 2 planes of 5 satellites each. The orbital planes are both circular and inclined at 45°. The satellites communicate with 12 gateways (ground stations), called Satellite Access Nodes (SANs), spread around the world. Each SAN (Fig. 9.8) is equipped with five antennas to track as many satellites as might be in view at any given time. Due to the higher altitude of the constellation, each satellite is able to cover around 25% of Earth's surface at any given time. The air interface is very similar to the terrestrial cellular IS-136 (D-AMPS) standard, in that it is an FDMA/TDMA scheme (QPSK modulation) that can support as many as 40 time slots (nominal-rate users) on five subcarriers within a 156-kHz channel bandwidth. The fact that the channel bandwidth is reasonably wide (approximately 156 kHz) is the reason that the system is readily modified to accept GPRS-like (and even EDGE-like) waveforms. There is, therefore, scope to provide packet-based medium data rate services in the ICO system, and this is one of the "future-proofing" characteristics that were designed in.

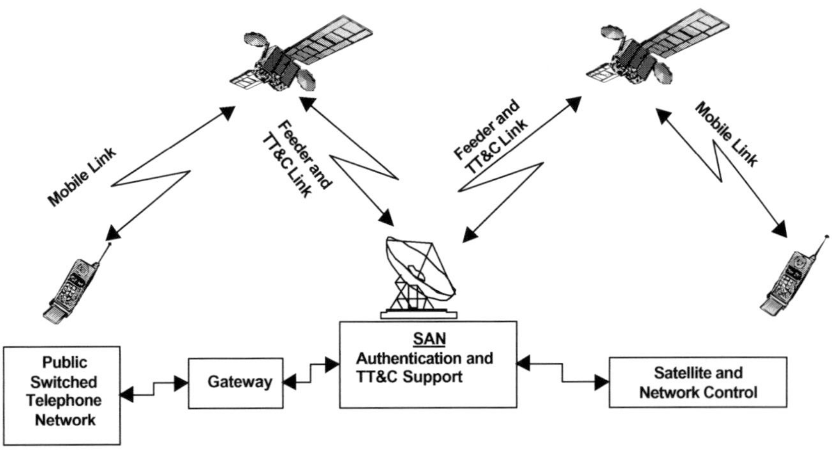

FIGURE 9.8 Generalized ICO system architecture.

Mobile downlinks are at L-band (1980 to 2010 MHz) while the uplinks are at S-band (2170 to 2200 MHz). The aggregated signals, shipped between the satellites and the SANs, are transmitted at C-band (5187 to 5237 MHz up, and 7018 to 7068 MHz down).

The ICO Network Management Center (NMC) is located in Tokyo, Japan, while the Satellite Control Center (SCC) has been placed in Uxbridge, England. The SCC monitors the health and status of the spacecraft, and also takes care of any orbital adjustments that might become necessary in the course of the mission life.

The ICO spacecraft design is fairly sophisticated. Moderately large (2600 kg and consuming around 8700 W), the satellite is based on the HS 601 design, which has been extensively deployed for GEO missions. There are two mobile link antennas, one for transmit and the other for receive, each measuring around 2 m in diameter. These antennas are direct radiating arrays driven by a sophisticated digital beamformer (DBF), which allows the antenna patterns (cells) to be dynamically shaped in order to respond to changing loading needs. This design also means that complexity (cost) has been added to the spacecraft, and that it must be supported by an intricate calibration infrastructure as well. Each antenna subsystem consists of 127 radiating elements and forms around 163 beams that, together, cover a ground surface diameter of around 12,900 km. The system frequency reuse factor is about four times with this design. Each satellite has switching capacity for around 6000 voice channels, although power constraints limit the actual capacity to around 4500 circuits. On the other hand, given the system modifications in process under McCaw's direction, the system is poised to enter the packet switched regime. Entirely different capacity calculations are possible under such an operation environment.

ICO's development has been a mixed bag, and apart from McCaw and Chandra's involvement, the value proposition is uncertain. Factors include:

- Higher orbit, slower movement, fewer hand-offs: this characteristic makes for easier tracking and allows certain simplifications in the ground infrastructure.
- Smaller constellation (with respect to LEO): fewer satellites are built and launched, but each satellite is heavier and more complex.
- Wider field of view: higher orbit sees more of Earth, and can cover oceanic regions (a mixed blessing).
- Fewer gateways required: infrastructure cost is lower than LEO, but higher than a GEO system.
- IS-136 (D-AMPS)-like air interface: this allows adaptation of standard terrestrial cellular hardware and easier manufacture of multimode handsets.

ICO has clearly had a tough time getting started, and for a time it appeared that the system would suffer the same fate as Iridium. The system is relatively expensive (included a significant amount of innovative technology). Again, estimates vary, but the overall system cost around \$3 billion to \$5 billion in its original state. McCaw's modifications will add additional cost, and he has invited other investors to participate, but his Eagle River organization is financially powerful enough to drive progress forward on its own if need be. With the further innovations to wireless packet data support, the future of the ICO system will be interesting to watch as it unfolds.

9.4.2.2 Ellipso[10,20,21]

Ellipso has also been placed in the class of the "Big LEOs." The system orbit design is unique, to the point of actually having been patented (U.S. patent #5,582,367). Due to the altitude (mean altitude, in the case of the elliptic planes) we class the Ellipso system here as a MEO system.

Mobile Communications Holdings, Inc. owns Ellipso with major contractors Lockheed Martin (ground) and Harris (space) supplying equipment to the system. Services are provided globally, though biased to favor populated areas, through satellites in three orbital planes; two elliptical ones that are called "Borealis" and one circular called "Concordia." The orbits are optimized to provide regional coverage proportional to the distribution of population on the surface of Earth. The Borealis orbits are elliptical, sun-synchronous, inclined at 116°, and each contain five satellites. These orbits each have a perigee at

520 km and an apogee at 7846 km. The Concordia orbit, on the other hand, is equatorial, circular, and has seven equally spaced satellites. Concordia's altitude is 8060 km. Both of these orbits are well within the band separating the two Van Allen radiation belts.

Ellipso's guiding philosophy is to perform all system trades with an eye toward the lowest end cost to the subscriber. Its services (including voice, messaging, positioning, and Internet access) are targeted toward "everyman." In other words, in contrast to the target customers of Iridium and ICO, Ellipso wants to reach deeper into the market and not just focus on the affluent, globe-trotting businessman.

Like the ICO system, Ellipso provides a mobile user downlink at L-band (1610 to 1621.35 MHz) with the uplink placed at S-band (2483.5 to 2500 MHz). Feeder (ground station) communication, on the other hand, is done with the uplink at Ku-band (15450 to 15650 MHz) and the downlink at high C-band, nearly X-band (6875 to 7075 MHz). Each satellite forms a cell beam pattern consisting of 61 spot beams incorporating a high degree of frequency reuse (by way of cell isolation and orthogonal coding) across the coverage pattern. The system air interface is based on third-generation (3G) wideband CDMA (W-CDMA) with an occupied bandwidth of 5 MHz. This is a technology that is only just starting to be deployed in the terrestrial cellular world at the time of this writing. Consequently, the Ellipso design exhibits a tremendous degree of forethought and "future-proof" planning. Because of its W-CDMA infrastructure, Ellipso is poised to launch services with a 3G infrastructure in place and is, consequently, ready to provide high data rate (up to 2 Mb/s) packet-based services.

A general diagram of the overall Ellipso system is shown in Fig. 9.9. The central System Coordination Center takes care of the system network planning and monitoring. The Ground Control Stations (GCS) provide the gateway function as the interface to the communication signals going to, and coming from, the mobile stations via the satellites. Regional Network Control Stations provide local network control functions and collection of billing records. Associated with each GCS is an Ellipso Switching Office (ESO) that provides the interface between the PSTN and the Ellipso system. The ESO, additionally houses the HLR and VLR, and takes care of the user authentication function.

All Ellipso satellites are of identical design, regardless of the orbit into which they are placed. Simplicity is the driver behind the satellite design as well. They are straightforward bent-pipe translators with separate transmit and receive, direct radiating, fixed beam-formed phased-array antennas with 127 radiating elements in each planar array. Each satellite is of medium size with a mass of around 700 kg which, in turn, keeps launch costs down.

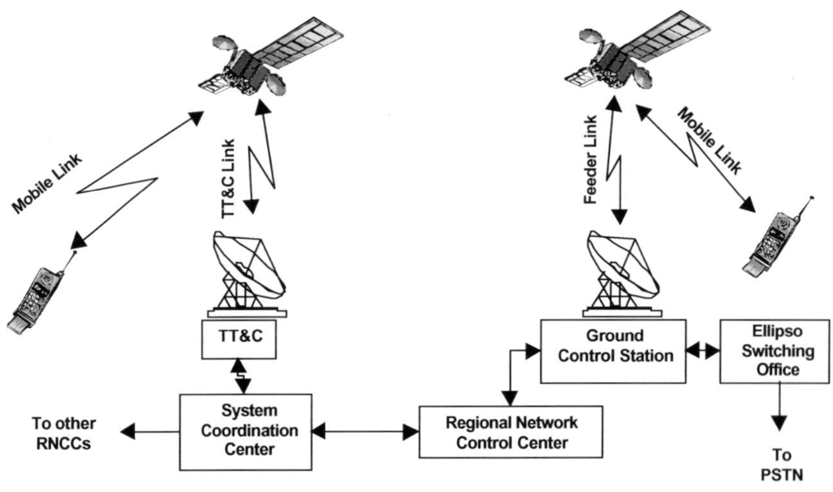

FIGURE 9.9 General diagram of the Ellipso MEO system.

The success of Ellipso has yet to be seen at this juncture. The system does, however, contain many of the features important to an attractive value proposition. This is another system that will be interesting to watch as development unfolds. Some of Ellipso's more interesting value features include:

- Low system cost: the system designers are fanatical about keeping costs down, as they are keenly aware of the connection between system deployment cost and the ultimate price point that can be offered to the subscriber.
- Wideband 3G-compatible air interface: a very forward-looking design feature. The system will be ready for enhanced 3G services, including always connected, high-speed packet-based Internet access. This feature has the potential of being very attractive to subscribers.
- Complexity on the ground: the space segment is designed as simply as practical in the overall system context. Complexity is kept on the ground where system upgrades are more readily accomplished.
- Low price point and large target market: the designers are targeting average consumers and not focusing on the lower quantity of affluent business travelers.
- Revenue area concentration: system design focuses on areas on Earth where the bulk of the people are. Resources are not wasted over large ocean regions and areas of sparse population.
- Phased deployment: the characteristics of the Ellipso orbits are such that service can be initiated with a partially deployed constellation of satellites. In effect, early revenues can be generated that will help to pay for the remainder of the system deployment.

9.4.3 GEO

Two of the most advanced GEO-based cellular satellite systems are the ACeS and the Thuraya systems. Both of these systems are slated to provide commercial service offerings in the 2000 to 2001 time frame.

9.4.3.1 ACeS[10,22]

ACeS (Asia Cellular Satellite) is a GEO cellular satellite system conceived, designed, and backed by a partnership consisting of P.T. Pasifik Satelit Nusantara (of Indonesia), the Philippine Long Distance Telephone Company (PLDT), Lockheed Martin Global Telecommunications (LMGT), and Jasmine International Overseas Company, Ltd. (of Thailand).

The ACeS primary target market is the Southeast Asian area comprising the 5000 islands of Indonesia in the south, Northern China in the north, Pakistan in the west, and Japan in the east (see Fig. 9.10). The coverage area encompasses some three billion people, many who have little or no access to a wired communication infrastructure. The Indonesian archipelago, for instance, is an expanse of islands that stretches some 4000 miles from east to west. It is not difficult to imagine the tremendous challenge of building a

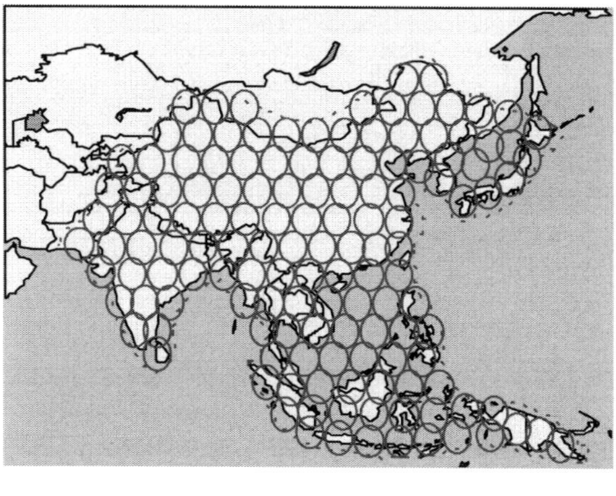

FIGURE 9.10 ACeS coverage and beam.

wired infrastructure to interconnect such a country. As such, the ACeS service is focused on a tightly defined market region, widely recognized as a rapidly expanding industrial world sector. While the typical ACeS user is perceived to be an active business traveler, the pricing of the planned services is expected to be at a level well within the reach of middle-class business people. Consequently, ACeS has a large addressable user population. This approach stands in stark contrast to some systems that target the high-end traveling businessman and seek to provide complete global coverage. In itself, a global coverage system, like some low Earth orbiting (LEO) systems, requires a large quantity of satellites (along with sophisticated hand-off and traffic management methods) and has an attendant high implementation price tag. For instance, the Iridium system is reported to have cost some $7 billion, about an order of magnitude more than the final cost of the ACeS system.

FIGURE 9.11 AceS.

ACeS is designed to operate in a clearly defined geographic coverage area; therefore, a regional GEO-based satellite system was chosen. Such a well-defined coverage area ensures that a maximum amount of precious satellite resources is concentrated on the desired revenue-producing areas. The satellite air interface standard was based on the ubiquitous Global System for Mobile Communication (GSM) terrestrial cellular standard in order to take advantage of its feature-rich suite of services, as well as the availability of a large quantity of standard supporting hardware. A GSM-based system also eases the integration of terrestrial hardware (e.g., dual-mode ACeS-GSM handsets) and facilitates intersystem roaming (based on GSM Subscriber Identity Module [SIM] cards). A mutually beneficial cooperative effort between LMGT and Ericsson (a world-class supplier of mobile communication equipment), assured the definition of an optimally tailored AIS. Further, Earth–satellite links were conservatively designed to ensure good service to disadvantaged users; frequently dropped calls are death to service acceptance and customer loyalty. An essential component of customer acceptance, facilitated by the strong links, is a handset form factor that is comparable to what is expected in modern cellular handsets. Figure 9.11 shows a picture of the ACeS handset in the form factor to be used at service launch. Other terminal types are also in development. Additionally, low cost of both service (airtime) and equipment is essential to customer uptake and heavy system use. Again, the modest deployment cost of a regional GEO system implementation aids in keeping the costs low.

The ACeS system has two main components, viz. the ground segment and the space segment, where the ground segment is further subdivided into the back-haul and control function (implemented at C-band) and the L-band user link function (see Fig. 9.12). There is one Satellite Control Facility (SCF) for each spacecraft. The Network Control Center (NCC) provides the overall control and management of the ACeS system, including such functions as resource management, call setup and teardown, call detail records, and billing support (customer management information system). Regional gateways, operated by the various National Service Providers (NSPs) manage the subset of system resources as allocated by the NCC. These gateways also provide the local interface and billing to the actual system users, and also provide connectivity between ACeS users and the wired infrastructure (Public Switched Telephone Network, Private Networks, and/or the Public Land Mobile Network). The user segment consists of handheld, mobile, or fixed terminals. These terminals can be configured to provide basic digital voice, data, and fax services. Further, since the system is based on GSM at the physical layer (in particular, 200-kHz Time Division Multiple Access, or TDMA, channels), it is future proofed in the sense that it will support General Packet Radio System (GPRS) and Enhanced Data for GSM Evolution (EDGE) upgrades with *no change required in the space segment.* This feature is a very important characteristic of the ACeS system.

The ACeS spacecraft, dubbed *Garuda-1*, is one of a family of Lockheed Martin modular spacecraft in the A2100-series (see Fig. 9.13). In particular, *Garuda-1* is an A2100AXX, one of the largest models. This spacecraft is three-axis stabilized, with a bus subsystem that has been applied across numerous other spacecraft programs (e.g., Echostar, GE, LMI, and others) and, consequently, has a significant amount of application heritage. The payload is designed with two separate L-band antennas on the user link side. One antenna is dedicated to the transmit function (forward link) with the other, naturally, for receive (return

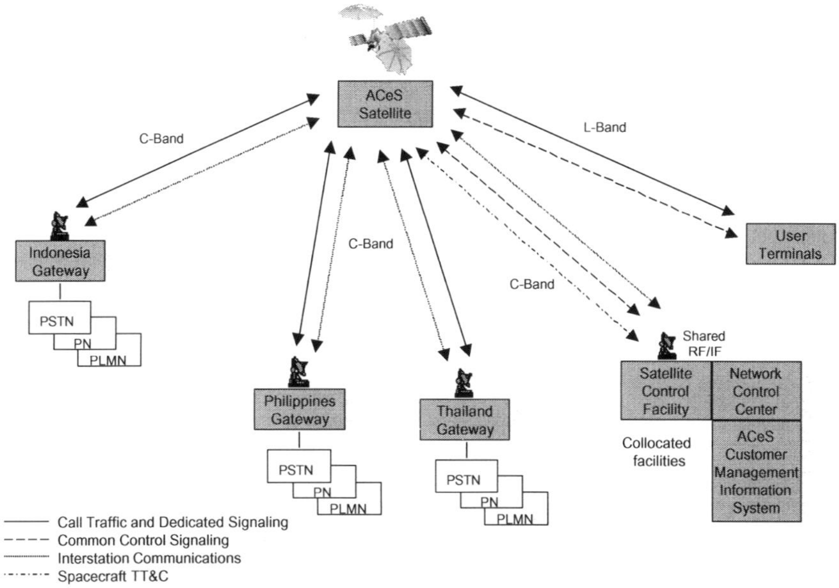

FIGURE 9.12 The ACeS system functions.

FIGURE 9.13 ACeS spacecraft: deployed.

link). This separation of antenna functions was done to minimize the probability of receive-side interference from passive intermodulation (PIM) in the transmit side. Large (12-m) projected antenna apertures (two of them) provide the underpinnings for strong user links, crucial to reliable call completion under a variety of user circumstances. The result is high aggregate effective isotropic radiated power (EIRP), at 73 dBW, and high G/T, at +15.3 dB/K.

Figure 9.14 illustrates the major block functions of the *Garuda-1* communication subsystem (CSS). As noted earlier, the user link is closed in the conventional mobile satellite system L-band. Narrow beams (140) are formed with a low risk, low power analog beam-forming network (BFN) approach, in both forward and return directions. This beam design gives rise to a composite pattern that provides 20-times frequency reuse, thereby efficiently conserving valuable spectrum. A beam congruency system (BCS) operates in conjunction with ground beacons to ensure proper overlap of the corresponding transmit and receive beams. The L-band transmit amplification subsystem is implemented with a distributed set of Butler matrix based amplifier blocks called matrix power amplifiers (MPA). The MPA construct provides equal loading for all amplifiers in the block, which, in turn, minimizes phase and amplitude variation across the block. Small

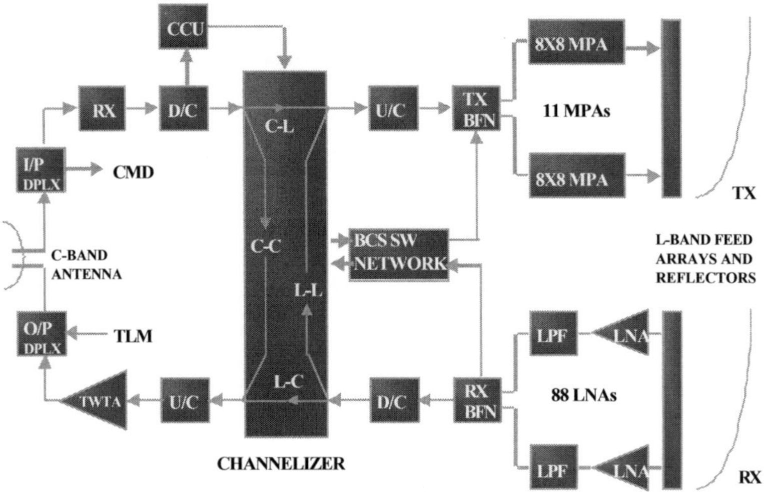

FIGURE 9.14 ACeS payload block diagram.

variations are key to good isolation between beams, giving good frequency reuse performance and, hence, maximum traffic capacity.

At the heart of the CSS sits the digital channelizer. This channelizer performs the important function of filtering the TDMA traffic channels and, also, of routing the individual traffic bursts as needed (particularly for the case of direct mobile-to-mobile communication).

The balance of the CSS (shown to the left of the channelizer) is the C-band back-haul transmit and receive equipment. This equipment is, largely, heritage being very similar to that used on previous direct broadcast and fixed service satellites.

The ACeS system has several important characteristics that hallmark a successful system. Among these characteristics are:

- Low infrastructure cost: being a GEO system, once the satellite is on-station, instant connectivity is possible. The entire system, both space and ground segments, are reported to be under $1 billion (about an order of magnitude less than most Big LEO systems).
- Low targeted service cost: low infrastructure cost allows lower price points to be charged while still allowing the business to turn a profit. This approach also means that a greater market can be attracted for the service.
- Well-researched target market: the Asia-Pacific region includes a multitude of islands that are not well connected by any kind of terrestrial infrastructure. This area is also one of high industrial growth which, in turn, needs good communication for support.
- Aesthetically pleasing UT: the satellite bears the burden of providing the link margin, so that the UT can be sized similar to terrestrial cellular telephones. On the downside, complexity is added to the space segment thereby adding cost to the satellite. Since only one satellite is involved, this added complexity is easier to bear.
- Well-chosen air interface: the ACeS air interface is based around the GSM standard, implying the ability to directly use terrestrial hardware in the terrestrial equipment (thereby reducing development cost).
  - Wideband: 200-kHz channels (the GSM standard) will support future 2.5G services (GPRS and EDGE) without modification to the space segment. This is a design forethought that helps to future-proof the system.

- Adjacent service compatibility: the adjacent terrestrial cellular services are largely based on GSM. Consequently, multimode UTs can be provided in a cost-effective manner.

9.4.3.2 Thuraya[10,23,24]

A partnership group led by Etisalat (Emirates Telecommunication Corporation) of the United Arab Emirates owns the Thuraya system. The objective of the system is to provide regional cellular satellite service to an area that includes Continental Europe, Northern Africa, the Middle East, and India. The Thuraya coverage area is adjacent to the ACeS coverage area, and also shares many of the same design features seen in that system. The target subscriber population includes national and regional roamers in an area (desert) that is not well served by terrestrial cellular service. Types of services to be offered include voice, facsimile, low-rate data, short messaging, and position determination. Major suppliers of the system components include Hughes Space and Communications (space segment), Hughes Network Systems (ground infrastructure), and Ericsson (network switching equipment).

The Thuraya system has one GEO satellite positioned at 44° E longitude (a second satellite is, reportedly, in process with the intended placement at 28.5° E). Mobile users connect with the Thuraya system through handheld UTs operating at L-band (1626.5- to 1660.5-MHz uplink and 1525- to 1559-MHz downlink). The aggregated signals destined for connection to the public wired infrastructure (PSTNs, etc.) are connected between the satellite and the Gateway stations at C-band (6425 to 6725 MHz up and 3400 to 3625 MHz down). The air interface is similar to that developed by Hughes Network Systems for the ICO system. That is, it is similar to the terrestrial cellular IS-136 (D-AMPS) system with Offset QPSK modulation, although the higher protocol elements are GSM compatible. Consequently, the Thuraya system is GSM compatible on a network and service level. Figure 9.15 shows the main elements that comprise the system.

The Gateway station serves as the main interface for communication signals into the public infrastructure. The system has a primary gateway that is located in Sharjah, UAE. Accommodation for several regional gateway stations is provided in the network infrastructure. The main functions performed by the gateway include user authentication (HLR and VLR), call control (setup and teardown), billing records, resource allocation, and roaming support. All interface conditioning required to connect to PSTNs, PLMNs, etc., are also handled in the gateway.

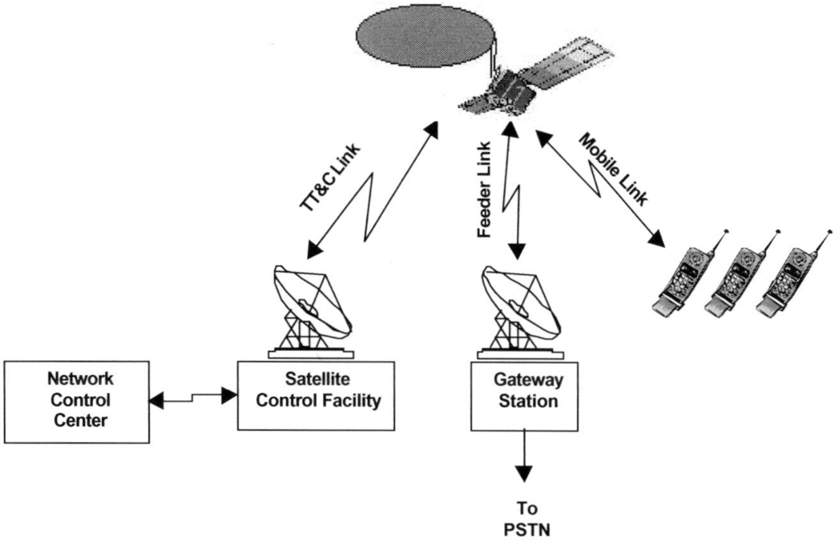

FIGURE 9.15 General diagram of the Thuraya system.

The Satellite Control Facility is actually classed as part of the space segment in that it handles the monitoring and control of the satellite. All telemetry is monitored in the SCF, and the required bus and payload commands are initiated at this facility. The Network Control Center, as the name implies, handles all of the network administration functions (routing, congestion control, and similar functions). Any payload commands required in this context are translated by the SCF and sent to the satellite.

The satellite is a very sophisticated element of the Thuraya system. It is a very large spacecraft (4500 kg) with a mission life of 12 years. The spacecraft is based on the well-known HS-601, but with several enhancements. For instance, solar concentrators are used on the solar arrays in order to enhance collection. This modification is done in order to help supply the large quantity of electrical power required (12.5 kW). The satellite is capable of switching about 13,750 simultaneous voice circuits. Mobile links interface to the ground through a single 12.25-m projected aperture reflector. Passive intermodulation (PIM) (i.e., unwanted mixing noise from the transmitter entering the receiver) is normally a central concern in systems such as these. Mitigation methods in other systems (e.g., Inmarsat-3 and ACeS) have included the use of two separate antenna apertures. The engineering of the Thuraya system has solved that problem, and only one large reflector is required (serving both transmit and receive functions). The 256 cellular beams are generated by a state-or-the-art digital beamformer, which allows dynamic beam-shaping as required to meet variations in traffic loading, as required. This digital beamformer, though adding a great deal of flexibility to the system, comes at a price of added complexity (cost), power consumption, and necessary dynamic calibration infrastructure. Some of the blocks of the communication subsystem of the Thuraya satellite are shown in Fig. 9.16.

As a GEO system, Thuraya has many of the same positive value characteristics previously described:

- Low infrastructure cost: Thuraya has the instant connectivity trait common to GEO systems where only one satellite is required. The system cost is reported to be around $1 billion.
- Low targeted service cost: low-price points are planned for system services (reportedly, on the order of $0.50 per minute). This approach also means that a greater market can be attracted for the service.
- Target market: the northern Africa area contains a lot of open territory (desert) where a significant industry is conducted (e.g., petroleum). Existing terrestrial infrastructure is, clearly, not adequate, so a good business opportunity exists. On the other hand the potential for generating significant revenue from European coverage area is questionable given the ubiquity of GSM in that region.
- Aesthetically-pleasing UT: Thuraya, with its 12.25-m antenna aperture, allows the UT to be sized similar to terrestrial cellular telephones. On the downside, complexity and cost accrues to the satellite as a result. As we noted in the ACeS case, only one satellite is involved so this added complexity is easier to bear.

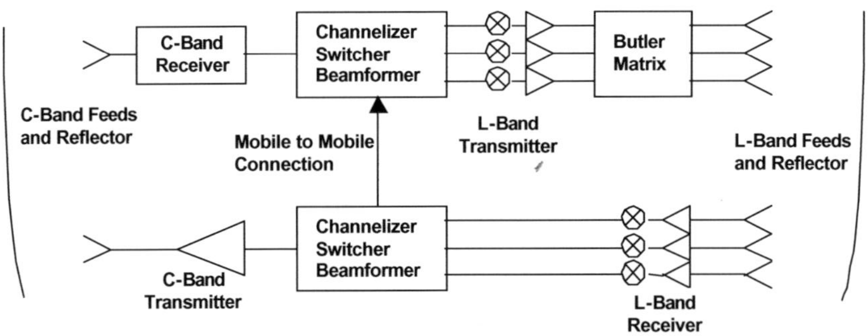

FIGURE 9.16 Major elements of the Thuraya communications payload.

- Air interface: Thuraya's air interface shares many of the characteristics of the D-AMPS IS-136 standard, implying the ability to directly use terrestrial hardware in the terrestrial equipment (thereby reducing development cost). It is also relatively wideband (around 156 kHz) and is, therefore, suited to support 2.5G services as they become available.
- GSM network infrastructure: the network services are based upon the GSM model and can support GSM services as a result.

9.5 Trends

Satellite cellular service developments have, by and large, mimicked the advancements in terrestrial cellular,[25-27] albeit with a predictable delay. Clearly, the satellite service infrastructure takes longer to develop and field than its terrestrial counterpart. Cellular satellite service providers have continued to look to terrestrial developments for the "next step." Activities are underway to adapt existing systems, and to incorporate future enhancements, to effectively support data transmission. The immense data communication infrastructure demands seen today are being fueled (akin to gasoline being fire-hosed onto a blaze!) by the explosive growth of the Internet. Cellular satellite system developers aim to have a part in the ongoing explosion. Internet traffic, from the user's perspective, is primarily a bursty form of communication based on packet switching. Circuit switched systems are not efficient in bursty packet mode, so cellular satellite systems are being adapted to communicate in packet mode and at higher data rates. Mobile data communications will require packet data access on the order of 100 kb/s or higher. GPRS, in GSM, accommodates dynamically adjustable rates and, in eight-slot full-rate mode will support a peak rate on the order of 115 kb/s. If EDGE is included (with adaptive coding and its 8PSK modulation approach), it will squeeze 384 kb/s into a 200-kHz GSM channel. Plainly, cellular satellite systems based on GSM will readily be able to support GPRS and EDGE.

3G W-CDMA terrestrial systems are being deployed in order to support mobile data rates as high as 2 Mb/s. One cellular satellite system (Ellipso) is already designed to provide such support. Others may follow. Teledesic, the wideband system owned by Bill Gates and Craig McCaw, is a LEO system specifically designed to provide high-speed data to the home fixed locations), but it is not a cellular satellite system under the criteria we have used here.

Other terrestrial enhancements are in the works, and it is likely that their incorporation into the cellular satellite world will occur at an accelerated pace. These enhancements include:

- WAP: the Wireless Application Protocol, which is a protocol stack specifically designed to allow "thin clients" (limited capability devices like cellular telephones) to take greater advantage of the Internet. WAP allows a more streamlined approach for cell phones to receive and transmit e-mail, browse Internet Websites, interact with corporate enterprise structures (Intranet services), and other well-established Internet-based activities.
- Bluetooth: a wireless picocell communication method that allows direct synchronization between Bluetooth enabled devices. Initially proposed as a method for interconnecting personal computers, printers, data devices, cell phones, and digital assistants over a spread-spectrum link in the unlicensed ISM band (Industrial, Scientific, and Medical band; the same band used by microwave ovens, for example). It is obvious to see how the use of Bluetooth can be extended to other applications, for example, as an intermediate link between a cellular satellite and a user who might be shopping in a mall.

These types of enhancements will only serve to heighten the need for, and the growth of cellular satellite systems.[28] It is clear that future applications are only limited by the imagination of entrepreneurs and system designers.

References

1. Quality of Service Forum Web page, http://www.QoSforum.com
2. Butash, T. C., Lockheed Martin Manassas, private communication, 2000. I am grateful to Dr. Butash for his clear insight into all-digital satellite system assessment.
3. Matolak, D., Lockheed Martin Global Telecommunications, private communication, 1998.
4. Kadowaki, N., et al., ATM transmission performance over the trans-Pacific HDR satcom link, *Proc. Second International Workshop on Satellite Communications in the Global Information Infrastructure*, 63, 1997.
5. Fitch, M., ATM over satellite, *Proc. Second International Workshop on Satellite Communications in the Global Information Infrastructure*, 67, 1997.
6. Falk, A., TCP over satellite, *Proc. Second International Workshop on Satellite Communications in the Global Information Infrastructure*, 74, 1997.
7. Johnson, Lt. G.W., and Wiggins, ET1 M.D., Improved Coast Guard communications using commercial satellites and WWW technology, *Proc. Fifth International Mobile Satellite Conference*, 519, 1997.
8. Logsdon, T., *Mobile Communication Satellites*, McGraw-Hill, Inc., New York, New York, 1995, 131.
9. Stern, P. and Mauricio, P. The exploration of the Earth's magnetosphere, *NASA Web tutorial*, 1998, http://www-spof.gsfc.nasa.gov/Education
10. Miller, B., Satellites free the mobile phone, *IEEE Spectrum*, 26, March, 1998.
11. eGlobal Report, March 2000, http://www.emarketer.com
12. Lonergan, D., Strategy Analytics, Inc. Web-based report, January, 2000, http://www.strategyanalytics.com
13. Elizondo, E. et al., Success criteria for the next generation of space based multimedia systems, *Satellite Communications Symposium of the International Astronautical Federation*, Session 33, September 1998.
14. Maine, K. et al., Overview of Iridium satellite network, Motorola Satellite Communications Division, Chandler, AZ, 1995.
15. Lloyd's Satellite Constellations, *Big LEO Tables*, http://www.ee.surrey.AC.uk/Personal/L.Wood/constellations/tables/
16. Dietrich, F.J., The Globalstar satellite cellular communication system design and status, *Proc. Fifth International Mobile Satellite Conference*, 139, 1997.
17. Wilkinson, G., Satellite companies face murky fate after Iridium's demise, *Total Telecom*, 24, March 2000.
18. ICO Global Communications, http://www.icoglobal.com/
19. Foley, T., Mobile & satellite: McCaw unveils WAP strategy for new ICO, *Communications Week International*, 03, April 2000.
20. Draim, J.E. et al., Ellipso — An affordable global, mobile personal communications system, *Proc. Fifth International Mobile Satellite Conference*, 153, 1997.
21. Ellipso Web site, http://www.ellipso.com
22. Nguyen, N.P. et al., The Asia Cellular Satellite System, *Proc. Fifth International Mobile Satellite Conference*, 145, 1997.
23. Alexovich, A. et al., The Hughes geo-mobile satellite system, *Proc. Fifth International Mobile Satellite Conference*, 159, 1997.
24. Thuraya Web site, http://www.thuraya.com
25. The future: Down the road for PCS computing, *PCS Data Knowledge Site*, Intel, 1998, http://www.pcsdata.com/future.htm
26. Route to W-CDMA, *Ericsson Mobile Systems*, Ericsson, December, 1998, http://www.ericsson.com/wireless/products
27. Cadwalader, D., Toward global IMT-2000, *Ericsson Wireless NOW!*, January, 1998, http://www.ericsson.com/WN/wn1-98/imt2000.html
28. Tuffo, A.G., From the 'outernet' to the Internet, *18th AIAA International Communications Satellite Systems Conference*, April, 2000.

10

Electronic Navigation Systems

Benjamin B. Peterson
U.S. Coast Guard Academy

In an attempt to treat electronic navigation systems in a single chapter, one must either treat lightly or completely ignore many aspects of the subject. Before going into detail on some specific topics, I would like to point out what is and is not in this chapter and why, and where interested readers can go for missing topics or more details. The chapter begins with fairly brief descriptions of the U.S. Department of Defense (DoD) satellite navigation system, NAVSTAR Global Positioning System (GPS), its Russian counterpart, GLObal NAvigation Satellite System, (GLONASS), and LORAN-C. For civil use of GPS and GLONASS there are numerous existing and proposed augmentations to improve accuracy and integrity which will be mentioned briefly. These include maritime Differential GPS, the FAA Wide Area Augmentation System (WAAS), and Local Area Augmentation System (LAAS) for aviation. The second half of the chapter looks in detail at how position is determined from the measurements from these systems, the relationships between geometry, and the statistics of position and time errors. It also considers how redundant information may be optimally used and how information from multiple systems can be integrated.

GPS and its augmentations were chosen because they already are and will continue to be the dominant radio navigation technologies well into the next century. GLONASS could potentially become significant by itself and potentially integrated with GPS. LORAN-C is included while other systems were not for a variety of reasons. The combination of its long history and military significance dating back to World War II, and its large user base, has resulted in a wealth of research effort and literature. Because the basic principle of LORAN-C (i.e., the measurement of relative times of arrival of signals from precisely synchronized transmitters) is the same as that of GPS, it is both interesting and instructive to analyze the solution for position of both systems in a unified way. While the future of LORAN-C in the United States is uncertain

0-8493-1732-0/03/$0.00+$1.50

[DoD/DoT, 1999], it is expanding in Europe and Asia and may remain significant on a worldwide basis for many years. Because LORAN-C has repeatable accuracy adequate for many applications, since it is more resistant to jamming, has failure modes independent of those of GPS, and since its low frequency signals penetrate locations like urban canyons and forests better than those of GPS, analysis of its integration with GPS is felt useful and is included.

Of other major systems, VHF Omnidirectional Range (VOR), Instrument Landing System (ILS), Distance Measuring Equipment (DME), and Microwave Landing System (MLS) are relatively shorter range aviation systems that are tentatively planned to be phased out in favor of GPS/WAAS/LAAS beginning in 2008 [DoD/DoT, 1999]. Due to space and because their principles of operation differ from GPS, they are not included. The U.S. Institute of Navigation in its 50th Anniversary issue of Navigation published several excellent historical overviews. In particular, see Enge [1995] for sections on VOR, DME, ILS, and MLS, and Parkinson [1995] for the history of GPS.

Recently, there has been a significant and simultaneous reduction in the printing of U.S. Government documents and increased distribution of this information via the Internet. Details on Internet and other sources of additional information are included near the end.

10.1 The Global Positioning System (NAVSTAR GPS)

GPS is a U.S. DoD developed, worldwide, satellite-based radio navigation system that will be the DoD's primary radio navigation system well into the 21st century. GPS Full Operational Capability (FOC) was declared on July 17, 1995 by the Secretary of Defense and meant that 24 operational satellites (Block II/IIA) were functioning in their assigned orbits and the constellation had successfully completed testing for operational military functionality.

GPS provides two levels of service — a Standard Positioning Service (SPS) and a Precise Positioning Service (PPS). SPS is a positioning and timing service, which is available to all GPS users on a continuous, worldwide basis. SPS is provided on the GPS L1 frequency, which contains a coarse acquisition (C/A) code and a navigation data message. The current official specifications state that SPS provides, on a daily basis, the capability to obtain horizontal positioning accuracy within 100 m (95% probability) and 300 m (99.99% probability), vertical positioning accuracy within 140 m (95% probability), and timing accuracy within 340 ns (95% probability). For most of the life of GPS, the civil accuracy was maintained at approximately these levels through the use of Selective Availability (SA) or the intentional degradation of accuracy via the dithering of satellite clocks. In his Presidential Decision Directive in 1996, President Clinton committed to terminating SA by 2006. On May 1, 2000, in a White House press release [White House, 2000] the termination of SA was announced and a few hours later it was turned off. Work is starting on a new civil GPS signal specification with revised accuracy levels. Very preliminary data indicate 95% horizontal accuracy of approximately 7 m will be possible under some conditions. When SA was on, it was the dominant error term; accuracy for all receivers with a clear view of the sky was easily predictable and independent of other factors. Now, it is expected that terms such as multipath and ionospheric delay, which vary greatly with location, time, and receiver and antenna technology will dominate, and exact prediction of error statistics will more difficult.

The GPS L1 frequency also contains a precision (P) code that is not a part of the SPS. PPS is a highly accurate military positioning, velocity, and timing service which is available on a continuous, worldwide basis to users authorized by the DoD. PPS is the data transmitted on GPS L1 and L2 frequencies. PPS is designed primarily for U.S. military use and is denied to unauthorized users by the use of cryptography. Officially, P-code-capable military user equipment provides a predictable positioning accuracy of at least 22 m (2 drms) horizontally and 27.7 m (2 sigma) vertically, and timing/time interval accuracy within 90 ns (95% probability).

The GPS satellites transmit on two L-band frequencies: L1 = 1575.42 MHz and L2 = 1227.6 MHz. Three pseudorandom noise (PRN) ranging codes are in use giving the transmitted signal direct sequence, spread-spectrum attributes. The coarse/acquisition (C/A) code has a 1.023 MHz chip rate, a period of one millisecond (ms), and is used by civil users for ranging and by military users to acquire the P-code. Bipolar

Phase-Shift Key (BPSK) modulation is utilized. The transmitted PRN code sequence is actually the Modulo-2 addition of a 50 Hz navigation message and the C/A code. The SPS receiver demodulates the received code from the L1 carrier, and detects the differences between the transmitted and the receiver-generated code. The SPS receiver uses an exclusive or truth table, to reconstruct the navigation data, based upon the detected differences in the two codes. Ward [1994–1995] contains an excellent description of how receivers acquire and track the transmitted PRN code sequence.

The precision (P) code has a 10.23 MHz rate, a period of seven days, and is the principle navigation ranging code for military users. The Y-code is used in place of the P-code whenever the anti-spoofing (A-S) mode of operation is activated. *Anti-spoofing* (A-S) guards against fake transmissions of satellite data by encrypting the P-code to form the Y-code. A-S was exercised intermittently through 1993 and implemented on January 31, 1994. The C/A code is available on the L1 frequency and the P-code is available on both L1 and L2. The various satellites all transmit on the same frequencies, L1 and L2, but with individual code assignments.

Each satellite transmits a navigation message containing its orbital elements, clock behavior, system time, and status messages. In addition, an almanac is provided that gives the approximate data for each active satellite. This allows the user set to find all satellites once the first has been acquired. Tables 10.1 and 10.2 include examples of ephemeris and almanac information for the same satellite at approximately the same time.

The nominal GPS constellation is composed of 24 satellites in 6 orbital planes, (4 satellites in each plane). The satellites operate in circular 20,200-km altitude (26,570-km radius) orbits at an inclination angle of 55° and with a 12-h period. The position is therefore the same at the same sidereal time each day (i.e., the satellites appear four minutes earlier each day).

The GPS Control segment consists of five Monitor Stations (Hawaii, Kwajalein, Ascension Island, Diego Garcia, Colorado Springs), three Ground Antennas, (Ascension Island, Diego Garcia, Kwajalein), and a Master Control Station (MCS) located at Falcon AFB in Colorado. The monitor stations passively

TABLE 10.1 Ephemeris Information

Satellite Ephemeris Status for PRN07
Ephemeris Reference Time = 14:00:00 02/10/1995, Week Number = 0821
All Navigation Data is Good, All Signals OK
Code on L2 Channel = Reserved, L2 P Code Data = On
Fit Interval = 4 hours, Group Delay = 1.396984e-09 s
Clock Correction Time = 14:00:00 02/10/1995, af0 = 7.093204e-04 s
af1 = 4.547474e-13 s/s, af2 = 0.000000e+00 s/s2
Semi-Major Axis = 26560.679 km, Eccentricity = 0.007590
Mean Anom = –23.119829, Delta n = 2.655361e-07, Perigee = –144.510162
Inclination = 55.258496, Inclination Dot = 1.100943e-08
Right Ascension = –10.880005, Right Ascension Dot = –4.796266e-07
Crs = 5.3750e+01 m, Cuc = 1.6691e-04, Crc = 2.7141e+02 m, Cic = 0.0000e+00

TABLE 10.2 Almanac Information

SATELLITE ALMANAC STATUS FOR PRN07
Almanac Validity = Valid Almanac at 09:06:07 01/10/1995
Semi-Major Axis = 26560.168 km
Eccentricity = 0.007590
Inclination = 55.258966
Right Ascension = –10.829880
Right Ascension Rate = –4.655885e-07/s
Mean Anomaly = –172.426664
Argument of Perigee = –144.590529
Clock Offset = 7.095337e-04 s, Clock Drift = 0.000000e+00 s/s
Navigation Health Status Is: All Data OK
Signal Health Status Is: All Signals OK

track all satellites in view, accumulating ranging data. This information is processed at the MCS to determine satellite orbits and to update each satellite's navigation message. Updated information is transmitted to each satellite via the Ground Antennas.

The GPS system uses time of arrival (TOA) measurements for the determination of user position. A precisely timed clock is not essential for the user because time is obtained in addition to position by the measurement of TOA of four satellites simultaneously in view. If altitude is known (e.g., for a surface user), then three satellites are sufficient. If a stable clock (say, since the last complete coverage) is keeping time, then two satellites in view are sufficient for a fix at known altitude. If the user is, in addition, stationary or has a known speed then, in principle, the position can be obtained by the observation of a complete pass of a single satellite. This could be called the "transit" mode, because the old Transit system used this method. In the case of GPS, however, the apparent motion of the satellite is much slower, requiring much more stability of the user clock.

The current (May 2000) GPS constellation consists of 28 satellites counting 27 operational and one recently launched and soon to become operational. Figures 10.1 and 10.2 illustrate the history of the constellation including both the limiting dates of the operational periods, individual satellites, and the total number available as a function of time.

10.1.1 GPS Augmentations

Many augmentations to GPS exist to improve accuracy and provide integrity. These are both government operated (for safety of life navigation applications) and privately operated. This section will only deal with government operated systems. The recent termination of Selective Availability has sparked some debate on the need for GPS augmentation, but the most prevalent views are that stand-alone GPS will not satisfy all accuracy and integrity requirements, and that some augmentation will continue to be necessary. Since the dominant post-SA errors vary much more slowly than SA-induced errors did, the information bandwidth requirements to meet specified accuracy levels have certainly been reduced. Since integrity requirements specify maximum times to warn users of system faults, these warning times and time to first fix specifications will now drive information bandwidth requirements.

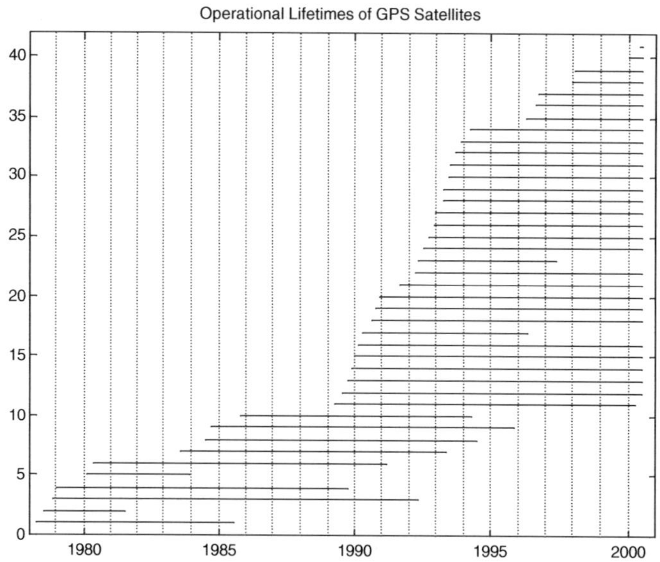

FIGURE 10.1 Operational lifetimes of GPS satellites.

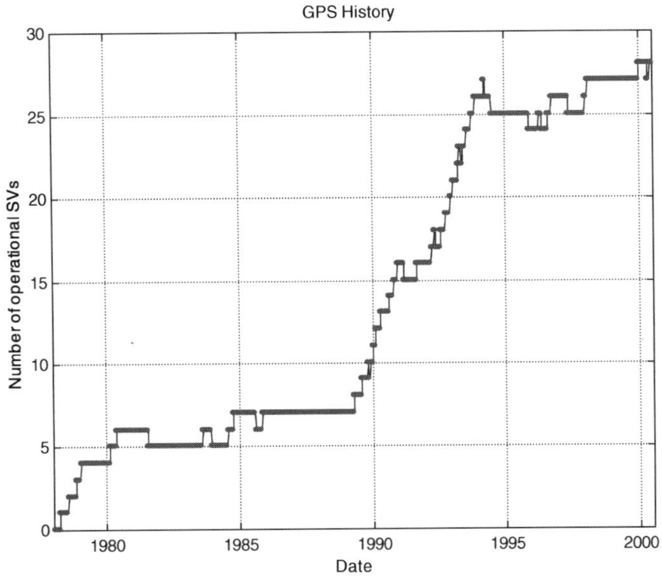

FIGURE 10.2 GPS history.

The simplest and first to become operational were maritime Differential GPS stations. Existing radio beacons in the 285- to 325-kHz band were converted to transmit differential corrections via a Minimum-Shift Keyed (MSK) modulation scheme. A base station receiver in a fixed, known location measures pseudorange errors relative to its own clock and known position. Messages containing the time of observation, these pseudorange errors, and their rates of change are then transmitted at either 100 or 200 Bd. Parity bits are added, but no forward error correction is used. If the user receiver notes a data error, it merely waits for the next message. The user receiver uses the corrections extrapolated ahead in time based on the age of the correction and its rate of change. For a description of the message format the reader is referred to RTCM [1992].

With SA on, satellite clock dithering was the dominant error and was common to all users. It is also assumed that over the few hundred kilometers or less where the signal can be received, that errors such as ephemeris and ionospheric and tropospheric delay are correlated between base station and user receiver as well and can be considerably reduced. The algorithms to predict these delays are disabled in both the base station and user receivers.

Because the corrections are measured relative to the base station clock, they are relative and not absolute corrections. This means they must have corrections for all satellites used for a position and all of these corrections must be from the same base station. For fixed position, precise time users, DGPS provides no improvement in accuracy. For moving users, their solution for time will track the base station clock, and since this base station receiver is in a fixed known location, the time solution in the moving receiver will see improved accuracy as well.

The U. S. Coast Guard operates maritime DGPS along the coasts and rivers. In addition, the system is being expanded inland in support of the Federal Railroad Administration in the National DGPS project. Many foreign governments worldwide operate compatible systems primarily for maritime applications. RTCM type DGPS messages are also transmitted by modulating LORAN-C transmitters in Europe. Additional details on this system known as EUROFIX are included in the later section on LORAN-C.

Three Satellite Based Augmentation Systems (SBASs) for aviation users are in various stages of development. These include:

1. The Wide Area Augmentation System (WAAS) by the U.S. FAA
2. The European Geostationary Navigation Overlay System (EGNOS) jointly by the European Union, the European Space Agency (ESA), and EUROCONTROL
3. MTSAT Satellite Based Augmentation System (MSAS) by the Japan Civil Aviation Bureau (JCAB)

These systems are intended for the enroute, terminal, non-precision approach, and Category I (or near Category I) precision approach phases of flight. All of these systems are designed to be compatible, and to the level of detail in this chapter, are equivalent. For additional details on all three systems the reader is referred to Walter [1999] and to RTCA [1999] for the WAAS signal specification. In SBASs, corrections need to be applied over the very large geographic area in which the satellite signal can be received. In this case the errors due to satellite ephemeris and ionospheric delay are not the same for all users and cannot be combined into one overall correction. Separate messages are provided for satellite clock corrections, vector corrections of satellite position and velocity, integrity information, and vertical ionospheric delay estimates for selected grid points. These grid points are at 5° increments in latitude and longitude, except larger increments occur in extreme northern or southern latitudes. The user receiver calculates its estimates of ionospheric delay by first determining the ionospheric pierce point of the propagation path from satellite to user receiver, interpolating between grid points, and then correcting for slant angle.

These corrections are or will be transmitted from geostationary satellites at GPS L1 frequencies with signal characteristics similar to GPS. The SBASs will provide additional ranging signals and transmit their own ephemeris information to improve fix availability. Message symbols at 500 symbols per second will be modulo 2 added to a 1023-b PRN code. The baseline data rate is 250 b/s. This data rate is ½ convolutional encoded with a Forward Error Correction code resulting in 500 symbols per second. The data will be sent in 250-b blocks or one block per second. The data block contains an 8-b preamble, a 6-b message type, a 212-b message, and 24 b of CRC parity.

For Category II and III precision approach, the U.S. FAA will implement the Local Area Augmentation System (LAAS). The transmitted data will include pseudorange correction data, integrity parameters, approach data, and ground station performance category. The broadcast will be in the 108- to 117.95-MHz band presently used for VOR and ILS systems.

The modulation format is a differentially encoded, eight-phase-shift-keyed (D8PSK) scheme. The broadcast uses an eight time slot per half second, fixed-frame, time division multiple access (TDMA) structure. The total data rate of the system is 31,500 b/s. After the header and cyclic redundancy check (CRC), the effective rate of the broadcast is 1776 b (222 bytes) of application data per time slot, or a total of 16 × 1776 b/s equaling 28,416 application data bits per second. For additional information on LAAS the reader is referred to Braff [1997] and Skidmore [1999].

10.2 Global Navigation Satellite System (GLONASS)

GLONASS is the Russian parallel to GPS and has its origins in the mid-1970s in the former Soviet Union. Like GPS, the system has been primarily developed to support the military, but in recent years has been broadened to include civilian users. In September 1993, Russian President Boris Yeltsin officially proclaimed GLONASS to be an operational system and the basic unit of the Russian Radionavigation Plan. The well-publicized political, economic, and military uncertainties within Russia have undoubtedly hampered the implementation and maintenance of the system. These issues combined with the poor reliability record of the early satellites, raised questions relative to its eventual success. Figures 10.3 and 10.4 show data on the operational life of those satellites that became operational from the first one in 1982 through the present. It is also believed that attempts were made to place approximately 10 or more other satellites into operation resulting in failure for an number of reasons [Dale, 1989]. However, in the mid-1990s, the reliability of the satellites considerably improved, and many more satellites were put in orbit such that in early 1996 there were 24 operational satellites. The triple launch on December 30, 1998 has been the only launch since late 1995, and the constellation has degraded to only 10 operational

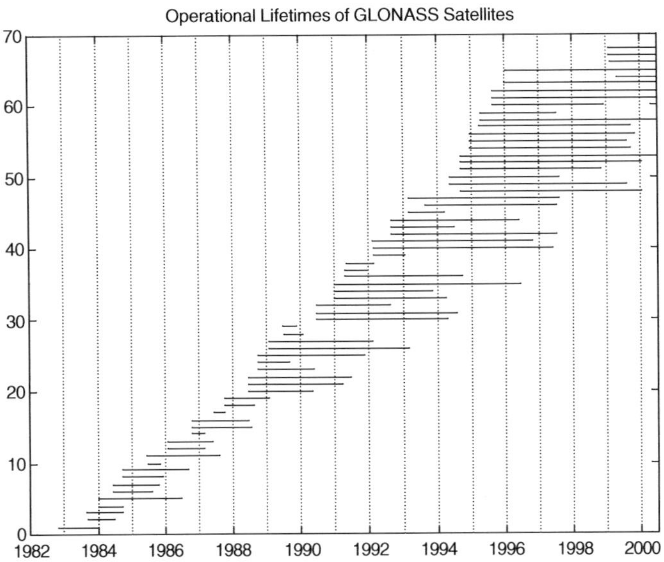

FIGURE 10.3 Operational lifetimes of GLONASS satellites.

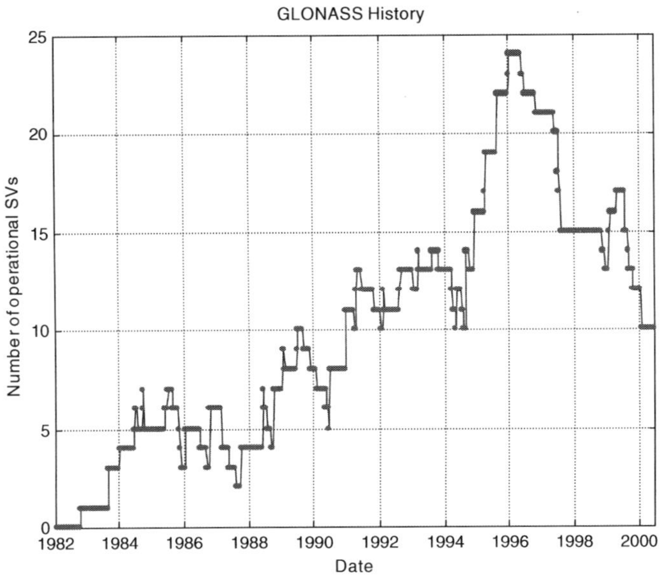

FIGURE 10.4 GLONASS history.

satellites, again raising concerns on the future of the system. Publicly, the Russian Federation remains fully committed to the system. To date, GLONASS receivers have been quite expensive and have only been produced in limited quantities.

While similar in many respects to GPS, GLONASS does have important differences. While all GPS satellites transmit at the same frequencies and are distinguished by their PRNs, all GLONASS satellites transmit the same PRN and are distinguished by their frequencies. The L1 transmitted frequencies in MHz are given by 1602 + 0.5625 × Channel Number, which extends from 1602.5625 to 1615.5 MHz. L2 frequencies are seven-ninths L1 and in MHz are given by 1246 + 0.4375 × Channel Number or from 1246.4375 to 1256.5 MHz. GLONASS satellites opposite each other in the same orbit plane have been assigned the same channel number and frequency to limit spectrum use. Like GPS, the C/A code is transmitted on L1 only, and P code on both L1 and L2. The C/A code is 511 chips long at 511 kb/s for a length of 1 ms. The P code is at 5.11 Mb/s. Like GPS, the actual modulation is the Modulo-2 addition of the PRN sequence and 50 b/s data and BPSK modulation is used.

The GLONASS planes have a nominal inclination of 64.8° compared to 55° for GPS, which gives slightly better polar fix geometry at the cost of fix geometry at lower latitudes. The 24 slots are in three planes of 8 slots each. The orbital altitude is 25,510 km or 1050 km less than GPS. The orbit period is 11 hours, 15 minutes.

Rather than transmitting ephemeris parameters as in GPS, GLONASS satellites transmit actual position, velocity, and acceleration in Earth-centered, Earth-fixed (ECEF) coordinates that are updated on half-hour intervals. The user receiver integrates using Runge-Kutta techniques for other times. All monitor sites are within the former Soviet Union, which limits accuracy of both ephemeris and time and would delay user notification of satellite failures. The system produces both high accuracy signals for Russian military use only and lesser accuracy for civilian use. The civilian use signals are not degraded to the same extent as GPS was with SA, which resulted in significantly better accuracy than civil GPS during the period GLONASS had a complete constellation. Military accuracy is classified. For more details the reader is referred to the GLONASS Interface Control Document [CSIC, 1998] available at http://www.rssi.ru/sfcsic/sfcsic_main.html

10.3 LORAN-C History and Future

Early in World War II, both the U.S. and Great Britain recognized the need for an accurate, long range, radio navigation system to support military operations in Europe. As a result the British developed Gee and the U.S. developed LORAN (Long Range Navigation). Both were pulsed, hyperbolic systems with time differences between master and secondary transmitters measured by an operator matching envelopes on a delayed sweep CRT. Gee operated at several carrier frequencies from 20 to 85 MHz and had a pulse width of 6 μs. Standard LORAN (or LORAN-A) operated at 1.95 MHz and had a pulse width of 45 μs. The first LORAN-A chain was completed in 1942 and by the end of the war 70 stations were operating. For details on the World War II LORAN effort see Pierce [1948].

It was recognized that lower frequencies would propagate longer distances, and near the end of the war testing began on a 180-kHz system. The pace of development slowed considerably after the war. The band from 90 to 110 kHz was established by international agreement for long-range navigation. In 1958, system tests started on the first LORAN-C chain, which consisted of stations at Jupiter, Florida, Carolina Beach, North Carolina, and Martha's Vineyard, Massachusetts.

Over the next two decades LORAN C coverage was expanded by the U.S. Coast Guard in support of the DoD to much of the U.S. (including Alaska and Hawaii), Canada, northwest Europe, the Mediterranean, Korea, Japan, and Southeast Asia. The Southeast Asia chain ceased operations with the fall of South Vietnam in 1975. In 1974, LORAN-C was designated as the primary means of navigation in the U.S. Coastal Confluence Zone (CCZ), the cost of civilian receivers declined sharply, and the civil maritime use of the system became very large. In the late 1980s at the request of the Federal Aviation Administration, the Coast Guard built four new stations in the mid-continent and established LORAN-C coverage for the entire continental U.S.

Other countries are developing and continuing LORAN-C to meet their future navigational needs. Many of the recent initiatives have taken place as a result of the termination of the U.S. DoD requirement for overseas LORAN-C. This need came to an end as of December 31, 1994. With the introduction of GPS, many countries have decided that it is in their own best interests not to have their navigational needs met entirely by a U.S. DoD-controlled navigation system. Many of these initiatives have resulted in multilateral agreements between countries, which have common navigational interests in those geographic areas where LORAN-C previously existed to meet U.S. DoD requirements (e.g., northern Europe and the Far East). The countries of Norway, Denmark, Germany, Ireland, the Netherlands, and France established a common LORAN-C system under the designation of the Northwest European LORAN-C System (NELS). Recently the governments of Italy and the United Kingdom have applied for membership. This system presently comprises eight stations forming four chains. In conjunction with this system, respective foreign governments took over operation of the former USCG stations of B0 and Jan Mayen, Norway; Sylt, Germany; and Ejde, Faroe Islands (Denmark) as of December 31, 1994. The two French stations formerly operated by the French Navy in the rho-rho mode were reconfigured and included and two new stations in Norway were constructed. A planned ninth station at Loop Head, Ireland has yet to be constructed. The former USCG stations at Angissoq, Greenland and Sandur, Iceland were closed.

An important feature of NELS is the transmission of DGPS corrections by pulse position modulation. This concept, called EUROFIX, was developed by Professor Durk Van Willigen and his students at Delft University of Technology in the Netherlands. The last six pulses in each group are modulated in a three-state pattern of prompt and 1 μs advance or retard. Of the possible 729 combinations, 142 of which are balanced, 128 balanced sequences are used to transmit a 7-b word. Extensive Reed Solomon error correction is used resulting in a very robust data link, although at a somewhat lower data rate than the MSK beacons. With Selective Availability on, the horizontal accuracy was 2 to 3 m, 2 drms, or slightly worse than conventional DGPS due to the temporal decorrelation of SA. With SA off, the performance should be virtually comparable. EUROFIX has been operating at the Sylt, Germany transmitter since February 1997; expansion to other NELS transmitters is planned for the near future. In a joint FAA/USCG effort, preliminary testing is underway of transmitting WAAS data via a much higher data rate LORAN communications channel. If LORAN-C survives in the long term, it is likely to be both as a navigation system and as a communications channel to enhance the accuracy and integrity of GPS.

In the Far East, an organization was formed called the Far East Radionavigation Service (FERNS). This organization consists of the countries of Japan, the Peoples Republic of China, the Republic of Korea, and the Russian Federation. Japan took over operation of the former USCG stations in its territory and they are currently being operated by the Japanese Maritime Safety Agency (JMSA). In the Mediterranean Sea area, the four USCG stations were turned over to the host countries. The two stations in Italy, Sellia Marina and Lampedusa, are currently being operated. The stations at Kargaburun, Turkey and Estartit, Spain remain off air. Saudi Arabia and India are currently each operating two LORAN chains.

There is ongoing debate at high levels concerning the future of LORAN-C in the United States. According to the 1999 Federal Radionavigation Plan [DoD/DoT, 1999], "While the administration continues to evaluate the long term need for continuation of the LORAN-C navigation system, the Government will operate the LORAN-C system in the short term. The U. S. Government will give users reasonable notice if it concludes that LORAN-C is not needed or is not cost effective, so that users will have the opportunity to transition to alternative navigational aids."

10.3.1 LORAN Principles of Operation

Each chain consists of three or more stations, including a master and at least two secondary transmitters. (Each master-secondary pair enables determination of one LOP, and two LOPs are required to determine a position. The algorithms to convert these LOPs into latitude and longitude are discussed in a later section.) Each LORAN-C chain provides signals suitable for accurate navigation over a designated geographic area termed a *coverage area.*

The stations in the LORAN chain transmit in a fixed sequence that ensures that Time Differences (TDs) between receiving master and secondary can be measured throughout the coverage area. The length of time in microseconds over which this sequence of transmissions from the master and the secondaries takes place is termed the Group Repetition Interval (GRI) of the chain. All LORAN-C chains operate on the same frequency (100 kHz), but are distinguished by the GRI of the pulsed transmissions.

The LORAN-C system uses pulsed transmission, nine pulses for the master and eight pulses for the secondary transmissions. Figure 10.5 shows this overall pulse pattern for the master and three secondary transmitters (X, Y, and Z). Coding delay is the time between when a secondary receives the master signal and when it transmits. The time differences measured on the baseline extension beyond the secondary

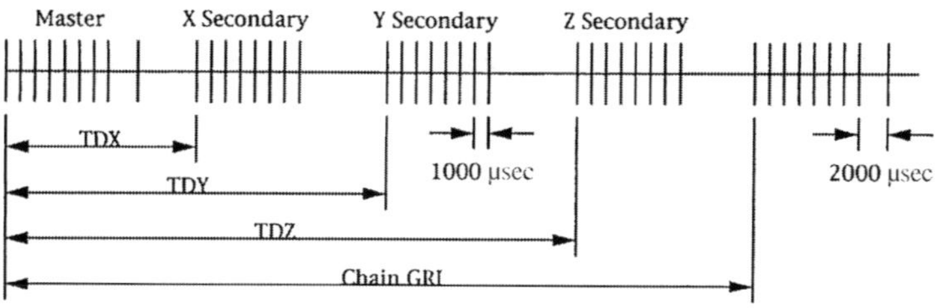

FIGURE 10.5 LORAN-C pulse pattern.

should be coding delay. Emission delay is the difference in time of transmission between master and secondary and is the sum of coding delay and baseline length (converted to time). Shown in Fig. 10.6 is an exploded view of the LORAN-C pulse shape. It consists of sine waves within an envelope referred to as a t-squared pulse. (The equation for the envelope is also included in Fig. 10.6.) This pulse will rise from zero amplitude to maximum amplitude within the first 65 μs and then slowly trails off or decays over a 200 to 300 μs interval. The pulse shape is designed so that 99% of the radiated power is contained within the allocated frequency band for LORAN-C of 90 to 110 kHz. The rapid rise of the pulse allows a receiver to identify one particular cycle of the 100 kHz carrier. Cycles are spaced approximately 10 μs apart. The third cycle of this carrier within the envelope is used when the receiver matches the cycles. The third zero crossing (termed the positive 3rd zero crossing) occurs at 30 μs into the pulse. This time is both late enough in the pulse to ensure an appreciable signal strength and early enough in the pulse to avoid sky wave contamination from those sky waves arriving close after the corresponding ground wave.

Within each pulse group from the master and secondary stations, the phase of the radio frequency (RF) carrier is changed systematically from pulse-to-pulse in the pattern shown in Fig. 10.6 and Table 10.3. This procedure is known as phase coding. The patterns A and B alternate in sequence. The pattern of phase coding differs for the master and secondary transmitters. Thus, the exact sequence of pulses is actually matched every two GRIs, an interval known as a phase code interval (PCI).

Phase coding enables the identification of the pulses in one GRI from those in an earlier or subsequent GRI. Just as selection of the pulse shape and standard zero crossing enable rejection of certain early sky waves interfering with the same pulse, phase coding enables rejection of late sky waves interfering with the next pulse. Since 7 of the 14 pulses that come immediately before another pulse have the same phase code and 7 are different, late sky waves interfering with the next pulse will average to zero. Because the master and secondary signals have different phase codes, the LORAN receiver can distinguish between them.

10.4 Position Solutions from Radio Navigation Data

In general, closed-form solutions that allow direct calculation of position from radio navigation observables do not exist. However, the inverse calculations can be done. Given one's position and clock offset, one can predict GPS or GLONASS pseudoranges or LORAN TDs or TOAs. These expressions are nonlinear but can be linearized about a point and the problem solved by iteration. The basic approach can be summarized by:

1. Assume a position.
2. Calculate the observables one should have seen at that point.
3. Calculate rate of change of observables (partial derivatives) as position is varied.
4. Compare calculated to measured observables.
5. Compute a new position based on differences between calculated and measured observables.
6. Depending on exit criteria, either exit algorithm or return to Step 2 using the calculated position as the new assumed position.

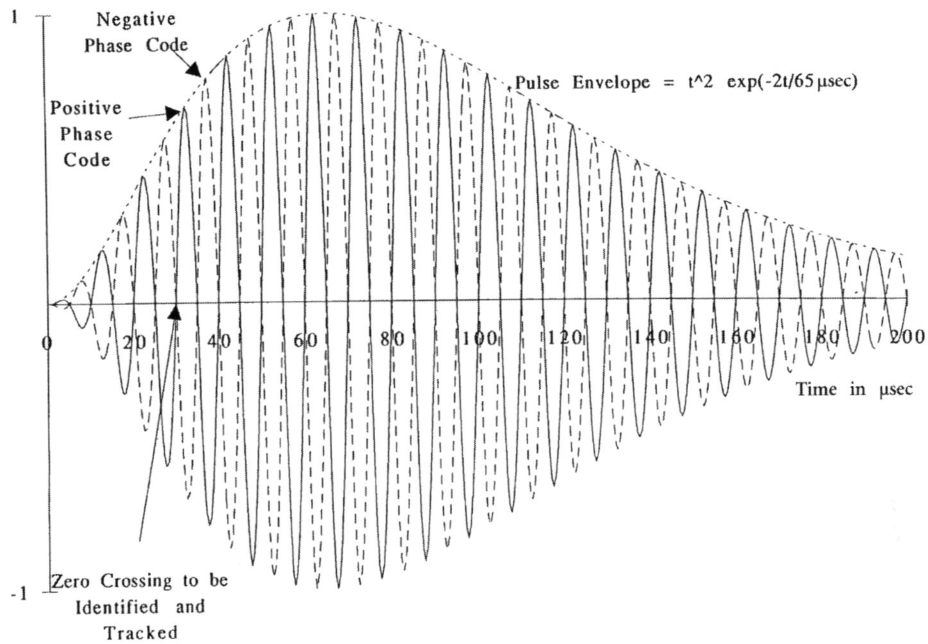

FIGURE 10.6 Ideal LORAN pulse.

TABLE 10.3 LORAN-C Phase Codes

GRI Interval	Master	Secondary
A	++−−+−+−+	+++++−−+
B	+−−+++++−	+−+−++−−

We will look at the details of each of these steps and the expected statistics of the results for a number of cases. We will start with the example of processing three LORAN TOAs (or two TDs), but the basic principles apply to GPS as well. We will then consider overdetermined solutions and the techniques for assuring the solutions are optimal in some sense. We will also consider integrated solutions and Kalman filter based solutions.

For LORAN the calculated Times of Arrival (TOAs) are given by:

For Master $\quad TOA_m = d_i$,
and for secondaries $\quad TOA_i = d_i + ED_i$, etc.

where d_m and d_i are the propagation times from the master and i^{th} secondary stations to the assumed position and ED_i is Emission Delay. (Note: Secondary stations are designated by the letters V, W, X, Y, and Z, but because we will also want to use x and y for position variables, in order to avoid confusion, we will use integer subscripts to denote secondaries.) The LORAN propagation times are the sums of three terms:

1. *Primary Factor* (PF) or the geodesic distance to the station divided by a nominal phase velocity. See WGA [1982, 57–61] for an algorithm that will work to distances up to 3000 nautical miles.
2. *Secondary Factor* (SF) which takes into account the additional phase lag due to the signal following a curved Earth surface over an all-seawater path (see COMDT(G-NRN) [1992] II-14 for formula).

3. *Additional Secondary Factor* (ASF) which takes into account the difference between the actual path traveled and an all-seawater path. This factor can be calculated based on conductivity maps, or can be obtained by tables (from the U. S Defense Mapping Agency) or from charts published by the Canadian government. The latter two are based both on calculations and observations.

A typical LORAN receiver measures time differences (TDs) vs. TOAs and the receiver need not necessarily solve for time. In GPS or GLONASS, since the transmitters move rapidly with time, in order to know their location it is essential that an explicit solution for time be made. In GPS each satellite transmits ephemeris parameters which, when substituted in standard equations [NATO, 1991, A-3-26 and 27], give the satellite location as a function of time expressed in Earth-centered, Earth-fixed (ECEF) coordinates. In this system, the positive x-axis goes from Earth's center to 0° latitude and longitude, the positive y-axis goes from Earth's center to 0° latitude and 90°E longitude, and the positive z-axis goes from Earth's center to 90°N latitude. For GLONASS, the satellites transmit their location, velocity, and acceleration, in ECEF coordinates, and validate at half-hour intervals. The user is expected to numerically integrate the equations on motion to obtain position for a particular time.

The receiver measures the time of arrival of the satellite signal, and since the signal is tagged precisely in time, the difference in time of transmission and time of arrival (converted to distance) is a pseudorange. "Pseudo" is used as the time of arrival relative to the receiver clock, which cannot be assumed exact but must be solved for as part of the solution. These pseudoranges are then corrected for both ionospheric and tropospheric delay. In civil C/A code receivers, ionospheric delay parameters are transmitted and the delay calculated via an algorithm [NATO, 1991, p. A-6-31]. For dual frequency receivers, the ionospheric delay is assumed to be inversely proportional to frequency and the delay determined from the difference in pseudoranges at the two frequencies. For differential GPS, ionospheric delay is assumed to be the same at the reference station and the user's location, and the algorithm is disabled at both locations.

10.4.1 Solution Using Two TDs (LORAN only)

TDs at the assumed position are calculated ($\mathrm{TD_p}$) and subtracted from observations ($\mathrm{TD_o}$), i.e.,

$$\mathrm{TD_i} = \mathrm{TOA_i} - \mathrm{TOA_m}$$

$$\Delta\mathrm{TD} = \begin{bmatrix} \Delta\mathrm{TD_1} \\ \Delta TD_2 \end{bmatrix} = \mathrm{TD_o} - \mathrm{TD_p}$$

This difference in TDs can be linearly related to a difference in position $\Delta\mathrm{P} = \begin{bmatrix} \Delta x \\ \Delta y \end{bmatrix}$ by

$$\begin{bmatrix} \sin(\phi_m) - \sin(\phi_1) & \cos(\phi_m) - \cos(\phi_1) \\ \sin(\phi_m) - \sin(\phi_2) & \cos(\phi_m) - \cos(\phi_2) \end{bmatrix} \begin{bmatrix} \Delta x \\ \Delta y \end{bmatrix} = c \begin{bmatrix} \Delta TD_1 \\ \Delta TD_2 \end{bmatrix}$$

where ϕ_m, ϕ_1, and ϕ_2 are the azimuths to the transmitters measured in a clockwise direction from the north or positive y direction as shown in Fig. 10.7. While exact calculations taking into account the ellipsoidal nature of the shape of Earth must be used in the calculation of predicted TDs, only approximate expressions based on a spherical Earth are necessary to determine azimuths. The expression above can be expressed in matrix form:

$$\mathrm{A_{TD}}\Delta\mathrm{P} = \mathrm{c}\,\Delta\mathrm{TD}$$

and solved by:

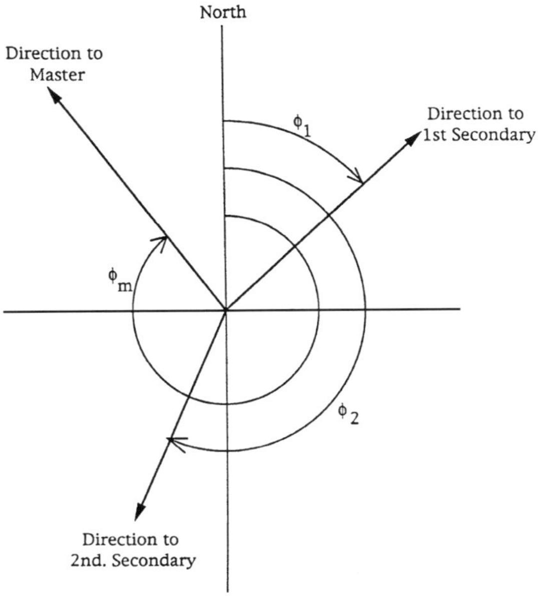

FIGURE 10.7 Definition of LORAN azimuths.

$$\Delta P = c\, A_{TD}^{-1}\, \Delta TD$$

At this point a new assumed position would be calculated (using an appropriate conversion of meters to degrees); the predicted TDs at that location determined; and if they were within a selected tolerance of the measured, the algorithm would stop. Otherwise a new solution would be determined. An alternative criteria based on the movement of the assumed position, $|\Delta P|$, can be used as well and will be more appropriate in an over-determined solution because exact match of TDs is not possible. Figures 10.8 and 10.9 show typical convergence of this type of algorithm. Since the algorithm is based on the assumption that the TD grid is linear in the vicinity of the assumed position, we need to limit the step size so that the position does not jump to radically different geometry in one step. As can be seen in Fig.10.8 all lines of constant TD between two transmitters are closed curves on Earth's surface. Two of these curves intersect, either do not intersect, or intersect exactly twice. Since we assume we are starting with a set of TDs that exist, the latter applies. The initial assumed position is what determines which of these two ambiguous positions the algorithm will converge to. Essentially since the algorithm is based on the gradient of the TDs with respect to change in position and that gradient changes sign on the baseline extension, the position will move to the solution it can reach without crossing a baseline extension. The direction of the gradient, and hence the final solution also changes as the lines of constant TD become parallel as at 30°N and 10°E in Fig. 10.9.

Frequently, as shown in the example in Fig. 10.8, one of these ambiguous positions is well beyond the range the signals can be received and can easily be eliminated. This is not true when one of the solutions is near the baseline extension as shown in Fig. 10.9. In this example, it is necessary to have prior knowledge that the actual position is north or south of the Master station.

10.4.2 Solution Using TOAs

This method yields a solution for receiver clock bias (modulo one Phase Code Interval) and even though not generally implemented in present receivers, is presented for a number of reasons:

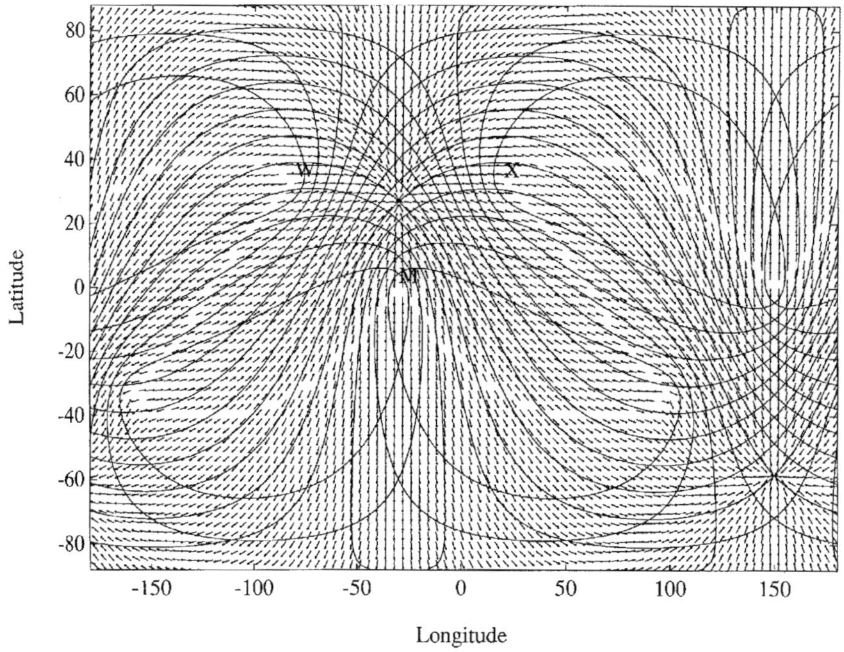

FIGURE 10.8 Convergence of position solution algorithm.

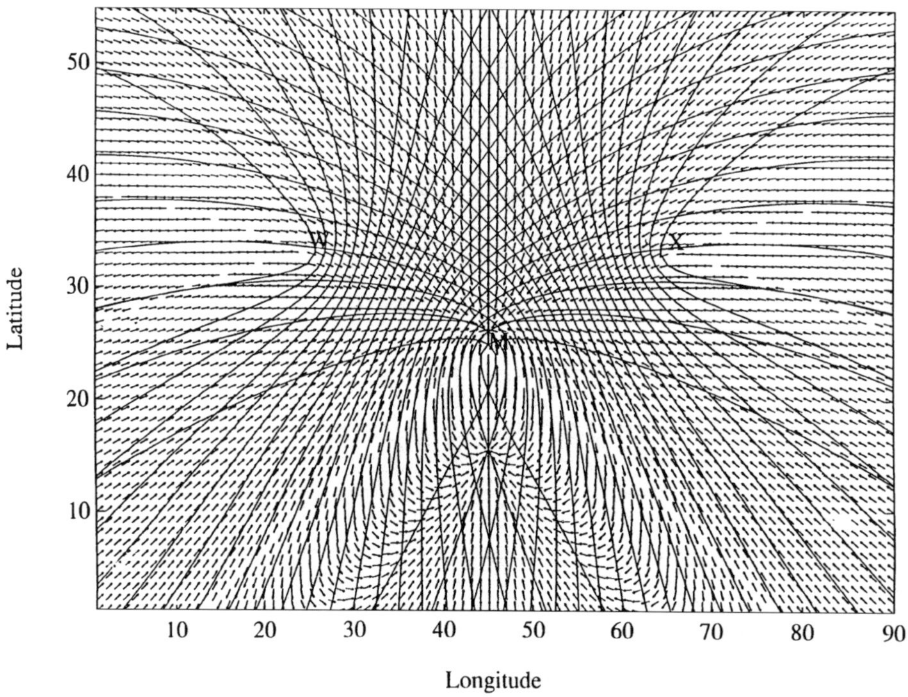

FIGURE 10.9 Example of position solution with close by ambiguous position.

1. It is functionally equivalent to the GPS position solution but is presented first because it is a two- vs. three-dimensional and is somewhat easier to visualize.
2. For the exactly determined (two-TD or three-TOA) case it is exactly equivalent to the TD solution and therefore we can extend results from TOA or pseudorange analysis to the TD case. Early LORAN TD analysis was typically done in scalar form before computer programs such as spreadsheets and MatLab™ made matrix calculations trivial. GPS fix analysis has typically been expressed in matrix form resulting is simpler expressions.
3. For over-determined solutions, since we can assume TOA errors are statistically independent, but cannot make the same assumption for TD errors, the solution is slightly easier to implement and analyze.
4. Solutions using Kalman filters, integrated GPS/LORAN, or a precise clock with only two TOAs, will typically be based on TOAs vs. TDs.

The vector difference between observed (TOA_o) and predicted (TOA_o) TOAs are calculated via:

$$\Delta TOA = TOA_o - TOA_p$$

This equation is replaced by:

$$-\begin{bmatrix} \sin(\phi_m) & \cos(\phi_m) & 1 \\ \sin(\phi_1) & \cos(\phi_1) & 1 \\ \sin(\phi_2) & \cos(\phi_2) & 1 \end{bmatrix} \begin{bmatrix} \Delta x \\ \Delta y \\ c\Delta t \end{bmatrix} = c \begin{bmatrix} \Delta TOA_m \\ \Delta TOA_1 \\ \Delta TOA_2 \end{bmatrix}$$

or:

$$\mathrm{a}_{\mathrm{TOA}} \Delta \mathrm{P} = \mathrm{c}\, \Delta \mathrm{TOA}$$

and solved by:

$$\Delta P = c\, A_{TOA}^{-1}\, \Delta TOA$$

It is useful to think of the sines and cosines in the matrix as the cosines of the angles between the positive direction of the coordinate axes and the directions to the transmitters. This concept can be easily extended to the three-dimensional case in GPS or GLONASS solutions. For a normal GPS-only solution it is usually more convenient to express both satellite and user positions in an ECEF coordinate system. In this case the direction cosines are merely the difference of the particular coordinate divided by the range to each satellite. After position solution, an algorithm is used to convert from (xyz) to latitude, longitude, and altitude. An alternative that will prove useful for meaningful Dilution of Precision (DOP) calculations or for integrated GPS/LORAN receivers is to use a east/north/altitude coordinate system and to solve for the azimuth (AZ) and elevation (EL) of the satellites. In this case the direction cosines for each satellite become:

East: sin(AZ) cos(EL)
North: cos(AZ) cos(EL)
Altitude: sin(EL)

10.5 Error Analysis

With only three TOAs this solution will be exactly the same as the two-TD solution above. In both of these solutions after the algorithm has converged, we can consider ΔTOA or ΔTD as measurement errors and ΔP as the position error due to these errors. The covariance of the errors in position (and in time for TOA processing) can be expressed as functions of the direction cosine matrix and the covariance of the TDs or TOAs.

$$\Delta P\ \Delta P^T = c^2 A_{TD}^{-1} \Delta TD\ \Delta TD^T \left(A_{TD}^{-1}\right)^T$$

$$\mathrm{cov}(\Delta P) = E\{\Delta P\ \Delta P^T\} = c^2 A_{TD}^{-1} E\{\Delta TD\ \Delta TD^T\} \left(A_{TD}^{-1}\right)^T = c^2 A_{TD}^{-1} \mathrm{cov}(\Delta TD) \left(A_{TD}^{-1}\right)^T$$

$$\mathrm{cov}(\Delta P) = c^2 A_{TOA}^{-1} \mathrm{cov}(\Delta TOA) \left(A_{TOA}^{-1}\right)^T$$

Given that the covariances of the measurements are known it becomes a simple task to evaluate the covariance of the position errors. We will first assume each TOA measurement is uncorrelated, i.e.,

$$\mathrm{cov}(\Delta TOA) = \begin{bmatrix} \sigma_m^2 & 0 & 0 \\ 0 & \sigma_1^2 & 0 \\ 0 & 0 & \sigma_2^2 \end{bmatrix}$$

and

$$\mathrm{cov}(\Delta TD) = \begin{bmatrix} \sigma_m^2 + \sigma_1^2 & \sigma_m^2 \\ \sigma_m^2 & \sigma_m^2 + \sigma_2^2 \end{bmatrix}$$

For simplicity we will also assume each TOA measurement has the same variance (σ_{TOA}^2). While this assumption is not in general valid, particularly for LORAN-C, it is useful in that it allows us to separate out purely fix geometry from signal-in-space issues. Under this assumption:

$$\mathrm{cov}(\Delta TOA) = \sigma_{TOA}^2 I$$

and

$$\mathrm{cov}(\Delta TD) = \sigma_{TOA}^2 \begin{bmatrix} 2 & 1 \\ 1 & 2 \end{bmatrix}$$

For the particular case of the same number of measurements and unknowns, these both yield the same results:

$$\mathrm{cov}(\Delta P) = \begin{bmatrix} \sigma_x^2 & \sigma_{xy} & c\sigma_{xt} \\ \sigma_{xy} & \sigma_y^2 & c\sigma_{yt} \\ c\sigma_{xt} & c\sigma_{yt} & c^2\sigma_t^2 \end{bmatrix} = c^2 \sigma_{TOA}^2 A_{TOA}^{-1} \left(A_{TOA}^{-1}\right)^T = c^2 \sigma_{TOA}^2 \left[A_{TOA}^T A_{TOA}\right]^{-1}$$

What is significant is that the matrix $G = [A_{TOA}^T A_{TOA}]^{-1}$ is a dimensionless multiplier that relates our ability to measure the range to a transmitter (LORAN station or GPS satellite). In GPS, [NATO GPS 1991] $\sqrt{\mathrm{trace}(G)}$ is defined as the Geometric Dilution of Precision (GDOP), which includes the time term. [NATO GPS 1991]. In LORAN-C, since convention has been not to explicitly solve for time, GDOP is normally considered as $\sqrt{G_{11} + G_{22}}$, which only contains horizontal position terms and is equivalent to Horizontal Dilution of Precision (HDOP) in GPS. Since we want to consider both LORAN and GPS in a consistent, integrated way, we will use HDOP and reserve GDOP for expressions including time term. The scalar expression for this quantity is

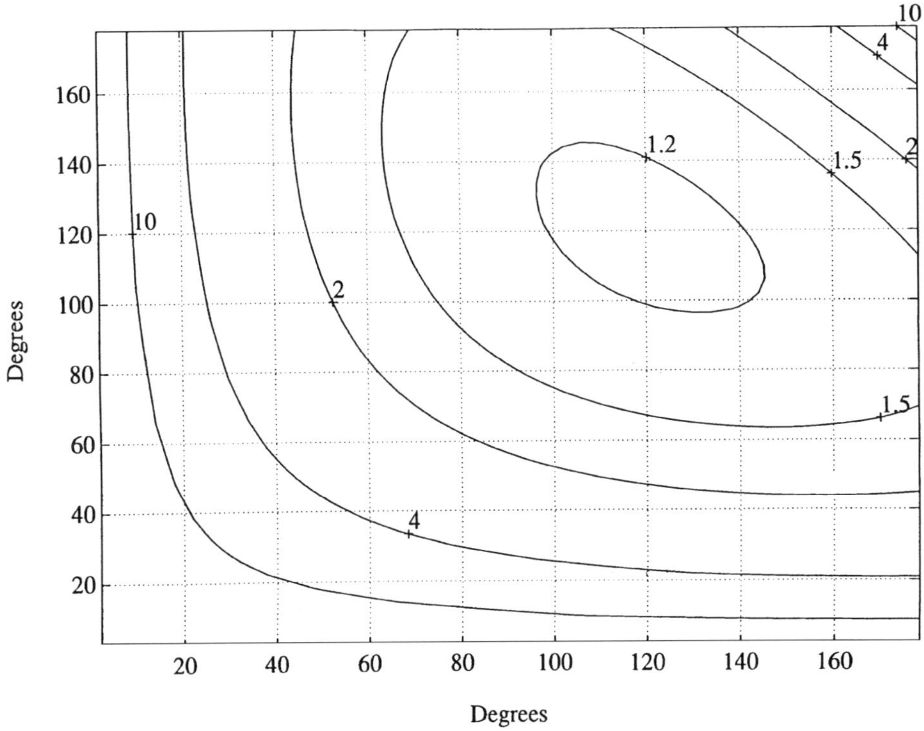

FIGURE 10.10 HDOP as function of angles from master to secondaries.

$$HDOP = \frac{1}{\sqrt{2}\sin(C)}\sqrt{\frac{1}{\sin^2(A/2)} + \frac{1}{\sin^2(B/2)} + \frac{\cos(C)}{\sin(A/2)\sin(B/2)}}$$

where A and B are the angles from the Master to each secondary and $C = \frac{A+B}{2}$. Figure 10.10 shows contours of constant HDOP as functions of these two angles. While the horizontal accuracy in one dimension is well defined ($\sigma_x^2 = G_{11}\, c^2\, \sigma_{TOA}^2$, for example), the two-dimensional accuracy is slightly more complex. Generally the terms *distance root mean square* (drms) and *2 drms accuracy* are used to refer to the one-sigma and two-sigma accuracies. For example:

$$\text{2 drms horizontal accuracy} = 2\ HDOP\ c\ \sigma_{TOA}$$

A circle of this radius should contain between 95 and 98.2% of the fixes depending on the eccentricity of the error ellipse describing the distribution of the fixes. For example, if $\sigma_x = \sigma_y$ and $\sigma_{xy} = 0$, the error ellipse becomes a circle and one of radius 2 drms contains [1 – exp(–4)] = 98.2%. In the other extreme, when the ellipse collapses to a line ($\sigma_{xy} \cong \pm\, \sigma_x\sigma_y$) the 2 drms circle contains the probability within two standard deviations of the mean of a scalar normal variable or 95.4%.

The relationship between geometry and accuracy can best be explained via numerical example. Shown below are spreadsheet calculations for a Master and two secondaries at 0°, 80°, and 300°, respectively.

	Bearing	East(Sine)	North(Cosine)	Time
Master	0	0	1	1
Sec. #1	80	0.9848	0.1736	1
Sec. #2	300	–0.8660	0.5000	1
		INVERSE(A^{T}A)		
TOA		0.7122	0.6748	–0.4047
		0.6748	3.5258	–1.9937
		–0.4047	–1.9937	1.4616
TD		sin(M) –sin(S)	cos(M)–cos(S)	
		–0.9848	0.8264	
		0.8660	0.5000	
		R(TD)		
		2	1	
		1	2	
		INV(A) R INV(A^T)		
		0.7122	0.6748	
		0.6748	3.5258	

Assuming σ_{TOA} = 100 ns or equivalently c σ_{TOA} = 30 m, we can make specific calculations regarding repeatable accuracy. The standard deviations in the east and north directions and time are:

$$EDOP = \sqrt{0.7122} \quad \sigma_x = 30 \text{ m } EDOP = 25.3 \text{ m}$$

$$NDOP = \sqrt{3.5258} \quad \sigma_y = 30 \text{ m } NDOP = 56.3 \text{ m}$$

$$TDOP = \sqrt{1.4616} \quad \sigma_t = 100 \text{ ns } TDOP = 120.9 \text{ ns}$$

$$HDOP = \sqrt{0.7122 + 3.5258} = 2.058$$

and the 2 drms accuracy = 2 × 30 m × 2.058 = 123.5 m.

10.6 Error Ellipses [Pierce, 1948]

Frequently the scalar 2 drms accuracy does not adequately describe the position accuracy and the distribution of the error vector provides useful additional information. The probability density of two jointly normal random variables is given by:

$$p(x, y) = \frac{1}{2\pi\sqrt{\sigma_x^2 \sigma_y^2 - \sigma_{xy}^2}} \exp\left[-\frac{\sigma_x^2 \sigma_y^2}{2(\sigma_x^2 \sigma_y^2 - \sigma_{xy}^2)}\left(\frac{x^2}{\sigma_x^2} - \frac{2xy\sigma_{xy}}{\sigma_x^2 \sigma_y^2} + \frac{y^2}{\sigma_y^2}\right)\right]$$

and the ellipses

$$\frac{x^2}{\sigma_x^2} - \frac{2xy\sigma_{xy}}{\sigma_x^2 \sigma_y^2} + \frac{y^2}{\sigma_y^2} = 2\left(1 - \frac{\sigma_{xy}^2}{\sigma_x^2 \sigma_y^2}\right)c^2$$

are curves of constant probability density such that the probability a sample lies within the ellipse is 1 – exp($-c^2$). For algebraic simplicity this equation will be rewritten:

$$Ax^2 + 2Hxy + By^2 = C$$

which can be rotated by angle (Ω) into a (ξ, η) coordinate system aligned with the ellipse axes;

$$\frac{\xi^2}{\alpha^2} + \frac{\eta^2}{\gamma^2} = C$$

where

$$\tan(2\Omega) = \frac{-2H}{B-A}$$

$$\alpha^2 + \gamma^2 = \frac{A+B}{AB-H^2}$$

$$\alpha^2 - \gamma^2 = \frac{(B-A)\sec(2\Omega)}{AB-H^2}$$

Taking the sum and difference of the last two expressions;

$$2\alpha^2 = \frac{A[1-\sec(2\Omega)] + B[1+\sec(2\Omega)]}{AB-H^2}$$

$$2\gamma^2 = \frac{A[1+\sec(2\Omega)] + B[1-\sec(2\Omega)]}{AB-H^2}$$

In the example above, $\Omega = -12.8°$. To draw an ellipse that contains $[1 - \exp(-4)] = 98.17\%$ of the samples, $\alpha = 63.4$ m and $\gamma = 162.8$ m. (Note: α is the length of the ellipse axis rotated by angle Ω from the original positive x-axis.) Figures 10.11 and 10.12 illustrate an example of 2000 fixes based on the above geometry and TOA statistics. In this example 1926 or 96.3% are contained in the 2 drms circle of radius 123.5 m.

Figure 10.13 is an illustration of error ellipses (not to scale) and lines of constant time difference for a typical triad. The main point is that when geometry becomes poor in the fringes of the coverage area, representing the accuracy by circles of increasing 2 drms radii does not truly reflect one's knowledge of accuracy. In many examples from both air and marine navigation, cross-track accuracy is far more important than along-track accuracy. Knowledge of the orientation of the error ellipse may be important.

In GPS or GLONASS, exactly the same error analysis and ellipses apply, the main difference being that the geometry at any particular point on Earth's surface has constantly changing geometry as satellites move and are added/subtracted from the available constellation. When we look at distributions of GPS or GLONASS fixes they tend to appear circularly symmetric, not because the instantaneous distributions are that way, but because the averaged distributions over long periods such as a day or longer become circularly symmetric.

10.7 Overdetermined Solutions

In the previous examples the issue of the distribution of measurement errors had no impact on the method used to find a solution, only on the error analysis of that solution. We assumed a normal

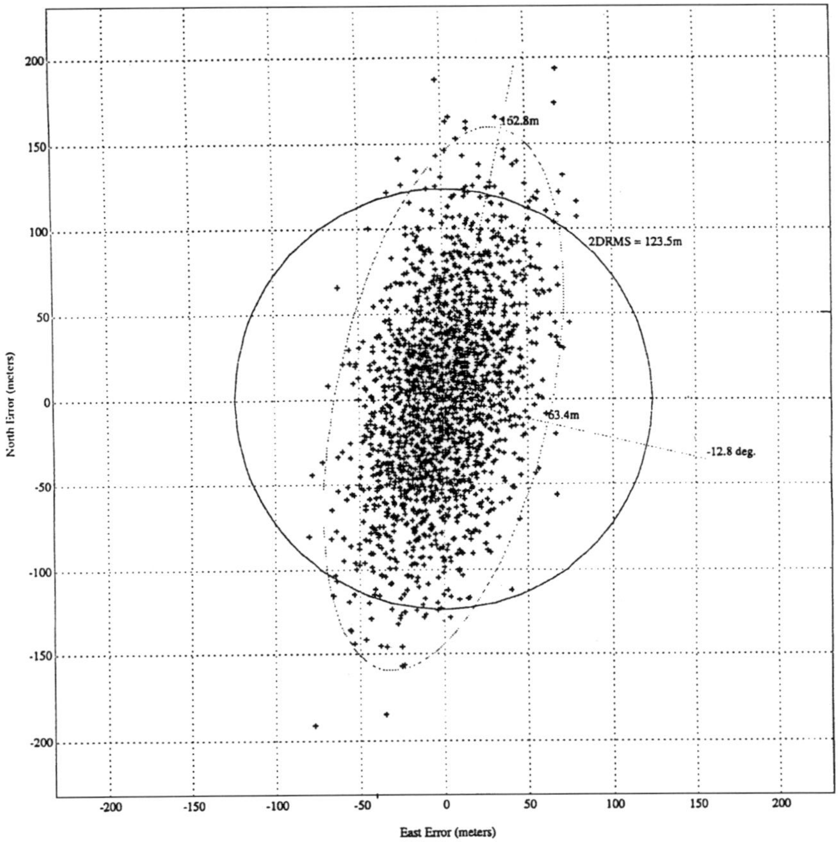

FIGURE 10.11 Error ellipse and distribution of fix errors.

distribution on these errors not necessarily because we could guarantee that distribution, but because it allowed us to make meaningful predictions. When the number of observables exceeds the number of unknowns, we have choices in how to process the observations and the optimum choice depends both on the distribution of our measurement errors and what parameter we are trying to optimize. Since the GPS space segment now has 28 satellites, commonly there are more than the minimum of 3 or 4 required.

If the errors in the measured pseudoranges or TOAs are statistically independent and have the same variance, then the solution that minimizes the sum of the squares of the pseudorange residuals is the linear least squares solution.

$$\Delta P = \left(A^T A\right)^{-1} A^T \, \Delta PR$$

where A is the matrix of direction cosines and ΔPR is the difference between measured and calculated pseudoranges. The covariance of the position error is now:

$$\text{cov}(\Delta P) = \left(A^T A\right)^{-1} A^T \, \text{cov}(\Delta PR) A \left(A^T A\right)^{-1} = \sigma_{PR}\left(A^T A\right)^{-1}$$

where $\text{cov}(\Delta P) = \sigma_{PR}$ I.

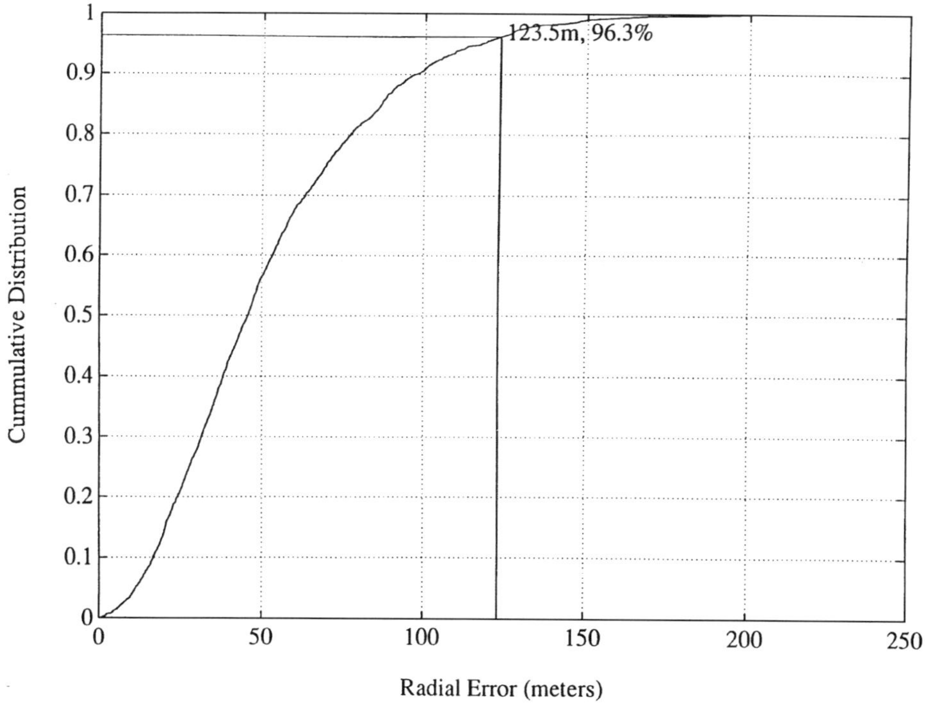

FIGURE 10.12

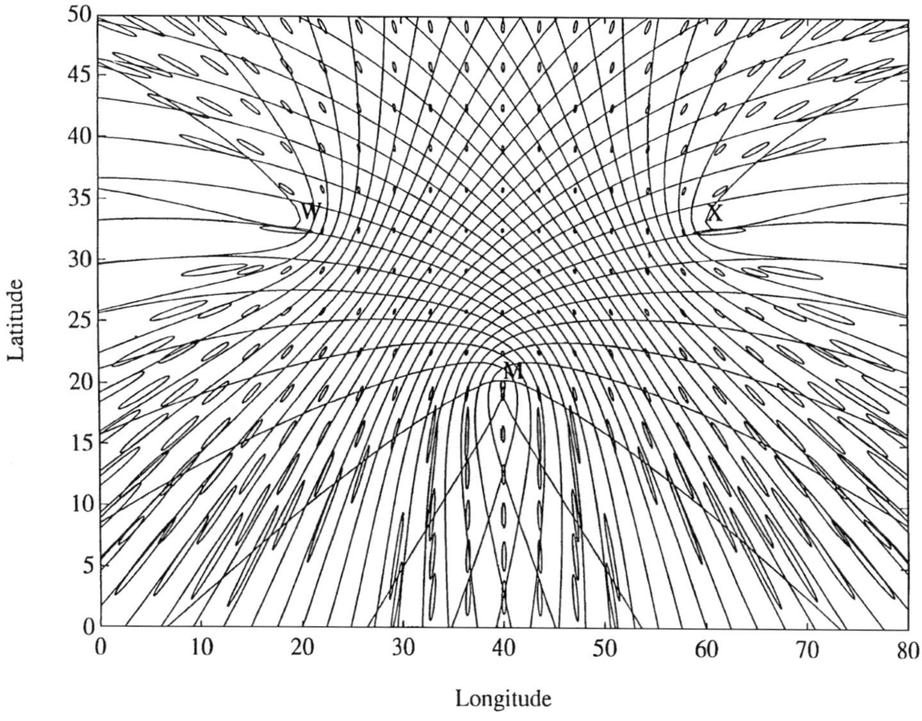

FIGURE 10.13 Lines of constant time difference and error ellipses for typical triad.

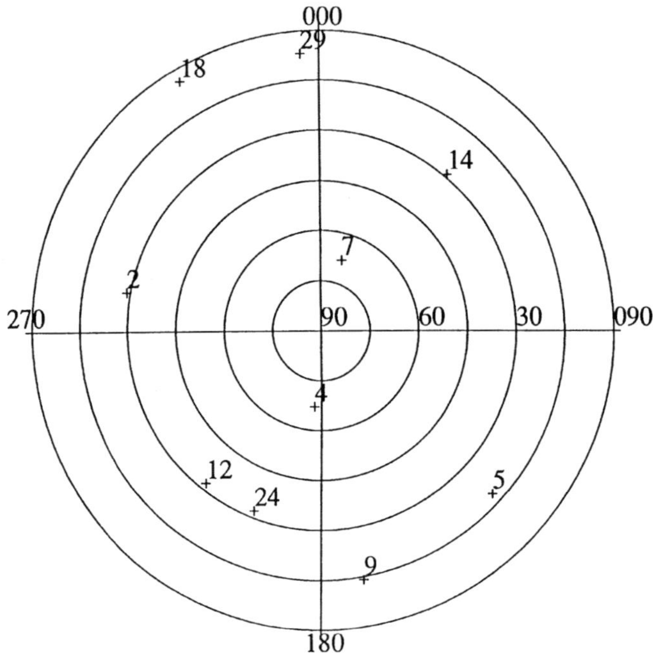

FIGURE 10.14 Azimuth and elevation plot of GPS satellites for 1400Z, August 28, 1995, New London, Connecticut.

Figure 10.14 shows the azimuth and elevation plot of GPS satellites for 1400Z, August 28, 1995 in New London, Connecticut, as a typical example. Eliminating the low elevation satellite (18) due to potential multipath and using the nine remaining listed in the table below, we can calculate expected accuracies.

SVN	Elevation	Azimuth
5	18°	317°
12	32°	232°
7	68°	73°
24	32°	249°
14	29°	50°
29	7°	94°
2	29°	169°
9	14°	280°
4	67°	265°

East	North	Altitude	Time
cos(EL)sin(AZ)	cos(EL)cos(AZ)	sin(EL)	
–0.6486	0.6956	0.309	1
–0.6683	–0.5221	0.5299	1
0.3582	0.1095	0.9272	1
–0.7917	–0.3039	0.5299	1
0.6700	0.5622	0.4848	1
0.9901	–0.0692	0.1219	1
0.1669	–0.8586	0.4848	1
–0.9556	0.1685	0.2419	1
–0.3892	–0.0341	0.9205	1
$G = \mathrm{inv}(A^TA)$			
0.2533	–0.0192	0.0223	0.0239
–0.0192	0.5248	0.1160	–0.0466
0.0223	0.1160	1.6749	–0.8404
0.0239	–0.0466	–0.8404	0.5380

$$GDOP = \sqrt{trace(G)} = 1.73$$

$$PDOP = \sqrt{G_{11} + G_{22} + G_{33}} = 1.57$$

$$HDOP = \sqrt{G_{11} + G_{22}} = 0.88$$

$$VDOP = \sqrt{G_{33}} = 1.29$$

Exactly what these DOPs imply in terms of accuracy depends on what version of GPS/DGPS service one is using.

For the Standard Positioning Service (SPS) with Selective Availability (SA) on, typically the value 32 m was used for the one sigma error. Since this value depended almost entirely on policy, advances in technology did not change its value. In Conley [1999] we can get a glimpse of what accuracies will be expected in the short term, and how they may improve in the future. They report less than 1.4 m rms for the on-orbit clocks and 59 cm for the ephemeris contribution to user range error (URE) in 1999. For SVN 43, the first operational Block IIR satellite, the combined clock and ephemeris errors were 70 cm RMS. This means that the errors in predicting ionospheric delay at about 5 m, the errors in predicting tropospheric delay at about 2 m, and multipath now dominate the statistics. DGPS will remove most of the residual errors due to satellite clock and ephemeris, and ionospheric and tropospheric delay, leaving multipath as the most common dominant error source. The multipath errors are highly dependent on receiver and antenna design and antenna placement. Preliminary data after the termination of SA, from fixed sites, with high quality receivers, and where some care has been taken to minimize multipath indicate 7 m 2 drms accuracy may be achievable. The performance on typical installations with low cost receivers on moving platforms will be considerably worse. In the GPS modernization program to be implemented over the next decade, the addition of C/A code on L2 and third civil frequency, will mean civil receivers will measure ionospheric delay via differential pseudorange as in dual frequency, military receivers. This combined with other improvements in technology is expected to improve civil GPS accuracy to better than 5 m 2 drms.

For Precise Positioning Service users, NATO GPS [1991] specifies 13.0 m (95%) URE, which would imply a standard deviation of 6.5 m assuming a normal distribution. The actual performance, while classified, is generally acknowledged to be considerably better than this 1991 value, and will continue improve with technology.

The negative correlation ($\rho = -0.8404/\sqrt{1.6749 * 0.5380} = -0.885$) between the altitude and time errors is common because the satellites used (by receivers on Earth's surface) all must lie within a cone starting at some (elevation mask) angle above the horizon. Figure 10.15 shows the error ellipse for the joint time and altitude distribution. By using the first (east), second (north), and fourth (time) columns in the matrix A above we can calculate the expected DOPs for a GPS receiver in a fixed (known) altitude mode. HDOP is virtually unchanged but TDOP improves from 0.73 to 0.34. Similarly, if we assume we know time accurately and are only solving for three-dimensional position, we use the first three columns and VDOP improves from 1.29 to 0.60. Realistically, we will not know time exactly because we use the GPS to get time. We can improve vertical accuracy somewhat by propagating clock bias ahead via a Kalman filter as described in a later section, but can never get to zero clock error and this VDOP.

10.8 Weighted Least Squares

When the covariance of the measurement error is not a constant times an identity matrix, the optimum solution is a weighted vs. linear least squares solution.

$$\Delta P = \left(A^T W A\right)^{-1} A^T W \Delta PR$$

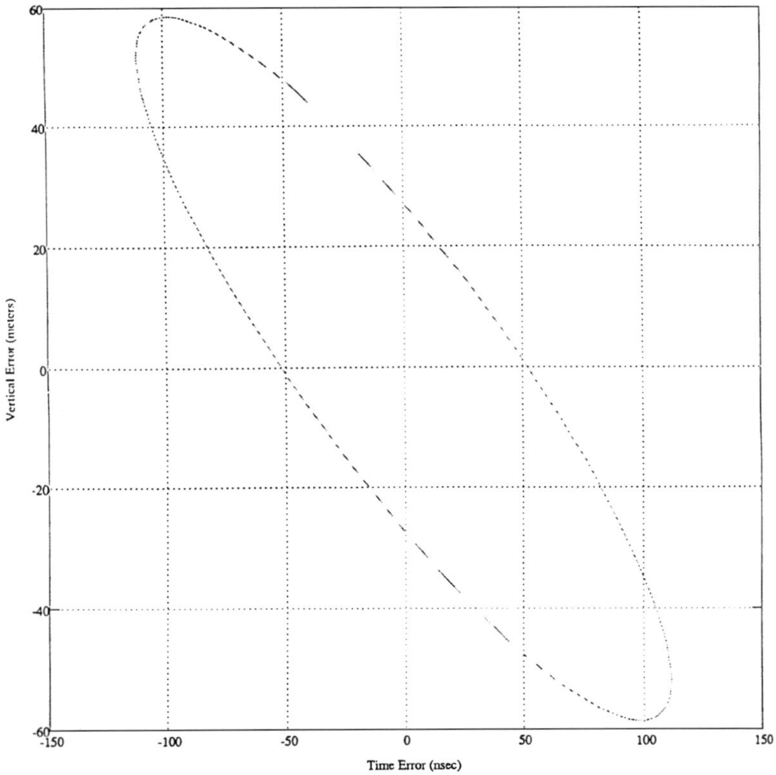

FIGURE 10.15 Typical joint distribution of GPS altitude and time errors.

Now the covariance of the error is given by

$$E\{\Delta P \Delta P^T\} = E\{(A^T W A)^{-1} A^T W \Delta PR \, \Delta PR^T W^T A (A^T W A)^{-1}\}$$

This error is minimized if

$$W = R^{-1}$$

This minimum error is given by

$$E\{\Delta P \, \Delta P^T\} = (A^T W A)^{-1}$$

For example, if we assume the standard deviation of each TOA measurement is the same for a three-TD fix we should use a weighting matrix given by

$$\mathrm{W} = \mathrm{R}^{-1} = \begin{bmatrix} \sigma_m^2 + \sigma_1^2 & \sigma_m^2 & \sigma_m^2 \\ \sigma_m^2 & \sigma_m^2 + \sigma_2^2 & \sigma_m^2 \\ \sigma_m^2 & \sigma_m^2 & \sigma_m^2 + \sigma_3^2 \end{bmatrix}^{-1} = \frac{1}{\sigma_{\mathrm{TOA}}} \begin{bmatrix} 0.75 & -0.25 & -0.25 \\ -0.25 & 0.75 & -0.25 \\ -0.25 & -0.25 & 0.75 \end{bmatrix}$$

When this matrix is multiplied times the TD observation vector, the correlation of the errors in the resulting vector is removed. It can be shown that weighted least squares processing of the TD measurements with the weighting matrix above is exactly equivalent to linear least squares processing of TOA measurements. For GPS, GLONASS, or DGPS there may be some advantages to weighted least squares processing. For example, low elevation satellites are subject to larger pseudorange errors due to multipath and ionosphere delay modeling errors. The typical approach is to reject satellites below some elevation mask threshold of 5 to 10°. An alternative method using low elevation satellites with lower weights may help eliminate occasional spikes in GDOP. In integrated GPS/GLONASS, weighted least squares would have advantages in assigning different weights according to the relative accuracy of the two system's pseudoranges.

In LORAN, TOAs do not have same variance, TDs are not independent, and the position solution should be designed accordingly. Figures 10.16 to 10.18 show the results of 1166 LORAN fixes over 6.5 h using the 9960 chain at a stationary location in New York Harbor. The LOCUS™ receiver used was modified to use a Cesium time standard and provided TOA data. The observed TOA standard deviations were:

Seneca, NY	(175 nm)	7.6 ns
Caribou, ME	(452 nm)	100.6 ns
Nantucket, MA	(186 nm)	8.6 ns
Carolina Beach, NC	(438 nm)	21.1 ns
Dana, IN	(618 nm)	60.0 ns

(*Note:* Even though distances are comparable, the Carolina Beach TOAs are much better than those of Caribou because the path is seawater vs. land. In addition, Carolina Beach TDs are monitored and controlled using data from nearby Sandy Hook, New Jersey, and Caribou is controlled using data from a monitor in Maine.)

The 2 drms repeatable accuracy is seen to be 21.8 m for linear least squares and 7.3 m for weighted least squares. Since for this particular data set the Caribou and Dana signals are virtually ignored (Caribou is weighted $(7.6/100.6)^2 = 0.0057$ relative to Seneca), comparable accuracy to weighted least squares would have been obtained by totally ignoring Caribou and Dana data and doing three-TOA/two-TD fixes. In general using all data provides for a much more reliable solution. Figure 10.19 shows a histogram of time difference data of the 5930Y (Caribou/Cape Race) baseline as measured in New London, Connecticut. What is seen for marginal signals like Cape Race at 875 nm from New London, is frequent cycle errors resulting in 10 and 20 μs TD errors. Any kind of standard least squares algorithm will not work. Assuming an over-determined solution, there are two basic alternatives; first, a maximum likelihood algorithm that takes into account the potential for local maxima in the likelihood function due to cycle errors, and second, Receiver Autonomous Integrity Monitoring (RAIM). In RAIM, a fix using all measurements is calculated and the residuals are compared to a threshold. For RAIM to be valid, assuming N measurements, each of the N fixes available from each subset of N-1 measurements must have acceptable geometry. For GPS, if we have two or more measurements than unknowns, the potential exists to identify and remove bad data via residual analysis on these N over-determined fixes. Since for LORAN, cycle errors have predictable behavior, we may be able to identify and remove them with only one measurement more than the number of unknowns.

10.9 Kalman Filters

This section points out the basic principles of Kalman filters and specifically how they can be used in radio navigation systems. All of our analysis above has assumed we process each set of data independently. We know that parameters such as clock frequency and our velocity can only change at finite rates and therefore we ought to be able to do better using all past measurements. When we integrate LORAN TOAs

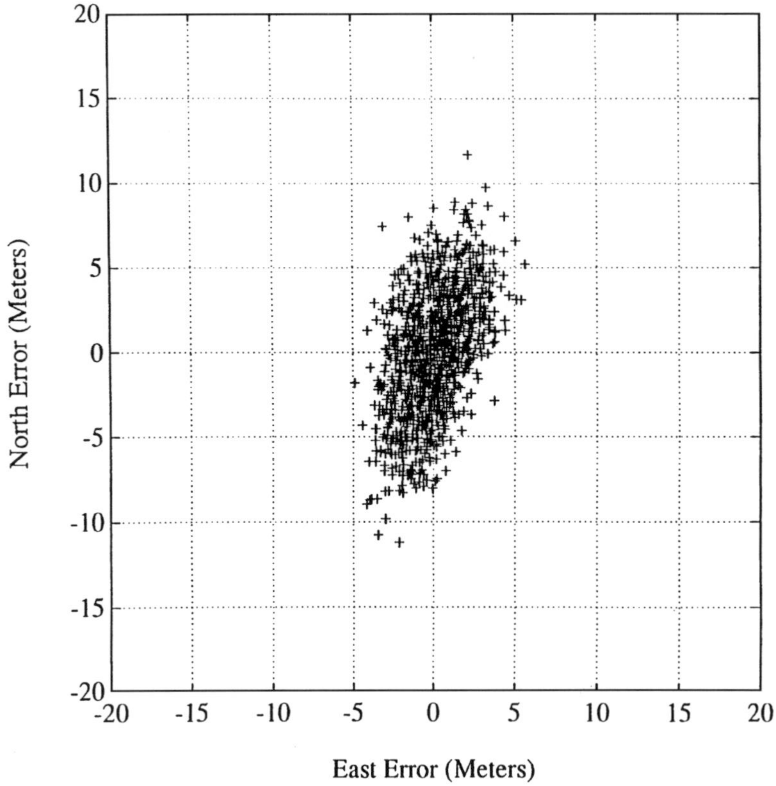

FIGURE 10.16

or GLONASS pseudoranges with GPS pseudoranges, we may not be able to assume the various systems are exactly synchronized in time. Therefore, not only can we solve for one time term but also we know the drift between systems is extremely slow and need to somehow exploit that knowledge. Kalman filters allow us to use past measurements together with a priori knowledge of platform and clock dynamics to obtain optimum values of position and velocity. Typically, Kalman filters have found extensive use in the integration of inertial and electronic navigation systems. Here we will concentrate on stand-alone electronic navigation systems. The reader is referred to more extensive references such as Brown [1992] for more detail.

Figure 10.20 shows the basic structure of a discrete Kalman filter. The process starts with initial estimates of the state [X(0)] and the error covariance [P(0)]. The other variables are:

R(k) is the covariance of the measurement errors. (They are assumed white; this is not valid for SA-dominated pseudorange errors, which are correlated over minutes. Strictly speaking this correlation should be modeled via additional state variables in system model, but normally is not.)

H(k) is matrix of direction cosines and ones (as A above) that relate pseudorange or TOA errors to positions and clock bias and Doppler's to velocity and frequency errors.

z(k) is pseudorange (or TOA) and Doppler measurements.

K(k) is Kalman gains.

Φ(k) is state transition matrix.

Q(k) is covariance of noise driving state changes. This matrix controls allowed accelerations, clock frequency, and phase variations. Smaller Q(k) will result in smaller estimates of state error covariance and less weighting of measurements. Larger Q(k) will result in faster system response to measurements. Conventional wisdom suggests using larger Q when uncertain of the exact model.

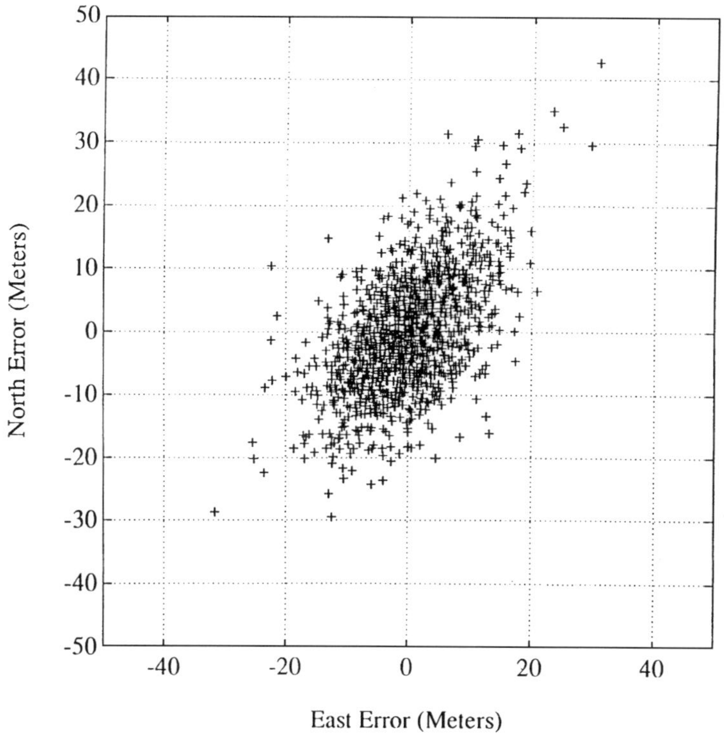

FIGURE 10.17

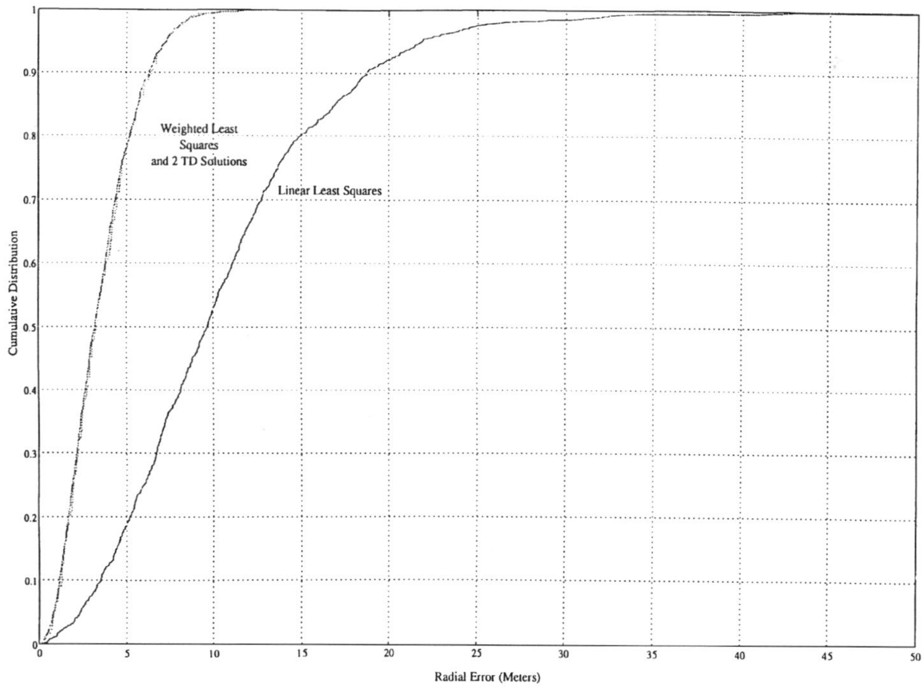

FIGURE 10.18

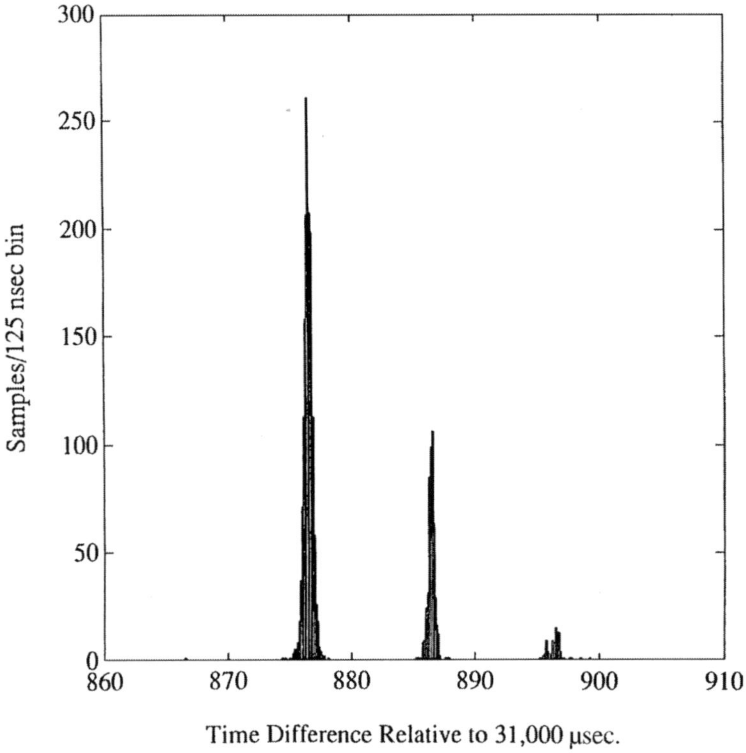

FIGURE 10.19 Histogram of time difference data of the 5930Y (Caribou/Cape Race) baseline measured in New London, Connecticut.

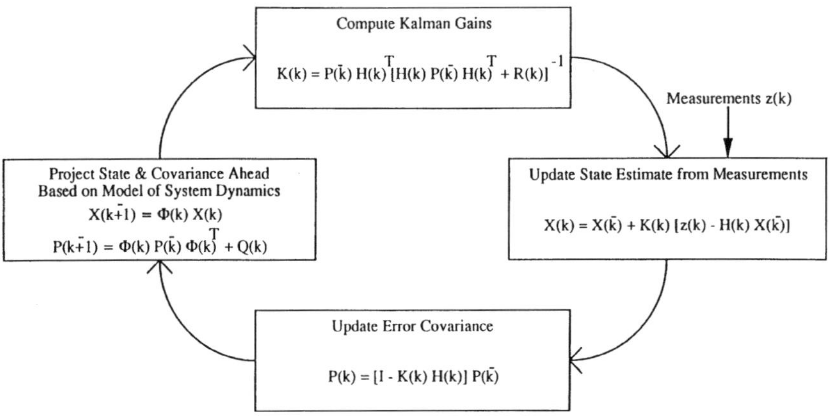

FIGURE 10.20 Kalman filter structure.

A typical state space model for stand-alone GPS would have eight states, the spatial coordinates and their velocities, and the clock offset and frequency. The individual pseudorange measurements can be processed sequentially, which means that the Kalman gains can be calculated as scalars without the need for matrix inversions. There is no minimum number of measurements required to obtain an updated position estimate. The measurements are processed in an optimum fashion and if not enough for good geometry, the estimate of state error variance [P(k)] will grow. If two satellites are available, the clock bias terms are just propagated forward via the state transition matrix.

The structure is ideal for integrated systems where the offset between GPS and GLONASS time may only be known approximately initially, but is known to be virtually constant. This offset would then be estimated as a state variable with some initial uncertainty but with very small Q(k) driving changes. The same would be true for LORAN/GPS time offsets and the concept could be extended to ASFs for individual stations as well. The ASFs would be represented by very slowly varying states to be estimated. If GPS pseudoranges became unavailable such as in an urban canyon, in a mountain valley, or under a heavy foliage cover, the LORAN would now be well calibrated.

Defining Terms

Accuracy: The degree of conformance between the estimated or measured position and/or velocity of a platform at a given time and its true position or velocity. Radio navigation system accuracy is usually presented as a statistical measure of system error and is specified as follows:

Predictable: The accuracy of a radio navigation system's position solution with respect to the charted solution. Both the position solution and the chart must be based upon the same geodetic datum.

Repeatable: The accuracy with which a user can return to a position whose coordinates have been measured at a previous time with the same navigation system.

Relative: The accuracy with which a user can measure position relative to that of another user of the same navigation system at the same time.

Availability: The availability of a navigation system is the percentage of time that the services of the system are usable. Availability is an indication of the ability of the system to provide usable service within the specified coverage area. Signal availability is the percentage of time that navigational signals transmitted from external sources are available for use. Availability is a function of both the physical characteristics of the environment and the technical capabilities of the transmitter facilities.

Block I and Block II Satellites: The Block I is a GPS concept validation satellite; it does not have all of the design features and capabilities of the production model GPS satellite, the Block II. The FOC 24 satellite constellation is defined to consist entirely of Block II/IIA satellites.

Coordinated Universal Time (UTC): UTC, an atomic time scale, is the basis for civil time. It is occasionally adjusted by 1-s increments to ensure that the difference between the uniform time scale, defined by atomic clocks, does not differ from Earth's rotation by more than 0.9 s.

Differential: A technique used to improve radio navigation system accuracy by determining positioning error at a known location and subsequently transmitting the determined error, or corrective factors, to users of the same radio navigation system, operating in the same area.

Dilution of Precision (DOP): The magnifying effect on radio navigation position error induced by mapping ranging errors into position and time through the position solution. The DOP may be represented in any user local coordinate desired. Examples are HDOP for local horizontal, VDOP for local vertical, PDOP for all three coordinates, TDOP for time, and GDOP for position and time.

Distance Root Mean Square (drms): The root-mean-square value of the distances from the true location point of the position fixes in a collection of measurements. As used in this chapter 2 drms is the radius of a circle that contains at least 95% of all possible fixes that can be obtained with a system at any one place. Actually, the percentage of fixes contained within 2 drms varies between approximately 95.5 and 98.2%, depending on the degree of ellipticity of the error distribution.

Hyperbolic Navigation System: A navigation system that produces hyperbolic lines of position (LOPs) through the measurement of the difference in times of reception (or phase difference) of radio signals from two or more synchronized transmitters.

Integrity: Integrity is the ability of a system to provide timely warnings to users when the system should not be used for navigation.

Multipath Transmission: The propagation phenomenon that results in signals reaching the receiving antenna by two or more paths. When two or more signals arrive simultaneously, wave interference results. The received signal fades if the wave interference is time varying or if one of the terminals is in motion.

Pseudorange: The difference between the ranging signal time of reception (as defined by the receiver's clock) and the time of transmission contained within the satellite's navigation data (as defined by the satellite's clock) multiplied by the speed of light.

Receiver Autonomous Integrity Monitoring (RAIM): A technique whereby a civil GPS receiver/processor determines the integrity of the GPS navigation signals without reference to sensors or non-DoD integrity systems other than the receiver itself. This determination is achieved by a consistency check among redundant pseudorange measurements.

Selective Availability (SA): The denial of full GPS accuracy to civil users by manipulating navigation message orbit data (epsilon) and/or satellite clock frequency (dither).

References

Braff, R., Description of the FAA's Local Area Augmentation System (LAAS), *Navigation*, 44, 4, 411–424, Winter 1997.

Brown, R. G. and Hwang, P. Y. C., *Introduction to Random Signals and Applied Kalman Filtering,* 2nd Ed., John Wiley & Sons, New York, 1992.

Conley, R. and Lavrakas, J. W., The World after Selective Availability, *Proceedings of ION GPS '99, 14–17,* 1353–1361, Nashville, TN, September 1999.

CSIC, *GLONASS Interface Control Document, Version 4.0,* The Ministry of Defence of the Russian Federation Coordination Scientific Information Center, 1998. Available in electronic form at http://www.rssi.ru/SFCSIC/SFCSIC_main.html

Dale, S, Daly, P., and Kitching, I., Understanding signals from GLONASS navigation satellites, *Int. J. Satellite Commun.,* 7, 11–22, 1989.

U.S. Departments of Defense and Transportation, (DoD/Dot, 1994) *1994 Federal Radionavigation Plan,* U.S. Departments of Defense and Transportation, NTIS Report DOT-VNTSC-RSPA-95-1/DoD-4650.5, 1995. Available in electronic form (Adobe Acrobat) from USCG NAVCEN (www.navcen,uscg.mil).

Enge, P., Swanson, E., Mullin, R., Ganther, K., Bommarito, A., and Kelly, R., Terrestrial radionavigation technologies, *Navigation*, 42, 1, 61–108, Spring, 1995.

Feairheller, S., The Russian GLONASS system, a U.S. Air Force study, *Proceedings of Institute of Navigation GPS-94,* Salt Lake City, UT, September 1994.

NATO Navstar GPS Technical Support Group (NATO GPS 1991), *Technical Characteristics of the Navstar GPS,* June 1991.

Parkinson, B., Stansell, T., Beard, R., and Gromov, K., A history of satellite navigation, *Navigation*, 42, 1, 109–164, Spring 1995.

Pierce, J. A., Mckenzie, A. A., and Woodward, R. H., Eds., *LORAN*, MIT Radiation Laboratory Series, McGraw-Hill, New York, 1948.

RTCA, *Minimum Aviation Performance Standards for Local Area Augmentation System,* RTCA/DO-245, September 1998.

RTCA, *Minimum Operational Performance Standards for GPS/Wide Area Augmentation System Airborne Equipment*, RTCA/DO 229B, October 1999.

RTCM Special Committee 104, *RTCM Recommended Standard for Differential Navstar GPS Service,* Version 2.0, December 10, 1992.

Skidmore, T. A., Nyhus, O. K., and Wilson, A. A., An overview of the LAAS VHF data broadcast, *Proceedings of ION GPS '99, 14–17,* 671–680, Nashville, TN, September 1999.

U.S. Air Force, GPS Joint Program Office, (USAF JPO, 1995) *Global Positioning System, Standard Positioning Service, Signal Specification,* June 1995. Available in electronic form (Adobe Acrobat) from USCG NAVCEN (www.navcen,uscg.mil).

U. S. Coast Guard, (USCG, 1992), *LORAN C User Handbook,* COMDTPUB P16562.6, COMDT(G-NRN), 1992. Available in electronic form from USCG NAVCEN (www.navcen,uscg.mil).

Walter, T. and Bakery El-Arini, M., Eds., *Selected Papers on Satellite based Augmentation Systems (SBASs),* VI in the GPS Series, Institute of Navigation, Alexandria, VA, 1999.

Ward, P. W., GPS receiver RF interference monitoring, mitigation and analysis techniques, *Navigation,* 41, 4, 367–391, Winter 1994–1995.

White House Press Secretary, Statement by the President Regarding the United States Decision to Stop Degrading Global Positioning System Accuracy, White House, May 1, 2000.

Wild Goose Association (WGA, 1982), On the calculation of geodesic arcs for use with LORAN, *Wild Goose Association Radionavigation Journal,* 57–61, 1982.

Further Information

Government Agencies

1. **U.S. Coast Guard Navigation Information Service (NIS)**
   Address:Commanding Officer
   USCG Navigation Center
   7323 Telegraph Road
   Alexandria, VA 22315
   Internet:http://www.navcen.uscg.mil/navcen.htm
   Phone:(703) 313-5900 (NIS Watchstander)

Formerly the GPS Information Center (GPSIC), NIS has been providing information since March 1990. The NIS is a public information service. At the present time, there is no charge for the information provided. Hours are continuous: 24 hours a day, 7 days a week, including federal holidays.

The mission of the Navigation Information Service (NIS) is to gather, process, and disseminate timely GPS, DGPS, Omega, LORAN-C status, and other navigation related information to users of these navigation services. Specifically, the functions to be performed by the NIS include the following:

Provide the Operational Advisory Broadcast Service (OAB).
Answer questions by telephone or written correspondence.
Provide information to the public on the NIS services available.
Provide instruction on the access and use of the information services available.
Maintain tutorial, instructional, and other relevant handbooks and material for distribution to users.
Maintain records of GPS, DGPS, and LORAN-C broadcast information, and databases or relevant data for reference purposes.
Maintain bibliography of GPS, DGPS, and LORAN-C publications.

The NIS provides a watchstander to answer radio navigation user inquiries 24 h a day, seven days a week and disseminates general information on GPS, DGPS, Loran-C, and Radiobeacons through various mediums.

2. **NOAA, National Geodetic Survey**
   Address: 1315 East-West Highway, Station 09202
   Silver Spring, MD 20910
   Phone: (301) 713-3242
   Fax: (301) 713-4172
   Monday through Friday, 7:00 A.M.–4:30 P.M., Eastern Time
   Internet: http://www.ngs.noaa.gov/

NOAA is modernizing the Nation's Spatial Reference System, providing horizontal and vertical positions for navigation and engineering purposes. By implementing a new system called CORS (Continuously Operating Reference Stations), NOAA will continuously record carrier phase and pseudorange measurements for all GPS (Global Positioning System) satellites at each CORS site. A primary objective of CORS is to monitor a particular site, which is determined to millimeter accuracy and provide local users a tie to the National Spatial Reference System. NOAA will also use the CORS sites along with other globally distributed tracking stations in computing precise GPS orbits.

NOAA is also the federal agency responsible for providing accurate and timely Global Positioning System (GPS) satellite ephemerides ("orbits") to the general public. The GPS precise orbits are derived using 24-h data segments from the global GPS network coordinated by the International Geodynamics GPS Service (IGS). The reference frame used in the computation is the International Earth Rotation Service Terrestrial Reference Frame (ITRF). In addition, an informational summary file is provided to document the computation and to convey relevant information about the observed satellites, such as maneuvers or maintenance. The orbits generally are available two to six days after the date of observation.

Also available:

- Software programs to compute, verify, or adjust field surveying observations; convert coordinates from one geodetic datum to another; or assist in other specialized tasks utilizing geodetic data.
- Listings of publications available on geodesy, mapping, charting, photogrammetry, and related topics, such as plane coordinate systems and numerical analysis.
- Information on NGS programs to assist users of geodetic data, including technical workshops and the geodetic advisor program.
- Information on survey data recently incorporated into NSRS.
- Information on the location of NGS field parties.

3. **U.S. Naval Observatory (USNO)**
   Internet:http://www.usno.navy.mil/

USNO is the official source of time used in the United States. USNO timekeeping is based on an ensemble of cesium beam and hydrogen maser atomic clocks. USNO disseminates information on time offsets of the world's major time services and radio navigation systems.

4. **USSPACECOM GPS Support Center**
   300 O'Malley
   Suite 41
   Schriever AFB, CO 80912-3041
   http://www.peterson.af.mil/usspace/gps_support

The **GPS Support Center** (GSC) is the DoD's focal point for operational issues and questions concerning military use of GPS. The GSC is responsible for:

a. Receiving reports and coordinating responses to radio frequency interference in the use of GPS in military operations;
b. Providing prompt responses to DoD user problems or questions concerning GPS;
c. Providing official USSPACECOM monitoring of GPS performance provided to DoD users on a global basis;
d. Providing tactical support for planning and assessing military missions involving the use of GPS.

As the DoD's focal point for GPS operational matters, the GSC serves as U.S. Space Command's interface to the civil community, through the U.S. Coast Guard's Navigation Center and Federal Aviation Administration's National Operations Command Center.

5. **The Ministry of Defence of the Russian Federation**
   Coordination Scientific Information Center
   Internet: http://www.rssi.ru/SFCSIC/SFCSIC_main.html

The mission of the Coordination Scientific Information Center is to plan, manage, and coordinate the activities on

- Use of civil-military space systems (navigation, communications, meteorology, etc.);
- Realization of Russian and international scientific and economic space programs;
- Realization of programs of international cooperation;
- Conversional use of military space facilities, as well as to provide the scientific-informational, contractual, and institutional support of these activities.

11

Microwave and Radio Frequency (RF) Avionics Applications

James L. Bartlett
Rockwell Collins

The term *avionics* was originally coined by contracting *avia*tion and electr*onics*, and has gained widespread usage over the years. Avionics differs from most of the other wireless applications in several important areas. Avionics applications typically require functional integrity and reliability that are orders of magnitude more stringent than most commercial wireless applications. The rigor of these requirements is matched or exceeded only by the requirements for space or certain military applications. The need for this is readily understood when one compares the impact (pun intended) of a failure of a cellular phone call, and a failure of an aircraft instrument landing system on final approach during reduced visibility conditions. Avionics must function in environments that are more severe than most other wireless applications. Extended temperature ranges, high vibration levels, altitude effects (including corona and high-energy particle upset (known as single event upset) of electronics are all factors that must be considered in the design of avionics products. Quantities for the avionics market are typically very low when compared to commercial wireless applications (e.g., the number of cell phones manufactured every single working day far exceeds the number of aircraft that are manufactured in the world in a year). Wireless systems for avionics applications cover an extremely wide range in a number of dimensions, including frequency, system function, modulation type, bandwidth, and power. Due to the number of systems aboard a typical aircraft, electromagnetic interference (EMI) and electromagnetic compatibility (EMC) between systems is a major concern, and EMI/EMC design and testing is a major factor in the flight certification testing of these systems.

11.1 Communications Systems, Voice and Data

Very low frequency (VLF) is not used in civil aviation, but is used for low data rate transmission to and from strategic military platforms, due to its ability to transmit reliably worldwide, including water penetration adequate to communicate with submerged submarines. These are typically very high power systems, with 100 to 200 kW of transmit power.

High-frequency (HF) communication from 2 to 30 MHz has been used since the earliest days of aviation, and continues in use on modern aircraft, both civil and military. HF uses single sideband, suppressed carrier modulation, with a relatively narrow modulation bandwidth of about 2.5 kHz, typically several-hundred-watt transmitters, and is capable of worldwide communications. Because propagation

0-8493-1732-0/03/$0.00+$1.50

conditions vary with frequency, weather, ionosphere conditions, time of day, and sun spot activity, the whole HF band will generally not be available for communications between any two points on Earth at any given time, but some portion will be suitable. Establishing a usable HF-link frequency between two stations is an essential part of the communications protocol in this band, and until recently, required a fair amount of time. Modern HF systems have Automatic Link Establishment (ALE) software and hardware to remove that burden from the operator. Prior to the advent of communications satellites, HF was the only means of communicating with an aircraft for large parts of transoceanic flights.

Very high frequency (VHF) communication as used in the aviation world comprises two different frequency bands. The military uses 30 to 88 MHz, with narrowband frequency modulation (FM), primarily as an air-to-ground link for close air support. RF power for this usage ranges from 10 to 40 W. Civil and military aircraft both use the 116- to 136-MHz band with standard double sideband amplitude modulation (AM) for air traffic control (ATC) purposes, typically with 10 to 30 W of carrier power. Although this band has been in use for decades, the technical requirements for radios in this band have not remained constant. The proliferation of potential interfering sites and competition for spectrum space has led to adapting ever narrower channel spacings, and increased dynamic range requirements. An aircraft operating near an airport in an urban environment frequently has to fly near a commercial FM station operating at several tens of kilowatts in the 88- to 108-MHz band, without compromising communications capability. Currently, VHF ATC radios operate with both 25- and 8.33-KHz channel spacings. Additionally, this band is being used for Addressing and Reporting System (ACARS) and data-link purposes, known as VHF data link (VDL) with differing modes of operation, depending on the required data and throughput required. The ACARS function, also known as VDL mode A, uses an AM minimum-shift keying modulation at 2.4 kb/s. VDL modes 2 and 3 are the highest data rate modes currently defined, achieving 31.5 kb/s using a differential eight-phase shift-keying (D8PSK) modulation scheme in a 25-kHz channel. These two modes differ in the networking scheme, with mode 2 using a Collision Sense, Multiple Access (CSMA) network protocol, and mode 3 using a Time Division Multiple Access (TDMA) network. Both VDL mode 2 and 3 operate in an air-to-ground mode only. VDL mode 4 utilizes a Gaussian frequency-shift-keying modulation at 19.5 kb/s in a self-organizing TDMA (STDMA) network.

Ultra high frequency (UHF) communication as used in the aviation world actually bridges the VHF and UHF regions, operating from 225 to 400 MHz. Various radios in use here range in power output from 10 to 100 W of carrier power for FM, and 10 to 25 W of AM carrier power. This band is used for military communications, and many waveforms and modulation formats are in use here, including AM, FM, and a variety of pulsed, frequency hopping, antijam waveforms using various protocols and encryption techniques.

Satellite communication (SatCom) is used for aircraft data-link purposes, as well as for communication by the crew and passengers. Passenger telephones use this service. Various satellites are used, including INMARSAT, Aero-H, and Aero-I, all operating at L-band.

Data links are a subset of the overall communication structure, set up to transmit digital data to and from aircraft. They cover an extremely wide range of operating frequencies, data rates, and security features. Data-link applications may share use of an existing radio, or use a specialized transceiver, dedicated to the application. Examples include the ARINC Communication ACARS, which uses the existing VHF Comm radio on civil aircraft, and the Joint Tactical Information Distribution System (JTIDS), which is a secure, fast-hopping L-band system that uses a dedicated transceiver.

Data-link usage is an application area that has been dominated by military applications, but is now used for an increasing number of commercial applications. Virtually all the network management and spread-spectrum methods coming into use for personal communications were originally pioneered by military secure data-link applications. Code Division Multiple Access (CDMA), TDMA, Biphase-Shift Keying (BPSK), Quadrature Phase-Shift Keying (QPSK), Minimum-Shift Keying (MSK), Continuous Phase-Shift Keying (CPSM), Gaussian Minimum-Shift Keying (GMSK), various error detecting and correcting codes, along with interleaving and various forms of data redundancy have all found applications in military data-link applications. A great deal of effort has been devoted to classes of orthogonal code generation and secure encryption methods for use in these systems. It is common to find systems using a variety of techniques together to enhance their redundancy and antijam nature.

One system, for example, uses the following techniques to increase the antijam capabilities of the system. Messages are encoded with an error detecting and correcting code to provide a recovery capability for bit errors. The message is then further encoded using a set of orthogonal CCSK code symbols representing a smaller number of binary bits (i.e., 32 chips representing 5 binary bits), giving some level of CDMA spreading, which further increases the probability of successful decoding in the face of jamming. The result is then multiplied with a pseudorandom encryption code, further pseudorandomizing the information in a CDMA manner. The resultant bit stream is then modulated on a carrier in a CPSM format and transmitted symbol by symbol, with each symbol transmitted on a separate frequency in a pulsed, frequency-hopped TDMA sequence determined by another encryption key. The message itself may also be parsed into short segments and interleaved with other message segments to time spread it still further, and repeated with different data interleaving during different time slots in the TDMA network arrangement to provide additional redundancy. Such systems have a high degree of complexity, and a large throughput overhead when operating in their most secure modes, but can deliver incredible robustness of message delivery in the face of jamming or spoofing attempts by an adversary. Additionally, these systems are incredibly difficult to intercept successfully and decode, which is frequently just as important to the sender as the successful delivery of the message to its intended user.

11.2 Navigation and Identification Systems

Navigation and identification functions comprise a large share of the avionics aboard a typical aircraft. There are systems used for enroute navigation, including Long-Range Radio Navigation (LORAN), Automatic Direction Finder (ADF), Global Positioning System (GPS), Global Orbital Navigation Satellite System (GLONASS), Distance Measuring System (DME), and Tactical Air Navigation (TACAN). There are also systems used for approach navigation, including Marker Beacon, VHF Omnidirectional Range (VOR), Instrument Landing System (ILS), Microwave Landing System (MLS), Radar Altimeter, and Ground Collision Avoidance System (GCAS). Finally, there are systems used for identification of aircraft and hazards, including Air Traffic Control Radar Beacon System (ATCRBS), Mode S, Identification Friend or Foe (IFF), Traffic Collision Avoidance System (TCAS), and radar.

Commercial airborne radar systems are typically operated in either C-band, around 5 GHz, or more commonly at X-Band at 9.33 GHz. Their primary purpose is weather surveillance and hazard avoidance. Many use Doppler techniques with color-coded displays to quantify and clearly display the hazard level in a particular storm front, allowing the pilot to make better weather avoidance decisions. The operating principles of radar are not discussed in this chapter.

Current trends in avionics radio navigation systems mirror those of consumer systems in providing multiple capabilities in a single unit. They are referred to as Multi Mode Receivers (MMR), and may combine VOR, ILS, Marker Beacon, GPS, and MLS functions. They are installed as doubly or triply redundant systems, just as their predecessors are, but still only replicating one system, not several as before.

Many of the radio navigation systems rely on varying complexity schemes of spatial modulation for their function. In almost all cases, there are multiple navaids functional at a given airport ground station, including Marker Beacons, a DME or TACAN, a VOR, and an ILS or MLS.

In a typical example, the VOR system operates at VHF (108 to 118 MHz) to provide an aircraft with a bearing to the ground station location in the following manner. The VOR transmits continuously at its assigned frequency, with two antennas (or the electronically combined equivalent), providing voice transmission and code identification to ensure that the aircraft is tracking the proper station. The identification is in Morse code, and is a two- or three-letter word repeated with a modulation tone of 1020 Hz. The transmitted signal from the VOR station contains a carrier that is amplitude modulated at a 30-Hz rate by a 30-Hz variable signal, and a 9960-Hz subcarrier signal that is frequency modulated between 10440 and 9480-Hz by a 30-Hz reference signal. The reference signal is radiated from a fixed omnidirectional antenna, and thus contains no time varying spatial modulation signal. The variable signal is radiated from a rotating (electrically or mechanically), semidirectional element driven at 1800 r/min or 30 Hz, producing a spatial AM at 30 Hz. The phasing of the two signals is set to be in phase at magnetic North, 90° out of phase at magnetic East, 180° when South, and 270° when West of the VOR station. VOR receivers function by receiving both signals, comparing their phase, and displaying the bearing to the station to the pilot. Directly above a VOR station is an area that has no AM component. This cone-shaped area is referred to as the "cone of confusion," and most VOR receivers contain delays to prevent "invalid data" flags from presentation during the brief flyover time.

The Instrument Landing System (ILS) contains three functions: localizer, glideslope, and marker beacon. Three receivers are used in this system: a VHF localizer, a UHF glide slope, and a VHF marker beacon. Their functions are described in the following paragraphs.

The marker beacon receiver functions to decode audio and provide signaling output to identify one of three marker beacons installed near the runway. The outer beacon is typically about 5 miles out, the middle is approximately 3500 ft., and the inner is just off the runway end. These beacons radiate a narrow beamwidth, 75-MHz signal in a vertical direction, and each has a different distinct modulation code so the receiver can identify which one it is flying over.

The localizer transmitter is located at the far end of the runway, and radiates two intersecting lobes in the 108- to 112-MHz frequency band along the axis of the runway, with the left lobe modulated at 90 Hz, and the right lobe modulated at 150 Hz. The installation is balanced so that the received signal is equally modulated along the centerline of the runway. The total beamwidth is approximately 5° wide. The localizer receiver uses this modulation to determine the correct approach path in azimuth. This pattern also extends beyond the departure end of the runway, and is called the back course. In the case of an aircraft flying an inbound back course, the sensed modulations are reversed, and there are no marker beacons or glide slope signals available on the back course. In the event of a missed approach and go-around, the side lobes of the localizer antenna pattern could present erroneous or confusing bearing data and validity flags, so most installations utilize a localizer go-around transmitter on each side of the runway that radiates a signal directed outward from the side of the runway that is modulated with a go-around level modulation that identifies the signal and masks the side lobe signals.

The glide slope transmitter is located close to the near end of the runway, and similarly radiates two intersecting lobes. These are one on top of the other, in the 329- to 335-MHz frequency band along the axis of, and at the angle of, the desired approach glide slope, usually approximately 3°. The upper lobe is modulated at 90 Hz, and the lower lobe is modulated at 150 Hz. The installation is balanced so that the received signal is equally modulated along the centerline of the glide slope. The total beamwidth is approximately 1.4° wide. The localizer receiver uses this modulation to determine the correct glide slope path. The combination of localizer and glide slope modulations for the approach is illustrated in Figure 11.1. The back course, marker beacons, and go-around patterns are deleted for simplicity of visualization.

The addition of MLS technology to an MMR provides flight crews with an alternative to existing ILS capability wherever MLS is installed. MLS technology improves margins of safety around airports in highly developed areas, especially when weather conditions degrade and category II capability is required to maintain airport capacity. MLS is used in conjunction with the Distance Measuring Equipment, Precision (DME/P) to determine slant range measurements, while the MLS provides the angular data needed to determine present position relative to the runway. One of the major advantages that MLS has over the current VHF/UHF ILS system is the relative ease and cost of an installation. Because an ILS uses the ground in front of the glide slope antenna to form the beam, a large area must be level. Frequently, the cost of ground preparation for an ILS site exceeds the equipment costs. Additionally, an ILS is sensitive to nearby reflecting surfaces that can reflect and distort the glide slope beam. Consequently, aircraft and other vehicles must be held a long distance from the takeoff threshold to stay clear of the critical reflection area. MLS, operating at 5 GHz, does not suffer nearly as severely from these effects, and is much faster and easier to install and calibrate. Multipath distortions and errors at 5 GHz can be at high levels, and

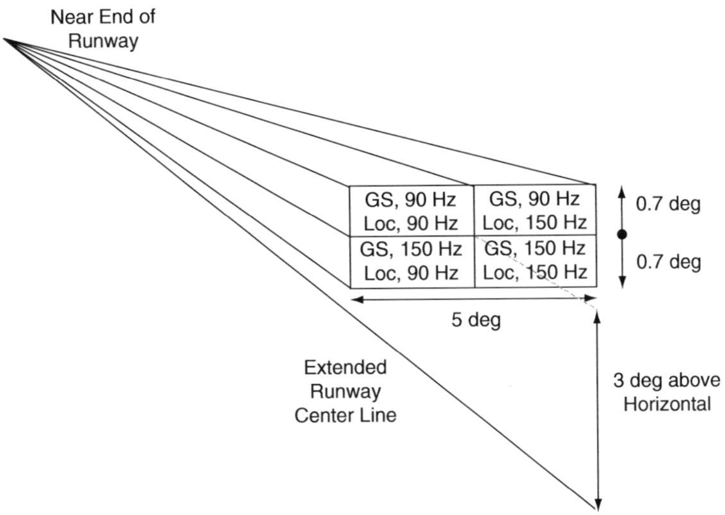

FIGURE 11.1 Glide slope and localizer modulation.

require much smaller reflective surfaces to create, but they also are varying over very small distances, allowing easier receiver designs to accommodate. The relatively easy, fast setup allows MLS to be used by the military to provide a portable precision approach capability at otherwise unimproved airstrips.

MLS operates on 200 channels in a band from 5.03 to 5.09 GHz using a scanning beam in both elevation and azimuth which the aircraft can use to determine its position relative to the desired glide slope. The MLS signal structure is a series of transmissions in sequence from the different antennas and is very flexible, allowing azimuth elevation, and back azimuth to be transmitted in any order. The entire sequence takes approximately 75 ms to complete, and includes an azimuth scan, three elevation scans, and a back azimuth scan. Thus, the azimuth data are updated at ~ 13.5 Hz, and elevation data are updated at three times that, or ~ 40 Hz. The MLS transmits a broad sector beam that contains a differential phase-shift keying (DPSK) modulated preamble to provide the initial timing mark as well as information concerning the antenna function and offset from the runway. MLS angle measurements are based on a Time Reference Scanning Beam (TRSB) system that uses the timing between received pulses of a beam in both elevation and azimuth, which is scanned back and forth across the coverage area to extract angular data. In either case, the radiated beams are narrow, fan-shaped beams, scanned at 20,000° per second. The azimuth beam coverage is set at installation, and can cover up to ± 60°. A typical installation is set to ±40°, whereas the minimum is ±10°. The azimuth coverage does not have to be separate around the runway centerline. For example, it could be set to +10°, and – 40° to avoid mountains or some other obstruction. Figure 11.2 is a pictorial representation of the azimuth scanning beam format.

The elevation beam is set to scan from a minimum of 0.9° to a maximum of 15°. Typical fixed wing aircraft glide slopes on approach are around 3°. Figure 11.3 shows the scanning beam in elevation.

As stated earlier, the time difference between the received pulses on the "to" scan and the "fro" scan is used to calculate the angle relative to the MLS transmitter and hence the airport centerline. Figure 11.4 shows the principle of TRSB angular measurement in azimuth, with elevation measured in the same manner.

Note that the scanned beam does not have to be modulated at all. The receiver uses amplitude information only to establish the timing between the pulses, and calculates the aircraft's angular position from the timing between received pulses. When this information is combined with slant range from the DME/P system, the aircraft's current position is fully established. Like the current ILS installations, there is a provision included to cover the back side of the runway for takeoffs, missed landings, etc. Unlike the older VHF/UHF ILS system, which requires a straight-in path down the glide slope, MLS can support

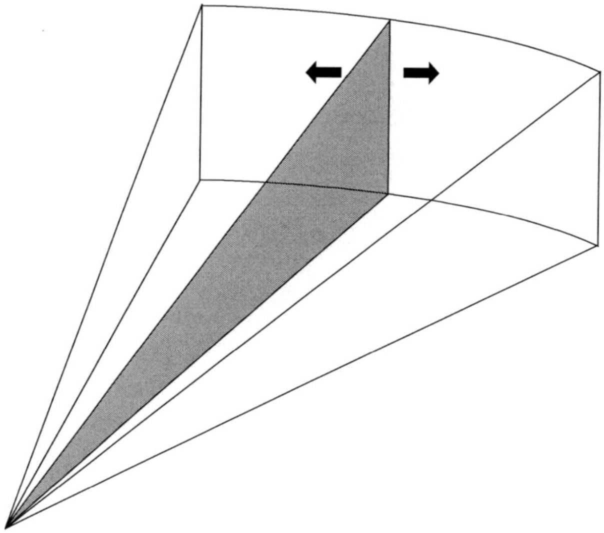

FIGURE 11.2 Azimuth scanning beam format.

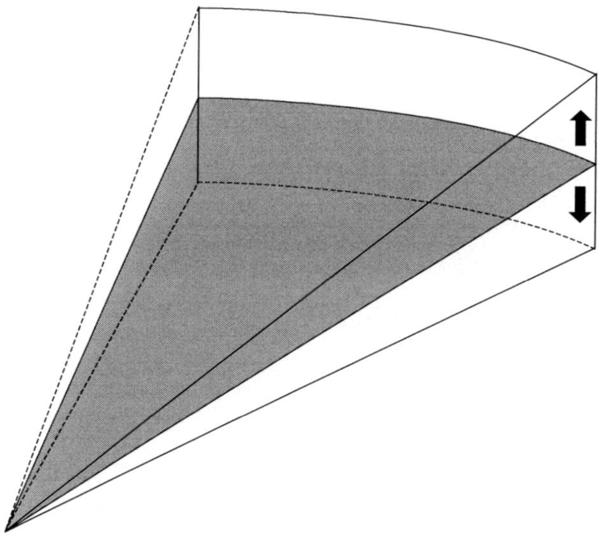

FIGURE 11.3 Scanning beam in elevation.

arbitrarily curved approach paths. This is useful in maintaining separation distances in highly congested approach environments, and is especially useful for the military where mountains, structures, or enemy activity near the airfield might preclude a classic straight-in glide path.

GPS is coming online as a viable approach navigation aid. The installation of Wide Area Augmentation Service (WAAS) and Local Area Augmentation Service (LAAS) has enhanced the accuracy of civil GPS to the point where it is now useful for approach navigation service. Both of these services transmit corrections to GPS to allow significant improvements in position accuracy. These augmentations are also at higher levels, and not as easily jammed (deliberately or accidentally) as the very low level GPS signal is. This and the advances in GPS receivers that track many satellites simultaneously have allowed integrity monitoring to be built into the system that can meet the integrity requirements of approach navigation.

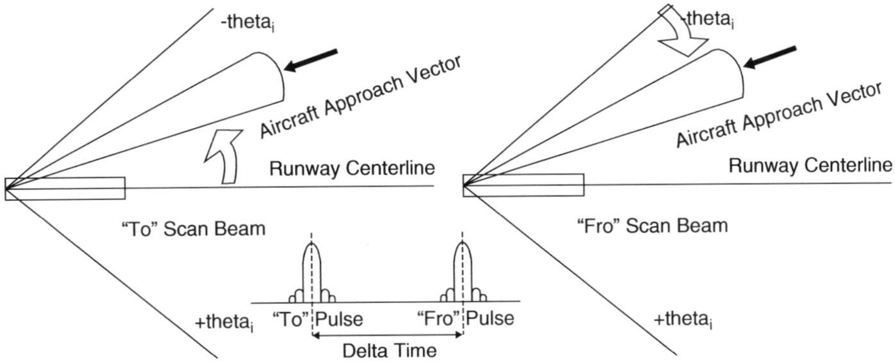

FIGURE 11.4 Principle of TRSB angular measurement in azimuth.

Another example of an avionics system in widespread use that has significant RF content is the Traffic Alert and Collision Avoidance System (TCAS). TCAS functions in close harmony with the Air Traffic Control Radar Beacon System (ATCRBS) at 1030 and 1090 MHz, which has been in operation for several decades. The ATCRBS system comprises a system of ground-based interrogators and airborne responders, which serve to provide the ATC system with the necessary information to safely manage the air space.

There are several interrogation and reply modes incorporated into the ATCRBS/TCAS system. Mode A and mode C interrogations use a combination of pulse position and width coding, and replies are in a pulse position/timing format. The mode A response identifies the aircraft responding. Aircraft equipped with mode C transponders interrogated by the ATC Secondary Surveillance Radar system respond with a coded pulse train that includes altitude information, allowing automated altitude tracking by the ATC system. TCAS incorporates mode C and mode select (or mode S as it is more commonly known), which incorporates a data-link capability, and interrogates these transponders in other aircraft. Mode S uses pulse position coding for interrogations and replies, and DPSK to send data blocks containing 56 or 112 data chips. All modes of transponder responses to an interrogation are designed to be tightly controlled in time so that slant range between the interrogator and responder are easily calculated. Altitude is directly reported in both mode C and mode S responses as well. Specific details of all the allowable interrogation, reply, and data transmissions would take up far too much room to include in this chapter. As with most essential avionics equipment, passenger aircraft may have two TCAS systems installed for redundancy.

TCAS provides for a surveillance radius of 30 miles or greater and displays detected aircraft in four categories to the pilot: other, proximate, traffic advisories (TA), and resolution advisories (RA). Each has its own unique coding on the display. Intruders classified as "other" are outside of 6 nautical miles (nmi), or more than ±1200 ft. of vertical separation, with projected flight paths not taking them closer than these thresholds. A "proximate" aircraft is within 6 nmi, but with a flight path not representing a potential conflict. TCAS provides the pilot with audible advisory messages for the remaining two categories; TA for "proximate" aircraft when they close to within 20 to 48 s of Closest Point of Approach (CPA) announced as "Traffic, Traffic," and RA for aircraft when the flight path is calculated to represent a threat of collision. RAs are given 15 to 35 s prior to CPA, depending on altitude. RAs are limited to vertical advisories only; no turns are ever commanded. RAs may call for a climb, a descent, or a maneuver to maintain or increase the rate of climb or descent. RAs are augmented by displaying a green arc on the vertical speed indicator, which corresponds to the range of safe vertical speeds.

TCAS antennas are multielement, multioutput port antennas, matched quite closely in phase and gain so that the receiver may determine the angle of arrival of a received signal. With altitude, slant range, and angle of arrival of a signal known, the present position, relative to own aircraft is computed. Based on changes between replies, the CPA is computed, and the appropriate classification of the threat is made. If both aircraft are equipped with mode S capability, information is exchanged and RAs are coordinated to ensure that both aircraft do not make the same maneuver, requiring a second RA to resolve.

From an RF and microwave point of view, there are several challenging aspects to a TCAS design.

1. The antenna requires four elements located in close proximity within a single enclosure that radiate orthogonal cardiod patterns matched in gain and phase quite closely (i.e., 0.5 dB and 5°, typically). With the high amount of mutual coupling present, this presents a challenging design problem.
2. Because of the high traffic density that exists within the U.S. and Europe, the potential for multiple responses and interference between aircraft with the same relative angle to the interrogating aircraft but with different ranges is significant.

To prevent this, a "whisper shout" algorithm has been implemented for TCAS interrogations. This involves multiple interrogations, starting at low power, with each subsequent interrogation 1 dB higher in power. This requires control of a high power (typically in excess of 1 kW) transmitter output over a 20-dB range, with better than 1 dB linearity. Attaining this linearity with high-speed switches and attenuators capable of handling this power level is not an easy design task. The whisper shout pulse sequence is illustrated in Figure 11.5.

Each interrogation sequence consists of four pulses, S1, P1, P3, and P4. Each pulse is 0.8 υSec in width, with a leading edge spacing of 2.0 υSec for pulses S1 to P1 and P3 to P4. There is a 21 υSec spacing from P1 to P3 (compressed in the figure). Each interrogation sequence can be increased in power level in 1-dB steps over a 20-dB range. A transponder only replies to pulse sequences that meet the timing requirements, and whose amplitude is such that S1 is below the receiver noise floor and P1 is above a minimum trigger level. This effectively limits the number of responses to the whisper shout sequence from any aircraft to about two, thus limiting the inter-aircraft interference.

11.3 Passenger Business and Entertainment Systems

Entertainment is the latest application of microwave technology to the aircraft industry. Although the traditional definition of avionics has referred to cockpit applications, in recent years there has been a move to receive direct broadcast TV services and provide Internet access services to passengers via individual displays at each seat. These applications require steerable, aircraft-mounted antennas at X- or Ku-bands capable of tracking the service provider satellites; low noise down-converters; some sort of closed loop tracking system, typically utilizing the aircraft's inertial navigation system; a receiver or receivers with multiple tunable output channels; and the requisite control at each seat to select the desired channel. Thus, such a system is much more complex and expensive than a household direct broadcast system.

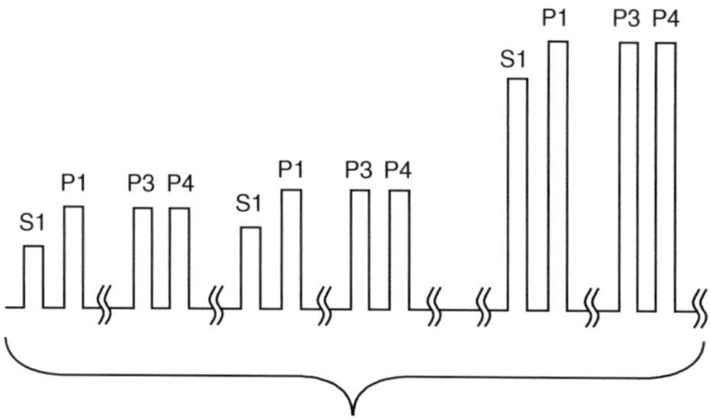

FIGURE 11.5 Mode C interrogation whisper shout sequence (all call).

11.4 Military Systems

It is beyond the scope of this chapter to discuss military systems as separate entities. Military aircraft utilize all the avionics systems listed earlier (with the exception of entertainment), and depending on the mission of the individual airframe, may also contain systems to spoof, jam, or just listen to other users of any or all those systems. The radar systems on military aircraft may be much more powerful and sophisticated than those required for civil use and serve a variety of weapons or intelligence gathering functions in addition to the basic navigation and weather-sensing functions of civil aircraft. In fact, virtually all the functions used by civil aircraft may be exploited or denied for a military function, and systems exist on various platforms to accomplish just that.

12

Continuous Wave Radar

James C. Wiltse
Georgia Institute of Technology

Continuous wave (CW) radar employs a transmitter that is on all or most of the time. Unmodulated CW radar is very simple and is able to detect the **Doppler-frequency shift** in the return signal from a target that has a component of motion toward or away from the transmitter. While such a radar cannot measure range, it is used widely in applications such as police radars, motion detectors, burglar alarms, proximity fuzes for projectiles or missiles, illuminators for semiactive missile guidance systems (such as the Hawk surface-to-air missile), and scatterometers (used to measure the scattering properties of targets or clutter such as terrain surfaces) [Nathanson, 1991; Saunders, 1990; Ulaby and Elachi, 1990].

Modulated versions include frequency-modulated (FM/CW), interrupted frequency-modulated (IFM/CW), and phase-modulated. Typical waveforms are indicated in Fig. 12.1. Such systems are used in altimeters, Doppler navigators, proximity fuzes, over-the-horizon radar, and active seekers for terminal guidance of air-to-surface missiles. The term *continuous* is often used to indicate a relatively long waveform (as contrasted to pulse radar using short pulses) or a radar with a high duty cycle (for instance, 50% or greater, as contrasted with the typical duty cycle of less than 1% for the usual pulse radar). As an example of a long waveform, planetary radars may transmit for up to 10 h and are thus considered to be CW [Freiley et al., 1992]. Another example is interrupted CW (or **pulse-Doppler**) radar, where the transmitter is pulsed at a high rate for 10 to 60% of the total time [Nathanson, 1991]. All of these modulated CW radars are able to measure range.

The first portion of this section discusses concepts, principles of operation, and limitations. The latter portion describes various applications. In general, CW radars have several potential advantages over pulse radars. Advantages include simplicity and the facts that the transmitter leakage is used as the local oscillator, transmitter spectral spread is minimal (not true for wide-deviation FM/CW), and peak power is the same as (or only a little greater than) the average power. This latter situation means that the radar is less detectable by intercepting equipment.

The largest disadvantage for CW radars is the need to provide antenna isolation (reduce spillover) so that the transmitted signal does not interfere with the receiver. In a pulse radar, the transmitter is off before the receiver is enabled (by means of a duplexer and/or receiver-protector switch). Isolation is frequently obtained in the CW case by employing two antennas, one for transmit and one for reception. When this is done, there is also a reduction of close-in clutter return from rain or terrain. A second

0-8493-1732-0/03/$0.00+$1.50

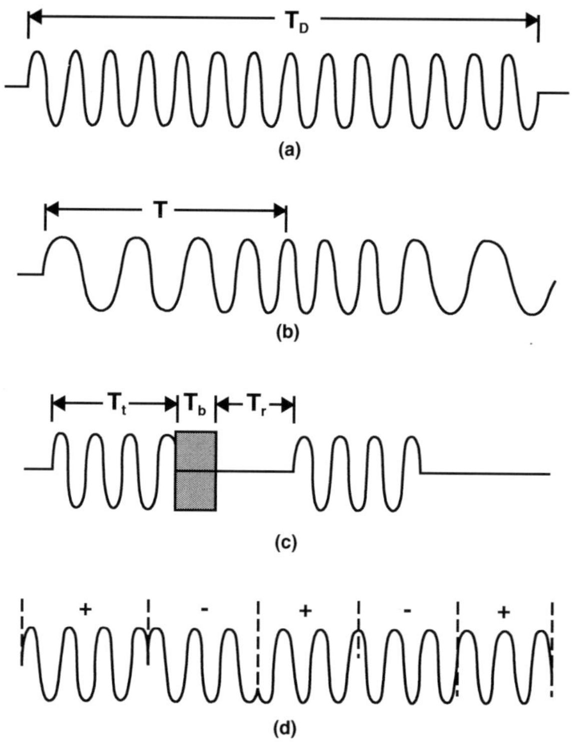

FIGURE 12.1 Waveforms for the general class of CW radar: (a) continuous sine wave CW; (b) frequency modulated CW; (c) interrupted CW; (d) binary phase-coded CW. (*Source:* F. E. Nathanson, *Radar Design Principles*, New York: McGraw-Hill, 1991, p. 450. With permission.)

disadvantage is the existence of noise sidebands on the transmitter signal which reduce sensitivity because the Doppler frequencies are relatively close to the carrier. This is considered in more detail below.

12.1 CW Doppler Radar

If a sine wave signal were transmitted, the return from a moving target would be Doppler-shifted in frequency by an amount given by the following equation:

$$f_d = \frac{2v_r f_T}{c} = \text{Doppler frequency} \tag{12.1}$$

where f_T = transmitted frequency; c = velocity of propagation, 3×10^8 m/s; and v_r = radial component of velocity between radar and target.

Using Eq. (12.1) the Doppler frequencies have been calculated for several speeds and are given in Table 12.1.

As may be seen, the Doppler frequencies at 10-GHz (X-band) range from 30 Hz to about 18 kHz for a speed range between 1 and 600 mi/h. The spectral width of these Doppler frequencies will depend on target fluctuation and acceleration, antenna scanning effects, frequency variation in oscillators or components (for example, due to microphonism from vibrations), but most significantly, by the spectrum of the transmitter, which inevitably will have noise sidebands that extend much higher than these Doppler

TABLE 12.1 Doppler Frequencies for Several Transmitted Frequencies and Various Relative Speeds (1 m/s = 2.237 mi/h)

Microwave	Relative Speed			
Frequency—f_T	1 m/s	300 m/s	1 mi/h	600 mi/h
3 GHz	20 Hz	6 kHz	8.9 Hz	5.4 kHz
10 GHz	67 Hz	20 kHz	30 Hz	17.9 kHz
35 GHz	233 Hz	70 kHz	104 Hz	63 kHz
95 GHz	633 Hz	190 kHz	283 Hz	170 kHz

frequencies, probably by orders of magnitude. At higher microwave frequencies the Doppler frequencies are also higher and more widely spread. In addition, the spectra of higher frequency transmitters are also wider, and, in fact, the transmitter noise-sideband problem is usually worse at higher frequencies, particularly at millimeter wavelengths (i.e., above 30 GHz). These characteristics may necessitate frequency stabilization or phase locking of transmitters to improve the spectra.

Simplified block diagrams for CW Doppler radars are shown in Fig. 12.2. The transmitter is a single-frequency source, and leakage (or coupling) of a small amount of transmitter power serves as a local oscillator signal in the mixer. This is called homodyning. The transmitted signal will produce a Doppler-shifted return from a moving target. In the case of scatterometer measurements, where, for example, terrain reflectivity is to be measured, the relative motion may be produced by moving the radar (perhaps on a vehicle) with respect to the stationary target [Wiltse et al., 1957]. The return signal is collected by the antenna and then also fed to the mixer. After mixing with the transmitter leakage, a difference frequency will be produced which is the Doppler shift. As indicated in Table 12.1, this difference is apt to range from low audio to over 100 kHz, depending on relative speeds and choice of microwave frequency. The Doppler amplifier and filters are chosen based on the information to be obtained, and this determines the amplifier bandwidth and gain, as well as the filter bandwidth and spacing. The transmitter leakage may include reflections from the antenna and/or nearby clutter in front of the antenna, as well as mutual coupling between antennas in the two-antenna case.

The detection range for such a radar can be obtained from the following [Nathanson, 1991]:

$$R^4 = \frac{\overline{P}_T G_T L_T A_e L_R L_p L_a L_s \delta_T}{(4\pi^2) k T_s b(S/N)} \tag{12.2}$$

where R = the detection range of the desired target.
$\overline{P}_T$ = the average power during the pulse.
G_T = the transmit power gain of the antenna with respect to an omnidirectional radiator.
L_T = the losses between the transmitter output and free space including power dividers, waveguide or coax, radomes, and any other losses not included in A_e.
A_e = the effective aperture of the antenna, which is equal to the projected area in the direction of the target times the efficiency.
L_R = the receive antenna losses defined in a manner similar to the transmit losses.
L_p = the beam shape and scanning and pattern factor losses.
L_a = the two-way-pattern propagation losses of the medium; often expressed as $\exp(-2\propto R)$, where $\propto$ is the attenuation constant of the medium and the factor 2 is for a two-way path.
L_s = signal-processing losses that occur for virtually every waveform and implementation.
δ_T = the radar cross-sectional area of the object that is being detected.
k = Boltzmann's constant (1.38×10^{-23} W-s/K).
T_s = system noise temperature.
b = Doppler filter or *speedgate* bandwidth.
S/N = signal-to-noise ratio.

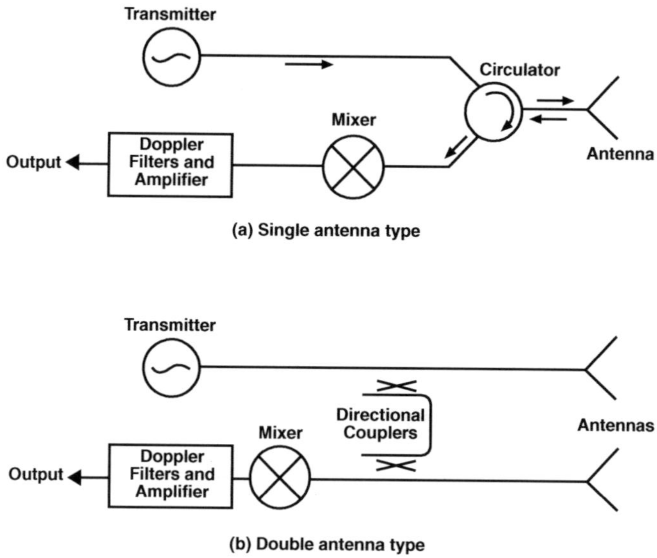

FIGURE 12.2 Block diagrams of CW-Doppler radar systems: (a) single antenna type; (b) double antenna type.

S_{min} = the minimum detectable target-signal power that, with a given probability of success, the radar can be said to *detect*, *acquire*, or *track* in the presence of its own thermal noise or some external interference. Since all these factors (including the target return itself) are generally noise-like, the criterion for a detection can be described only by some form of probability distribution with an associated probability of detection P_D and a probability that, in the absence of a target signal, one or more noise or interference samples will be mistaken for the target of interest.

While the Doppler filter should be a matched filter, it usually is wider because it must include the target spectral width. There is usually some compensation for the loss in detectability by the use of post-detection filtering or integration. The S/N ratio for a CW radar must be at least 6 dB, compared with the value of 13 dB required with pulse radars when detecting steady targets [Nathanson, 1991, p. 449].

The Doppler system discussed above has a maximum detection range based on signal strength and other factors, but it cannot measure range. The rate of change in signal strength as a function of range has sometimes been used in fuzes to estimate range closure and firing point, but this is a relative measure.

12.2 FM/CW Radar

The most common technique for determining target range is the use of frequency modulation. Typical modulation waveforms include sinusoidal, linear sawtooth, or triangular, as illustrated in Fig. 12.3. For a linear sawtooth, a frequency increasing with time may be transmitted. Upon being reflected from a stationary point target, the same linear frequency change is reflected back to the receiver, except it has a time delay that is related to the range to the target. The time is $T = (2R)/c$, where R is the range. The received signal is mixed with the transmit signal, and the difference or beat frequency (F_b) is obtained. (The sum frequency is much higher and is rejected by filtering.) For a stationary target this is given by

$$F_b = \frac{4R}{c} \cdot \Delta F \cdot F_m \tag{12.3}$$

where ΔF = frequency deviation and F_m = modulation rate.

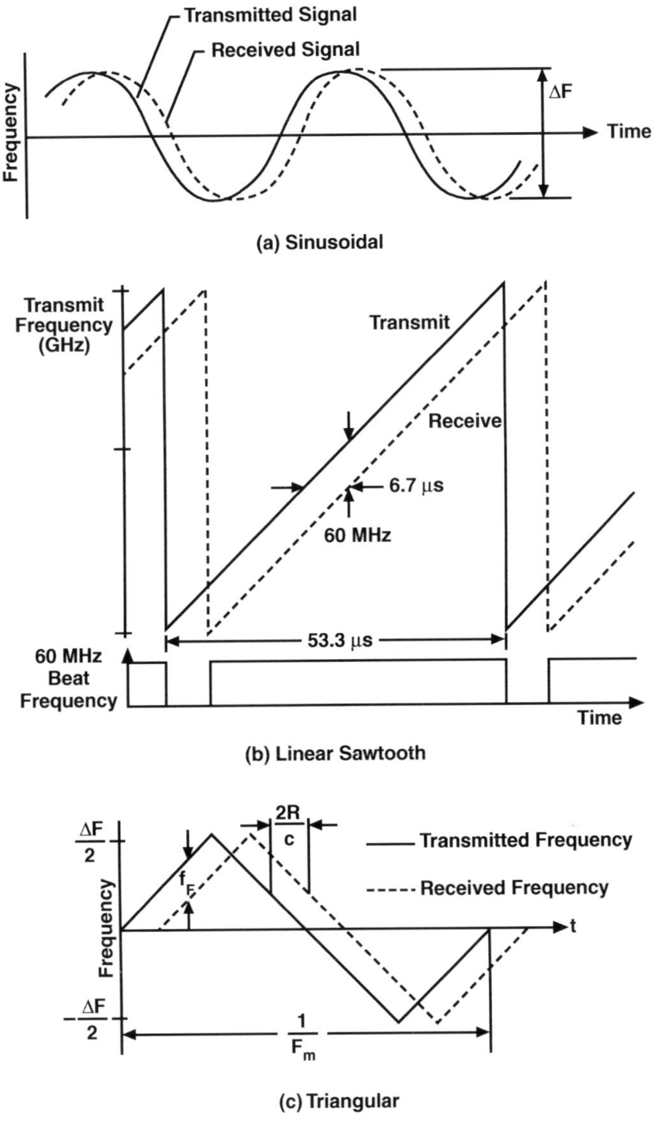

FIGURE 12.3 Frequency vs. time waveforms for FM/CW radar: (a) sinusoidal, (b) linear sawtooth, (c) triangular modulations.

The beat frequency is constant except near the turnaround region of the sawtooth, but, of course, it is different for targets at different ranges. (If it is desired to have a constant intermediate frequency for different ranges, which is a convenience in receiver design, then the modulation rate or the frequency deviation must be adjusted.) Multiple targets at a variety of ranges will produce multiple-frequency outputs from the mixer and frequently are handled in the receiver by using multiple range-bin filters.

If the target is moving with a component of velocity toward (or away) from the radar, then there will be a Doppler frequency component added to (or subtracted from) the difference frequency (F_b), and the Doppler will be slightly higher at the upper end of the sweep range than at the lower end. This will introduce an uncertainty or ambiguity in the measurement of range, which may or may not be significant, depending on the parameters chosen and the application. For example, if the Doppler frequency is low (as in an altimeter) and/or the difference frequency is high, the error in range measurement may be tolerable. For the symmetric triangular waveform, a Doppler less than F_b averages out, since it is higher on one-half of a cycle and lower on the other half. With a sawtooth modulation, only a decrease or increase is noted, since the frequencies produced in the transient during a rapid flyback are out of the receiver passband. Exact analyses of triangular, sawtooth, dual triangular, dual sawtooth, and combinations of these with noise have been carried out by Tozzi [1972]. Specific design parameters are given later in this chapter for an application utilizing sawtooth modulation in a **missile terminal guidance seeker**.

For the case of sinusoidal frequency modulation the spectrum consists of a series of lines spaced away from the carrier by the modulating frequency or its harmonics. The amplitudes of the carrier and these sidebands are proportional to the values of the Bessel functions of the first kind (J_n, $n = 0, \dots 1, \dots 2, \dots 3, \dots$), whose argument is a function of the modulating frequency and range. By choosing a particular modulating frequency, the values of the Bessel functions and thus the characteristics of the spectral components can be influenced. For instance, the signal variation with range at selected ranges can be optimized, which is important in fuzes. A short-range dependence that produces a rapid increase in signal, greater than that corresponding to the normal range variation, is beneficial in producing well-defined firing signals. This can be accomplished by proper choice of modulating frequency and filtering to obtain the signal spectral components corresponding to the appropriate order of the Bessel function. In a similar fashion, spillover and/or reflections from close-in objects can be reduced by filtering to pass only certain harmonics of the modulating frequency (F_m). Receiving only frequencies near $3F_m$ results in considerable spillover rejection, but at a penalty of 4 to 10 dB in signal-to-noise [Nathanson, 1991].

For the sinusoidal modulation case, Doppler frequency contributions complicate the analysis considerably. For details of this analysis the reader is referred to Saunders [1990] or Nathanson [1991].

12.3 Interrupted Frequency-Modulated CW (IFM/CW)

To improve isolation during reception, the IFM/CW format involves preventing transmission for a portion of the time during the frequency change. Thus, there are frequency gaps, or interruptions, as illustrated in Fig. 12.4. This shows a case where the transmit time equals the round-trip propagation time, followed by an equal time for reception. This duty factor of 0.5 for the waveform reduces the average transmitted power by 3 dB relative to using an uninterrupted transmitter. However, the improvement in the isolation should reduce the system noise by more than 3 dB, thus improving the signal-to-noise ratio [Piper, 1987]. For operation at short range, Piper points out that a high-speed switch is required [1987]. He also points out that the ratio of frequency deviation to beat frequency should be an even integer and that the minimum ratio is typically 6, which produces an out-of-band loss of 0.8 dB.

IFM/CW may be compared with pulse compression radar if both use a wide bandwidth. Pulse compression employs a “long” pulse (i.e., relatively long for a pulse radar) with a large frequency deviation or “chirp.” A long pulse is often used when a transmitter is peak-power limited, because the longer pulse produces more energy and gives more range to targets. The frequency deviation is controlled in a predetermined way (frequently a linear sweep) so that a matched filter can be used in the receiver. The large time-bandwidth product permits the received pulse to be compressed in time to a short pulse in order to make an accurate range measurement. A linear-sawtooth IFM/CW having similar pulse length, frequency deviation, and pulse repetition rate would thus appear similar, although arrived at from different points of view.

12.4 Applications

Space does not permit giving a full description of the many applications mentioned at the beginning of this chapter, but several will be discussed.

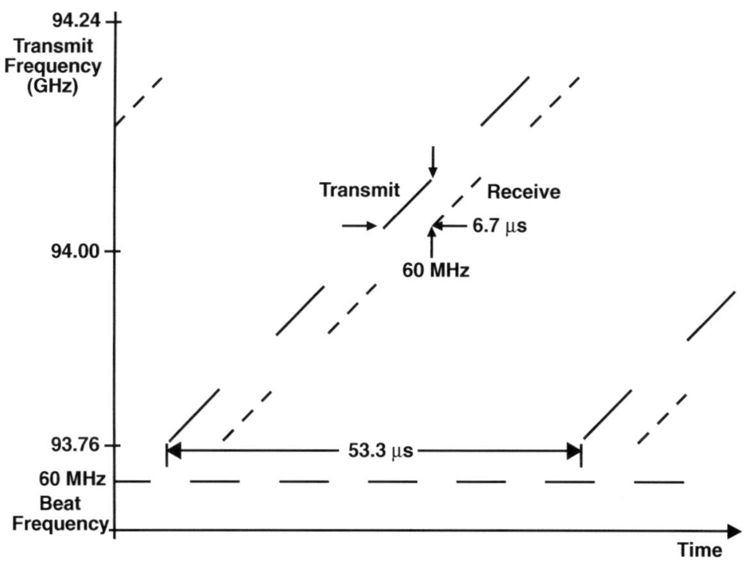

FIGURE 12.4 Interrupted FM/CW waveform. (*Source:* S.O. Piper, "MMW seekers," in *Principles and Applications of Millimeter Wave Radar,* N. Currie and C.E. Brown, Eds., Norwood, MA: Artech House, 1987, p. 683. With permission.)

12.4.1 Radar Proximity Fuzes

Projectiles or missiles designed to be aimed at ships or surface land targets often need a height-of-burst (HOB) sensor (or target detection device) to fire or fuze the warhead at a height of a few meters. There are two primary generic methods of sensing or measuring height to generate the warhead fire signal. The most obvious, and potentially the most accurate, is to measure target round-trip propagation delay employing conventional radar ranging techniques. The second method employs a simple CW Doppler radar or variation thereof, with loop gain calibrated in a manner that permits sensing the desired burst height by measurement of target return signal amplitude and/or rate of change. Often the mission requirements do not justify the complexity and cost of the radar ranging approach. Viable candidates are thus narrowed down to variations on the CW Doppler fuze.

In its simplest form, the CW Doppler fuze consists of a fractional watt RF oscillator, homodyne detector, Doppler amplifier, Doppler envelope detector, and threshold circuit. When the Doppler envelope amplitude derived from the returned signal reaches the preset threshold, a fire signal is generated. The height at which the fire signal occurs depends on the radar loop gain, threshold level, and target reflectivity. Fuze gain is designed to produce the desired height of burst under nominal trajectory angle and target reflectivity conditions, which may have large fluctuations due to glint effects, and deviations from the desired height due to antenna gain variations with angle, target reflectivity, and fuze gain tolerances are accepted. A loop gain change of 6 dB (2 to 1 in voltage), whether due to a change in target reflection coefficient, antenna gain, or whatever, will result in a 2 to 1 HOB change.

HOB sensitivity to loop gain factors can be reduced by utilizing the slope of the increasing return signal, or so-called rate-of-rise. Deriving HOB solely from the rate-of-rise has the disadvantage of rendering the fuze sensitive to fluctuating signal levels such as might result from a scintillating target. The use of logarithmic amplifiers decreases the HOB sensitivity to the reflectivity range. An early (excessively high) fire signal can occur if the slope of the signal fluctuations equals the rate-of-rise threshold of the fuze. In practice a compromise is generally made in which Doppler envelope amplitude and rate-of-rise contribute in some proportion of HOB.

Another method sometimes employed to reduce HOB sensitivity to fuze loop gain factors and angle of fall is the use of FM sinusoidal modulation of suitable deviation to produce a range correlation function comprising the zero order of a Bessel function of the first kind. The subject of sinusoidal modulation is quite complex, but has been treated in detail by Saunders [1990, pp. 1422–1446 and 144.41]. The most important aspects of fuze design have to do with practical problems such as low cost, small size, ability to stand very high-g accelerations, long life in storage, and countermeasures susceptibility.

12.4.2 Police Radars

Down-the-road police radars, which are of the CW Doppler type, operate at 10.25 (X-Band), 24.150 (K-band), or in the 33.4- to 36.0-GHz (Ka-band) range, frequencies approved in the United States by the Federal Communications Commission. Half-power beamwidths are typically in the 0.21 to 0.31 radian (12 to 18°) range. The sensitivity is usually good enough to provide a range exceeding 800 m. Target size has a dynamic range of 30 dB (from smallest cars or motorcycles to large trucks). This means that a large target can be seen well outside the antenna 3-dB point at a range exceeding the range of a smaller target near the center of the beam. Range is not measured. Thus there can be uncertainty about which vehicle is the target. Fisher [1992] has given a discussion of a number of the limitations of these systems, but in spite of these factors probably tens of thousands have been built.

The transmitter is typically a Gunn oscillator in the 30- to 100-mW power range, and antenna gain is usually around 20 to 24 dB, employing circular polarization. The designs typically have three amplifier gains for detection of short, medium, or maximum range targets, plus a squelch circuit so that sudden spurious signals will not be counted. Provision is made for calibration to assure the accuracy of the readings. Speed resolution is about 1 mi/h. The moving police radar system uses stationary (ground) clutter to derive the patrol car speed. Then closing speed, minus patrol car speed, yields target speed.

The limitations mentioned about deciding which vehicle is the correct target have led to the development of laser police radars, which utilize much narrower beamwidths, making target identification much more accurate. Of course, the use of microwave and laser radars has spawned the development of automotive radar detectors, which are also in wide use.

12.4.2.1 Altimeters

A very detailed discussion of FM/CW altimeters has been given by Saunders [1990, pp. 14.34–14.36], in which he has described modern commercial products built by Bendix and Collins. The parameters will be summarized below and if more information is needed, the reader may want to turn to other references [Saunders, 1990; Bendix Corp., 1982; Maoz et al., 1991; and Stratakos, 2000]. In his material, Saunders gives a general overview of modern altimeters, all of which use wide-deviation FM at a low modulation frequency. He discusses the limitations on narrowing the antenna pattern, which must be wide enough to accommodate attitude changes of the aircraft. Triangular modulation is used, since for this waveform the Doppler averages out, and dual antennas are employed. There may be a step error or quantization in height (which could be a problem at low altitudes), due to the limitation of counting zero crossings. A difference of one zero crossing (i.e., 1/2 Hz) corresponds to 3/4 m for a frequency deviation of 100 MHz. Irregularities are not often seen, however, since meter response is slow. Also, if terrain is rough, there will be actual physical altitude fluctuations. Table 12.2 shows some of the altimeter parameters. These altimeters are not acceptable for military aircraft, because their relatively wide-open front ends make them potentially vulnerable to electronic countermeasures. A French design has some advantages in this respect by using a variable frequency deviation, a difference frequency that is essentially constant with altitude, and a narrowband front-end amplifier [Saunders, 1990].

12.4.2.2 Doppler Navigators

These systems are mainly sinusoidally modulated FM/CW radars employing four separate downward looking beams aimed at about 15° off the vertical. Because commercial airlines have shifted to nonradar forms of navigation, these units are designed principally for helicopters. Saunders [1990] cites a particular example of a commercial unit operating at 13.3 GHz, employing a Gunn oscillator as the transmitter,

TABLE 12.2 Parameters for Two Commercial Altimeters

Modulation Frequency	Frequency Deviation	Prime Power	Weight (pounds)	Radiated Power
Bendix ALA-52A	150 Hz	130 MHz	30 W	11[a]
Collins ALT-55	100 kHz	100 MHz	8	350 mW

[a]Not including antenna and indicator.

with an output power of 50 mW, and utilizing a 30-kHz modulation frequency. A single microstrip antenna is used. A low-altitude equipment (below 15,000 ft.), the unit weighs less than 12 pounds. A second unit cited has an output power of 300 mW, dual antennas, dual modulating frequencies, and an altitude capability of 40,000 ft.

12.4.2.3 Millimeter-Wave Seeker for Terminal Guidance Missile

Terminal guidance for short-range (less than 2 km) air-to-surface missiles has seen extensive development in the last decade. Targets such as tanks are frequently immersed in a clutter background that may give a radar return that is comparable to that of the target. To reduce the clutter return in the antenna footprint, the antenna beamwidth is reduced by going to millimeter wavelengths. For a variety of reasons the choice is usually a frequency near 35 or 90 GHz. Antenna beamwidth is inversely proportional to frequency, so in order to get a reduced beamwidth we would normally choose 90 GHz; however, more deleterious effects at 90 GHz due to atmospheric absorption and scattering can modify that choice. In spite of small beamwidths, the clutter is a significant problem, and in most cases signal-to-clutter is a more limiting condition than signal-to-noise in determining range performance. Piper [1987] has done an excellent job of analyzing the situation for 35- and 90-GHz pulse radar seekers and comparing those with a 90-GHz FM/CW seeker. His FM/CW results will be summarized below.

In his approach to the problem, Piper gives a summary of the advantages and disadvantages of a pulse system compared to the FM/CW approach. One difficulty for the FM/CW can be emphasized here. That is the need for a highly linear sweep, and, because of the desire for the wide bandwidth, this requirement is accentuated. The wide bandwidth is desired in order to average the clutter return and to smooth the glint effects. In particular, glint occurs from a complex target because of the vector addition of coherent signals scattered back to the receiver from various reflecting surfaces. At some angles the vectors may add in phase (constructively) and at others they may cancel, and the effect is specifically dependent on wavelength. For a narrowband system, glint may provide a very large signal change over a small variation of angle, but, of course, at another wavelength it would be different. Thus, very wide bandwidth is desirable from this smoothing point of view, and typical numbers used in millimeter-wave radars are in the 450- to 650-MHz range. Piper chose 480 MHz.

Another trade-off involves the choice of FM waveform. Here the use of a triangular waveform is undesirable because the Doppler frequency averages out and Doppler compensation is then required. Thus the sawtooth version is chosen, but because of the large frequency deviation desired, the difficulty of linearizing the frequency sweep is made greater. In fact many components must be extremely wideband, and this generally increases cost and may adversely affect performance. On the other hand, the difference frequency (F_b) and/or the intermediate frequency (F_{IF}) will be higher and thus farther from the carrier, so the phase noise will be lower. After discussing the other trade-offs, Piper chose 60 MHz for the beat frequency.

With a linear FM/CW waveform, the inverse of the frequency deviation provides the theoretical time resolution, which is 2.1 ns for 480 MHz (or range resolution of 0.3 m). For an RF sweep linearity of 300 kHz, the range resolution is actually 5 m at the 1000-m nominal search range. (The system has a mechanically scanned antenna.) An average transmitting power of 25 mW was chosen, which was equal to the average power of the 5-W peak IMPATT assumed for the pulse system. The antenna diameter was 15 cm. For a target radar cross section of 20 m^2 and assumed weather conditions, the signal-to-clutter and signal-to-noise ratios were calculated and plotted for ranges out to 2 km and for clear weather or 4 mm/h of rainfall. The results show that for 1-km range the target-to-clutter ratios are higher for the FM/CW case than the pulse system in clear weather or in rain, and target-to-clutter is the determining factor.

12.5 Summary Comments

From this brief review it is clear that there are many uses for CW radars, and various types (such as fuzes) have been produced in large quantities. Because of their relative simplicity, today there are continuing trends toward the use of digital processing and integrated circuits. In fact, this is exemplified in articles describing FM/CW radars built on single microwave integrated circuit chips [Maoz et al., 1991; Chang et al., 1995; Haydl et al., 1999; Menzel et al., 1999].

Defining Terms

Doppler-frequency shift: The observed frequency change between the transmitted and received signal produced by motion along a line between the transmitter/receiver and the target. The frequency increases if the two are closing and decreases if they are receding.

Missile terminal guidance seeker: Located in the nose of a missile, a small radar with short-range capability which scans the area ahead of the missile and guides it during the terminal phase toward a target such as a tank.

Pulse Doppler: A coherent radar, usually having high pulse repetition rate and duty cycle and capable of measuring the Doppler frequency from a moving target. The radar has good clutter suppression and thus can see a moving target in spite of background reflections.

References

Bendix Corporation, Service Manual for ALA-52A Altimeter; Design Summary for the ALA-52A, Bendix Corporation, Ft. Lauderdale, FL, May 1982.

K.W. Chang, H. Wang, G. Shreve, J.G. Harrison, M. Core, A. Paxton, M. Yu, C.H. Chen, and G.S. Dow, Forward-looking automotive radar using a W-band single-chip transceiver, *IEEE Transactions on Microwave Theory and Techniques,* 43, 1659–1668, July 1995.

Collins (Rockwell International), ALT-55 Radio Altimeter System; Instruction Book, Cedar Rapids, IA, October 1984.

P.D. Fisher, Improving on police radar, *IEEE Spectrum,* 29, 38–43, July 1992.

A.J. Freiley, B.L. Conroy, D.J. Hoppe, and A.M. Bhanji, Design concepts of a 1-MW CW X-band transmit/receive system for planetary radar, *IEEE Transactions on Microwave Theory and Techniques,* 40, 1047–1055, June 1992.

W.H. Haydl et al., Single-chip coplanar 94 GHz FMCW radar sensors, *IEEE Microwave and Guided Wave Lett.,* 9, 73–75, February 1999.

B. Maoz, L.R. Reynolds, A. Oki, and M. Kumar, FM-CW radar on a single GaAs/AlGaAs HBT MMIC chip, *IEEE Microwave and Millimeter-Wave Monolithic Circuits Symposium Digest,* 3–6, June 1991.

W. Menzel, D. Pilz, and R. Leberer, A 77-GHz FM/CW radar front-end with a low-profile low-loss printed antenna, *IEEE Trans. Microwave Theory and Techniques,* 47, 2237–2241, December 1999.

F.E. Nathanson, *Radar Design Principles,* New York: McGraw-Hill, 1991, 448–467.

S.O. Piper, MMW seekers, in *Principles and Applications of Millimeter Wave Radar,* N.C. Currie and C.E. Brown, Eds., Norwood, MA: Artech House, 1987, chap. 14.

W.K. Saunders, CW and FM radar, in *Radar Handbook,* M.I. Skolnik, Ed., New York: McGraw-Hill, 1990, chap. 14.

G.E. Stratakos, P. Bougas, and K. Gotsis, A low cost, high accuracy radar altimeter, *Microwave J.,* 43, 120–128, February 2000.

L.M. Tozzi, Resolution in frequency-modulated radars, Ph.D. thesis, University of Maryland, College Park, 1972.

F.T. Ulaby and C. Elachi, *Radar Polarimetry for Geoscience Applications*, Norwood, MA: Artech House, 1990, 193–200.

J.C. Wiltse, S.P. Schlesinger, and C.M. Johnson, Back-scattering characteristics of the sea in the region from 10 to 50 GHz, *Proceedings of the IRE*, 45, 220–228, February 1957.

Further Information

For a general treatment, including analysis of clutter effects, Nathanson's [1991] book is very good and generally easy to read. For extensive detail and specific numbers in various actual cases, Saunders [1990] gives good coverage. The treatment of millimeter-wave seekers by Piper [1987] is excellent, both comprehensive and easy to read.

13

Pulse Radar

Melvin L. Belcher, Jr.
Georgia Institute of Technology

Josh T. Nessmith
Georgia Institute of Technology

13.1 Overview of Pulsed Radars

13.1.1 Basic Concept of Pulse Radar Operation

The basic operation of a pulse radar is depicted in Fig. 13.1. The radar transmits pulses superimposed on a radio frequency (RF) carrier and then receives returns (reflections) from desired and undesired scatterers. Scatterers corresponding to desired targets may include space, airborne, and sea and/or surface-based vehicles. They can also include Earth's surface and atmosphere in remote sensing applications. Undesired scatterers are termed *clutter.* Clutter sources include Earth's surface, natural and man-made discrete objects, and volumetric atmospheric phenomena such as rain and birds. Short-range/low-altitude radar operation is often constrained by clutter since the multitude of undesired returns masks returns from targets of interest such as aircraft. Conversely, volumetric atmospheric phenomena may be considered as targets for weather radar systems.

The range, azimuth angle, elevation angle, and range rate can be directly measured from a return to estimate target metrics, position, and velocity; and to support tracking. Signature data to support noncooperative target identification or environmental remote sensing can be extracted by measuring the amplitude, phase, and polarization of the return.

Pulse radar affords a great deal of design and operational flexibility. Pulse duration, pulse rate, and pulse bandwidth can be tailored to specific applications to provide optimal performance. Modern computer-controlled multiple-function radars exploit this capability by choosing the best waveform from a repertoire for a given operational mode and interference environment automatically.

0-8493-1732-0/03/$0.00+$1.50

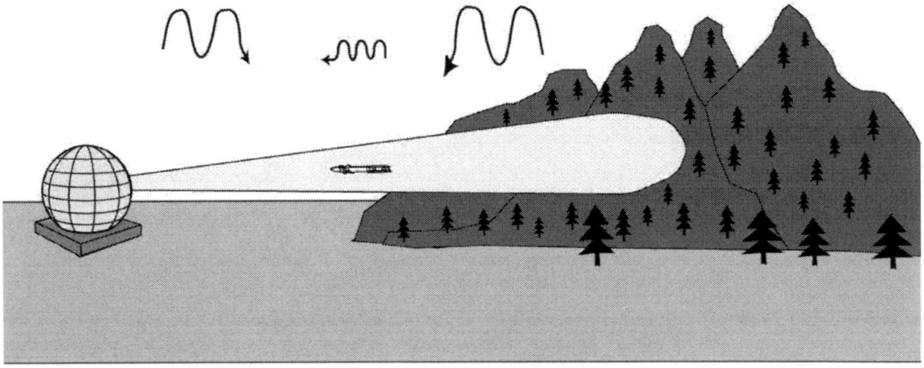

FIGURE 13.1 Pulse radar.

13.1.2 Radar Applications

The breadth of pulse radar applications is summarized in Table 13.1 in terms of operating frequencies. Radar applications can also be grouped into search, track, and signature measurement applications. Search radars are used for surveillance tracking but have relatively large range and angle errors. The search functions favor broad beamwidths and low bandwidths in order to efficiently search over a large spatial volume. As indicated in Table 13.1, search is preferably performed in the lower frequency bands. The antenna pattern is typically narrow in azimuth and has a cosecant pattern in elevation to provide acceptable coverage from the horizon to the zenith.

Tracking radars are typically characterized by a narrow beamwidth and moderate bandwidth in order to provide accurate range and angle measurements on a given target. The antenna pattern is a pencil beam with approximately the same dimensions in azimuth and elevation. Track is usually conducted at the higher frequency bands in order to minimize the beamwidth for a given antenna aperture area. After each return from a target is received, the range and angle are measured and input into a track filter. Track filtering smooths the data to refine the estimate of target position and velocity. It also predicts the target's flight path to provide range gating and antenna pointing control to the radar system.

Signature measurement applications include remote sensing of the environment as well as the measurement of target characteristics. In some applications, synthetic aperture radar (SAR) imaging is conducted from aircraft or satellites to characterize land usage over broad areas. Moving targets that present changing aspect to the radar can be imaged from airborne or ground-based radars via inverse synthetic aperture radar (ISAR) techniques. As defined in the subsection Resolution and Accuracy, cross-range resolution improves with increasing antenna extent. SAR/ISAR effectively substitutes an extended observation interval over which coherent returns are collected from different target aspect angles for a large antenna structure that would not be physically realizable in many instances.

In general, characterization performance improves with increasing frequency because of the associated improvement in range, range rate, and cross-range resolution. However, phenomenological characterization to support environmental remote sensing may require data collected across a broad swath of frequencies.

A multiple-function **phased-array** radar generally integrates these functions to some degree. Its design is usually driven by the track function. Its operational frequency is generally a compromise between the lower frequency of the search radar and the higher frequency desired for the tracking radar. The degree of signature measurement implemented to support such functions as noncooperative target identification depends on the resolution capability of the radar as well as the operational user requirements. Multiple-function radar design represents a compromise among these different requirements. However, implementation constraints, multiple-target handling requirements, and reaction time requirements often dictate the use of phased array radar systems integrating search, track, and characterization functions.

TABLE 13.1 Radar Bands

Band	Frequency Range	Principal Applications
HF	3–30 MHz	Over-the-horizon radar
VHF	30–300 MHz	Long-range search
UHF	300–1000 MHz	Long-range surveillance
L	1000–2000 MHz	Long-range surveillance
S	2000– 4000 MHz	Surveillance Long-range weather characterization Terminal air traffic control
C	4000–8000 MHz	Fire control Instrumentation tracking
X	8–12 GHz	Fire control Air-to-air missile seeker Marine radar Airborne weather characterization
Ku	12–18 GHz	Short-range fire control Remote sensing
Ka	27– 40 GHz	Remote sensing Weapon guidance
V	40–75 GHz	Remote sensing Weapon guidance
W	75–110 GHz	Remote sensing Weapon guidance

13.2 Critical Subsystem Design and Technology

The major subsystems making up a pulse radar system are depicted in Fig. 13.2. The associated interaction between function and technology is summarized in this subsection.

13.2.1 Antenna

The radar antenna function is to first provide spatial directivity to the transmitted EM wave and then to intercept the scattering of that wave from a target. Most radar antennas may be categorized as mechanically scanning or electronically scanning. Mechanically scanned reflector antennas are used in applications where rapid beam scanning is not required. Electronic scanning antennas include phased arrays and frequency scanned antennas. Phased array beams can be steered to any point in their field of view, typically within 10 to 100 μs, depending on the latency of the beam-steering subsystem and the switching time of the phase shifters. Phased arrays are desirable in multiple function radars since they can interleave search operations with multiple target tracks.

There is a Fourier transform relationship between the antenna illumination function and the far-field antenna pattern analogous to spectral analysis. Hence, tapering the illumination to concentrate power near the center of the antenna suppresses side lobes while reducing the effective antenna aperture area. The phase and amplitude control of the antenna illumination determines the achievable side lobe suppression and angle measurement accuracy.

Perturbations in the illumination due to the mechanical and electrical sources distort the illumination function and constrain performance in these areas. Mechanical illumination error sources include antenna shape deformation due to sag and thermal effects as well as manufacturing defects. Electrical illumination error is of particular concern in phased arrays where sources include beam steering computational error and phase shifter quantization. Control of both the mechanical and electrical perturbation errors is the key to both low side lobes and highly accurate angle measurements. Control denotes that either tolerances are closely held and maintained or that there must be some means for monitoring and correction. Phased arrays are attractive for low side lobe applications since they can provide element-level phase and amplitude control.

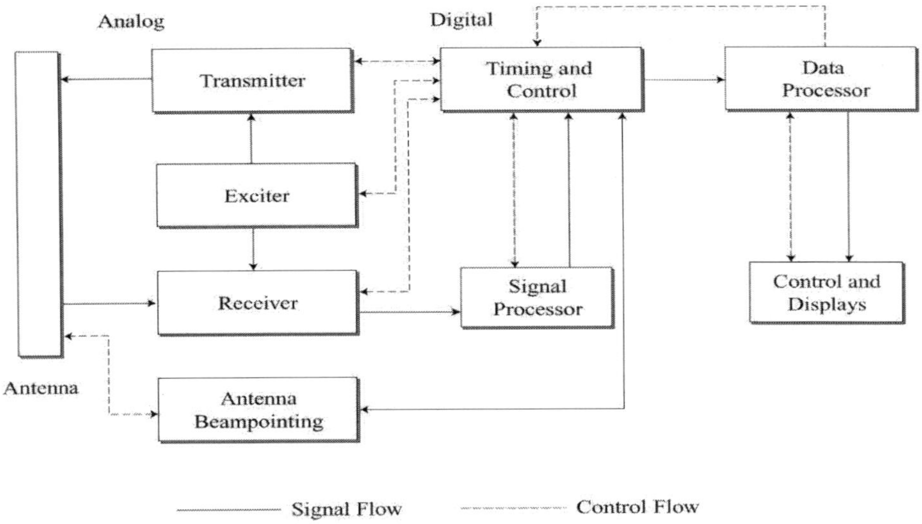

FIGURE 13.2 Radar system architecture.

13.2.2 Transmitter

The transmitter function is to amplify waveforms to a power level sufficient for target detection and estimation. There is a general trend away from tube-based transmitters toward solid-state transmitters. In particular, solid-state transmit/receive modules appear attractive for constructing phased array radar systems. In this case, each radiating element is driven by a module that contains a solid-state transmitter, phase shifter, low-noise amplifier, and associated control components. Active arrays built from such modules appear to offer significant reliability advantages over radar systems driven from a single transmitter. Microwave tube technology offers substantial advantages in power output over solid-state technology. However, there is a strong trend in developmental radars toward use of solid-state transmitters due to production base as well as performance considerations.

13.2.3 Receiver and Exciter

This subsystem contains the precision timing and frequency reference source or sources used to derive the master oscillator and local oscillator reference frequencies. These reference frequencies are used to downconvert received signals in a multiple-stage superheterodyne architecture to accommodate signal amplification and interference rejection. Filtering is conducted at the carrier and intermediate frequencies in processing to reject interference outside the operating band of the radar. The receiver front end is typically protected from overload during transmission through the combination of a circulator and a transmit/receive switch.

The exciter generates the waveforms for subsequent transmission. As in signal processing, the trend is toward programmable digital signal synthesis because of the associated flexibility and performance stability.

13.2.4 Signal and Data Processing

Digital processing is generally divided between two processing subsystems (i.e., signals and data), according to the algorithm structure and throughput demands. Signal processing includes pulse compression, Doppler filtering, and detection threshold estimation and testing. Data processing includes track filtering, user interface support, and such specialized functions as electronic counter-countermeasures (ECCM) and built-in test (BIT), as well as the resource management process required to control the radar system.

The signal processor is often optimized to perform the repetitive complex multiply-and-add operations associated with the fast Fourier transform (FFT). FFT processing is used for implementing **pulse compression** via fast convolution and for Doppler filtering. Pulse compression consists of matched filtering on receive to an intrapulse modulation imposed on the transmitted pulse. As delineated subsequently, the imposed intrapulse bandwidth determines the range resolution of the pulse while the modulation format determines the suppression of the waveform matched-filter response outside the nominal range resolution extent. Fast convolution consists of taking the FFT of the digitized receiver output, multiplying it by the stored FFT of the desired filter function, and then taking the inverse FFT of the resulting product. Fast convolution results in significant computational saving over performing the time-domain convolution of returns with the filter function corresponding to the matched filter. The signal processor output can be characterized in terms of range gates and Doppler filters corresponding approximately to the range and Doppler resolution, respectively.

In contrast, the radar data processor typically consists of a general-purpose computer with a real-time operating system. Fielded radar data processors range from microcomputers to mainframe computers, depending on the requirements of the radar system. Data processor software and hardware requirements are significantly mitigated by off-loading timing and control functions to specialized hardware. This timing and control subsystem typically functions as the two-way interface between the data processor and the other radar subsystems. The increasing inclusion of BIT (built-in-test) and built-in calibration capability in timing and control subsystem designs promises to result in significant improvement in fielded system performance. The trend is toward increasing use of commercial off-the-shelf digital processing elements for radar applications and tighter integration of the signal and data processing functions.

13.3 Radar Performance Prediction

13.3.1 Radar Line-of-Sight

With the exception of over-the-horizon (OTH) radar systems, which exploit either sky-wave bounce or ground-wave propagation modes and sporadic ducting effects at higher frequencies, surface and airborne platform radar operation is limited to the refraction-constrained line of sight. Atmospheric refraction effects can be closely approximated by setting Earth's radius to 4/3 its nominal value in estimating horizon-limited range. The resulting line-of-sight range is depicted in Fig. 13.3 for a surface-based radar, an airborne surveillance radar, and a space-based radar.

As evident in the plot, airborne and space-based surveillance radar systems offer significant advantages in the detection of low-altitude targets that would otherwise be masked by earth curvature and terrain features from surface-based radars. However, efficient clutter rejection techniques must be used in order to detect targets since surface clutter returns will be present at almost all ranges of interest.

13.3.2 Radar Range Equation

The radar range equation is commonly used to estimate radar system performance, given that line-of-sight conditions are satisfied. This formulation essentially computes the signal-to-noise ratio (S/N) at the output of the radar signal processor. In turn, S/N is used to provide estimates of radar detection and position measurement performance as described in subsequent subsections. S/N can be calculated in terms of the number of pulses coherently integrated over a single coherent processing interval (CPI) using the radar range equation such that

$$S/N = \frac{P_D A T_p N_p \sigma}{(4\pi)^2 R^4 L_t L_{rn} L_{sp} kT_s} \tag{13.1}$$

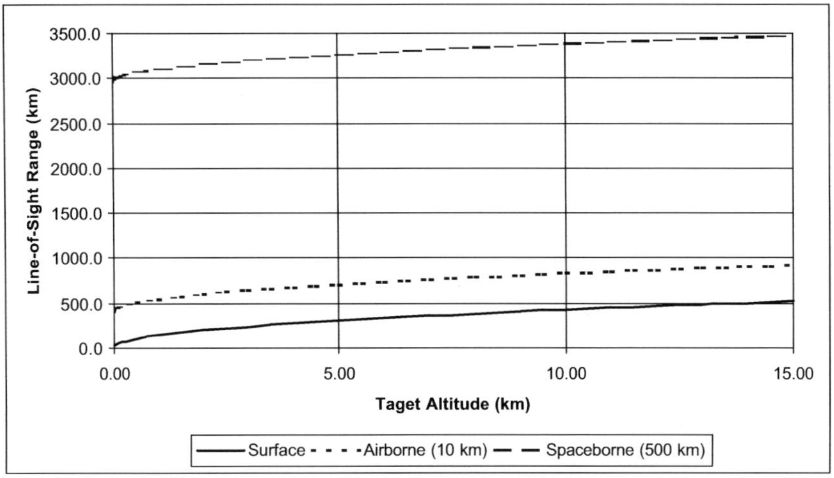

FIGURE 13.3 Maximum line-of-sight range for a surface-based radar, an airborne surveillance radar, and a space-based radar.

where P is peak transmitter power output, D is directivity of the transmit antenna, A is effective aperture area of the receive antenna in meters squared, T_p is pulse duration, σ is **radar cross section** in square meters, N_p is the number of coherently integrated pulses within the coherent processing interval, R is range to target in meters, L_t is system ohmic and nonohmic transmit losses, L_{rn} is system nonohmic receive losses, L_{sp} is signal processing losses, k is the Boltzmann constant (1.38×10^{-23} degrees K), and T_s is system noise temperature, including receive ohmic losses (Kelvin).

At X-band and above it may also be necessary to include propagation loss due to atmospheric absorption [Blake, 1986]. This form of the radar range equation is applicable to radar systems using pulse compression or pulse Doppler waveforms as well as the unmodulated single-pulse case. In many applications, average power is a better measure of system performance than peak power since it indicates the S/N improvement achievable with pulse integration over a given interval of time. Hence, the radar range equation can be modified such that

$$S/N = \frac{P_a D A T_c \sigma}{(4\pi)^2 R^4 L_t L_{rn} L_{sp} k T_s} \tag{13.2}$$

where P_a is average transmitter power and T_c is coherent processing interval (CPI).

The portion of time over which the transmitter is in operation is referred to as the radar duty cycle. The average transmitter power is the product of duty cycle and peak transmitter power. Duty cycle ranges from less than 1% for typical **noncoherent** pulse radars to somewhat less than 50% for high pulse repetition frequency (PRF) pulse Doppler radar systems. The CPI is the period over which returns are collected for **coherent** processing functions such as pulse integration and Doppler filtering. The CPI can be estimated as the product of the number of coherently integrated pulses and the interval between pulses. Noncoherent pulse integration is less efficient and alters the statistical character of the signal and interference.

13.3.3 Antenna Directivity and Aperture Area

The directivity of the antenna is

$$D = \frac{4\pi A\eta}{\lambda^2} \tag{13.3}$$

where η is aperture efficiency and λ is radar carrier wavelength. Aperture inefficiency is due to the antenna illumination factor.

The common form of the radar range equation uses power gain rather than directivity. Antenna gain is equal to the directivity divided by the antenna losses. In the design and analysis of modern radars, directivity is a more convenient measure of performance because it permits designs with distributed active elements, such as solid-state phased arrays, to be assessed to permit direct comparison with passive antenna systems. Beamwidth and directivity are inversely related; a highly directive antenna will have a narrow beamwidth. For typical design parameters,

$$D = \frac{10^7}{\theta_{az}\,\theta_{el}} \tag{13.4}$$

where θ_{az} and θ_{el} are the radar azimuth and elevation beamwidths, respectively, in milliradians. The directivity then gives the power density relative to an isotropic radiator.

13.3.4 Radar Cross-Section

In practice, the *radar cross section* (RCS) of a realistic target must be considered a random variable with an associated correlation interval. Targets are composed of multiple interacting scatters so that the composite return varies in magnitude with the constructive and destructive interference of the contributing returns. The target RCS is typically estimated as the mean or median of the target RCS distribution. The associated correlation interval indicates the rate at which the target RCS varies over time. RCS fluctuation degrades target detection performance at moderate to high probability of detection.

The median RCS of typical targets is given in Table13.2. The composite RCS measured by a radar system may be composed of multiple individual targets in the case of closely spaced targets such as a bird flock.

13.3.5 Loss and System Temperature Estimation

Sources of *S*/*N* loss include ohmic and nonohmic (mismatch) loss in the antenna and other radio frequency components, propagation effects, signal processing deviations from matched filter operation, detection thresholding, and search losses. Scan loss in phased array radars is due to the combined effects of the decrease in projected antenna area and element mismatch with increasing scan angle.

Search operations impose additional losses due to target position uncertainty. Because the target position is unknown before detection, the beam, range gate, and Doppler filter will not be centered on the target return. Hence, straddling loss will occur as the target effectively straddles adjacent resolution cells in range and Doppler. Beam shape loss is a consequence of the radar beam not being pointed directly at the target so that there is a loss in both transmit and receive antenna gain. In addition, detection threshold loss associated with radar system adaptation to interference must be included [Nathanson, 1991].

System noise temperature estimation corresponds to assessing the system thermal noise floor referenced to the antenna output. Assuming the receiver hardware is at ambient temperature, the system noise temperature can be estimated as

$$T_s = T_a + 290\left(L_{ro}\,F - 1\right. \tag{13.5}$$

where T_a is the antenna noise temperature, L_{ro} is receive ohmic losses, and F is the receiver noise figure.

TABLE 13.2 Median Target RCS (m^2)

Carrier Frequency, GHz	1–2	3	5	10	17
Aircraft (nose/tail avg.)					
Small propeller	2	3	2.5		
Small jet (Lear)	1	1.5	1	1.2	
T38-twin jet, F5	2	2–3	2	1–2/6	
T39-Sabreliner	2.5		10/8	9	
F4, large fighter	5–8/5	4–20/10	4	4	
737, DC9, MD80	10	10	10	10	10
727, 707, DC8-type	22–40/15	40	30	30	
DC-10-type, 747	70	70	70	70	
Ryan drone				2/1	
Standing man (180 lb)	0.3	0.5	0.6	0.7	0.7
Automobiles	100	100	100	100	100
Ships-incoming ($\times 10^4$ m^2)					
4K tons	1.6	2.3	3.0	4.0	5.4
16K tons	13	18	24	32	43
Birds					
Sea birds	0.002	0.001–0.004	0.004		
Sparrow, starling, etc.	0.001	0.001	0.001	0.001	0.001

Note: Slash marks indicate different set.

Source: F.E. Nathanson, *Radar Design Principles,* 2nd ed., New York: McGraw-Hill, 1991. With permission.

In phased-array radars, the thermodynamic temperature of the antenna receive beam former may be significantly higher than ambient, so a more complete analysis is required. The antenna noise temperature is determined by the external noise received by the antenna from solar, atmospheric, Earth surface, and other sources.

Table 13.3 provides typical loss and noise temperature budgets for several major radar classes. In general, loss increases with the complexity of the radar hardware between the transmitter/receiver and the antenna radiator. Reflector antennas and active phased arrays impose relatively low loss, while passive array antennas impose relatively high loss.

13.3.6 Resolution and Accuracy

The fundamental resolution capabilities of a radar system are summarized in Table 13.4. In general, there

TABLE 13.3 Typical Microwave Loss and System Temperature Budgets

	Mechanically Scanned Reflector Antenna	Electronically Scanned Slotted Array	Solid-State Phased Array
Nominal losses			
Transmit loss, L_t (dB)	1	1.5	0.5
Nonohmic receiver loss, L_r (dB)	0.5	0.5	0.1
Signal processing loss, L_{sp} (dB)	1.4	1.4	1.4
Scan loss (dB)	N/A	N/A	30 log [cos (scan angle)]
Search losses, L_{DS}			
Beam shape (dB)	3	3	3
Range gate straddle (dB)	0.5	0.5	0.5
Doppler filter straddle (dB)	0.5	0.5	0.5
Detection thresholding (dB)	1	1	1
System noise temperature (kelvin)	500	600	400

TABLE 13.4 Resolution and Accuracy

Dimension	Nominal Resolution	Noise-Limited Accuracy
Angle	$\frac{\alpha\lambda}{d}$	$\frac{\alpha\lambda}{dK_m\sqrt{2S/N}}$
Range	$\frac{\alpha C}{2B}$	$\frac{\alpha C}{2BK_i\sqrt{2S/N}}$
Doppler	$\frac{\alpha}{\text{CPI}}$	$\frac{\alpha}{\text{CPI}\,K_i\sqrt{2S/N}}$
SAR/ISAR	$\frac{\alpha\lambda}{2\Delta\theta}$	$\frac{\alpha\lambda}{2\Delta\theta K_i\sqrt{2S/N}}$

Note: α, taper broadening factor, typically ranging from 0.89 (unweighted) to 1.3 (Hamming); d, antenna extent in azimuth/elevation; B, waveform bandwidth; K_m, monopulse slope factor, typically on the order of 1.5; Ki, interpolation factor, typically on the order of 1.8; $\Delta\theta$, line-of-sight rotation of target relative to radar over CPI.

is a trade-off between main lobe resolution corresponding to the nominal range, Doppler, and angle resolution; and effective dynamic range corresponding to suppression of sidelobe components. This is evident in the use of weighting to suppress Doppler side bands and angle side lobes at the expense of broadening the main lobe and S/N loss.

Cross range denotes either of the two dimensions orthogonal to the radar line of sight. Cross-range resolution in real-aperture antenna systems is closely approximated by the product of target range and radar beamwidth in radians. Attainment of the nominal ISAR/SAR cross-range resolution generally requires complex signal processing to generate a focused image, including correction for scatterer change in range over the CPI.

The best accuracy performance occurs for the case of thermal noise-limited error. The resulting accuracy is the resolution of the radar divided by the square root of the *S*/*N* and an appropriate monopulse or interpolation factor. In this formulation, the single-pulse *S*/*N* has been multiplied by the number of pulses integrated within the CPI as indicated in Eqs. (13.1) and (13.2).

In practice, accuracy is also constrained by environmental effects, target characteristics, and instrumentation error as well as the available *S*/*N*. Environmental effects include multipath and refraction. Target glint is characterized by an apparent wandering of the target position because of coherent interference effects associated with the composite return from the individual scattering centers on the target. Instrumentation error is minimized with alignment and calibration but may significantly constrain track filter performance as a result of the relatively long correlation interval of some error sources.

13.3.7 Radar Range Equation for Search and Track

The radar range equation can be modified to directly address performance in the two primary radar missions: search and track.

Search performance is basically determined by the capability of the radar system to detect a target of specific RCS at a given maximum detection range while scanning a given solid angle extent within a specified period of time. *S*/*N* can be set equal to the minimum value required for a given detection performance, S/N^*r, while R can be set to the maximum required target detection range, R_{max}. Manipulation of the radar range equation results in the following expression:

$$\frac{P_a A}{L_t L_r L_{sp} L_{os} T_s} \geq \left(\frac{S}{N}\right) \frac{R_{max}^4 \Omega}{\sigma T_{fs}} \cdot 16k \tag{13.6}$$

where Ω is the solid angle over which search must be performed (steradians), T_{fs} is the time allowed to search Ω by operational requirements, and L_{os} is the composite incremental loss associated with search.

The left-hand side of the equation contains radar design parameters, while the right-hand side is determined by target characteristics and operational requirements. The right-hand side of the equation is evaluated to determine radar requirements. The left-hand side of the equation is evaluated to determine if the radar design meets the requirements.

The track radar range equation is conditioned on noise-limited angle accuracy as this measure stresses radar capabilities significantly more than range accuracy in almost all cases of interest. The operational requirement is to maintain a given data rate track providing a specified single-measurement angle accuracy for a given number of targets with specified RCS and range. Antenna beamwidth, which is proportional to the radar carrier wavelength divided by antenna extent, impacts track performance since the degree of S/N required for a given measurement accuracy decreases as the beamwidth decreases. Track performance requirements can be bounded as

$$\frac{P_a A^3}{\lambda^4 L_t L_r L_{sp} T_s} k_m^2 \eta^2 \geq 5k \frac{r N_t R^4}{\sigma \sigma_\theta^2} \tag{13.7}$$

where r is the single-target track rate, N_t is the number of targets under track in different beams, σ_θ is the required angle accuracy standard deviation (radians), and σ is the RCS. In general, a phased array radar antenna is required to support multiple target tracking when $Nt > 1$.

Incremental search losses are suppressed during single-target-per-beam tracking. The beam is pointed as closely as possible to the target to suppress beam shape loss. The tracking loop centers the range gate and Doppler filter on the return. Detection thresholding loss can be minimal since the track range window is small, though the presence of multiple targets generally mandates continual detection processing.

13.4 Radar Waveforms

13.4.1 Pulse Compression

Typical pulse radar waveforms are summarized in Table 2.6. In most cases, the signal processor is designed to closely approximate a matched filter. As indicated in Table 13.5, the range and Doppler resolution of any match-filtered waveform are inversely proportional to the waveform bandwidth and duration, respectively. Pulse compression, using modulated waveforms, is attractive since S/N is proportional to pulse duration rather than bandwidth in matched filter implementations. Ideally, the intrapulse modulation is chosen to attain adequate range resolution and range sidelobe suppression performance while the pulse duration is chosen to provide the required sensitivity. Pulse compression waveforms are characterized as having a time bandwidth product (TBP) significantly greater than unity, in contrast to an unmodulated pulse, which has a TBP of approximately unity.

13.4.2 Pulse Repetition Frequency

The radar system pulse repetition frequency (PRF) determines its ability to unambiguously measure target range and range rate in a single CPI as well as determining the inherent clutter rejection capabilities of the radar system. In order to obtain an unambiguous measurement of target range, the interval between

TABLE 13.5 Selected Waveform Characteristics

	Comments	Time Bandwidth Product	Range Side Lobes (dB)	S/N Loss (dB)	Range/Doppler Coupling	ECM/EMI Robustness
Unmodulated	No pulse compression	~1	Not applicable	0	No	Poor
Linear frequency modulation	Linearly swept over bandwidth	>10	Unweighted: –13.5 Weighted: >– 40[a]	0 0.7–1.4	Yes	Poor
Nonlinear FM	Multiple variants specific	Waveform specific	Waveform specific	0	Waveform	Fair
Barker	N-bit biphase	≤13 (N)	–20 log(N)	0	No	Fair
LRS	N-bit biphase	~N; > 64/pulse[a]	~ –10 log (N)	0	No	Good
Frank	N-bit polyphase (N = integer^2)	~N	~ –10 log ($\pi^2 N$)	0	Limited	Good
Frequency coding	N subpulses noncoincidental in time and frequency	~N^2	Waveform specific • Periodic • Pseudorandom	Waveform specific 0.7–1.40 0		

[a] Constraint due to typical technology limitations rather than fundamental waveform characteristics.

radar pulses (1/PRF) must be greater than the time required for a single pulse to propagate to a target at a given range and back. The maximum unambiguous range is then given by $C/(2 \cdot \text{PRF})$ where C is the velocity of electromagnetic propagation.

Returns from moving targets and clutter sources are offset from the radar carrier frequency by the associated Doppler frequency. As a function of range rate, $R\cdot$, the Doppler frequency, f_D, is given by $2R\cdot/\lambda$. A coherent pulse train samples the return Doppler modulation at the PRF. Most radar systems employ parallel sampling in the in-phase and quadrature baseband channels so that the effective sampling rate is twice the PRF. The targets return is folded in frequency if the PRF is less than the target Doppler.

Clutter returns are primarily from stationary or near-stationary surfaces such as terrain. In contrast, targets of interest often have a significant range rate relative to the radar clutter. Doppler filtering can suppress returns from clutter. With the exception of frequency ambiguity, the Doppler filtering techniques used to implement pulse Doppler filtering are quite similar to those described for CW radar in Chapter 12. Ambiguous measurements can be resolved over multiple CPIs by using a sequence of slightly different PRFs and correlating detections among the CPIs.

13.4.3 Detection and Search

Detection processing consists of comparing the amplitude of each range gate/Doppler filter output with a threshold. A detection is reported if the amplitude exceeds that threshold. A false alarm occurs when noise or other interference produces an output of sufficient magnitude to exceed the detection threshold. As the detection threshold is decreased, both the detection probability and the false alarm probability increase. S/N must be increased to enhance detection probability while maintaining a constant false alarm probability.

As noted in the subsection on Radar Cross Section, RCS fluctuation effects must be considered in assessing detection performance. The Swerling models, which use chi-square probability density functions (PDFs) of 2° and 4° of freedom (DOF), are commonly used for this purpose. The Swerling 1 and 2 models are based on the 2 DOF PDF and can be derived by modeling the target as an ensemble of independent scatterers of comparable magnitude. This model is considered representative of complex targets such as aircraft. The Swerling 3 and 4 models use the 4 DOF PDF and correspond to a target with a single dominant scatterer and an ensemble of lesser scatterers. Missiles are sometimes represented by Swerling 2 and 4 models. The Swerling 1 and 3 models presuppose slow fluctuation such that the target RCS is constant from pulse to pulse within a scan. In contrast, the RCS of Swerling 2 and 4 targets is modeled as independent on a pulse-to-pulse basis.

Single-pulse detection probabilities for nonfluctuating, Swerling 1/2, and Swerling 3/4 targets are depicted in Fig. 13.4. This curve is based on a typical false alarm number corresponding approximately to a false alarm probability of 10^{-6}. The difference in S/N required for a given detection probability for a fluctuating target relative to the nonfluctuating case is termed the fluctuation loss.

The detection curves presented here and in most other references presuppose noise-limited operation. In many cases, the composite interference present at the radar system output will be dominated by clutter returns or electromagnetic interference such as that imposed by hostile electronic countermeasures. The standard textbook detection curves cannot be applied in these situations unless the composite interference is statistically similar to thermal noise with a Gaussian PDF and a white power spectral density. The presence of non-Gaussian interference is generally characterized by an elevated false alarm probability. Adaptive detection threshold estimation techniques are often required to search for targets in environments characterized by such interference [Nathanson, 1991].

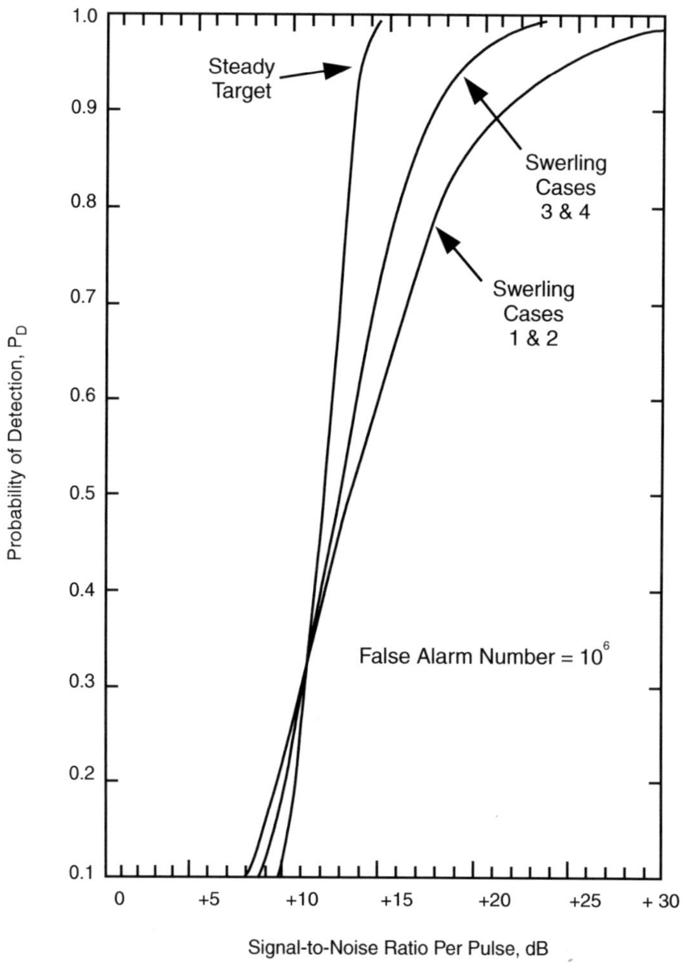

FIGURE 13.4 Detection probabilities for various target fluctuation models. (*Source:* F. E. Nathanson, *Radar Design Principles*, 2nd ed., New York: McGraw-Hill, 1991, p. 91. With permission.)

13.5 Estimation and Tracking

13.5.1 Measurement Error Sources

Radars measure target range and angle position and, potentially, Doppler frequency. Angle measurement performance is emphasized here since the corresponding cross-range error dominates range error for most practical applications. Target returns are generally smoothed in a tracking filter, but tracking performance is ultimately determined by the measurement accuracy and associated error characteristics of the subject radar system. Radar measurement error can be characterized as indicated in Table 13.6.

The radar design and the alignment and calibration process development must consider the characteristics and interaction of these error components. Integration of automated techniques to support alignment and calibration is an area of strong effort in modern radar design that can lead to significant performance improvement in fielded systems.

As indicated previously, angle measurement generally is the limiting factor in measurement accuracy. Target azimuth and elevation position are primarily measured by a monopulse technique in modern radars though early systems used sequential lobing and conical scanning. Specialized monopulse tracking radars utilizing reflectors have achieved instrumentation and S/N angle residual systematic error as low as 50 μrad. Phased-array antennas have achieved a random error of less than 60 μrad, but the composite systematic residual errors remain to be measured. The limitations are primarily in the tolerance on the phase and amplitude of the antenna illumination function.

Figure 13.5 shows the monopulse beam patterns. The first is the received sum pattern that is generated by a feed that provides the energy from the reflector or phased array antenna through two ports in equal amounts and summed in phase in a monopulse comparator shown in Fig. 13.6. The second is the difference pattern generated by providing the energy through the same two ports in equal amounts but taken out with a phase difference of π radians, giving a null at the center. A target located at the center of the same beam would receive a strong signal from the sum pattern with which the target could be detected and ranged. The received difference pattern would produce a null return, indicating the target was at the center of the beam. If the target were off the null, the signal output or difference voltage would be almost linear and proportional to the distance off the center (off-axis), as shown in Fig. 13.5. This output of the monopulse processor is the real part of the dot product of the complex sums and the difference signals divided by the absolute magnitude of the sum signal squared, i.e.,

$$e_d = \mathrm{Re}\left[\frac{\Sigma \cdot \Delta}{|\Sigma|^2}\right] \tag{13.8}$$

The random instrumentation measurement errors in the angle estimator are caused by phase and amplitude errors of the antenna illumination function. In reflector systems, such errors occur because

TABLE 13.6 Radar Measurement Error

Random errors	Those errors that cannot be predicted except on a statistical basis. The magnitude of the random error can be termed the *precision* and is an indication of the repeatability of a measurement.
Bias errors	A systematic error, whether due to instrumentation or propagation conditions. A nonzero mean value of a random error.
Systematic error	An error, whose quantity can be measured and reduced by calibration.
Residual systematic error	Those errors remaining after measurement and calibration. A function of the systematic and random errors in the calibration process.
Accuracy	The magnitude of the rms value of the residual systematic and random errors.

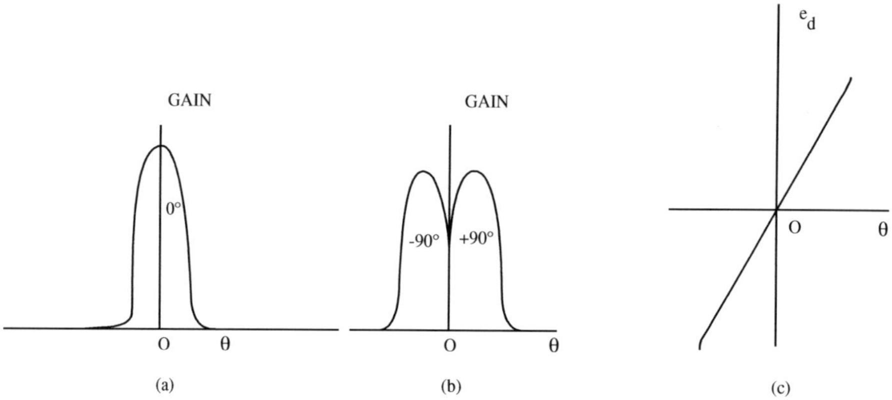

FIGURE 13.5 Monopulse beam patterns and difference voltage: (a) sum (S); (b) difference (D); (c) difference voltage.

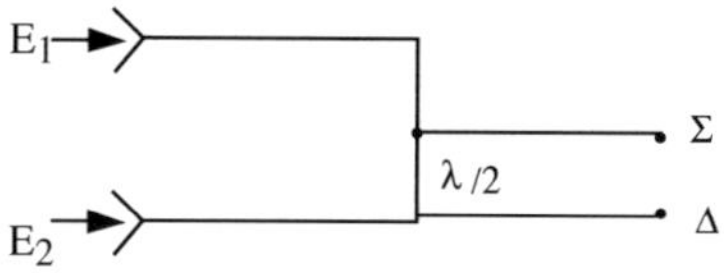

FIGURE 13.6 Monopulse comparator.

of the position of the feedhorn, differences in electrical length between the feed and the monopulse comparator, mechanical precision of the reflector, and its mechanical rotation. In phased array radars, these errors are a function of the phase shifters, time delay units, and combiners between the antenna elements and the monopulse comparator as well as the precision of the array. Although these errors are random, they may have correlation intervals considerably longer than the white noise considered in the thermal-noise random error and may depend upon the flight path of the target. For a target headed radially from or toward the radar, the correlation period of angle-measurement instrumental errors is essentially the tracking period. For crossing targets, the correlation interval may be pulse to pulse.

As in the estimate of range, the propagation effects of refraction and multipath also enter into the tracking error. The bias error in range and elevation angle by refraction can be estimated as

$$\begin{aligned} \Delta R &= 0.007\,N_s \text{cosecant}\, E_o \,(\text{meters}) \\ \Delta E_o &= N_s \cot E_o \,(\mu\text{rad}) \end{aligned} \tag{13.9}$$

where N_s is the surface refractivity and E_o is the elevation angle [Barton and Ward, 1984].

One can calculate the average error in multipath. However, one cannot correct for it as in refraction since the direction of the error cannot be known in advance unless there are controlled conditions such as in a carefully controlled experiment. Hence, the general approach is to design the antenna sidelobes to be as low as feasible and accept the multipath error that occurs when tracking close to the horizon. There has been considerable research to find means to reduce the impact, including using very wide bandwidths to separate the direct path from the multipath return as well as specialized track filtering techniques that accommodate multipath effects.

13.5.2 Tracking Filter Performance

Target tracking based on processing returns from multiple CPIs generally provides a target position and velocity estimate of greater accuracy than the single-CPI measurement accuracy delineated in Table 13.4. In principle, the error variance of the estimated target position with the target moving at a constant velocity is approximately $4/n \cdot \sigma_m^2$ where n is the number of independent measurements processed by the track filter and *sm* is the single measurement accuracy. In practice, the variance reduction factor afforded by a track filter is often limited to about an order of magnitude because of the reasons summarized in the following paragraphs.

Track filtering generally provides smoothing and prediction of target position and velocity via a recursive prediction-correction process. The filter predicts the target's position at the time of the next measurement based on the current smoothed estimates of position, velocity, and possibly acceleration. The subsequent difference between the measured position at this time and the predicted position is used to update the smoothed estimates. The update process incorporates a weighting vector that determines the relative significance given the track filter prediction versus the new measurement in updating the smoothed estimate.

Target model fidelity and adaptivity are fundamental issues in track filter mechanization. Independent one-dimensional tracking loops may be implemented to control pulse-to-pulse range gate positioning and antenna pointing. The performance of one-dimensional polynomial algorithms, such as the alpha–beta filter, to track targets from one pulse to the next and provide modest smoothing is generally adequate. However, one-dimensional closed-loop tracking ignores knowledge of the equations of motion governing the target so that their smoothing and long-term prediction performance is relatively poor for targets with known equations of motion. In addition, simple one-dimensional tracking-loop filters do not incorporate any adaptivity or measure of estimation quality.

Kalman filtering addresses these shortcomings at the cost of significantly greater computational complexity. Target equations of motion are modeled explicitly such that the position, velocity, and potentially higher-order derivatives of each measurement dimension are estimated by the track filter as a state vector. The error associated with the estimated state vector is modeled via a covariance matrix that is also updated with each iteration of the track filter. The covariance matrix determines the weight vector used to update the smoothed state vector in order to incorporate such factors as measurement *S/N* and dynamic target maneuvering.

Smoothing performance is constrained by the degree of *a priori* knowledge of the targets kinematic motion characteristics. For example, Kalman filtering can achieve significantly better error reduction against ballistic or orbital targets than against maneuvering aircraft. In the former case the equations of motion are explicitly known, while the latter case imposes motion model error because of the presence of unpredictable pilot or guidance system commands. Similar considerations apply to the fidelity of the track filters model of radar measurement error. Failure to consider the impact of correlated measurement errors may result in underestimating track error when designing the system.

Many modern tracking problems are driven by the presence of multiple targets which impose a need for assigning measurements to specific tracks as well as accommodating unresolved returns from closely spaced targets. Existing radars generally employ some variant of the nearest-neighbor algorithm where a measurement is uniquely assigned to the track with a predicted position minimizing the normalized track filter update error. More sophisticated techniques assign measurements to multiple tracks if they cannot clearly be resolved or make the assignment on the basis on several contiguous update measurements.

Defining Terms

Coherent: Integration where magnitude and phase of received signals are preserved in summation.

Noncoherent: Integration where only the magnitude of received signals is summed.

Phased array: Antenna composed of an aperture of individual radiating elements. Beam scanning is implemented by imposing a phase taper across the aperture to collimate signals received from a given angle of arrival.

Pulse compression: The processing of a wideband, coded signal pulse, of initially long time duration and low-range resolution, to result in an output pulse of time duration corresponding to the reciprocal of the bandwidth.

Radar cross section (RCS): Measure of the reflective strength of a radar target; usually represented by the symbol σ, measured in square meters, and defined as 4π times the ratio of the power per unit solid angle scattered in a specified direction of the power unit area in a plane wave incident on the scatterer from a specified direction.

References

Barton, D.K. and Ward, H.R., *Handbook of Radar Measurement,* Artech, Dedham, MA, 1984.
Blake, L.V., *Radar Range-Performance Analysis,* Artech, Dedham, MA, 1986.
Eaves, J.L. and Reedy, E.K., Eds., *Principles of Modern Radar,* Van Nostrand, New York, 1987.
Morris, G.V., *Airborne Pulsed Doppler Radar,* Artech, Dedham, MA, 1988.
Nathanson, F.E., *Radar Design Principles,* 2nd ed., McGraw-Hill, New York, 1991.

Further Information

Skolnik, M.I., Ed., *Radar Handbook,* 2nd ed., McGraw-Hill, New York, 1990.
IEEE Standard Radar Definitions, IEEE Standard 686-1990, April 20, 1990.

14

Electronic Warfare and Countermeasures

Robert D. Hayes
RDH Incorporated

Electronic warfare (EW) in general applies to military actions. Specifically, it involves the use of electromagnetic energy to create an advantage for the friendly side of engaged forces while reducing the effectiveness of the opposing hostile armed forces.

Electronic warfare support measures (ESM) or electronic warfare support (ES) are the actions and efforts taken to listen, to search space and time, to intercept and locate the source of radiated energy; and to analyze, identify, and characterize the electromagnetic energy. Tactical employment of forces is established and plans are executed for electronic countermeasures.

Electronic countermeasures (ECM) or electronic attack (EA) are those actions taken to reduce the enemy's effective use of the electromagnetic spectrum. These actions may include jamming, electronic deception, false targets, chaff, flares, transmission disruption, and changing tactics.

Electronic counter-countermeasures (ECCM) or electronic protection (EP) are actions taken to reduce the effectiveness of enemy used ECM to enhance the use of the electromagnetic spectrum for friendly forces. These actions are of a protective nature and are applied to tactical and strategic operations across the equipment used for sonar, radar, command, control, communications, and intelligence.

14.1 Radar and Radar Jamming Signal Equations

Electronic jamming is a technique where a false signal of sufficient magnitude and bandwidth is introduced into a receiving device so as to cause confusion with the expected received signal and create a loss of information. For a communication receiver, it simply overpowers the radio transmission; for radar, the jamming signal overrides the radar echo return from an object under surveillance or track. The jamming equation is the same for both communications and radar systems. In the case of a radar jammer, there are several operational scenarios to consider.

The radar jammer signal level must be determined for the various applications and jammer location. When the jammer is located on the target (self-protection), the jammer signal level must exceed the effective radar cross section of the target to form an adequate screen; when the jammer is located away

0-8493-1732-0/03/$0.00+$1.50

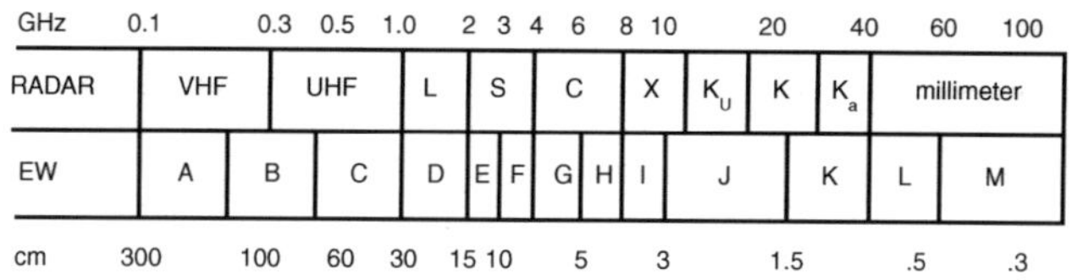

FIGURE 14.1 Frequency designation for radar and EW bands.

from the vehicle under protection (stand-off jammer), the jammer should simulate the expected radar cross section observed at a different range and different aspect.

The monostatic radar received signal power equation representing the signal received at an enemy detection radar, and the received jamming power transmitted toward the same enemy detection radar, are presented in the equations given in Table 14.1. Each quantity in equations is defined in this table. These quantities are dependent on electrical, polarization, material, shape, roughness, density, and atmospheric parameters.

Background interference is encountered in every branch of engineering. The received radar/communication signal power will compete with interfering signals which will impose limitations on performance of the detecting electronic system. There are many types of interference. A few are channel cross talk, nonlinear harmonic distortion, AM and FM interference, phase interference, polarization cross talk, and noise. The common limiting interference in electronic systems is noise.

There are many types of noise. Every physical body not at absolute zero temperature emits electromagnetic energy at all frequencies, by virtue of the thermal energy the body possesses. This type of radiation is called thermal noise radiation. Johnson noise is a result of random motion of electrons within a resistor; semiconductor noise occurs in transistors; shot noise occurs in both transmitter and receiver vacuum tubes as a result of random fluctuations in electron flow. These fluctuations are controlled by a random mechanism and, thus, in time are random processes described by the Gaussian distribution. The signal is random in phase and polarization, and usually is broad in frequency bandwidth. The average noise power is given by

$$P_N = kT\,B$$

where

P_N = power, watts
k = Boltzmann constant
T = temperature, absolute degrees

Noise jamming has been introduced into EW from several modes of operation. *Spot noise jamming* requires the entire output power of the jammer to be concentrated in a narrow bandwidth ideally identical to the bandwidth of the victim radar/communication receiver. It is used to deny range, conversation, and sometimes angle information. *Barrage noise jamming* is a technique applied to wide frequency spectrum to deny the use of multiple channel frequencies to effectively deny range information, or to cover a single wideband system using pulse compression or spread-spectrum processing. *Blinking* is a technique using two spot noise jammers located in the same cell of angular resolution of the victim system. The jamming transmission is alternated between the two sources causing the radar antenna to oscillate back and forth. Too high a blinking frequency will let the tracker average the data and not oscillate, while too low a frequency will allow the tracker to lock on one of the jammers.

In free space the radar echo power returned from the target varies inversely as the fourth power of the range, whereas the power received from a jammer varies as the square of the range from the jammer to the victim radar. There will be some range where the energy for the noise jammer is no longer great enough to hide the friendly vehicle. This range of lost protection is called the burn-through range.

TABLE 14.1

Monostatic Radar Equation	Jamming Radar Equation
$P_r = \frac{P_t G^2 \sigma^2 \lambda^2 F^4 E}{(4\pi)^3 R^4 L_t L_r} e^{-2\alpha R}$	$P_r = \frac{P_t G_t G_r \lambda^2 F^2 E}{(4\pi)^2 R^2 L_t L_r} e^{-\alpha R}$

Quantity	*Dependent Function*
P_t = transmitter power, W	ΔF = frequency spectrum ψ = phase, coherent
L_t = loss, transmitter line	Equipment path
L_r = loss, receiver line	Equipment path
E = receiver gain	Processing techniques
G = antenna gain	p = polarization λ = wavelength, RF carrier l = length w = width c = curvature ϕ = elevation angle θ = azimuth angle
σ = point target RSC, m^2	p λ l w c ϕ θ r = rotation s = surface roughness m = motion ϵ = dielectric constant μ = permeability
σ_0 = area extended target RCS, m^2/m^2	p λ ϕ θ s m ϵ τ = pulse length T = temperature d = density
σ_u = volume extended target RCS, m^2/m^3	p λ ϕ θ s m ϵ

Quantity	*Dependent Function*
σ_u = volume extended target RCS, m^2/m^3	p μ τ T d
α_0 = clear air loss, Np	λ T d P = pressure ρ = gas constant
α = rain, dust, snow, hail, etc., loss, Np	p = polarization λ c m m ϵ T d P ρ n = number of particles, or rain rate R = range, path length β = differential polarization phase shift
F = electric field propagation factor—multipath	p λ c ϕ θ R s ϵ μ ψ β
R = range, path length, m	n = index of refraction spherical geometry

Antenna gain = $G = \frac{28,000}{\theta\phi}$

where $\theta = 75\frac{\lambda}{\omega}$

and $\phi = 75\ \frac{\lambda}{l}$, in degrees

In both EW technology and target identification technology, the received energy, as compared to the received power, better defines the process of signal separation from interference. The received radar energy will be given by

$$S = P\tau$$

where τ is the transmitted pulse length. When the radar receiver bandwidth is matched to the transmitter pulse length, then

$$S = \frac{P_r}{B_r}$$

where B_r is the bandwidth of radar receiver. The transmitted jamming energy will be given by

$$J = \frac{P_j}{B_j}$$

where B_j is the bandwidth of jamming signal.

If for a moment, the system losses are neglected, the system efficiency is 100%, and there are negligible atmospheric losses, then the ratio of jamming energy received to radar energy

$$\frac{J}{S} = \frac{4\pi P_j G_j G_r \lambda^2 F_j^2 R_r^4 B_r}{P_r G_r^2 \sigma \lambda^2 F_r^4 R_j^2 B_j}$$

for the stand off jammer, and

$$\frac{J}{S} = \frac{4\pi P_j G_j B_r R^2}{P_r G_r \sigma F^2 B_j}$$

for the self-protect jammer. For whatever camouflage factor is deemed sufficient (example $J/S = 10$), the the burn-through range can be determined from this self-protect jamming equation.

14.1.1 Radar Receiver Vulnerable Elements

Range gate pull-off (RGPO) is a deception technique used against pulse tracking radars using range gates for location. The jammer initially repeats the skin echo signal with minimum time delay at a high power to capture the receiver automatic gain control circuitry. The delay is progressively changed, forcing the tracking gates to be pulled away from (walked off) the skin echo. Frequency memory loops (FMLs) or transponders provide the variable delay. The deceptive pulse can be transmitted before the radar pulse is received if the pulse repetition rate is stored and repeated at the proper time. In this technique the range gate can be either pulled off or pulled in.

One implementation is shown in Fig. 14.2. A split gate (early/late gates) range tracking scheme has two gates that are positioned by an automatic tracking loop so that equal energy appears in each gate. The track point is positioned on the target centroid. First, there is an initial dwell at the target range long enough for the tracker to capture the radar receiver. The received echo has S energy. Then, a jamming signal is introduced which contains J amount of energy. The energy in the early and late gates are compared and the difference is used to reposition the gates so as to keep equal energy in each gate. In this manner, the range gate is continually moved to the desired location. When RGPO is successful, the jamming signal does not compete with the target's skin return, resulting with an almost infinite J/S ratio.

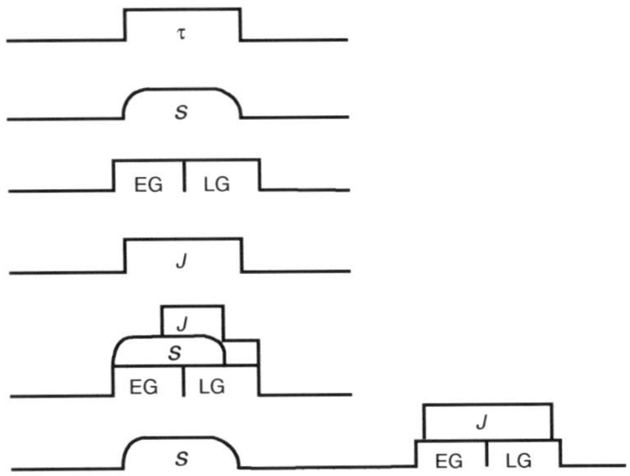

FIGURE 14.2 Range gate pull-off concept.

Velocity gate pulloff (VGPO) is used against the radar's received echo modulation frequency tracking gates in a manner analogous to the RGPO used against the received echo signal amplitude tracking gates. Typically in a Doppler or moving target indicator (MTI) system there are a series of fixed filters or a group of frequency tracking filters used to determine the velocity of a detected moving target. By restricting the band width of the tracking filters, the radar can report very accurate speeds, distinguish between approaching and receding targets, and reduce echos from wideband clutter. Typically, the tracking filters are produced by linearly sweeping across the frequency domain of interest to the detecting radar. The ECM approach to confuse this linear sweep is to transmit a nonlinear frequency modulated sweep which could be parabolic or cubic. The jamming signal will cause the Doppler detector to be walked off of the true frequency as shown in Fig. 14.3.

Another technique used to generate a frequency shift is to modulate the phase of the transmitted jamming signal. The instantaneous frequency of any modulated signal is directly related to the change in phase.

$$2\pi f = \frac{d\theta}{dt}$$

When the phase is given by $\theta = 2\pi f_c t + kt$ where f_c is the carrier frequency and k the change of phase with time, then

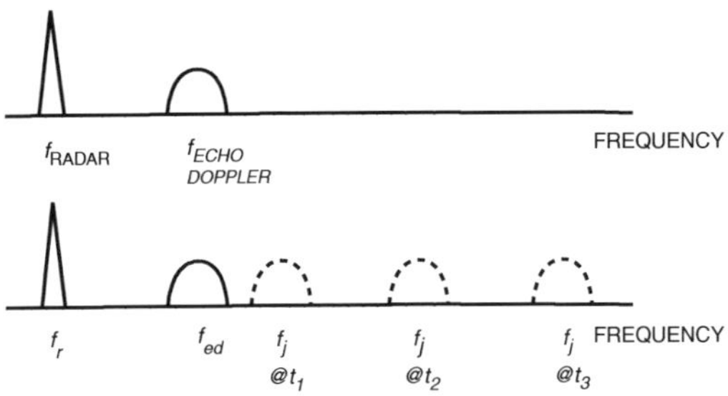

FIGURE 14.3 Frequency gate pull-off concept.

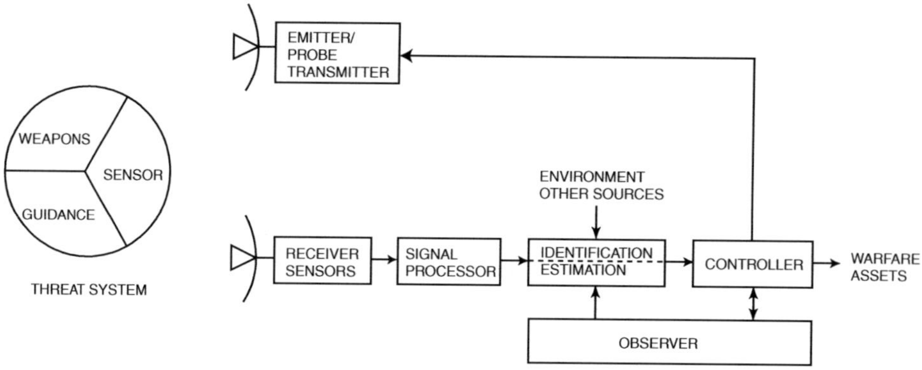

FIGURE 14.4 Surgical countermeasure block diagram.

$$\frac{d\theta}{dt} = 2\pi f_c + k$$

and

$$f = f_c + \frac{k}{2\pi}$$

This type of modulation has been referred to as serrodyne modulation and used with control voltage modulation of an oscillator or traveling wave tube (TWT).

An inverse-gain jammer is a repeater in which a signal is directed back toward the enemy radar to create false targets by varying the jammer transmitted signal level inversely with the magnitude of the received radar signal. To defeat conical scan trackers, the transponder signal has been modulated at the mutation of the scan rate to deceive angle tracking.

Most ECM techniques are open-loop control functions which add signals to the enemy radar return signal so as to confuse or deceive the weapon's ability to track. One program, named surgical countermeasures, employs closed-loop techniques to slice into the weapon tracking loop and control that loop from a stand off position. The basic elements of the EW system consist of the sensor, signal processor, controller, observer, and ECM output. In addition to being closed loop, the system is recursive and adaptable. The sensor observes the threat radar feeds the intercepted or reflected signals to the signal processor which will determine spectrum, carrier frequency, angle of arrival, pulse train, etc. and establish the electronic characteristic of the threat. The processed information is made available to the observer and the controller. The controller will generate ECM signals to probe, update, and fine-tune the techniques generator in the ECM output. The observer notes the effects of the jogging introduced in the threat radar tracker and updates the controlled tracking information so as not to cause a track break, or lock pull-off until the proper time in the engagement. The observer is the important feedback loop to determine the necessary refinements to previous observations and control equations.

The surgical countermeasure technique has been used successfully against both a conical-scan-on-receive-only (COSRO) tracker and in conjunction with a two-transmitter blinker ECM suite to cause break lock and transfer tracking see Timberlake [1986].

14.2 Radar Antenna Vulnerable Elements

Radar backscatter from a variety of antennas was measured by Hayes at Georgia Institute of Technology for the U.S. Air Force Avionics Laboratory, Wright-Patterson Air Force Base, Ohio [Hayes and Eaves, 1966].

Typical measured data are presented in Figs. 14.5 and 14.6 for an AN/APS-3 antenna. This antenna is an 18-in. parabolic dish with a horizontally polarized double-dipole feed built upon a waveguide termination for use at a wavelength of 3.2 cm with an airborne radar. The instrument measuring radar operated at a wavelength of 3.2 cm, and transmitted either horizontally or vertically polarized signals, while receiving simultaneously both horizontally and vertically polarized signals.

The APS-3 antenna was terminated with a matched waveguide load and the backscatter patterns first for horizontally polarized and vertically polarized echos obtained for both horizontal polarized transmissions, and second for vertically polarized transmissions. Then an adjustable short-circuit termination was placed at the end of the waveguide output port. The position of the waveguide termination greatly influenced the backscatter signal from the APS-3 antenna when the polarization of the interrogating measurement radar matched the operational horizontal polarization of the antenna. Tuning of the adjustable short circuit produced returns both larger and smaller than those returns recorded when the matched load was connected to the output port. The waveguide termination had very little effect on the backscatter when the interrogating transmission was vertical polarization (cross polarization). In addition to the APS-3, there were several other antennas and pyramidal horns (both in-band and out-of-band) investigated. The W shape in the center of the main beam is typical of backscatter patterns when the waveguide is terminated in a matched load, and when the interrogating measurement radar matched the operational frequency and polarization of the antenna under test.

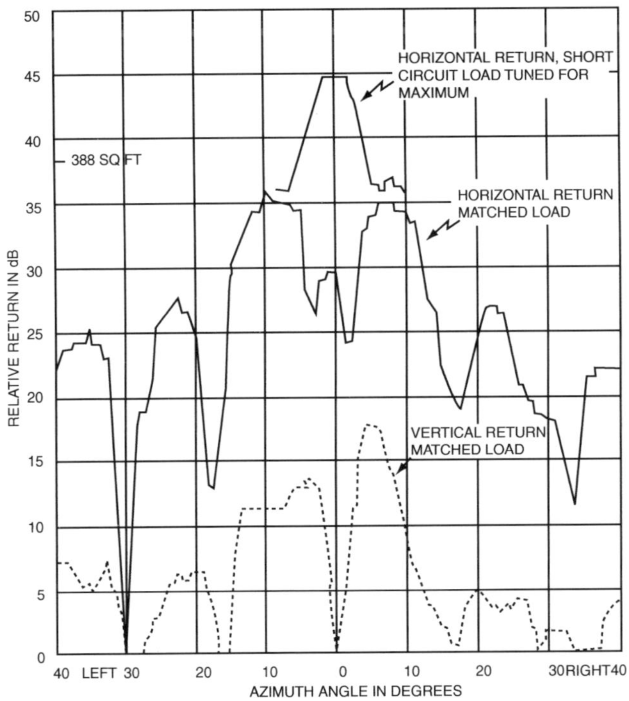

FIGURE 14.5 Backscatter from AN/APS-3 antenna, parallel polarization.

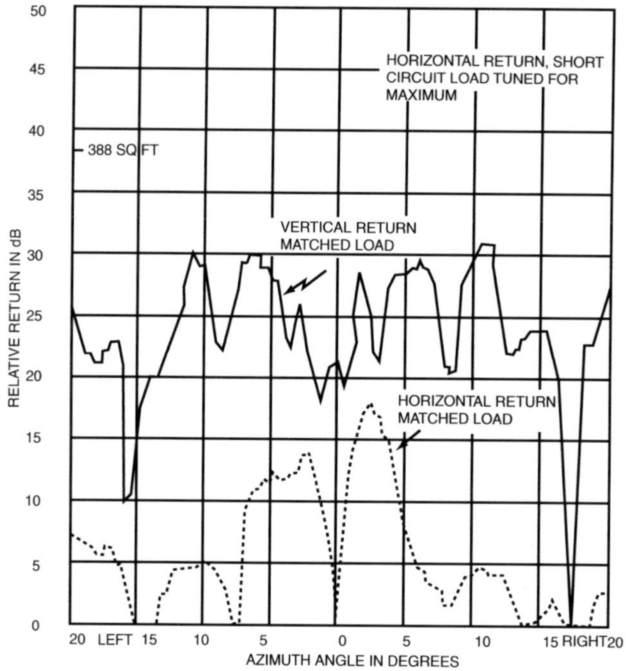

FIGURE 14.6 Backscatter from AN/APS-3 antenna, cross polarization.

Butler [1981] of Hughes Radar Systems has presented an analytical model of an antenna consisting of two scattering mechanisms. One component is the backscatter from the antenna surfaces (one-way scattering); and the second component is scattering from the feed system, which is coupled to the waveguide input/output port (round-trip scattering). It has been shown that the amplitude and phase relationships of the two scattering mechanisms will interact and can create this W shape in the equivalent main beam of an impedance-matched antenna when interrogated at the designed operating frequency and polarization.

A reflector antenna having a curved surface will generate a polarized signal, which is orthogonal to the polarization of the signal impinging upon the curved surface, when the impinging signal is not parallel to the surface. This new (cross-polarized) signal is a result of the boundary value requirement to satisfy a zero electrical field at the surface of a conductor. For a parabolic reflector antenna designed for operation with a horizontal polarization, there will be a vertically polarized component in the transmitted signal, the received signal, and the reflected signal. Curved surfaces produce cross-polarized components regardless of frequency and polarization. Thus, if horizontal polarized signals impinge upon a curved surface, vertical polarized signals as well as horizontal polarization are produced; vertical polarized impinging signals produced horizontal signals; right-circular polarized impinging signals produce a cross-polarized left-circular polarized signals, etc.

An example of parallel- and cross-polarized signals from a parabolic reflector antenna is shown in Fig. 14.7. These data are provided as a courtesy of Millitech Corporation. Please note that the antenna is designed to operate at 95 GHz, showing that the effect of cross-polarization will occur even at millimeter waves. The orthogonally polarized signals have four peaks (one in each quadrant), which are on the 45° lines to the principle axis, and are within the main beam of the principle polarized signal. The magnitude of these peaks is typically 13 to 20 dB below the peak of the principle polarized signal. The cross-polarized signal is theoretically zero on the two principle plane axis, so when perfect track lockup is made, cross-polarization will have no affect.

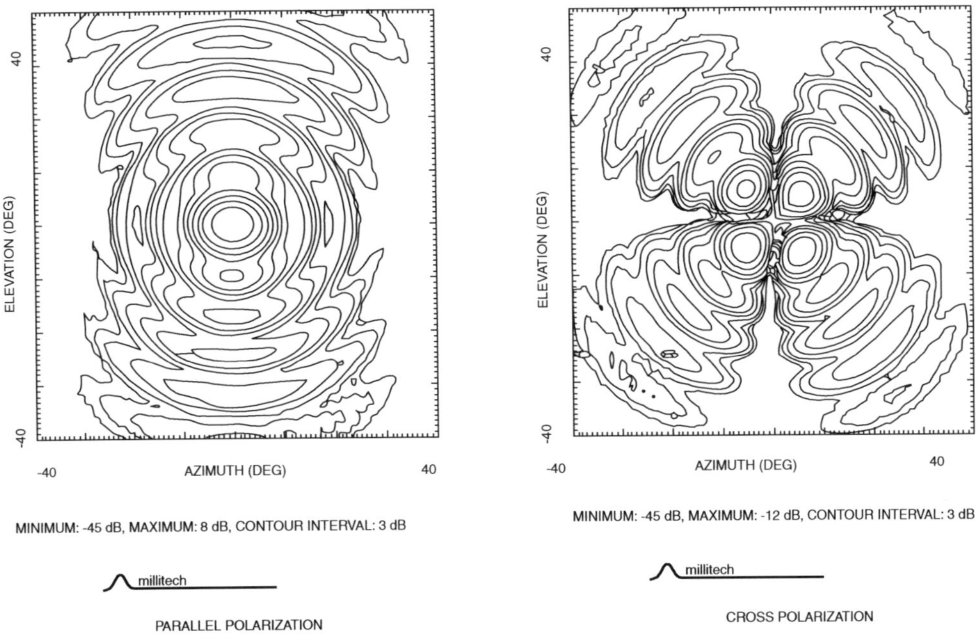

FIGURE 14.7 Antenna radiation patterns for parabola, conical feed, 94 GHz.

Injecting a cross-polarized signal into a system will have the same effect as entering a monopulse tracker. That is to say, the receiving system cannot detect the difference between parallel and cross-polarized signals and yet the cross-polarized pattern has nulls on the major axis, and peaks off boresight. When the cross-polarized signal dominates, the track should go to an angle about at the 10 dB level of the normal sum main beam.

Cross-eye is a countermeasure technique employed against conical scan and monopulse type angle-tracking radars. The basic idea is to intercept the enemy-propagated signal, phase shift the signal, and then retransmit the signal to make it appear that the signal represents a reflected return emanating from an object located at an angular position different from the true location. In generating the retransmitted signal, there are two basic approaches to signal level configuration: saturated levels and linear signal levels. A typical implementation of a four antenna cross-eye system is shown in Fig. 14.8. Let us assume that the two transmit antennas radiate about the same amount of power (slight difference could occur due to line loss, different amplifier gain, etc.). The signal received by antenna 2 is simply amplified with no phase shift and retransmitted out antenna 3, whereas the signal received by antenna 4 is amplified and phase shifted 180° before retransmitted out of antenna 1.

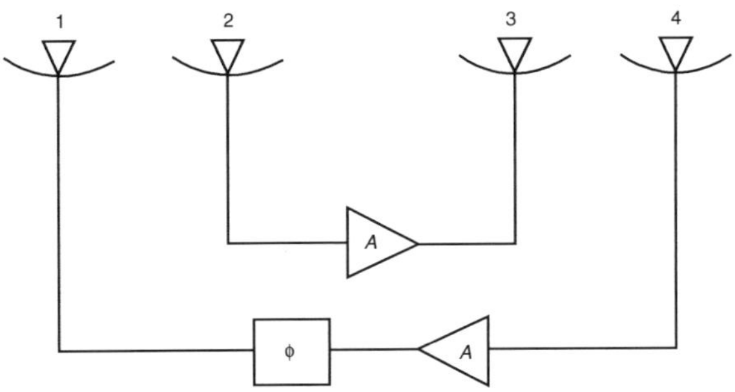

FIGURE 14.8 Cross-eye system block diagram.

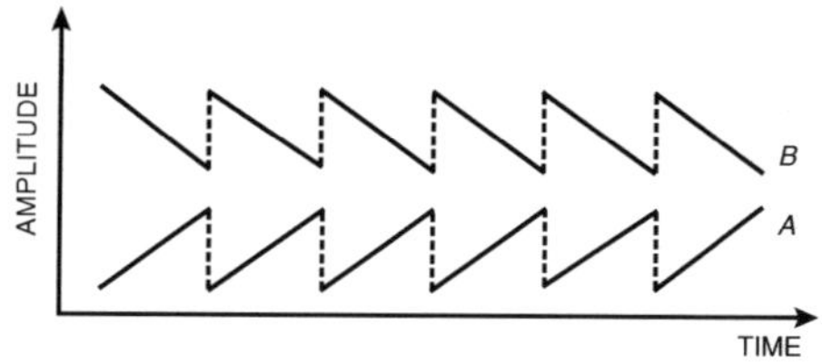

FIGURE 14.9 Saw-tooth modulation diagram for cross-eye.

The two wave fronts (generated by antennas 1 and 3) are of equal amplitude and 180° out of phase; thus their sum will be zero on boresight, while their differences will be finite. Radars, such as a monopulse tracker, which generate sum and difference signals will interpret the results of the two wave fronts as a tracking error. The track point will drift, moving the target off boresight in such a manner as to cause the radar lock-up point to be off target, where the difference signal becomes zero.

The maximum tracking error for a saturated cross-eye is about $0.5\theta_{3\,dB}$ where $\theta_{3\,dB}$ is the antenna half-power beamwidth of the sum pattern of the radar. This is the angular position where the difference pattern has peaks and zero slope.

The addition of sawtooth modulation to the cross-eye signal can produce multiple beamwidth tracking errors in excess of the $0.5\theta_{3\,dB}$. Figure 14.9 shows two sawtooth wave forms. The RF power from one antenna is linearly increased as a function of time, while the RF power from the other antenna is linearly decreased with time. At the end of the ramp, the process is repeated, creating the sawtooth. When the antenna is given sufficient momentum during the first half of the modulation cycle to enable the antenna to traverse the singularity at $0.5\theta_{3\,dB}$, tracking pull-off will occur. Successive periods of modulation can cause the monopulse tracker to lock onto successive lobes of the sum pattern.

In linear cross-eye, the amplifiers are never allowed to go into saturation. Linear cross-eye differs from saturated cross-eye in that the sum pattern is zero for all pointing angles, and the difference pattern is never zero. No stable track points exist with linear cross-eye and large tracking errors can occur, resulting in large angular breaklocks.

The two most common types of angle tracking used today are sequential lobing and simultaneous lobing.

A common implementation of sequential lobing is to rotate a single antenna lobe about the boresight axis in a cone; the signal amplitude will be modulated at the conical scan frequency. The direction off of boresight is determined by comparing the phase of the scan modulation with an internal reference signal. As the source is brought on to axis, the modulation will go to zero. The source of the RF signal can be either an active source such as a beacon or a communication transmitter, or the target scattering, as in a radar application.

In a typical radar, the conical scan rate is chosen early in the system design. The rate should be slow relative to the radar pulse repetition frequency (PRF), so as to have ample samples to determine the error amplitude. However, the scan rate must be high enough to track source motions in the expected dynamic engagement. The list on trade-offs puts the scan rates typically in the range of 10 to 1000 Hz, where the weight and size of the antenna play an important role.

The most effective countermeasure is to introduce an interference signal into the tracker that has the same frequency as the tracker's scan rate, but is 180° out of phase. This is simple to accomplish when the victim's scan rate is known to the ECM system. When the exact scan frequency is not known, then the ECM system must vary the jamming modulation over the expected range of the victim's operation with enough power and phase difference to cause mistracking. A technique known as swept square wave (SSW) has been successful in pulling radar conical scanners off track. The critical item is to keep within the audio bandwidth of the tracker control loop. There can be several variables in attempting to penetrate the tracking loop controls. The sweep frequency may be linear, nonlinear, or triangular, and the sweep time may have a varying time base and dwell times.

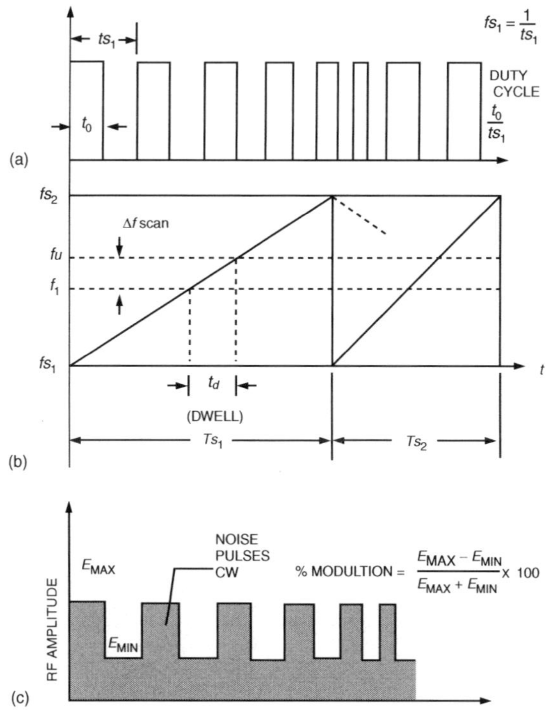

FIGURE 14.10 Swept square wave for conical scan pull-off.

For radar trackers which do not scan the transmitted beam, but instead scan a receiving antenna, the scanning rate is not usually known. It can be assumed with reasonably accurate results that the transmitted beam will be close to boresight of the conical-scan-on-receive-only (COSRO) tracker, so that a repeated RF signal directed back toward the illuminating beam will intercept the COSRO beam. This energy transmitted toward the victim radar should be at the same RF carrier frequency and amplitude modulated with sufficient noise spectrum so as to cover the audio bandwidth of the tracking control system. The use of pulsed and CW TWTs as jamming noise power sources is common.

In a radar system employing track while scan (TWS), the tracking function is performed in a signal processor/computer by keeping track of the reported position in range, azimuth, and elevation on a scan-to-scan basis. The antenna must move reasonably fast over the field-of-view in order to report target locations on a sufficiently timely basis. The antenna does not produce continued tracking, and thus does not remain fixed in azimuth or elevation for any long period of time. The countermeasure to a TWS radar is to move fast, turn fast, dispense chaff, noise jam, or some cover technique.

Simultaneous lobing or monopulse techniques, reduce the number of pulses required to measure the angular error of a target in two dimensions, elevation and azimuth. A minimum of three pulses are required in monopulse tracking, whereas more than four pulses are required for conical scan trackers. In the amplitude comparison monopulse, each pulse is measured in a sum and difference channel to determine the magnitude of the error off boresight, and then the signals are combined in a phase sensitive detector to determine direction. In a four horn antenna system, a sum pattern is generated for transmission and reception to determine range and direction. Two difference patterns, elevation and azimuth, are generated on reception for refining the error off boresight. The receiving difference patterns produce a passive receiving tracking system, and there is no scanning. Difference patterns are generated by two patterns (for each plane of tracking) offset in angle by the quantity squint angle, $u_s = \frac{\pi d}{\lambda} \sin \theta s$, and 180° phase difference, such that on boresight ($u = 0$) the difference pattern has a null, and this null is in the center of the sum pattern. The use of the function u instead of the angle θ keeps the offset functions mirror images. There is no absolute optimum value of squint angle since the desire may be to have the best linearity of the track angle output, or to have the maximum sensitivity, with all the possible illumination functions on the reflecting surface [Rhodes, 1980].

The monopulse tracker is almost completely immune to amplitude modulations.

The two beams which form the difference pattern (one beam +, and one beam –), cross each other at less than 3 dB below their peaks for maximum error slope-sum pattern product and to give increased angle tracking accuracy. This can be seen in Fig. 5.11 in Skolnik [1962] and in Fig. 6.1 in Rhodes [1980], where a comparison is shown between squint angle and the two difference beam magnitude crossover.

The difference pattern has a minimum on boresight which coincides with the peak of the sum pattern on boresight. The difference pattern is typically 35 to 40 dB below the sum pattern. These pattern values will typically improve the identity of the target location by 25:1 over a simple track-the-target in the main beam technique. Some ECM techniques are to fill the null of the difference channel so that the tracking accuracy is reduced from the 25:1 expected. If two transmitters are separated by the 3 dB width of the sum pattern (about the same as the peak of the two difference patterns), then a standard ECM technique of blinking of the two transmitters works very well.

Squint angle is a parameter within the control of the victim antenna designer; it may be chosen so as to optimize some desired function such as the best tracking linearity angle output, or maximum sensitivity on boresight. With the many options available to the antenna designer of the victim radar, the ECM designer will have to make some general assumptions. The assumption is that the squint angle off boresight is 0.4 of the half-power antenna beamwidth, given by setting the two difference patterns at a crossover of 2 dB, as shown in Fig. 14.11. (A squint of 0.5 the half-power angle is given by a 3 dB crossover). Two signals of equal amplitude (these may be radar backscatters or electronic sources, it makes no difference) and separated by 0.6 of the sum half-power beamwidth are introduced into the monopulse tracker. A typical monopulse tracker victim radar will seek the positive-going zero crossing of the error signal as a track point. In an example considering these system parameters, when the phase difference of these two signals is 90° or less, the track center position is between the two signals. When the phase difference between the two signals is between 90° and 180°, the positive going error signal is both to the left and the right outside the signal sources and the radar track will be outside of the sources. This can be seen in Fig. 14.12 (E. Rhodes, private communication, Georgia Institute of Technology). Neither target is sought, and wandering of the boresight center is expected. Complex targets such as multiscattering facet aircraft, can produce 180° differences by slight changes in angle of attack. At X-band,

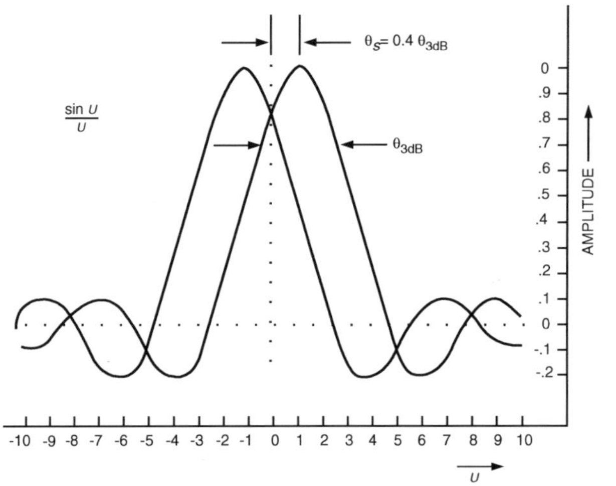

FIGURE 14.11 Squint angle monopulse $(\sin u)/u$ patterns.

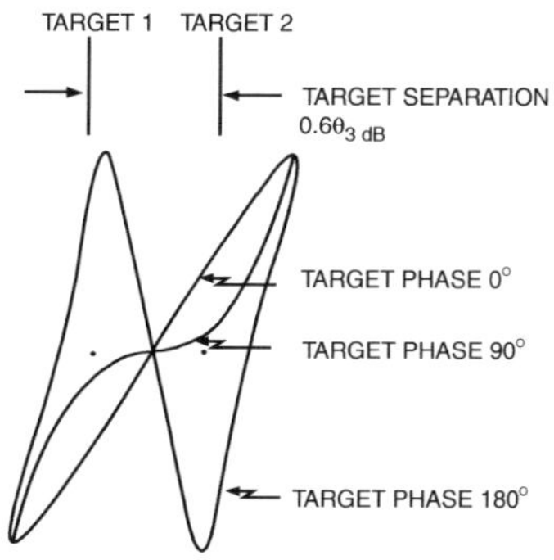

FIGURE 14.12 Two targets under track by monopulse antenna.

180° is only 1.5 cm. Thus, it is shown that in a buddy system, a monopulse radar tracker can be made to track off the two sources by slight changes in flight course.

14.3 Radar Counter-Countermeasures

The concept of counter-countermeasures is to prevent, or at least decrease, the detection, location, and tracking processes so that the mission can be accomplished before the victim group can defeat the attackers. Reducing the radar scattering cross section has produced real-time advantages in combat. This technique, commonly referred to as stealth, has become a part of all advanced systems. The concept is to reduce the cross section of the weapon's delivery system to such a low value that the detection range is so short that the defense system does not have time to react before the mission is completed.

The basic concepts of stealth are to make the object appear to have a characteristic impedance which is the same as free space, to absorb all energy which impinges upon our platform, or to reflect the energy in some obscure direction. This is not a new concept. The Germans were using "paints" and shaping techniques on their U-boats during WW II to reduce 3-GHz radar reflections. A semiflexible material was developed at the MIT Radiation Labs in the U.S. during WW II called HARP. The material consisted of copper, aluminum, and ferromagnetic material in a nonconduction rubber binder and made into cloth, sometimes with conductive backing. HARP-X was used at the X-band. Several techniques are appropriate in this process. First, the impinging signal can be reflected off axis so that no signals are returned in the direction of the transmitter/receiver, or multisignals are generated from multisurfaces which will be out of phase and appear to cancel in the direction of the transmitter/receiver. These techniques are accomplished by physical configuration and shaping of our platform. The second approach is to bend, absorb, or alter the propagation of the impinging wave in the material covering our platform. Radar absorbing material, (RAM), techniques are accomplished by material changes: typically, changing the complex dielectric constant, the dielectric-loss tangent, the magnetic permeability, and/or the magnetic-loss tangent. The Salisbury screen [Salisbury, 1952], a single layer of dielectric material backed by a reflecting surface, is often used as a model.

These devices create a resonant impedance to reduce reflections at multiple frequencies corresponding to multiple quarter-wavelengths of the dielectric material. Multilayers, or Jaumann absorbers, are generally more frequency broadband absorbers and may even be graded multidielectric or multimagnetic materials, and thus, increase the effectiveness at angles off of normal. Knott has shown that the thickness of the absorber can be reduced and the bandwidth considerable expanded if the sheets are allowed to have capacitive reactance in addition to pure resistance [Knott and Lunden, 1995].

The physics of RAM is well described by the Fresnel reflection and transmission equations. The material composite is changing daily, driven by application, and includes concerns such as weight, weather durability, flexibility, stability, life cycle, heat resistance, and cost. Manufacturers of RAM material such as Emerson and Cuming, Emerson Electric Rantic, Plussy, G.E.C., and Siemens should be consulted.

The application of electronic coding of the transmitter signal and new techniques in processing the received signals have led to a system concept referred to as low probability of intercept (LPI). There are many subtechnologies under this great umbrella, such as spread spectrum and cover pulses on transmission, pulse compression, frequency-coded and phase-coded CHIRP, frequency hopping, and variable pulse repetition frequencies. Barton [1975, 1977] has collected a number of papers describing the state of unclassified technology before 1977 in two ready references. A number of publications, primarily U.S. Naval Research Lab reports and IEEE papers, on coding techniques and codes themselves have been collected by Lewis et al. [1986].

In developing an LPI system concept, the first item will be to reduce the peak power on transmission so that the standoff detection range is greatly reduced. In order to have the same amount of energy on target, the transmitter pulse length must be increased. To have the same resolution for target detection, identification, and tracking, the new long pulse can be subdivided on transmission and coherently compressed to construct a much shorter pulse on reception. The compressed pulse produces an effective short pulse for greater range resolution and a greater peak signal which can exceed noise, clutter, or other interference for improved target detection.

One of the earliest pulse compression techniques was developed at the Bell Telephone Laboratories [Klauder et al., 1960; Rhodes, 1980]. The technique was called CHIRP and employs frequency modulation. In a stepped modulation, each separate frequency would be transmitted for a time that would correspond to a pulse length in a pulsed radar system. Upon reception, the separate frequencies are folded together to effectively form one pulse width containing the total energy transmitted. In a linear modulation, the frequency is continuous sweep over a bandwidth in a linear manner during the transmitted pulse length. The received signal can be passed through a matched filter, a delay line inversely related to frequency, an autocorrelater, or any technique to produce an equivalent pulse length equal to the reciprocal of the total bandwidth transmitted. The amplitude of this compressed signal is equal to the square root of the time-frequency-bandwidth product of the transmitted signal. The errors are limited to system component phase and frequency stabilities and linearities over the times to transmit and to receive. When using the Fourier transforms to transform from the frequency domain to the time and thus range domain, sidelobes are generated, and these side lobes appear as false targets in range. Side lobe suppression techniques are the same as employed in antenna side lobe suppression design.

One of the first biphase coded wave forms proposed for radar pulse compression was the Barker code. See for example, Farnett et al. [1970]. This code changes the phase of the transmitted signal 180° at a time span corresponding to a pulse length of a standard pulse-transmitter radar. The Barker codes are optimum in the sense that the peak magnitude of the autocorrelation function is equal to the length of the coded pulse (or the number of subpulses), and the side lobes are less than one. When the received signal is reconstructed from the phase domain, the range resolution is the same as the pulse length (or the length of a Barker subpulse) of the standard pulse-transmitter radar.

The maximum length of a Barker code is 13, and this code has side lobes of –22.3 dB. The phase will look like +++++−−++−+−+, or −−−−−++−−+−+−.

There are many phase codes discussed by Lewis et al. (1986), which are unique and thus have unique autocorrelation functions. The compression ratios can be very large and the sidelobes very low. The price to be paid is equipment phase stability in polyphase code implementation, length of time to transmit, minimum range due to the long transmitted pulse, and equipment for reception signal processing. The advantages gained are high range resolution, low peak power, forcing the observer to use more power in a noise mode because radar uses wider bandwidth spread over segments of narrow bandwidths.

Another technique found helpful in reducing interference is to change the transmitted frequency on a pulse-by-pulse basis. This frequency hopping is effective and does require that the receiver and local oscillator be properly tuned so that the receiver is matched to the return signal for efficient operation. When Doppler or moving target indicators are employed, then the frequency hopping is done in batches. There must be enough pulses at any one frequency to permit enough resolution in the Fourier transform (FT) or the proper number of filter banks to resolve the Doppler components. If the transmitter can handle the total bandwidth and keep each frequency separate so there is no cross talk in the FTs or in

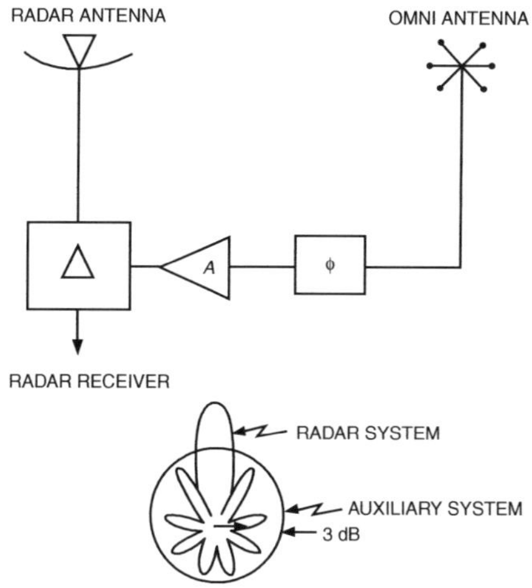

FIGURE 14.13 Side lobe interference reduction concept.

the filters, then this is very effective jamming reduction. Today's digital systems are set up for frequency groups like 32, 64, or 128; and then repeat the frequency hopping in the same frequency direction, or the reverse direction. This reduces velocity gate pull off.

If the countermeasure system has identified your pulse repetition frequency (PRF) and thus can lock on your radar's range gate, then there are several techniques to reduce the EW range gate pull-off. One technique is to place a noise pulse over the standard pulse. This noise pulse is longer than the system range resolution pulse and is not symmetric around the range pulse. Only your system should know were your own target identification pulse is in the chain of PRF pulses. Another technique that shows promise is to change the PRF after a batch of pulses have been transmitted. Typically the batch length is controlled in a digital manner. Again, as in the case of frequency hopping, PRF hopping still must permit processing in range for resolution, in frequency for Doppler, and in polarization (or other identifiers) for target identification.

So far as our system is concerned, there is no difference if a jamming or interference signal enters the main beam of our system antenna or a sidelobe of our system antenna. From the jammer's point of view, there needs to be 20 to 30 dB more power or 20 to 30 dB closer in range if the jammer is directed at a sidelobe instead of the main antenna lobe.

Suppression of side lobe jamming can be accomplished by employing omnidirectional auxiliary antennas having gains greater than the highest side lobe to be suppressed, and yet lower than the main beam of the antenna. The auxiliary antenna feeds a receiver identical to the receiver used by the radar. The detected output of the auxiliary receiver and the radar receiver are compared; and when the auxiliary receiver output is larger than the radar, then the radar is blanked in all range cells where this occurs. The auxiliary antenna should be placed as close as possible to the phase center of the radar antenna to ensure that the samples of the interference in the auxiliary antenna can be correlated with interference in the main radar antenna. This requires that the time of arrival of the signals to be compared must be less than the reciprocal of the smaller of the two bandwidths, in the radar receiver or the interference signal.

14.4 Chaff

Cylindrical metal rods scatter energy in the same manner as a dipole radiates energy. The same parameters affect the efficiency, beamwidth, loss, and spatial patterns. Thus, in this discussion of chaff as a scattering element, consideration will be given to rod length in terms of excitation frequency, length-to-diameter ratio, impedance or loss-scattering pattern as a function of polarization, and interaction (or shielding) between adjacent elements.

The aerodynamic characteristics are determined by material, density, shape, drag coefficient, and atmospheric parameters such as temperature, pressure, and relative humidity. In addition to these parameters, the movement of a rod is strongly influenced by local winds, both horizontal and vertical components, and by wind turbulence and shear.

When all of these variable parameters are defined, the radar scattering characteristic of a collection of rods forming a cloud of chaff can be calculated. The radar parameters of interest in this presentation are the scattering cross section from the cloud, the power spectral density to define the frequency spread, Doppler frequency components, polarization preference, signal attenuation through the cloud, bandwidth, and length of time that these clouds can be expected to exist.

From antenna theory, the tuned half-wave dipole produces the maximum scattering coefficient. There are many books for your reference reading (examples are Terman [1943]; Kraus [1950]; Thourel [1960]; Jasik [1961]; Eustace [1975–1991] and Van Vleck et al. [1947]).

When the dipole is observed broadside, and with a polarization parallel to the rod, then the maximum radar cross section is observed and has a value of

$$\sigma = 0.86\lambda^2$$

As the dipole tumbles and is observed over all possible orientation then, the average radar cross section has been calculated and found to be

$$\sigma = 0.18\lambda^2$$

Since the rod has a finite diameter, the maximum scattering cross section occurs for lengths slightly less than $\lambda/2$. A reasonable value of 0.48λ for the length will be assumed to compensate for the end effects of the rod. The first resonance is plotted as a function of the length-to-diameter ratio in the June 1976 edition of ICH [Eustace, 1975–1991]. The equations relating radar cross section to the rod length and diameter in terms of wavelength are complicated. The radar cross section has been shown to be directly related to the ratio of physical length-to-radar wavelength, and inversely related to the logarithm of the physical length-to-radius ratio in Van Vleck et al. [1947]. This article is a classic, and along with its references, is the place to begin.

The polarization of the observing radar and the alignment of the chaff material are important since horizontally aligned rods fall more slowly than vertically aligned rods. Under many applications, one end of the chaff filaments is weighted, causing these rods to be vertically aligned shortly after release. Thus, this enlarges the chaff cloud rapidly in the vertical elevation plane as well as making the radar cross section of the lower elevation portion of the cloud more susceptible to vertically polarized waves and the upper elevation section of the chaff cloud having a higher cross section to horizontally polarized waves.

Chaff elements have been constructed in all sizes and shapes. Early manufactured elements were cut from sheets of metal and were rectangular in cross section and cut to half-wave dipoles in length; some were bent in V shape, some were long to form ribbons, some were made as ropes, and some were cut from copper wire and some from aluminum. In the past decade or two, glass rods coated with aluminum have become very popular and this configuration will be assumed typical for this discussion. There has been a reasonable amount of experimental work and calculations made for dipoles constructed of both 1-mil- and 1/2-mil-diameter glass rods for use as chaff. The metal coating should be at least three skin depths at the designed frequency. One skin depth is defined as the thickness there the signal has decreased to $1/e$ of the original value.

There has been limited amounts of work done at 35 and 95 GHz as compared to much work done at lower frequencies. Consider that the wavelength at 35 GHz is only 0.857 cm and a half wavelength is only 0.1687 in. Thus, a cloud of dipoles will be a lot of very short elements of material. In order to have a reasonable length-to-diameter ratio and reduce end effects as well as to provide good electrical bandwidth, it will be assumed that 1/2-mil-diameter dipoles would be acceptable. From these dimensions, manufacturing of large bundles of these dipoles is not a trivial matter. Some areas of concern are: (1) the plating of the small diameter glass will not be simple and most likely will produce considerable variations in weight and *L*/*D*, with results in varying fall rates and changes in bandwidth of the scattering signal; (2) to consistently cut bundles of pieces of elements this short in length requires considerable ability; and (3) it is required that each dipole be independent and not mashed with several other dipoles or the result is reduction in radar cross section and bird nesting. These effects result in lost efficiency. One note: do shop around because technical and manufacturing capabilities are advancing at a rapid rate.

There are two types of antennas that are discussed in the literature on antennas: the thin-wire antenna, and the cylindrical antenna. The thin-wire antenna model, where assumed length-to-diameter is greater than 5000, is used to predict general scattering patterns and characteristic impedance as a function of rode length and frequency. The cylindrical antenna will show the effect of wire diameter which in turn reveals changes in the reactive portion of the terminal impedance and thus changes in gain, propagation and scattering diagrams, filling in of pattern nulls, reduction of side lobes, and wider frequency bandwidth. It has been shown that a thin-wire half-wave dipole, excited at the center, and at resonance, has an impedance in ohms of

$$Z = 73.2 + j42.5$$

When the length is not half-wave, and the diameter is not negligible, then to a first-order approximation

$$Z = 73.2 + j42.5 \pm j120\left[\ln\frac{L}{D} - 0.65\right]\cot\frac{2\pi L}{\lambda}$$

As is the response in all tuned circuits, the band-pass characteristics are primarily determined by the reactive elements as the frequency moves away from the resonant frequency. In low Q circuits, the real part of the impedance also has a notable effect on the total impedance, and thus, also on the band-pass characteristics. To a first-order effect, the real part of the half-wave dipole does not change drastically (less than 10%) from 73.2 Ω; however, the reactive part varies as (ln *L*/*D*). As shown by Van Vleck et al. [1947], the radar cross section of resonance is expressed in reciprocal terms of ln(rod length/rod radius). The effective half-power bandwidth of dipole scatters is thus usually referenced to the ratio of dipole length to dipole diameter as shown in Fig. 14.14. As it is desirable to have some bandwidth in television antennas, it is also desirable to have some bandwidth in the radar scattering from chaff, which leads to a measurable diameter of the rods or width of the ribbon strips used for chaff.

Consider an application at X-Band in which it is desirable to screen against a wideband pulse compression radar having a bandwidth of 1.5 GHz. The rods will be cut to 0.48λ. From Fig. 14.14, the value of *L*/*D* = 750,

PERCENT BANDWIDTH	LENGTH/DIAMETER
12	3000
13	1650
14	1050
15	740
16	520
17	400
18	320
19	260
20	220
21	180
22	150
23	140
24	120
25	110

FIGURE 14.14 Backscatter bandwidth response for chaff.

$$L/D = 750 = 0.48\lambda/D$$

where $D = 0.48 \times 3$ cm/750 = 0.00192 cm = 0.000756 in. = 0.756 mil.

Consider now the generation of a chaff cloud having a radar backscatter cross section per unit volume of –60 dB m^2/m^3. This value of cross section is comparable to the radar return from 10 mm/h of rain observed at X-Band [Currie, Hayes, and Trebets, 1992]. These small aluminum-coated glass cylinders fall very slowly in still air and have been known to stay aloft for hours. The movement is so strongly influenced by local wind conditions that meteorologists in several countries (including the U.S.S.R., Germany, and the U.S.) have used chaff clouds to determine wind profiles [Nathanson, 1969]. Local conditions give extreme wind variations even when the local wind structure is free from perturbations such as squalls, fronts, thunderstorms, etc. Some wind profile models include terrain surface material such as water, grass, dirt, and surface roughness; time of day; time of year; northern hemisphere; and other parameters shown to affect a particular area. At altitudes above 500 ft., the profile of wind has been shown to vary mostly in a horizontal manner, with little vertical constantly defined variations [AFCRC, 1960]. The strong jet streams at around 35,000 to 40,000 ft. define the topmost layer typical of zero water vapor levels, zero oxygen levels, definable temperature profiles, wind stability, and pressure profiles. This being the case, it behooves the EW engineer to apply local meteorological records whenever possible.

One-mil diameter aluminum-coated, glass-filament chaff, for example, will adjust to sharp wind gusts of up to 50 ft/s within 15 ms [Coleman, 1976–1977]. The vertical fall rate of chaff is proportional to the weight and air resistance; thus the important chaff parameters are diameter, length, and coating. A typical approach is to use a mixture of dipole diameter sizes so that a vertical cloud is generated from different fall rates. For the example above, the diameter mixture could be between 0.5- and 1.0-mil coated glass.

Consider a typical wind condition of a 5 knot horizontal wind, and a wind profile of 1/5 power relationship at a height below 500 ft. Coleman [1976–1977] has shown that a cloud mixture of 0.5 to 1.0 mil chaff would stay aloft for about 20 min when dispensed at 500 ft. The chaff cloud will smear in both elevation and in horizontal dimensions. The volume of the ellipsoid generated by the horizontal winds and the vertical fall rate is given by

$$\text{volume} = V = \frac{4}{3}\pi abc$$

In this example, α = 100 ft and $\beta = \chi$ = 300 ft. The resulting volume is 1,068,340 m^3.

The radar cross section of such a cloud is given by

$$\begin{aligned}\sigma_T &= V\sigma_V \\ &= \left(1.068 \times 10^6\right)\left(10^{-6}\right) \\ &= 1.068\,\text{m}^2\end{aligned}$$

This is not a strong target. But we know that 10 mm/h of rain is not a strong target at X-band. The big problem with rain is a deep volume producing a large amount of attenuation. You notice here that the chaff cloud is only 600 ft. across; this would not be a typical rain cloud. Perhaps the comparison should be for the chaff cloud to cover (hide) a 30-dB cross-section vehicle.

The number of dipoles in a cloud of chaff is given by

$$N = \frac{\sigma_T}{\sigma_d}$$

where σ_T is the total cross section, and σ_d is the cross section of each dipole

$$N = \frac{1000}{0.18(3/100)^2}, \quad \frac{\text{m}^2}{\text{m}^2}$$

$$= 6.17 \times 10^6$$

The density of dipoles is given by

$$\frac{N}{V} = \frac{6.17 \times 10^6}{1.068 \times 10^6}$$

$$= 5.78 \ \text{dipoles/m}^3$$

This is not a dense chaff cloud, and thus you can see the value of such a cloud to hide a good size target. Of course, there is a short fall; the cloud may last only 20 min and will drift several thousand feet in this time. This requires that there be proper area and time planning for chaff use. If the volume to be covered is larger than the 100 × 300 × 300 ft., then several clouds need to be dispersed.

In the *International Countermeasures Handbook* [EW Comm., 1975], the author has presented a graph to be used to determine the signal attenuation produced by a chaff cloud. The graph relates normalized two-way attenuation per wavelength squared to the number of half-way dipoles per cubic meter, as a function of the dipole orientation K relative to the radar polarization. When the dipoles are uniformly distributed and randomly oriented, K can be as low as 0.115; when the dipoles are aligned with the E-field, K = 1.0 and the loss is 10 dB higher; when the dipoles are almost cross-polarized to the radar, K = 0.0115 the attenuation will be 10 dB less than the random orientation. For K = 0.115, the attenuation relationship is given by

$$L/\lambda^2 = \text{N/m}^3$$

in units of dB/meter cube = number/cubic meter. Typical ranges of cloud density range from 10^{-7} to 10^{+2} dipoles per cubic meter. For the example given above, a cloud density of 5.78 dipoles/m^3 cube will give a value of $L/\lambda^2 = 5.78$ For a two-way loss per meter of chaff cloud, L = 0.0052 dB/m at λ = 0.03 m; giving 0.5 dB of attenuation in a 100-m cloud.

14.4.1 Expendable Jammers

Expandable jammers are a class of electronic devices which are designed to be used only once. These devices are typically released from chaff/flare dispensers much as chaff is deployed. The electronic package can be deployed as a free-falling object, or released with a parachute for slow fall rates. Such packages have been used as screening devices and as simulated false targets. The RF energy can be continuous, pulsed, or modulated to respond as a repeater or a false radar. Expandable jammers have been configured as spot noise jammers and have been deployed as barrage noise jammers depending upon the mission and victim radars.

Defining Terms

P = power, watts
k = Boltzmann constant = 1.38×10^{-23} joules/K
T = temperature = absolute temperature, K

S = energy = power x transmitted pulse length, joules
B = bandwidth, Hz
f = frequency, Hz
Hz = 1 cycle per second
dB = 10 times logarithm of power ratio
Z = impedance, ohms

References

AFCRC. 1960. *Geophysics Handbook,* Chap. 5. Air Force Cambridge Research Center, MA.

Barton, D.K. 1975. Pulse compression. *RADAR,* Vol. 3. Artech House, Norwood, MA.

Barton, D.K. 1977. Frequency agility and diversity. *RADAR,* Vol. 6. Artech House, Norwood, MA.

Butler, W.F. 1981. Antenna backscatter with applications to surgical countermeasures. Contract N00014-81-C-2313, Hughes Aircraft, September, El Segundo, CA.

Coleman, E.R. 1976–1977. *International Countermeasures Handbook.* EW Communications, Inc., Palo Alto, CA.

Currie, Hayes, and Trebets. 1992. *Millimeter-Wave Radar Clutter.* Artech House, Norwood, MA.

Eustace, H.F., ed., 1975–1991. *The Countermeasures Handbook,* EW Communications Inc., Palo Alto, CA.

EW Comm. 1975. *International Countermeasures Handbook,* June, EW Communications Inc., Palo Alto, CA, 227.

Farnett, Howard, and Stevens. 1970. Pulse-compression radar. In *Radar Handbook.* ed. Skolnik, Chap. 20. McGraw-Hill, New York.

Hayes, R.D. and Eaves, J.L. 1966. Study of polarization techniques for target exhancement. Georgia Institute of Technology Project A-871, March, AD 373522.

Jasik, H. 1961. *Antenna Engineering Handbook.* McGraw-Hill, New York.

Klauder, Prince, Darlington, and Albersheim, 1960. *Bell Syst. Tel. J.* 39(4).

Knott, E.F. and Lunden, C.D. 1995. The two-sheet capacitive Jaumann absorber. *IEEE Antennas Propag. Trans.* 43(11).

Kraus, J.D. 1950. *Antennas.* McGraw-Hill, New York.

Lewis, B.L., Kretschmer, F.F., Jr., and Shelton, W.W. 1986. *Aspects of Radar Signal Processing.* Artech House, Norwood, MA.

Nathanson, F.E. 1969. *Radar Design Principles.* McGraw-Hill, New York.

Rhodes, D.R. 1980. *Introduction to Monopulse.* Artech House, Norwood, MA.

Salisbury, W.W. 1952. Absorbent body for electromagnetic waves. U.S. Patent 2,599,944, June 10.

Skolnik, M.I. 1962. *Introduction to Radar Systems,* 2nd ed. McGraw-Hill, New York.

Terman, F.E. 1943. *Radio Engineers Handbook.* McGraw-Hill, New York.

Thourel, L. 1960. *The Antenna.* Chapman and Hall, New York.

Timberlake, T. 1986. *The International Countermeasure Handbook,* 11th ed. EW Communications Inc., Palo Alto, CA.

Van Vleck, Bloch, and Hamermesh. 1947. The theory of radar reflection from wires or thin metallic strips. *J. App. Phys.* 18 (March).

Further Information

Applied ECM, Vols. 1 and 2, 1978, EW Engineering, Inc., Dunn Loring, VA.

International Countermeasures Handbook, 11th ed. 1986, EW Communications, Inc.

Introduction to Radar Systems, 2nd ed. 1962, McGraw-Hill.

Modern Radar Systems Analysis, 1988, Artech House.

15

Automotive Radar

Madhu S. Gupta
San Diego State University

Scope: An automotive radar, as the name suggests, is any radar that has an application in automobiles and other autonomous ground vehicles. As a result, it represents a large and heterogeneous class of radars that are based on different technologies (e.g., laser, ultrasonic, microwave), perform different functions (e.g., obstacle and curb detection, collision anticipation, adaptive cruise control), and employ different operating principles (e.g., pulse radar, frequency-modulated continuous-wave [FMCW] radar, microwave impulse radar). This chapter is limited to microwave radars that form a commercially significant subset of automotive radars. Microwave radars have an advantage over the laser and infrared radars (or "lidar") in that they are less affected by the presence of precipitation or smoke in the atmosphere. They also have the advantage of being unaffected by air temperature changes that would degrade an ultrasonic radar.

Need: The need for automotive radars can be understood at three different levels. At the national level, the statistics on traffic fatalities, injuries, and property loss due to vehicle accidents, and estimates of their fractions that are preventable with technological aids, has encouraged the development of automotive radar. The economic value of those losses, when compared with the dropping cost of automotive radar, leads to a cost-benefit analysis that favors their widespread deployment. At the level of the automotive manufacturer, radar is another "feature" for the consumer to purchase that could be a possible source of revenue and competitive advantage. It is also a possible response to regulatory and public demands for safer vehicles. At the level of vehicle owners, automotive radar has an appeal as a safety device, and as a convenient, affordable gadget. Of greater practical importance is the

0-8493-1732-0/03/$0.00+$1.50

potential for radar to lower the stress in driving and decrease the sensory workload of the driver by taking over some of the tasks requiring attentiveness, judgment, and skill.

Antecedents: Lower frequency electronic systems have had a presence for decades in various automobile applications, such as entertainment, fuel injection, engine control, onboard diagnostics, antitheft systems, antiskid brakes, cruise control, and suspension control. The earliest microwave frequency products introduced in automobiles were speed radar detectors in the 1970s, followed by cellular telephones in the 1980s, and direct satellite receivers and GPS navigational aids in the 1990s. The use of microwave vehicular radars can also be traced back to the 1970s, when rear obstacle detection systems were first installed, mostly in trucks. Some of those radars evolved from simple motion detectors and security alarms that employed Gunn diodes and operated typically in the X-band around 10 GHz. The modern automotive radar owes its origin to three more recent technological advancements: low-cost monolithic microwave devices and hybrid or monolithic circuits, microprocessors, and digital signal processing.

Status: Automobile radars have been developed by a number of different companies in Europe, Japan, and the U.S., since the 1970s. Automotive radars for speed measurement and obstacle detection have been available for more than a decade as an aftermarket product, and have been installed primarily on trucks and buses in quantities of thousands. Only now (in 2000) are automotive radars becoming available as original equipment for installation on trucks, buses, and automobiles. However, other types of radars are still in the development, prototyping, field-testing, and early introduction stages. Some of the most important challenges for the future in the development of automotive radar include (1) development of more advanced human interface, (2) reduction of cost, (3) meeting user expectations, and (4) understanding the legal ramifications of equipment failure.

15.1 Classification

An automotive radar can be designed to serve a number of different functions in an automobile, and can be classified on that basis as follows.

1. *Speed Measuring Radar.* Vehicle speed is measured with a variety of types of speedometers, many of which are based on measuring the rate of revolution of the wheels, and are therefore affected by tire size and wheel slippage. By contrast, a speed measuring radar can determine the true ground speed of a vehicle, that may be required for the instrument panel, vehicle control (e.g., detection of skidding and antilock braking), and other speed-dependent functions.
2. *Obstacle Detection Radar.* Such a radar is essentially a vision aid for the driver, intended to prevent accidents by monitoring regions of poor or no visibility, and providing a warning to the driver. It can perform functions such as curb detection during parking, detection of obstacles in the rear of a vehicle while backing, and sensing the presence of other vehicles in blind zones (regions around the vehicle that are not visible to the driver via side- and rear-view mirrors) during lane changing.
3. *Adaptive Cruise Control Radar.* A conventional cruise control maintains a set vehicle speed regardless of the vehicular environment, and as a result, is ill suited in a heavy traffic environment where vehicles are constantly entering and leaving the driving lane. An adaptive cruise control, as the name suggests, adapts to the vehicular environment, and maintains a headway (clearance between a vehicle and the next vehicle ahead of it) that is safe for the given speeds of both vehicles.
4. *Collision Anticipation Radar.* This class of radars senses the presence of hazardous obstacles (other vehicles, pedestrians, animals) in the anticipated path of the vehicle that are likely to cause collision, given the current direction and speed of the vehicle. It is therefore potentially useful both under conditions of poor atmospheric visibility (e.g., in fog, sandstorm), as well as poor judgment on the part of the driver (unsafe headway, high speed). Its purpose may be to warn the driver, to deploy airbags or other passive restraints, or take over the control of vehicle speed.
5. *Other Vehicular Monitoring and Control Radars.* Many other vehicular control functions, such as vehicle identification, location, fleet monitoring, station keeping, guidance, and path selection,

can be performed with the aid of radar. Such radars may be placed on board the vehicle, or on the ground with the vehicle carrying a radar beacon or reflector, that may be coded to allow vehicle identification. Still other variations of automotive radars can be envisaged for special purposes, such as control of vehicles constrained to move along tracks, or joyride vehicles in an amusement park.

The following is a discussion of each of the first four types of radars, including their purpose, principle of operation, requirements imposed on the radar system and its constituent parts, limitations, and expected future developments. The emphasis is on the RF front end of the radar, consisting of the transmitter, the receiver, and the antenna. The adaptive cruise control and the collision anticipation radars share a number of characteristics and needs; they are therefore discussed together as forward-looking radars.

15.2 History of Automotive Radar Development

The idea of automotive radars is almost as old as the microwave radars themselves. The history of their development may be divided into four phases as follows.

1. *Conceptual Feasibility Phase.* The earliest known experiments, carried out in the late 1940s soon after the Second World War, involved a wartime radar mounted on top of an automobile, with the cost of the radar far exceeding that of the automobile. The idea that the radar could be used to control a vehicle was considered sufficiently novel that the U.S. Patent Office issued patents on that basis. An example of such early efforts is a patent (# 2,804,160), entitled "Automatic Vehicle Control System," issued to an inventor from Detroit, Michigan, that was filed in January 1954 and issued in August 1957.
2. *Solid-State Phase.* The presence of a microwave tube in the radar for generating the transmitted signal, and the high voltage supply they required, was known to be a principal bottleneck that impacted the size, cost, reliability, and safety of automotive radar. Therefore, the discovery of solid-state microwave sources in the form of two-terminal semiconductor devices (transferred-electron devices and avalanche transit-time diodes) in the mid-1960s led to a resurgence of interest in developing automotive radar. Since solid-state microwave devices were then one of the more expensive components of radar, the possibility of using the device, employed simultaneously as the source of the microwave signal and as a mixer by taking advantage of its nonlinearity, was considered another desirable feature. Such radars were the subject of numerous experimental and theoretical studies, and even reached the stage of limited commercial production. Radars based on BARITT diodes were deployed on trucks in the 1970s to warn the driver of obstacles in the rear.
3. *Monolithic microwave integrated circuits (MMIC) Phase.* The production of monolithic microwave integrated circuits (MMICs) in the 1980s made both the microwave devices, and the associated circuitry, sufficiently small and inexpensive that the weight and cost of microwave components in radar was no longer an issue. The prototype radars in this period were based both on Gunn diodes and on GaAs MMICs that employed metal semiconductor field-effect transistors (MESFETs) as the active device and source of microwaves. At the same time, the technology of microstrip patch antennas, that are particularly convenient for automotive use, became sufficiently developed. A variety of volume production problems had to be solved in this phase, including those of maintaining frequency stability and packaging, while keeping the cost down. The focus of development thus shifted to other aspects of the radar design problem: reliability of the hardware and its output data; operability of the radar under unfavorable conditions; affordability of the entire radar unit; and extraction of useful information from radar return by signal processing. Prototype units of various types were field tested, and deployed on a limited scale both for evaluation and feedback as well as in special markets such as in bus fleets.

4. *Product Development Phase.* Automotive radar became economically viable as a potential consumer product due to the decreasing ratio of the cost of the radar to the cost of the automobile. Moreover, the economic risk of radar development decreased due to the availability of specific frequency bands for this application, and the confidence resulting from the results of some field tests. As a result, automotive radar is now available as an "option" on some vehicles, and work continues on enhancing its capabilities for other applications, in human factors engineering, and for integration into the vehicle.

15.3 Speed-Measuring Radar

15.3.1 Operating Principle

The simplest method of measuring the true ground speed of a vehicle is by the use of a Doppler radar. In this radar, a microwave signal at frequency f is transmitted from on board the vehicle, aimed toward the ground; the signal scattered back by the ground is collected by a receiver also on board the vehicle. Given the large speed c of microwave signals, and the short distance between the antenna (typically mounted under the carriage) and the ground, the signal transit time is short. The returned signal is shifted in frequency due to Doppler effect, provided the vehicle and the scattering patch of the ground are moving with respect to each other, with a velocity component along the direction of propagation of the microwave signal. To ensure that, the microwave beam is transmitted obliquely toward the ground, making an angle q with respect to the horizontal, and in the plane formed by the ground velocity v of the vehicle and the perpendicular from the vehicle to the ground. Then the vehicle velocity component along the direction of propagation of the signal is v cos q, and the Doppler frequency shift is 2f(v/c)cos q, proportional both to the transmitted signal frequency f and to the vehicle velocity v. The Doppler shift frequency is extracted by mixing the returned signal with the transmitted signal, and carrying out filtering and signal processing. Typically (for a carrier frequency of 24 GHz and a tilt angle q = 30°), it lies in the range of 35 Hz to 3.5 kHz for vehicle speeds in the range of 1 to 100 mi/h.

15.3.2 Error Sources

Several sources of error in speed estimation can be identified from the above discussion.

1. *Vehicle Tilt.* Uneven loading of a vehicle, air pressure in the tires, and non-level ground can all contribute to a change in the tilt angle q of the beam, and hence in the estimated vehicle speed. A well-known technique for correcting this error is to employ a so-called Janus configuration in the forward and reverse directions (named after a Greek God with two heads). In this scheme, two microwave beams are transmitted from the transmitter, one in the forward and the other in the reverse direction, each making an angle q with the horizontal. The Doppler frequency shift has the same magnitude for each signal, but a tilt of the vehicle makes q one beam larger while simultaneously making the other smaller. The correct ground velocity, as well as the tilt angle of the vehicle, can be deduced from the sum and difference of the two Doppler shift frequencies.
2. *Nonzero Beam Width.* Any reasonably sized antenna will produce a beam of finite width, so that the angle q between the transmitted signal and the horizontal is not a constant. A spread in the values of q produces a corresponding spread in the values of the Doppler shift frequency.
3. *Vertical Vehicle Velocity.* Vibrations of the vehicle and uneven ground will cause the vehicle of have a velocity with respect to the ground in the vertical direction. This velocity also has a component in the direction of propagation of the microwave signal, and thus modulates the Doppler shift frequency.
4. *Surface Roughness.* Although surface roughness is essential for producing scattering in the direction of the receiver, it introduces errors because the surface variations appear as variations of q as well as of vehicle height, and therefore an apparent vertical vehicle velocity.

Extensive signal processing is required to extract an accurate estimate of vehicle velocity in the presence of these error sources. Current Doppler velocity sensors employ digital signal processing and are capable of determining the velocity with 99% certainty.

15.4 Obstacle-Detection Radar

15.4.1 Purpose

A driver is unable to view two principal areas around the vehicle that are a potential source of accidents: one is behind the vehicle, and the other is on the two sides immediately behind the driver. The need to view these areas arises only under specific conditions: when driving in reverse and when changing lanes. The exact boundaries of these areas depend on the style of vehicle, the placement and setting of the viewing mirrors, and the height and posture of the driver.

15.4.2 Mission Requirements

Since obstacle detection radar needs to operate over a small range (typically less than 10 m), cover a wide area, and does not need to determine the exact location of the obstacle in that area, its operating frequencies can be lower, where the antenna beamwidth is wide.

15.5 Adaptive Cruise Control Radar

15.5.1 Purpose

The adaptive cruise control (to be abbreviated hereafter as ACC) radar is so called because it not only controls the speed of a vehicle but also adapts to the speed of a vehicle ahead. The ACC radar controls the vehicle speed, subject to driver override, so as to maintain a safe distance from the nearest in-path vehicle ahead (the "lead" vehicle). If there are no lead vehicles within the stopping distance of the vehicle, the ACC functions as a conventional cruise control that maintains a fixed speed set by the driver. With lead vehicles present within the stopping distance, the system governs the acceleration and braking so as to control both the speed and the headway. Such radar has also been referred to by several other names such as intelligent cruise control (ICC), autonomous intelligent cruise control (AICC), and others.

15.5.2 Mission Requirements

First and foremost, the ACC radar must be capable of distinguishing between the closest vehicle ahead in the same lane and all other vehicles and roadside objects. As a result, a high accuracy in range and angular resolution is necessary. Second, it must acquire sufficient information to establish the minimum safe distance from the target vehicle, S_{min}. At the simplest level, S_{min} equals the stopping distance of the vehicle carrying the radar (the "host" vehicle), minus the stopping distance of the lead vehicle, along with an allowance for the distance traveled within the reaction time of the driver initiating the stopping action. If v_r and a_r are the velocity and deceleration of the host vehicle, v_t and a_t are those of the targeted lead vehicle, and T_r is the reaction time of the driver, then the minimum safe distance to the target can be estimated approximately as

$$S_{min} = \left(v_r^2/2a_r\right) - \left(v_t^2/2a_t\right) + v_r T_r$$

The distance calculated by this equation is subject to large uncertainty and additional safety margin, since the deceleration of each vehicle depends on the brake quality, road conditions, vehicle loading, and tire condition, while the reaction time depends on the driver's age, health, state of mind, training, and fatigue. However, the equation does show that to determine whether the following distance is safe requires not only the distance to the target but also the ground speed of the vehicle as well as the relative velocity with respect to the target. In particular, it is the radial component of the velocity that is pertinent, and the sign of the velocity is also important because it determines whether the vehicles are approaching or receding. This defines the minimum information that the ACC system must be designed to acquire.

15.6 Collision Anticipation Radar

15.6.1 Purpose

The purpose of collision anticipation radar is to sense an imminent collision. Several different variations of collision anticipation radars have been considered, differing in the use of the information gathered by the radar, and hence the definition of "imminent." For example, if the purpose of the radar is to initiate braking, the possibility of a collision must be sensed as early as possible to allow time for corrective action; if its purpose is to serve as a crash sensor for deploying an inflatable restraint system (commonly called airbag), only a collision that is certain is of interest. The different functions have been given various names, such as collision warning (CW), collision avoidance (CA), collision reduction (CR) radar, and others, but the nomenclature is not consistent and is sometimes based on marketing rather than on technical differences. The following discussion illustrates some applications.

15.6.2 Collision Warning Application

Traffic accident analyses show that a significant fraction of traffic accidents (30% of head-on collisions, 50% of intersection accidents, and 60% of rear-end collisions) can be averted if the drivers are provided an extra 0.5 s to react. The purpose of the collision warning radar is to provide such an advanced warning to the driver. In order to perform that function, the radar must resolve, classify, and track multiple targets present in the environment; collect range, speed, and acceleration information about individual targets; use the past and current information to predict the vehicular paths over a short interval; estimate the likelihood of an accident; and present the situational awareness information to the driver through some human interface. Moreover, these functions must be performed in real time, within milliseconds, and repeated several times per second for updating.

15.6.3 Crash Sensing Application

Passive restraints (such as airbags and seat belt tensioners) used to protect vehicle occupants from severe injuries, typically employ several mechanical accelerometers to sense the velocity change in the passenger compartment. A radar used as an electronic acceleration sensor can have a number of advantages, such as sophisticated signal processing, programmability to customize it for each vehicle structure, self-diagnosis and fault indication, and data recording for accident reconstruction. Since the passive restraints require only about 30 ms to deploy, the ranges and time intervals of interest are shorter than in collision warning applications.

15.6.4 Radar Requirements

In each case, the collision anticipation radar must detect objects in the forward direction (in the path of the vehicle), and acquire range and velocity data on multiple targets. In this respect, the radar function is similar to that of ACC radar, and there are many similarities in the design considerations for the RF front end of the two types of radars. The system requirements and radar architecture for the two are therefore discussed together in the following.

15.7 RF Front End for Forward-Looking Radars

ACC and collision anticipation radars have a number of similarities in areas they monitor, information about the vehicular environment they must acquire, and constraints under which they must operate. The major differences between them lie in the signal processing carried out on the radar return, the range of parameter values of interest, and the manner in which the collected information is utilized. From a user perspective, the primary difference between them stems from their roles: whereas the ACC radar is a "convenience" feature, a collision radar is thought as a "safety" device; consequently, the legal liability in case of malfunction is vastly different.

15.7.1 Radar Requirements

Some of the requirements to be satisfied by a forward-looking radar follow directly from its expected mission. The expected radar range is the distance to the lead vehicles of interest, and therefore lies in an interval of 3 m to perhaps as much as 200 m. The uncertainty in the measured value of range should not exceed 0.5 m. Since the lead vehicle is expected to be traveling in the same direction as the host vehicle, the maximum relative velocity of the vehicles can be expected to lie in the interval of +160 km/h to –160 km/h. If the permissible uncertainty in the measurement of this relative speed is 1% at the maximum speed, a speed measurement error of up to 1.5 km/h is acceptable. If the information must be refreshed upon a change of 2 m in the range even at the highest speed, the radar needs to update the range and speed information once every 50 ms.

15.7.2 Environmental Complexity

The conditions under which the forward-looking radars operate is made complex by four features of the roadway environment. First, the radar must operate under harsh ambient conditions with respect to temperature, humidity, mechanical vibration and acceleration, and electromagnetic interference. Second, it must operate in inclement weather, caused by rain, sleet, snow, hail, fog, dust storm, and smoke. Third, the roadside scene is rapidly changing, and includes a large number of both potentially hazardous and harmless objects, some moving and others stationary, and the radar must identify and discriminate between them in order to maintain a low probability of false alarm. Fourth, due to road curvature and steering, and tangential components of vehicle velocities, the discrimination between in-path and off-path objects becomes more involved, and requires computationally intensive prediction algorithms. Therefore, features like real-time signal processing, robust algorithms, and built-in fault detection are essential.

15.7.3 Frequency Selection

Although some of the earlier radars were designed for operation around 16 and 35 GHz, virtually all current developments employ V-band frequencies for forward-looking radars. Within the V-band, several different frequencies have been used in the past decade, including 77 GHz for U.S. and European systems, and 60 GHz in some Japanese systems. Three factors dictate the choice of millimeter wave range for this application. First, the range resolution of the radar is governed primarily by the bandwidth of the transmitted signal, and a resolution of the order of 1 m requires a minimum bandwidth of around 150 MHz. Such a large bandwidth is not available below the millimeter wave frequency range. Second, for a given performance level (such as antenna directivity), a higher frequency permits the use of smaller antennas and circuits, thereby reducing size, weight, and cost of the radar. Third, the higher atmospheric absorption of the millimeter wave signals is not a concern for short-range applications like automotive radars. However, the spray from other vehicles can impact the visibility of vehicles on the roadway at frequencies that lie in the water absorption band.

15.7.4 Signal Modulation

Although several different types of radar signal modulations have been evaluated over the years as possible candidates for forward-looking applications, most developers believe the frequency-modulated continuous-wave (FMCW) radar is the best overall choice. Some of the reasons for this choice of transmitted signal modulation include ease of modulation, simpler Doppler information extraction, higher accuracy for the short-range targets, and lower power rating for the active device generating the microwave signal for a given average transmitted power. In an FMCW radar, the frequency of the transmitted signal is changed with time in a prescribed manner (such as a linear ramp or a triangular variation), and the difference between the transmitted and returned signal frequencies is a measure of the range of the reflecting target.

15.7.5 Antenna Performance Requirements

Both the ACC and the collision anticipation radars require an antenna that meets several demanding specifications:

1. The size of the antenna should be small for cost and vehicle styling reasons.
2. The antenna beam should be narrow enough in azimuth that it can resolve objects in the driving lane from those in nearby lanes or adjacent to the roadway, as well as in elevation so as to distinguished on-road objects from overhead signs and bridges.
3. The side lobes of the antenna should be sufficiently low that small objects (like motorcycles) in the same lane are not masked by large objects (like trucks) in neighboring lanes.
4. The antenna should preferably be planar for ruggedness and ease of integration.

15.7.6 Choice of Antennas

Given the antenna beamwidth requirements and the limited physical dimensions permissible, some form of antenna scanning is needed for monitoring the relevant space, while at the same time maintaining the spatial discrimination. At least three choices are available for scanning. The first, and possibly the most versatile and powerful option is electronic scanning, using a phased array; with the presently available technologies, the cost of this option is prohibitive. Second, mechanically steered antennas have been developed that use reflectors. Third, synthetic aperture antenna techniques have been used that allow for sophisticated signal processing and information acquisition.

15.7.7 Signal Processing Needs

One of the most challenging aspects of automotive radar design, on which work presently continues, is the processing of the returned radar signals. Sophisticated signal processing algorithms have been employed to achieve many of the characteristics desired in radar performance, including the following:

1. Resolving small and large objects at different ranges and velocities;
2. Rejecting reflections from stationary objects;
3. Rejecting reflections from vehicles traveling in opposite directions;
4. Extracting target range despite road curvature;
5. Obtaining high performance despite low-cost components with low performance.

This software must be capable of simultaneously separating and tracking a dozen or more different targets within the field of view.

15.7.8 The Radar Assembly

The need for large volume production and cost considerations dictate the materials, technologies, and processes usable in the radar. The complete radar can be conceptually subdivided into two parts: the RF part including the antenna, transmitter, and receiver; and the baseband part, consisting of signal processing, power supply, microprocessors, displays, cables, and packaging and housing. The baseband part, as is typical of other automotive electronics, employs silicon chips, automated assembly methods such as flip-chip, and lightweight plastic housing. The cost of the front-end RF module at millimeter wave frequencies is usually minimized by use of a single hybrid assembly for the entire module, low-power Gunn devices as millimeter wave sources, silicon Schottky barrier diodes as mixers, and metalized injection-molded plastic waveguide cavities. More recently, as the cost of monolithic millimeter wave integrated circuit (MMIC) chips has decreased, a monolithic RF front end becomes cost effective, and can be integrated in a hybrid microstrip circuit. Chipsets for this purpose have been developed by a number of companies.

15.8 Other Possible Types of Automotive Radars

Several other types of radar architectures and types have been proposed for automotive applications, and have reached different stages of development. Some of the promising candidates are briefly summarized here.

1. *Noise Radar.* Given the large number of automotive radars that may be simultaneously present in a given environment, the need to minimize the likelihood of false alarms due to interference between them is important. Several different types of noise (or noise-modulated) radars have been advanced for this purpose. One proposed scheme employs a transmitted signal modulated by random noise, and a correlation receiver that determines the cross-correlation between the returned signal modulation and the transmitted signal modulation, to separate the returned signal from the noise due to the system, ambient, and other radars. In still other schemes, the transmitted signal can be broadband noise, or a CW signal modulated by a binary random signal.
2. *Micropower Impulse Radar (MIR).* Another class of radar that has been proposed for several automotive uses is the so-called micropower impulse radar (MIR). Its most distinguishing characteristic is that it transmits a very short pulse of electromagnetic energy, having a duration typically on the order of 0.1 ns, and a risetime measured in picoseconds. As a result, its spectrum occupies a bandwidth of several GHz, creating a so-called "ultra-wideband" (UWB) system. Consequently, to avoid interference problems, the transmitted power is kept very low, on the order of a microwatt. As a further consequence of that choice, it is necessary to integrate over a large number of pulses to improve the signal-to-noise ratio of the receiver; however, the pulse repetition rate can be random, so that multiple radars will not interfere with each other due to their distinctive pulse patterns. In addition, the receiver is gated in time, so that it receives radar signal echoes only over a narrow time window. The received signal is thus limited to echoes from targets lying at a pre-selected distance from the transmitter, allowing the echoes from smaller nearby targets to be distinguished in the presence of those from large faraway objects. Another consequence of the low power transmission is long battery life, and hence the small volume, weight, and low cost. Some consequences of ultra-wideband operation are high range resolution, substantial scattering cross section of targets at any observation angle, and signal penetration through dielectric materials.
3. *Interferometric Radar.* A radar system in which signals reflected from a target are simultaneously received by two physically separated antennas and are processed to determine the phase difference between them, can be used to estimate the target range over short distances. Such systems can employ a CW signal and a two-channel digital correlator, making them very simple, and can distinguish between approaching and receding targets. Since the range cannot be determined unambiguously from a known phase difference, the system is limited in its capability without signal processing to extract additional information from the returned signals.
4. *Cooperative Radar Systems.* Targets that modify the returned signal in known ways (e.g., by modulation or frequency translation, to allow it to be easily detected and distinguished against

other signals) are called cooperative targets. Radar systems with significantly higher performance capabilities can be developed if, as a result of common industry standards or regulatory directives, the vehicles incorporate appropriate means for enhancing returned radar signal, or carry a radar-interrogable distinguishing code, much like a license plate. In such a system, the ease of vehicle tracking can greatly improve the reliability and robustness of the radar system. The principal limitation of such sensors is their inability to handle targets with or without damaged reflectors, beacons, or other cooperative mechanisms.

15.9 Future Developments

Some of the major areas of emphasis in the future development of automotive radars are as follows.

1. *Radar Chipsets and MMICs.* With the frequency allocation and markets more definite, many manufacturers have developed chipsets for automotive radar that include GaAs MMIC chips for the front-end RF assembly. The availability of volume manufacturing capability, ease of incorporating additional functionality, and resulting cost reductions make this a promising approach.
2. *Phased-Array Antennas.* With presently available technologies, phased-array antennas are not viable candidates for consideration in automotive radar applications due to their high cost. When low-cost phased-array antennas become commercially available, automotive radar can be expected to undergo a significant transformation and to attain a multifunction capability.
3. *Signal Processing Capability.* Digital processing of complex returned signals makes possible the characterization of more complex vehicular environments, in which large numbers of objects can be tracked. Improved algorithms, and the ability to carry out extensive real-time processing with inexpensive processor chips, will allow the radar to serve more sophisticated functions.
4. *Human Interface Enhancement.* The willingness of a driver to relinquish partial control of the vehicle to the radar depends only partly on attaining a high reliability, robustness, and low false-alarm probability. User acceptance will depend strongly on the quality of human interface through which the information gathered by an automotive radar is presented and utilized.

References

History

1. H. P. Groll and J. Detlefsen, History of automotive anticollision radars, *IEEE Aerospace and Electronic Systems Magazine*, 12, 8, 15–19, August 1997.
2. D. M. Grimes and T. O. Jones, Automobile radar: A brief review, *Proc. IEEE*, 62, 6, 804–822, June 1974.
3. M. S. Gupta, et al., Noise considerations in self-mixing IMPATT-diode oscillators for short-range Doppler radar applications, *IEEE Transactions on Microwave Theory and Techniques*, MTT-22, 1, 37–43, January 1974.

Speed Measuring Radar

1. P. Heide, et al., A high performance multisensor system for precise vehicle ground speed measurement, *Microwave Journal*, 7, 22–34, July 1996.
2. P. Descamps, et al., Microwave Doppler sensors for terrestrial transportation applications, *IEEE Trans. Vehicular Technology*, 46, 1, 220–228, February 1997.
3. P. M. Schumacher, Signal processing enhances Doppler radar performance, *Microwaves & RF*, 30, 7, 79–86, June 1991.

Obstacle-Detection Radar

1. J. C. Reed, Side zone automotive radar, *IEEE Aerospace and Electronic Systems Magazine*, 13, 6, 3–7, June 1998.

Adaptive Cruise Control Radar

1. P. L. Lowbridge, Low cost millimeter-wave radar systems for intelligent vehicle cruise control applications, *Microwave Journal*, 38, 10, 20–33, October 1995.
2. L. H. Eriksson and S. Broden, High performance automotive radar, *Microwave Journal*, 39, 10, 24–38, October 1996.
3. M. E. Russel, et al., Millimeter wave radar sensor for automotive intelligent cruise control (ICC), *IEEE Transactions on Microwave Theory and Techniques*, 45, 12, 2444–2453, December 1997.
4. A. G. Stove, Automobile radar, *Applied Microwaves*, 5, 2, 102–115, Spring 1993.

Collision Anticipation Radar

1. C. D. Wang and S. Halajian, Processing methods enhance collision-warning systems, *Microwaves and RF*, 36, 3, 72–82, March 1997.

RF Front End for Forward-Looking Radars

1. W. H. Haydl, Single-chip coplanar 94-GHz FMCW radar sensors, *IEEE Microwave and Guided Wave Letters*, 9, 2, 73–75, February 1999.
2. D. D. Li, S. C. Luo, and R. M. Knox, Millimeter-wave FMCW radar transceiver/antenna for automotive applications, *Applied Microwaves and Wireless*, 11, 6, 58–68, June 1999.

Other Possible Types of Automotive Radars

1. S. Azevedo and T.E. McEwan, Micropower impulse radar: A new pocket-sized radar that operates up to several years on AA batteries, *IEEE Potentials*, 16, 2, 15–20, April–May 1997.
2. G. Heftman, Macroapplications seen for micro radar, *Microwaves and RF*, 39, 5, 39–40, May 2000.
3. B. M. Horton, Noise modulated distance measuring systems, *Proc. IRE*, 47, 5, 821–828, May 1959.
4. I. P. Theron, et al., Ultrawide-band noise radar in the VHF/UHF band, *IEEE Transactions on Antennas and Propagation*, 47, 6, 1080–1084, 1999.
5. A. Benlarbi and Y. Leroy, A novel short-range anticollision radar, *Microwave and Optical Technology Letters*, 7, 11, 519–521, August 5, 1994.
6. F. Sterzer, Electronic license plate for motor vehicles, *RCA Review*, 35, 2, 167–175, June 1974.

Future Developments

1. J.-E. Mueller, GaAs HEMT MMIC chip set for automotive radar systems fabricated by optical stepper lithography, *IEEE Journal of Solid-State Circuits*, 32, 9, 1342–1349, September 1997.
2. L. Verweyen, Coplanar transceiver MMIC for 77 GHz automotive applications based on a nonlinear design approach, *1998 IEEE Radio Frequency Integrated Circuits Symposium Digest*, Baltimore, MD, June 1998, 33–36.
3. T. Shimura, et al., 76 GHz Flip-Chip MMICs for automotive radars, *1998 IEEE Radio Frequency Integrated Circuits (RFIC) Symposium Digest*, Baltimore, MD, June 1998, 25–28.

16

New Frontiers for Radio Frequency (RF)/ Microwaves in Therapeutic Medicine

Arye Rosen
Drexel University

Harel D. Rosen
UMDNJ/Robert Wood Johnson Medical School

Stuart D. Edwards
Conway Stuart Medical, Inc.

The use of radio frequency (RF)/microwaves in therapeutic medicine has increased dramatically in the last few years. RF and microwave therapies for cancer in humans are well documented, and are presently used in many cancer centers. RF treatments for supraventricular arrhythmias, and more recently for ventricular tachycardia (VT) are currently employed by major hospitals. RF/microwaves are also used in human subjects for the treatment of benign prostatic hyperplasia (BPH), and have gained international approval, including approval by the United States Food and Drug Administration (FDA). In the last two years, several otolaryngological centers in the United States have been utilizing RF to treat upper airway obstruction and alleviate sleep apnea. Despite these advances, considerable efforts are being expended on the improvement of such medical device technology. Furthermore, new modalities such as microwave-aided liposuction, tissue anastomoses in conjunction with microwave irradiation in future endoscopic surgery, RF/microwaves for the enhancement of drug absorption, and microwave septic wound treatment are continually being researched.

0-8493-1732-0/03/$0.00+$1.50

16.1 RF/Microwave Interaction with Biological Tissue

Definitions: In this chapter, we detail two types of thermal therapies: RF and microwaves. We define RF as frequencies in the range between hundreds of kHz to a few MHz, and microwaves as those in the range of hundreds of MHz to about 10 GHz.

16.1.1 RF Energy[1-7]

The history of the effect of RF current on tissue began in the 1920s, with the work of W.T. Bovie who studied the use of RF for cutting and coagulating. The first practical and commercially available RF lesion generators for neurosurgery were built in the early 1950s by S. Aranow and B.J. Cosman at around 1 MHz.[1,2] The controlled brain lesions made had smooth borders, an immediate improvement over those obtained with DC. As far as the choice of frequency, Alberts et al.[3] have shown that frequencies of up to 250 kHz have stimulating effects on the brain. Thus, RF above 250 KHz was indicated.[4–7] The RF generator is a source of RF voltage between two electrodes. When the generator is connected to the tissue to be ablated, current will flow through the tissue between the active and dispersive electrodes. The active electrode is connected to the tissue volume where the ablation is to be made, and the dispersive electrode is a large-area electrode forcing a reduction in current density in order to prevent tissue heating. The total RF current, I_{RF} is a function of the applied voltage between the electrodes connected to the tissue and the tissue conductance. The heating distribution is a function of the current density. The greatest heating takes place in regions of the highest current density, J. The mechanism for tissue heating in the RF range of hundreds of kHz is primarily ionic. The electrical field produces a driving force on the ions in the tissue electrolytes, causing the ions to vibrate at the frequency of operation. The current density is $J = \sigma E$, where σ is the tissue conductivity. The ionic motion and friction heats the tissue, with a heating power per unit volume equal to J^2/σ. The equilibrium temperature distribution, as a function of distance from the electrode tip, is related to the power deposition, the thermal conductivity of the target tissue, and the heat sink, which is a function of blood circulation. The lesion size is, in turn, a function of the volume temperature. Many theoretical models to determine tissue ablation volume as a function of tissue type are available, but none is as good as actual data.

16.1.2 Microwave Energy

The need for accurate data on permittivity at microwaves and millimeter waves has long been recognized[8] and since 1980 many papers have appeared giving fairly extensive coverage of data up to 18 GHz.[9-13] The most recent tabulations of complex permittivity of various biological tissues are reported by Duck[14] and Gabriel et al.[15] Many of the applications and recent advances in the knowledge of the dielectric properties of tissues have been reviewed in the literature.[9,12,16,17] Since 1950, efforts have been directed toward characterization of a variety of tissues at microwave frequencies. Among many reported works in the literature Gabriel et al.[18] provide detailed measurements of a variety of tissues up to 20 GHz; they also fit the measured results to a Cole–Cole model with multiple relaxation time constants.[19] Knowledge of the dielectric properties of biological tissues and the basic physical properties of water in tissue at microwave frequencies is essential in order to predict the interaction between field and tissue, which in turn, provides the basis for some of the thermal applications of microwaves described in this chapter.

For sinusoidal fields of frequency f, the permittivity and conductivity are conveniently represented by a single parameter, the complex permittivity ε^*,

$$\varepsilon^* = \varepsilon' - j\varepsilon'' \tag{16.1}$$

where

$$\varepsilon'' = \frac{\sigma}{w\varepsilon_0} \tag{16.2}$$

and
$$w = 2\pi f \tag{16.3}$$

The real part ε' is referred to as the "relative permittivity" and is related to the energy storage. The imaginary part ε'' is called the "dielectric loss," and corresponds to the power absorption in terms of electromagnetic field interaction with matter; the former influences the phase of the transmitter wave, whereas the latter impacts its amplitude.

When a constant voltage is suddenly impressed on a system initially at equilibrium, for simple systems, the resulting response is usually found to be an exponential function of time. This response, for example, may be the charge buildup at an interface between two different dielectrics, or the alignment of dipoles with an applied electric field. When the voltage is removed, the system relaxes exponentially to its original state. In the general case of an alternating voltage of field, it can be shown[20–22] that the complex permittivity of such simple systems varies with frequency, and can be expressed in the form,

$$\varepsilon^*(w) = \varepsilon_\infty + \frac{\varepsilon_S - \varepsilon_\infty}{1 + j\omega\tau} \tag{16.4}$$

where τ is the time constant of the exponential relaxation process, ε_∞ is the permittivity at $\omega \ll 1/\tau$, and ε_s is the permittivity at $\omega \gg 1/\tau$. This is the Debye dispersion equation. The characteristic frequency f_c is defined as,

$$f_c = 1/2\pi\tau \tag{16.5}$$

By separating the Debye equation into real and imaginary parts,

$$\varepsilon' = \varepsilon_\infty + \frac{\varepsilon_S - \varepsilon_\infty}{1 + \left(\frac{f}{f_c}\right)^2} \tag{16.6}$$

$$\varepsilon'' = \frac{(\varepsilon_S - \varepsilon_\infty) f/f_c}{1 + \left(\frac{f}{f_c}\right)^2} \tag{16.7}$$

or

$$\sigma_d = \frac{(\varepsilon_S - \varepsilon_\infty) 2\pi\varepsilon_0 f^2}{f_c\left(1 + \left(\frac{f}{f_c}\right)^2\right)} \tag{16.8}$$

where the subscript d denotes the conductivity due to a Debye relaxation process. The measured conductivity, σ_m, will be higher if there are other loss mechanisms in the material, i.e.,

$$\sigma_m = \sigma_d + \sigma_S \tag{16.9}$$

where σ_s may be the conductivity due to ions and any other contributions at frequencies well below f_c.

If for $f \gg f_c$ we denote σ_d by σ_∞, then

$$(\sigma_\infty - \sigma_S) = 2\pi f_c (\varepsilon_S - \varepsilon_\infty)\varepsilon_0 \tag{16.10}$$

which states that the total change in conductivity is proportional to the total change in permittivity and to the characteristic frequency.

The dielectric properties of most materials are not exactly described by the Debye equation. The dielectric properties can then often be approximated empirically by the Cole–Cole equation[19,22]

$$\varepsilon^* = \varepsilon_\infty + \frac{\varepsilon_S - \varepsilon_\infty}{(1 + jw\tau)^{1-\alpha}} \tag{16.11}$$

It can be shown that this equation is valid for a distribution of relaxation times about a mean value. The Cole–Cole parameter α ranges from 0 to 1 and is an indication of the spread of relaxation times (for $\alpha = 0$, the Cole–Cole equation reduces to the Debye equation).

Depth of penetration (δ) of energy into tissue is defined as the distance in which the power density of a plane wave is reduced by the factor e^{-2}, which numerically is 0.135. Since the power density diminishes as e^{-aL} as the microwave energy travels into the lossy material, the attenuation factor (a) is inversely related to the depth of penetration: ($a = 1/\delta$). The generally accepted value for the depth of penetration of 2450-MHz energy into muscle tissue is 17 mm, and the corresponding attenuation constant is 0.118 for distances in millimeters. For a non-expanding plane wave front, the reduction of power density as it penetrates muscle tissue is shown in Fig. 16.1[71] (see also Table 16.1[70,74]). Figure 16.1 shows, for example, that 11% of the total input power is absorbed in the first millimeter of penetration, and that a total of 21% of the input is converted to heat in the first 2 mm of the tissue. Microwave antennas that are utilized in therapeutic medicine, however, are in the near field. The penetration depth is considerably lower depending on the antenna type).[23]

Previous studies[24] have shown that heat-induced damage to biological tissue is dependent on both the temperature and its duration. The temperature threshold for damage rises as the duration of exposure is shortened.

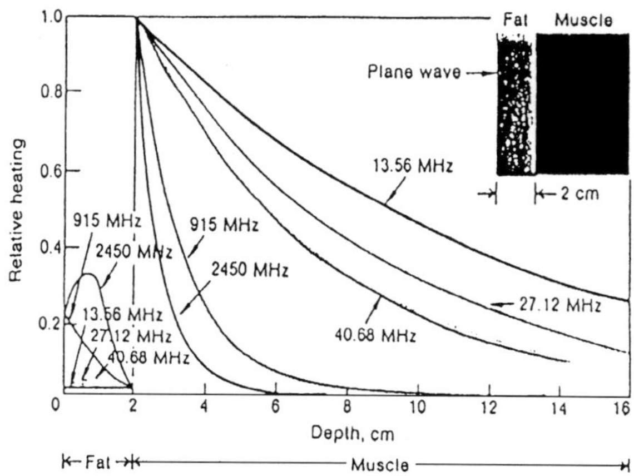

FIGURE 16.1 Calculated relative heating in fat and muscle as a function of distance for five frequencies. (After Paglione.[74])

TABLE 16.1 Relative Permittivity and Conductivity of Biological Media at Microwave Frequencies[70,74]

Frequency (MHz)	Wavelength (cm)	High Water Content Media		Low Water Content Media	
		c	σ (S/m)	c	σ (mS/m)
10	3,000	160	0.625	—	—
100	300	71.7	0.889	7.5	19.1–75.9
300	100	54	1.37	5.7	31.6–107
915	32.8	51	1.60	5.6	55.6–147
2,450	12.2	47	2.21	5.5	96.4–213
3,000	10	46	2.26	5.5	110–234
5,000	6	44	3.92	5.5	162–309
10,000	3	39.9	10.3	4.5	324–549

16.1.3 Test Fixture Structures for Biological Tissue Characterization[25]

A number of techniques have been established for microwave characterization of biological tissues. These techniques are subdivided in transverse electric and magnetic (TEM) transmission lines (i.e., coaxial lines) and non-TEM structures. The majority of published work deals with coaxial transmission lines either as a dielectric loaded[26,27] or an open-ended[28] coaxial line. In both approaches, the change in the terminating impedance causes change in the input reflection coefficient of the line. Different complex permittivity of the tissue under test causes change in the capacitance and conductance of termination, hence impacting amplitude and phase of the reflected wave. This technique is popular and is extensively developed by Burdette et al.[28] and Stuchly et al.[29]

Dielectric loaded waveguide structures such as circular or rectangular cross-section metallic waveguides are the second most popular structures for tissue characterization. Steel et al.[30,31] reported a technique for characterization of liquids and solids. For liquids measurements the sample is contained in a length of waveguide and a moving short circuit enables the liquid thickness to be varied. A microwave signal is applied to the sample, the modulus of the reflected signal is recorded as a function of sample length, and a least-squares curve-fitting analysis of data enables various parameters to be obtained. For measuring solid samples, an automated slotted line is used to record the standing wave ratio in front of the sample, the latter terminated by a short circuit.

The above techniques all use the reflecting wave from a terminating impedance and the transmission line theory to extract the electrical parameters of the tissues under test. On the other hand, methods exist that use tissue samples to change the resonance frequency and quality factor of a cavity resonator.[32] More accurate results can be achieved by using simpler setups. However, the resonance methods are only applicable to discrete frequency points corresponding to the resonance frequencies of the modes of interest. For instance, Land and Campbell[32] present a technique that uses simple formulas that relate the complex permittivity of a small piece of tissue to the change in resonance frequency and quality factor of a cylindrical cavity. The cavity is filled with PTFE and resonates for TM_{010} mode and has a diameter of 50.8 mm and a length of 7.6 mm. Three 1.5-mm diameter sample holes are provided in the cavity. The cavity resonates at 3.2 GHz. The microwave setup is very simple (i.e., signal generator, frequency meter, directional coupler, attenuator, diode detector, and voltmeter are used in the experiment).

16.1.4 Tissue Characterization through Reflection Measurements[25]

In order to characterize various biological tissues, three distinctive methods have been reported. The first method is based on voltage standing wave ratio measurements using slotted line waveguides.[31] The second approach is based on an impedance analyzer and is suitable for microwave frequencies.

Finally, the most popular approach in the last 20 years has been the use of a network analyzer.[28,33-38] With the advent of the automatic network analyzer (ANA) (e.g., Hewlett-Packard's HP8510, HP8753, HP8720), accurate measurements of scattering parameters have been extended to frequencies beyond

10 GHz (since 1984). The advantage of this technique is fast and accurate measurements of coaxial-based structures. The majority of attempts to accurately characterize complex permittivity of biological tissues in the last 10 years are based on these families of ANA.[18,38,39]

16.1.5 Microwave Antenna in Therapeutic Medicine: Issues

Biomedical antenna designs have typically addressed the applications of microwave hyperthermia for the treatment of malignant tumors, microwave catheter ablation for the treatment of cardiac arrhythmia, microwave balloon angioplasty, or microwave-assisted liposuction. The analytical basis for much of this work has been based on the lossy transmission line analysis developed by King et al.,[40,41] which allowed the calculation of input impedance and near fields for simple antenna geometries. Several researchers have refined this work to provide improved accuracy or wider applicability. For example, Iskander and Tumeh developed an iterative approach to designing multi-sectional antennas based on an improved King method[42] and have used this approach to compare different antennas.[43] Debicki and Astrahan developed correction factors to allow accurate modeling of the input impedance for electrically small multi-section antennas,[44] and Su and Wu have refined the King approach to determine the input impedance characteristics of coaxial slot antennas.[45] Casey and Bansal used a different approach than King to compute near fields of an insulated antenna using direct numerical integration of a surface integral.[46]

A problem inherent in many biomedical antenna designs is the effect of heating along the transmission line due to current flow on the outer conductor of the coaxial transmission line. In most applications this effect is undesirable since thermal energy is being delivered to healthy tissue outside the intended treatment area. Moreover, magnitude of this effect is a strong function of the insertion depth of the antenna. Similarly, the antenna input impedance also varies with insertion depth. Hurter et al.[67] proposed the use of a sleeve Balun to present a high impedance to the current on the outer conductor, thus concentrating the microwave energy at the antenna tip. Temperature profile measurements made in a phantom using a fiber-optic temperature probe clearly show improved localization of thermal energy delivery.

16.2 RF/Microwaves in Therapeutic Medicine

16.2.1 RF/Microwave Ablation for the Treatment of Cardiac Arrhythmias[47,48]

Cardiac arrhythmias can result from a variety of clinical conditions, but at their root is an abnormal focus, or pathway, of electrical activity. Abnormal sources of electrical activity most commonly occur at or above the AV node, and are thus deemed supraventricular tachyarrhythmias. Alternatively, abnormal ventricular foci cause ventricular tachycardia. The presence of abnormal conduction pathways can also result in an uncontrolled cycling of electrical activity resulting from retrograde signal conduction through the myocardium (reentry tachyarrhythmias). Reentry can occur within the AV node (AVNRT), or via accessory conduction pathways (AP). Regardless of the specific etiology, once the source of the arrhythmia has been identified, destruction of the abnormal cardiac tissue is curative. The goal of ablation is to modify the electrical system of the heart by converting electrically active cardiac tissue to electrically inactive scar tissue. The scar or lesion that forms then blocks the focus or accessory pathway and prevents the tachycardia. Various energy forms have been used to create such localized tissue injury, including direct current (DC), radiofrequency (RF), and microwave energy (Table 16.2).[47,48]

The clinical use of DC ablation dates back to 1982. An electrode catheter is placed at the desired location, and a DC shock is applied. Although complete ablation occurs in up to 65% of patients, DC ablation is fraught with complications. Hypotension, perforation, cardiac tamponade, embolization, pericarditis, and ventricular tachyarrhythmias have been reported in as many as 10% of patients. Mortality associated with DC ablation may be as high as 5% in some patient groups. RF ablation was developed with the hope of decreasing the risks associated with DC application. In RF ablation, lesion formation results from resistive tissue heating at the point of contact with the RF electrode (Fig. 16.2). This heating is thought to lead to coagulation necrosis and permanent tissue damage. If there is poor tissue contact,

TABLE 16.2 Energy Sources for Catheter Ablation[47]

	Direct Current	Radio Frequency	Microwave
Wave form	Monophasic, damped sinusoidal	Continuous unmodulated sinusoidal	N/A[a]
Frequency	DC	550–750 kHz	915, 2450 MHz
Voltage V	2000–3000 V	<100 V	N/A
Mechanism of injury	Passive heating, barotrauma, electric field effects	Resistive heating	Radiant heating
Sparking, barotrauma	Yes	No	No
General anesthesia	Yes	No	No
Lesion size	Moderate	Small	Unknown
Control of injury	Low	High	High

[a] N/A = data not available.

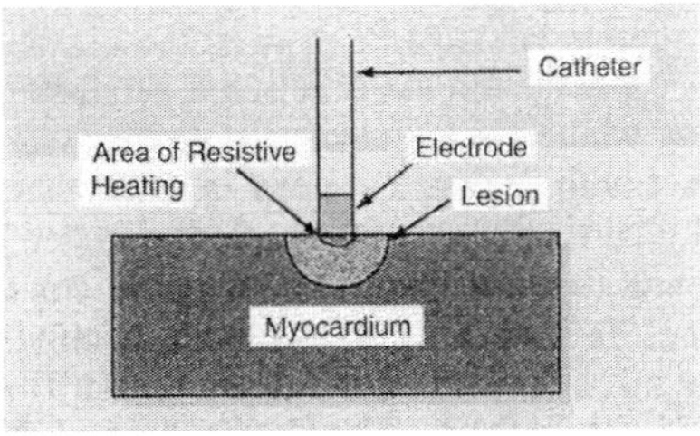

FIGURE 16.2 Mechanism of RF ablation. When RF current is delivered to the tip of a catheter electrode, resistive heating occurs along a small rim of tissue in direct contact with the electrode. A lesion is created as heat conducts passively away from this zone and the surrounding myocardium is heated to a temperature where cell death occurs (~50°C). Lesion size is therefore a function of the size of the electrode and the resulting temperature at the electrode–tissue interface.

RF current cannot be coupled to the underlying tissue, and the desired effect of tissue heating is lost. Overall success rates for RF ablation have been reported to be as high as 90% for AV junction ablation, and as high as 95% when applied to re-entry-mediated tachycardia. Furthermore, RF ablation has not been reported to result in serious side effects.

The first supraventricular tachycardias targeted for RF ablation were those associated with the Wolff–Parkinson–White syndrome. In this condition the anatomic basis for supraventricular tachycardias is an accessory connection or pathway that connects the atrium and ventricle outside the normal AV conduction pathway (Fig. 16.3). These accessory pathways cross between the atrium and the ventricle at the level of the mitral and tricuspid anulus. RF energy delivered to the mitral or tricuspid anulus either from the ventricular or atrial aspect can ablate these pathways (Fig. 16.4). Success rates of over 90% have been reported using either approach.

Recently, encouraging results in the treatment of ventricular arrhythmias occurring as a consequence of diffuse processes such as myocardial ischemia or infarction have been published. The search for ablation modalities capable of safely generating even larger lesions has spawned an interest in microwave ablation (Fig. 16.5). Unlike DC and RF techniques, which generate lesions of relatively limited size and penetration, microwave energy might allow for greater tissue penetration, and thus a greater volume of heating. Microwave ablation systems are currently being developed.

Microwave hyperthermia has been useful in radiation oncology for the treatment of various solid tumors.[49–51] The cardiac applications of this modality have only recently been explored. Microwave energy using either 915 or 2450 MHz has been studied in an attempt to enlarge myocardial lesions in catheter ablation.[52–54] Microwave energy is delivered down the length of a coaxial cable that terminates in an antenna capable of radiating the energy into tissue. Radiant energy will cause the water molecules in

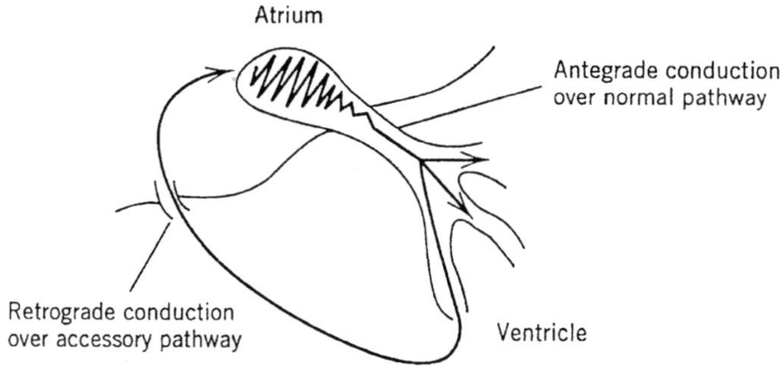

FIGURE 16.3 An arrhythmic circuit associated with the Wolff–Parkinson–White syndrome. In this syndrome, there is a connection between the atrium and ventricle outside the normal V nodal pathway (accessory pathway). A tachycardia circuit can develop if an impulse conducts antegrade (forward) via the normal V node pathway and is able to conduct retrograde (reverse) from the ventricle to the atrium via the accessory pathway. Catheter ablation successfully treats these arrhythmias because it interrupts accessory pathway conduction without interfering with normal AV nodal conduction.

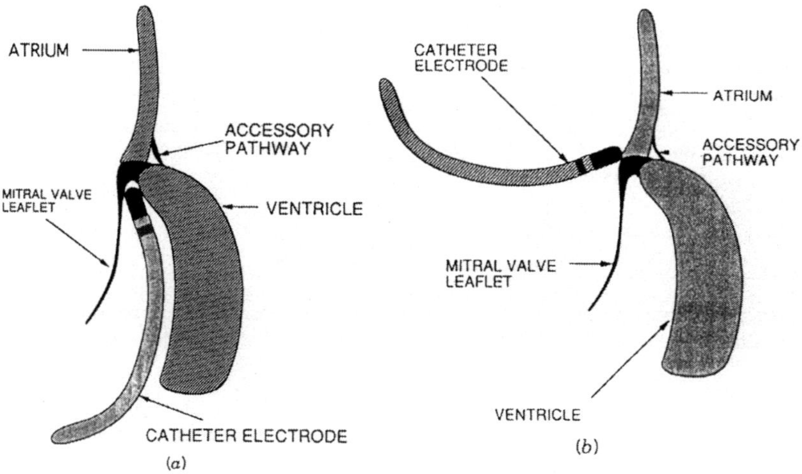

FIGURE 16.4 Diagrams of electrode positions used in RF catheter ablation of accessory pathways: (a) for the ventricular approach a catheter is passed retrograde across the aortic valve and positioned under the mitral leaflet; (b) for the atrial approach a catheter is passed across the interatrial septum (trans-septal catheterization) and positioned on top of the mitral valve leaflet. Electrical mapping confirms the site of the accessory pathway prior to the delivery of RF energy.

myocardial tissue to oscillate, producing tissue heating and cell death. The higher frequency of microwave energy allows for greater tissue penetration and theoretically a greater volume of heating than that possible with RF, which produces direct ohmic or resistive heating.

Wonnell and coworkers studied the effects of microwave energy for cardiac ablation using a helical antenna mounted on a coaxial cable (2.44 mm o.d.).[55] High-frequency current at 2450 MHz was delivered via the helical antenna into a tissue-equivalent phantom model. The temperature distribution profile was measured around the antenna as well as into surrounding volume (the depth of penetration). The volume of heating for the microwave catheter system was 11 times greater than that of an RF electrode catheter at the same surface temperature. In addition, the microwave catheter penetrated an area that was twice as large as that penetrated by the RF catheter. These data suggest that microwave energy will produce larger lesions than RF because a greater volume of tissue is being heated. An additional theoretical advantage of

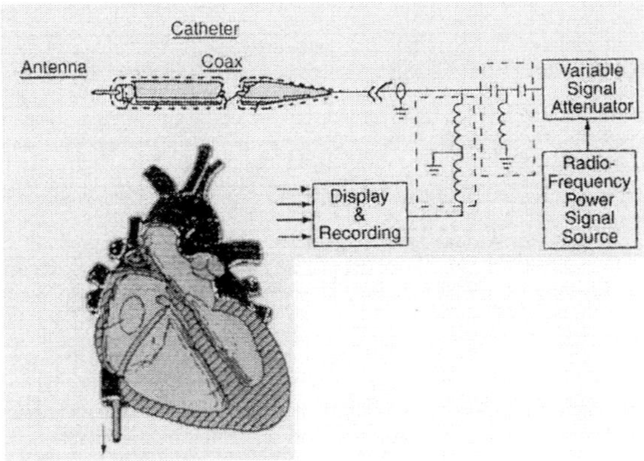

FIGURE 16.5 Microwave system used for myocardial tissue ablation.

the microwave system is that direct tissue contact is not crucial for tissue heating since heating occurs via radiation, and not via direct ohmic heating as seen with RF. Using this system, preliminary studies in six animals demonstrated that complete heart block could be achieved in all six animals by directing microwave energy (50 W at 2450 GHz for about 200 s) to the atrioventricular junction.

We evaluated helical and whip antenna designs in a tissue-equivalent phantom at 915 and 2450 MHz utilizing a coaxial cable (0.06 in o.d.).[56] All catheters were measured utilizing a network analyzer prior to placing them in the phantom model. Such analysis demonstrated the great variability in tuning of these microwave catheters. Microwave ablation catheters have suffered from imperfect tuning leading to inefficient radiation of energy. Consequently, there is generation of heat along the length of the catheter rather than radiation of energy into tissue. Little heating into tissue was observed in poorly tuned catheters. Such analysis underscores the critical importance of proper tuning of microwave catheters prior to any further studies.

A perfusion chamber containing a muscle-equivalent phantom was constructed and placed in a saline bath held at 37°C. The muscle-equivalent phantom consisted of TX150, polyethylene powder, NaCl, and water. Ablation catheters were placed on the surface of the phantom material. Temperature measurements were performed using a 12-channel Luxtron fiber-optic thermometry system. Probes were placed beneath the surface of the phantom. Saline at a constant temperature of 37°C was infused at a flow rate of 4 L/min across the surface of the phantom. This model simulates the heart where the phantom material has the dielectric properties of cardiac muscle and the saline properties of blood (Fig. 16.6).

Temperature curves were plotted from probes placed 1, 2.5, 5, and 7.5 mm from the point of maximal heating on the microwave catheter. Thermal profiling of these catheters demonstrated volume heating. Heating was proportional to power duration and to surface temperature. In addition to the volume heating, conductive heating was also present as a result of the increased temperature at the catheter–phantom interface.[56]

In vivo ablation using microwaves was performed on canine left ventricular myocardiums. A power of 80 W was delivered for a total of 5 min. Mean lesion size measured 435 ± 236 mm^3, which was similar in size to lesions created with small-tipped RF catheters. The microwave ablation catheters, as presently designed, were not capable of producing lesions larger than those produced by RF catheters.[56]

Practical problems remain to be solved before microwaves become a useful clinical energy source. These problems include (1) power loss in the coaxial cable, (2) resultant heating of the coaxial cable during power delivery that has led to a breakdown in the dielectric and catheter material, (3) inefficiency of the radiating antenna, and (4) lack of a unidirectional antenna that can radiate energy into tissue and not the circulating blood pool. At the present time microwave catheter systems are poorly efficient radiators of energy into cardiac tissue. These obstacles will have to be overcome before microwaves supplant radio frequency as the preferred energy source for cardiac ablation.

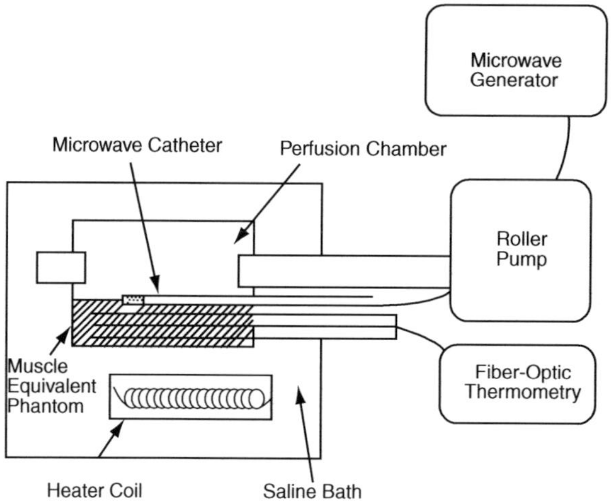

FIGURE 16.6 Flow-phantom model for cardiac ablation catheters.

16.2.2 RF/Microwave Treatment of BPH[57]

Benign prostatic hypertrophy (BPH) is an enlargement of the prostate gland which can lead to compression of the urethra, and thus cause urinary tract obstruction. The prostate gland is an organ at the base of the male bladder that surrounds the urethra and produces seminal fluid. Overgrowth of prostatic tissue leads to compression of the urethra. BPH is among the most common medical conditions affecting men over the age of 50. In fact, over 50% of men over 50 years of age have enlarged prostates. Symptoms of urinary tract obstruction (frequent urination, decreased urine flow, nocturia, dribbling, discomfort, pain) most commonly begin at 65 to 70 years of age.

Although drug therapy may be effective for patients with early stages of BPH, many men will need invasive intervention for relief of symptoms. Surgical excision of prostatic tissue has been the standard care for more advanced forms of BPH. Procedures such as prostatectomy and transurethral resection of the prostate, however, carry significant risks. To minimize hazards such as hemorrhage, coagulopathies, pulmonary emboli, bladder perforation, incontinence, infection, urethral stricture, retention of prostatic chips, infertility, and retrograde ejaculation, minimally invasive alternatives have been developed and are being investigated. Transurethral RF and microwave procedures are becoming promising alternatives to surgical intervention. The goal of therapy is to decrease the volume of prostatic tissue. RF Transurethral Needle Ablation (TUNA) (Fig. 16.7) involves the introduction of interstitial needle electrodes directly into prostatic tissue. This technology uses RF (460 kHz) with excellent control of the RF thermal energy applied to the tissue. The TUNA catheter used is 24.1 cm long and 21 French. Through the tip of the catheter, two needles (electrodes) oriented 40° apart can be deployed. The electrode–needles are shaped to facilitate passage through tissue. They are thin, and thus can be directed from the catheter through intervening tissue with a minimum of trauma to normal tissue. Each electrode–needle is enclosed within a longitudinally adjustable sleeve acting as a shield to prevent exposure of the tissue adjacent to the sleeve to the RF current, thus preserving the urethra by reducing the possibility of a rise in its temperature. The sleeve is also used to control the tissue interface, and therefore the ablation volume. Both the electrode–needle and the sleeve are locked into position. The TUNA catheter needle acts as the thermal electrode, and a grounding pad that is placed in back of the subject under treatment closes the RF circuit to the power supply (Fig. 16.8).

Thermocouples are located at the shield tip below each needle, and at the catheter bullet head (in order to record ablation temperature and prostatic urethral temperature), respectively. The RF unit (VidaMed, Inc.) includes an RF generator with the following readouts: RF power level, ablation time,

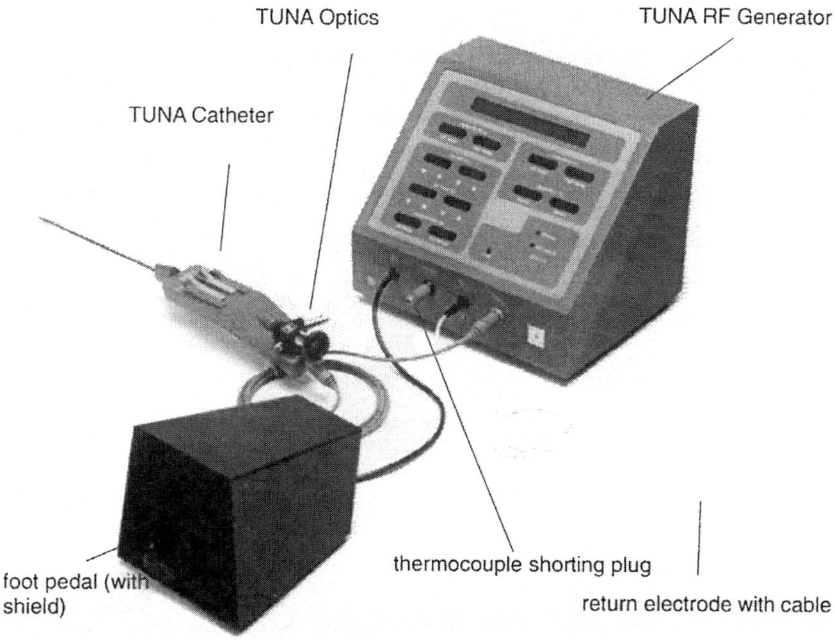

FIGURE 16.7 TUNA RF generator unit. (With permission from VidaMed, Inc.)

impedance, and six thermocouple readouts (Fig. 16.7). The TUNA catheter (Fig. 16.8a) includes direct fiber-optic vision, as well as provisions for introducing electrode–needles at various angles (Fig. 16.8b).

Transurethral Microwave Thermotherapy (TUMT) (Fig. 16.9a to c) has also shown promise as a therapeutic modality for the treatment of BPH. This technique uses a microwave delivery system housed within a transurethral catheter. Its goal is to selectively destroy prostatic tissue without damaging the urethral mucosa or structures surrounding the treatment area. At microwave frequencies, temperatures in the target tissue can be raised to as high as 45 to 70°C without damaging periprostatic tissue. TUMT is used routinely worldwide.

16.2.3 Microwave Balloon Catheter Techniques

16.2.3.1 Microwave Balloon Angioplasty[58,59]

Atherosclerosis, with its resultant occlusion of coronary blood flow, remains a leading cause of morbidity and mortality. For many patients with advanced disease, or in whom pharmacologic management has failed, percutaneous transluminal balloon angioplasty (PTCA) has offered an effective alternative to coronary bypass surgery. The efficacy of PTCA, however, has been limited by restenosis rates ranging from 17 to 47%, as well as by a risk of arterial dissection and/or thrombus formation. Furthermore, acute occlusion, resulting from elastic recoil at the angioplasty site, can occur in as many as 5% of patients undergoing PTCA. Such patients require emergency heart surgery. Microwave Balloon Angioplasty (MBA), the first microwave application in cardiology, was developed with the ultimate goal of decreasing both acute and long-term restenosis risks.

MBA, like PTCA, employs a balloon catheter that is advanced to the site of arterial stenosis. While PTCA uses only the pressure generated by balloon inflation to dilate the affected artery, MBA takes advantage of the volume heating properties of microwave emitters. In MBA, a microwave cable–antenna assembly is threaded through the catheter, with the antenna centered in the balloon portion of the catheter (Fig. 16.10). By heating the tissue as the balloon is inflated, it was hoped that a patent vessel would be created that would be resistant to both acute and chronic reocclusion. Early *in vivo* studies, at 2.45 GHz,

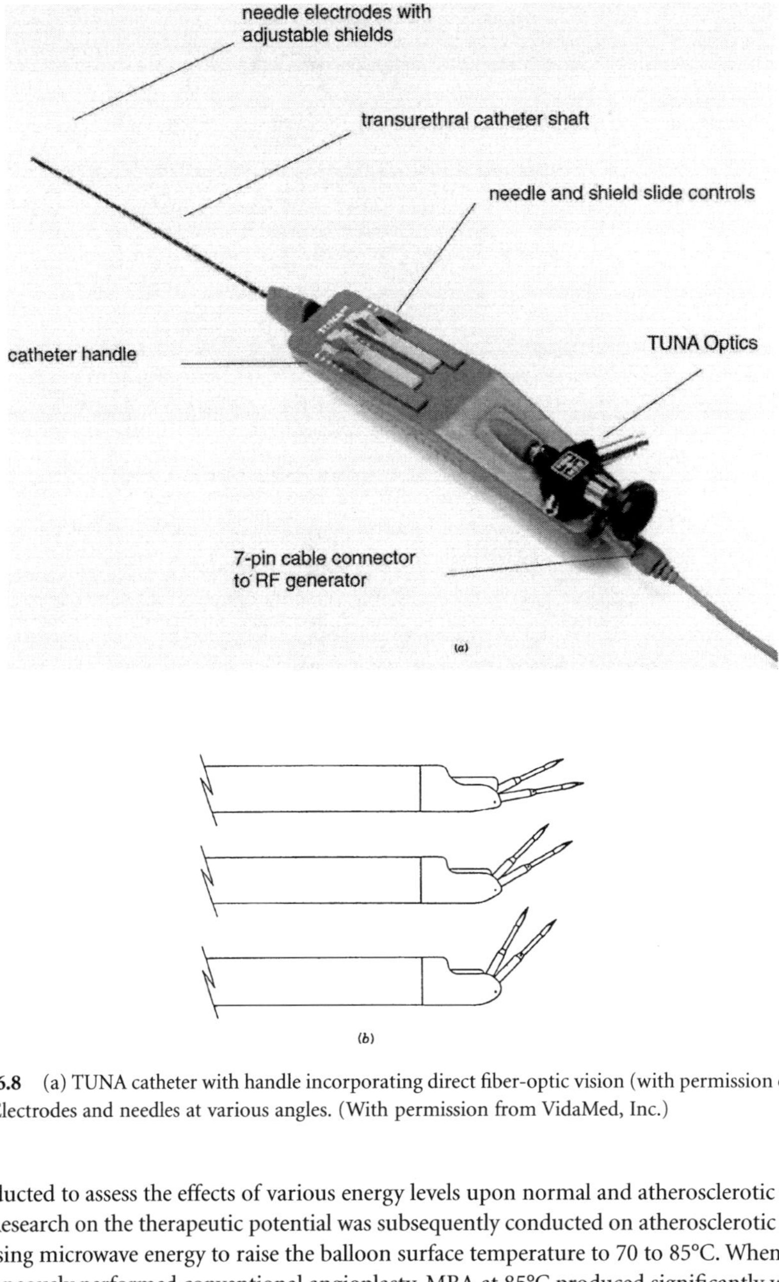

FIGURE 16.8 (a) TUNA catheter with handle incorporating direct fiber-optic vision (with permission of VidaMed, Inc.). (b) Electrodes and needles at various angles. (With permission from VidaMed, Inc.)

were conducted to assess the effects of various energy levels upon normal and atherosclerotic rabbit iliac arteries. Research on the therapeutic potential was subsequently conducted on atherosclerotic rabbit iliac arteries using microwave energy to raise the balloon surface temperature to 70 to 85°C. When compared to simultaneously performed conventional angioplasty, MBA at 85°C produced significantly wider luminal diameters, both immediately after angioplasty and 4 weeks after the procedure (Fig. 16.11). Further work, utilizing mongrel dogs with thrombin-induced coronary occlusion, has demonstrated the feasibility of MBA as a treatment modality for coronary thrombosis. MBA of such coronary thrombi in dogs resulted in patent vasculature with the added benefit of an organized and stabilized thrombus. Although the technique described was successful in animal studies, it has not yet found its way into clinical use. However, microwave balloon angioplasty has recently been suggested for applications in carotid stenosis and occlusions in peripheral circulation.

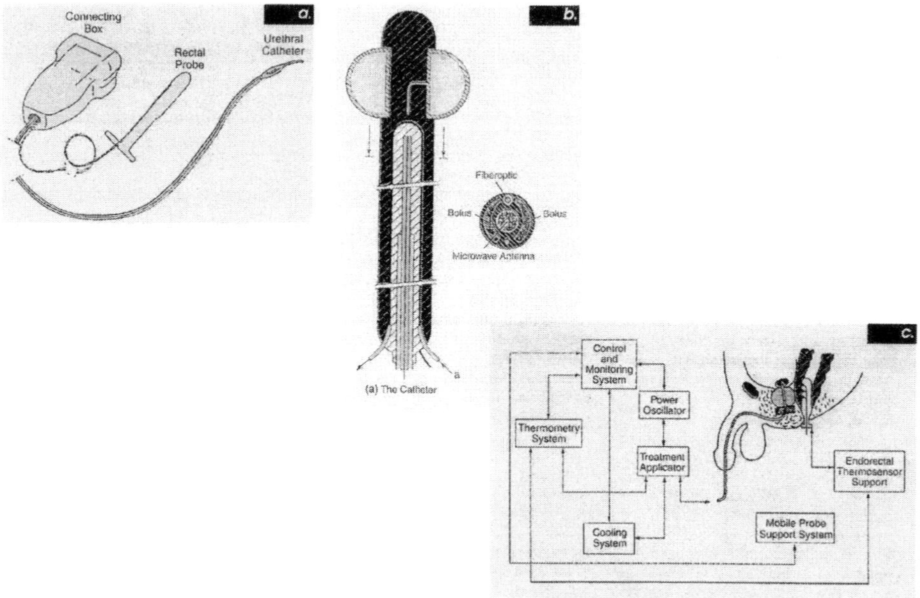

FIGURE 16.9 (a) Schematic representation of the treatment catheter; (b) cutaway view; (c) Prostatron treatment functional diagram. (With permission from Technemed Medical Systems.)

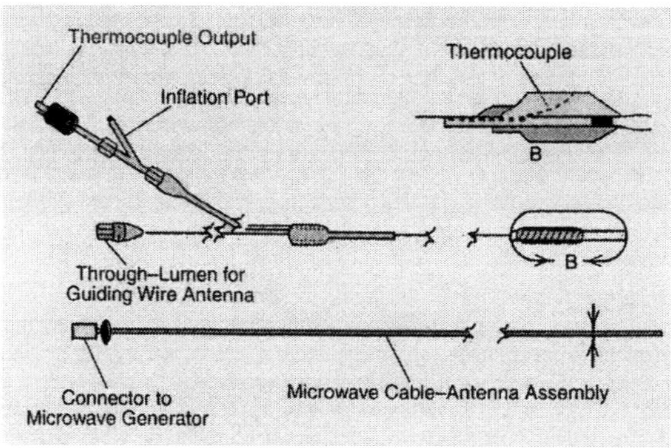

FIGURE 16.10 Microwave balloon angioplasty system.

16.2.3.2 Microwave Balloon Catheters in the Treatment of Benign Prostatic Hypertrophy[60]

Localized microwave hyperthermia has been used for more than a decade to treat cancer of the prostate and since 1985 to treat BPH. The initial hyperthermia treatments used microwave applicators that heated the prostate via the rectum, but today transurethral applicators are favored. Transurethral applicators are usually placed inside liquid-cooled catheters and the temperatures produced inside the treated prostate can be measured noninvasively with a microwave radiometer.

With balloon catheters it is possible to produce both high therapeutic temperatures throughout the prostate gland without causing tissue burning, and biological stents in the urethra in a single treatment session. Compared to conventional microwave catheters, the distances microwaves have to travel through the prostate to reach the outer surface of the gland are reduced by the use of balloon catheters, as is the radial spreading of the microwave energy (Fig. 16.12). Furthermore, compression of the gland tissues

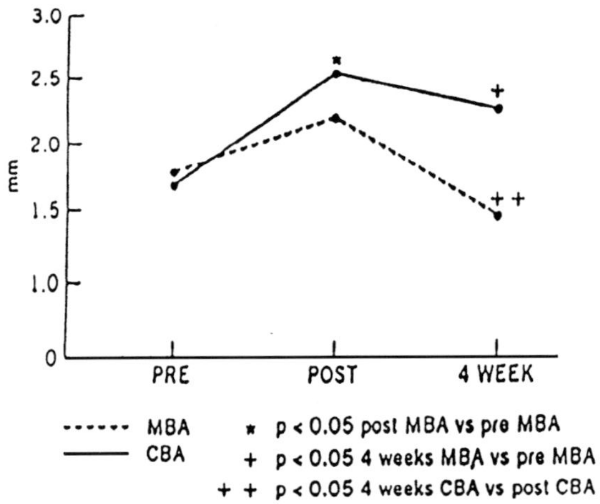

FIGURE 16.11 Microwave thermal angioplasty (85°C).

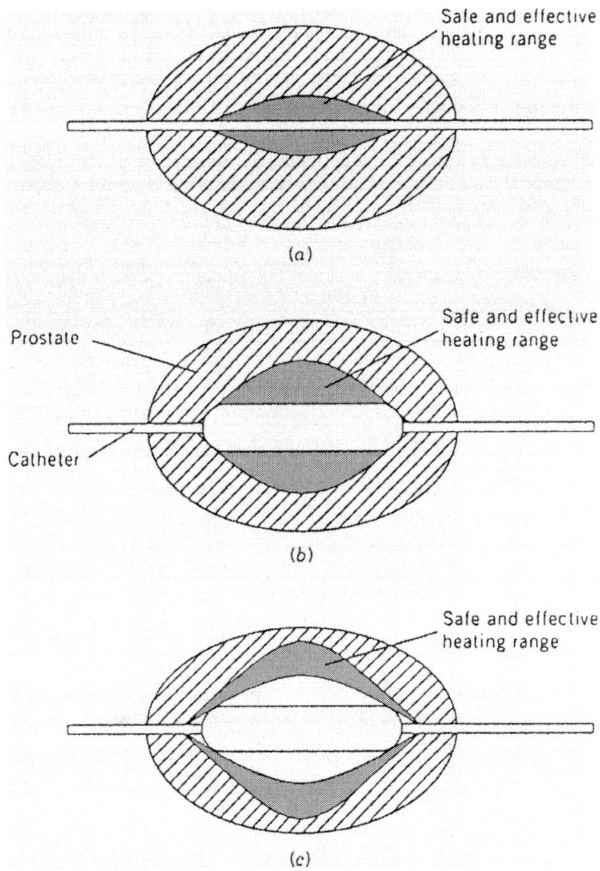

FIGURE 16.12 Safe and effective heating ranges in prostate glands with regular microwave catheters and with microwave balloon catheters. Microwave heating patterns: (a) with regular catheter; (b) with balloon catheter; (c) with balloon catheter and water cooling. (After Sterzer.[60])

reduces blood flow and its cooling effect within the gland. Also, since catheter balloons make excellent contact with the urethra, much better than do conventional catheters, the urethra is well cooled by the cooling liquid within the balloon, and is therefore well protected from thermal damage.

16.2.3.3 Microwave Balloon Catheters in the Treatment of Cancer[60]

Interstitial hyperthermia is usually combined with radiation therapy using radioactive seeds that are inserted into the tumor via the same tubing that is used for the hyperthermia (Fig. 16.13). A typical treatment sequence is brachytherapy (irradiation of the tumor with the seeds) followed by hyperthermia, followed again by brachytherapy. The interstitial hyperthermia enhances the efficacy of brachytherapy because (1) hyperthermia interferes with the repair of cells that have been sublethally damaged by the ionizing radiation, (2) cells in the S phase of the cell cycle and hypoxic tumor cells tend to be resistant to ionizing radiation but sensitive to heat, and (3) hyperthermia can be effective in oxygenating radiation-resistant hypoxic cells.

Interstitial arrays using conventional applicators are useful only for treating small tumors, because each interstitial applicator can heat and irradiate only a small volume of tissue, and the number of applicators that can be inserted into a tumor is limited because of their invasive nature. Interstitial applicators using balloons, on the other hand, can heat much larger volumes of tissues than can conventional interstitial applicators, making it possible to treat larger tumors. A catheter with a deflated balloon at its tip is inserted into the tumor volume to be heated, or where applicable, into a natural opening of the body such as the urethra, rectum, or vagina; the balloon is inflated, and radioactive seeds or a microwave antenna are inserted through the center lumen of the balloon catheter (Fig. 16.14).

16.2.4 RF in the Treatment of Obstructive Sleep Apnea[61,62,63]

Obstructive Sleep Apnea (OSA) is a disorder diagnosed when an individual's upper airway becomes intermittently blocked during sleep and breathing becomes interrupted. Approximately 20 million Americans are estimated to suffer from OSA, and over half of these are between the ages of 30 and 60 years. During sleep, there is a relaxation of the structures surrounding the pharynx/throat. Breathing becomes interrupted (apnea) when these anatomic structures relax in a position that occludes airflow. The most

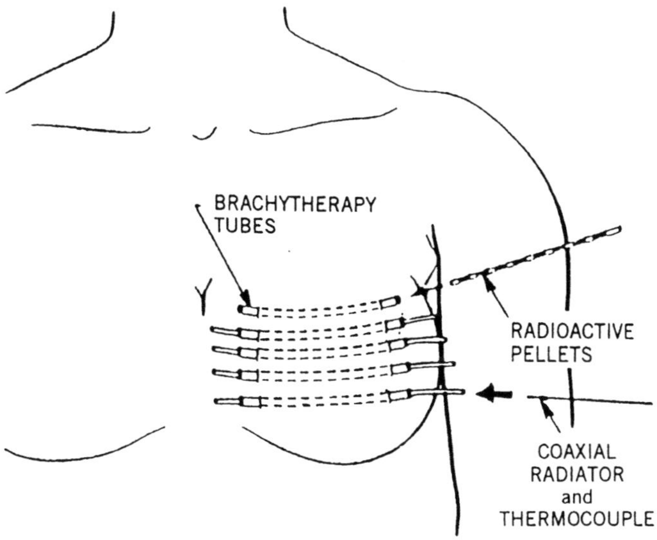

FIGURE 16.13 Interstitial hyperthermia combined with brachytherapy for treating breast cancer. (After Sterzer.[60])

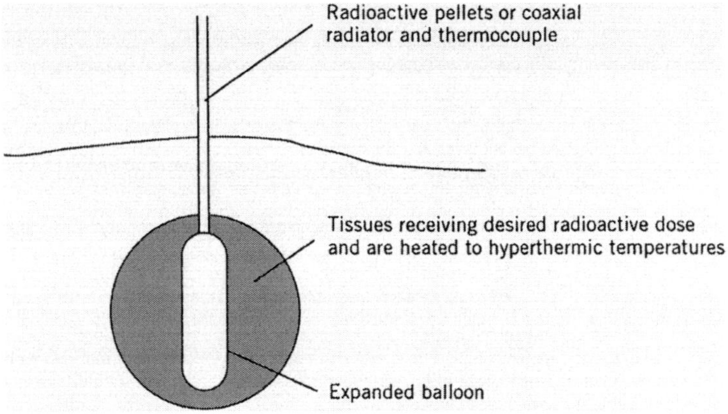

FIGURE 16.14 Interstitial microwave balloon catheter for combined localized hyperthermia and brachytherapy treatment of cancer. (After Sterzer.[60])

commonly involved structures include the soft palate, the base of the tongue, and the tonsils/adenoids. Enlarged turbinates within the nose can serve to further impede airflow.

OSA and its resultant interruption of normal sleep patterns have a wide range of clinical effects. Patients may experience daytime sleepiness, most hazardous while driving or during work. They may also exhibit personality changes, difficulty concentrating, memory difficulties, headaches, or sexual dysfunction. Sleep apnea is also associated with increased rates of systemic and pulmonary hypertension, stroke, heart failure, and myocardial infarction.

Treatment depends on the severity and frequency of symptoms. Some mild cases may be managed with weight loss alone. Often, however, further intervention is needed. Conventional management has relied upon dental appliances to maintain an open airway, ventilators to provide Continuous Positive Airway Pressure (CPAP), and attempts at surgical correction of the airway obstruction. Though effective, dental appliances and CPAP are both uncomfortable, and suffer from relatively low patient compliance rates (40 to 70%). Surgical correction may involve either excision of "excess tissue" (uvulopalatopharyngoplasty) or more involved maxillofacial surgery. Surgical cure rates have been reported to range between 30 and 75%.

Recently, Somnus Medical Technologies has developed an RF system (Somnoplasty™) that uses needle electrodes to create precise regions of submucosal tissue coagulation. Thus, both the tissue volume and its resulting airway obstruction are reduced. Applicator probes have been developed to target specific tissues including the base of the tongue (Fig. 16.15a), the uvula (Fig. 16.15b) and the soft palate (Fig. 16.15c), and the nasal turbinates (Fig. 16.15d). Somnoplasty is designed to be performed on an

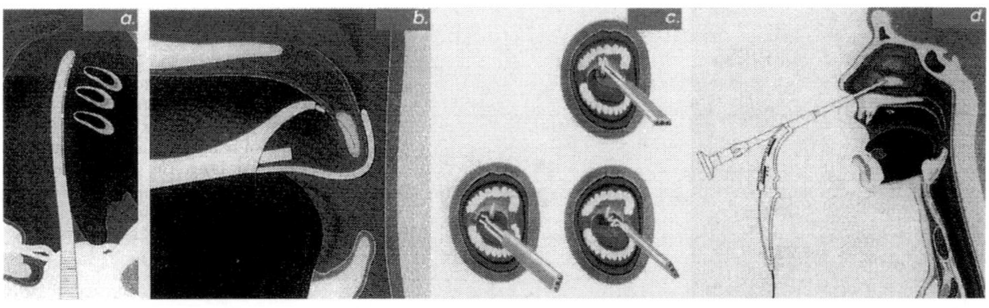

FIGURE 16.15 (**a**) Tongue Somnoplasty; (b) Uvula Somnoplasty; (c) Palatal Somnoplasty; (d) Turbinate Somnoplasty. (With permission from Somnus, Inc.)

outpatient basis, under local anesthesia, and is expected to boast such benefits as immediate results, little postoperative edema or discomfort, and no permanent scarring.

In the paper entitled "Radiofrequency Volumetric Reduction of the Tongue — A Porcine Pilot Study for the Treatment of Obstructive Sleep Apnea Syndrome," Powell et al.,[64] reported on the use of RF for the volumetric reduction of the tongue. Powell's three-stage pilot study investigated both the *in vitro* and *in vivo* effects of RF, delivered via a customized needle electrode. Volumetric measurements were performed using implanted ultrasonic crystals positioned around the treatment site. Changes in tissue volume could then be assessed both before and after the delivery of RF energy. To establish the feasibility of the technique, the initial stage of the project used two bovine tongues (*in vitro*). A single 0.05-in. diameter needle electrode delivered 30 kJ over a 20-min period at two sites per tongue. Volume reductions of between 12.8 and 26.7% were noted immediately after the procedure, with an additional 4% reduction noted after 4 h. The second stage was conducted using pigs, *in vivo*, and demonstrated that volume reduction increases as the amount of energy delivered is increased from 6.8 to 40 kJ. Finally, in the third stage, an *in vivo* porcine model was again used, this time assessing clinical efficacy of the procedure by measuring both tissue volume changes, and histological changes. RF tissue reduction was performed on 9 pigs, with 3 additional pigs serving as controls. An 0.035-in. diameter needle electrode was used to deliver 2.4 kJ over 6±1.20 min. Immediately after the procedure, a mean volume shrinkage of 7.02% was described. By 24 h after the procedure, edema resulted in a 4 to 6% increase in tissue volume, thus returning nearly to baseline volumes. Subsequently, however, a progressive volume reduction of up to 26.3% was identified over the following 10 days. Animals were sacrificed at between 1 h and 5 weeks after the procedure. Lesions were described as spherical, well-defined regions of tissue destruction, initially demonstrating edema and hemorrhage. As the lesion healed, scar formation occurred along with neovascularization. Tissue and vessels surrounding the lesion remained intact and viable. Given the seeming success of this technique in the animal model, RF tissue reduction may offer a promising alternative to the conventional management of obstructive sleep apnea.

16.2.5 Microwave-Aided Liposuction (MAL)[59,65]

Liposuction is used for aesthetic and reconstructive surgery. Its uses include the undermining of large flaps while preserving vascular attachments, removing lipomas, treating gynecomastia, and improving axillary hyperhidrosis. The application of microwave for aided liposuction may reduce some problems associated with standard mechanical liposuction, including blood loss, fluid shifts, and systemic effects.

Dry-technique liposuction vs. microwave-aided dry-technique liposuction — Preliminary work has been conducted, in swine, to compare the effects of the dry-technique liposuction vs. microwave-aided dry-technique liposuction. The "non-microwave" dry technique liposuction performed at the two cephalad sites yielded typical fat debris that grossly appeared to be mixed with a noticeable amount of blood. The "microwave" liposuction performed at the two caudal sites yielded fat that differed considerably in quality and texture from tissue extracted using the dry technique. The duration of microwave-aided suctioning appeared to be related to the histological changes observed in the subcutaneous fat derived from the caudal sites. The fat initially removed during the first 30 s grossly appeared similar to conventionally suctioned fat. However, the fat removed as the duration of microwave-aided liposuction increased from 30 s to 2 min appeared increasingly softened. The longest duration of microwave suctioning, from 2 to 4 min, yielded fat that grossly appeared to be fused into an opaque, amorphous melted state.

The tumescent technique for liposuction surgery — The "tumescent technique" of liposuction was introduced in 1986. Use of the Klein needle has allowed the anesthetic solution to be rapidly injected through the same incision used for liposuction, efficiently anesthetizing large subcutaneous areas, thus eliminating the need and risks of general anesthesia. Injection of a large volume of dilute lidocaine produces a swelling and firmness of the site to be aspirated, which greatly facilitates fat removal. The small (3 to 4 mm) cannulas produce less trauma and therefore result in less blood loss, bruising, and discomfort. The basic technique was later expanded, and much larger volumes of lidocaine were administered, resulting in the capability of aspirating significantly greater volumes of tissue with a minimum increase in blood loss. This was achieved with serum lidocaine levels well below the toxicity range. We believe that using microwave volume heating will further enhance and benefit the tumescent technique.

Tumescent-technique liposuction vs. microwave-aided tumescent liposuction — A similar protocol was followed at corresponding sites on the left side of the swine. The only modification was to employ tumescent liposuction instead of the dry technique that was used for sites on the right side. The solution used for tumescence consisted of 1000 cc of normal saline, combined with 60 cc of 1% lidocaine with epinephrine. Approximately 250 cc of this solution was infiltrated into each of the four sites prior to liposuction. The conventional "non-microwave" tumescent liposuction performed at the two cephalad sites yielded fat typically seen in such procedures; there was also less bleeding than seen with the dry technique.

Tumescent liposuction combined with microwaves between 30 and 40 W yielded a transformation in the fat suctioned, enabling easier fat removal with less bleeding in comparison to both conventional dry and tumescent liposuction without microwaves.

Cannula design — The cannula utilized was a Byron Accelerator III type cannula, which was modified to hold a microwave semirigid coaxial cable having a whip antenna at the distal end (Fig. 16.16). The tip of each cannula distal of the suction port was modified by removal of its metal tip, which was replaced with a dome made of plastic in order to facilitate microwave radiation. The suction port in the proximal end of the cannula handle was converted to accept the semirigid coaxial cable/antenna structure. Suction was effectuated through a new port installed in the cannula handle.

The system used in our preliminary experiments was designed for use at 2.45 GHz while immersed in a tissue phantom. With the modified liposuction cannula and antenna, we have measured return losses as low as –37 dB.

Antenna considerations[66] — The antenna design for Microwave-Aided Liposuction (MAL) presents some interesting challenges. The antenna must deliver microwave energy to heat the treated volume of fat. Unlike traditional biological antenna designs, the MAL antenna radiates in close proximity to a metallic cannula. The cannula serves two primary functions. First, the openings in the end of the cannula are sharp to facilitate fat removal via mechanical cutting. Second, removed fat is suctioned through the cannula. Clearly, the cannula imposes a more complex antenna geometry than exists for traditional biological antennas. Furthermore, as opposed to microwave hyperthermia application, for example, it

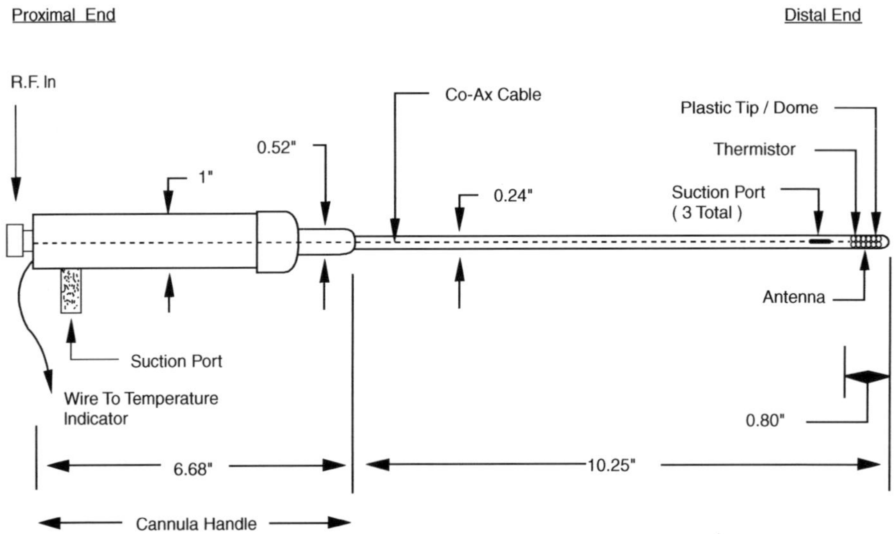

FIGURE 16.16 Microwave-aided liposuction cannula.

may actually be desirable to allow heating to occur along the transmission line since that heating will prevent the coagulation of fat being suctioned through the cannula. Analytical antenna design approaches become increasingly difficult under these design constraints. Therefore, it is appealing to consider the use of accurate simulation tools to design the antenna and to analyze the antenna performance. This approach was used by Labonte et al. who implemented a finite-element method (FEM) in the frequency domain to compare the near field radiation patterns of several types of antennas including the dielectric tip monopole, open tip monopole, and metal tip monopole.[67,68] In our future work, we propose to use a finite-element time domain (FDTD) code by REMCOM, Inc. to model the MAL antenna. This code will allow the computation of input impedance, near field values, and specific absorption rate (SAR).

16.2.6 Tissue Anastomoses Utilizing Biological Solder in Conjunction with Microwave Irradiation in Future Endoscopic Surgery[69]

Endoscopic surgery is revolutionizing many surgical procedures. For example, laparoscopic surgical procedures, particularly laparoscopic cholecystectomy, have gained widespread acceptance. Further expansion of the endoscopic approach is inevitable. Although minimal access surgery is advantageous to patients, the technical problems imposed by the limited access are pushing existing tissue closure technologies (mechanical stapling devices and hand-sewn sutures) to their limits. The laparoscopic closure of an incision made in the bile duct for removal of stones is an example of the shortcomings of current technologies. Closure of this incision with laparoscopically placed sutures is difficult and post-operative bile leakage may result. Mechanical stapling devices for this purpose are beyond currently available technology.

To enhance a tissue anastomosis with microwaves, the tissue temperature must be kept below the threshold for damage, while the biological solder is heated above 60°C. Microwave anastomosis may also prove useful for vascular repairs, for example. A microwave antenna can be positioned inside an artery, and solder (albumin) is then placed on the outside of the vessel and in any small gaps between the arterial segments undergoing repair. The successful results *in vitro* have encouraged the preliminary investigation in a rabbit model. Early results *in vivo*, however, have indicated the need for a dry environment. More research is needed to evaluate the full potential of the microwave anastomosis technique.

16.2.7 Nerve Ablation for the Treatment of Gastroesophageal Reflux Disease

Gastroesophageal reflux disease (GERD) results from the chronic backward flow of stomach contents into the esophagus. The acid, bile, and digestive enzymes cause irritation of the esophagus and symptoms of heartburn, regurgitation, chest pain, voice disorders, and swallowing problems.

Normally, the muscular valve (lower esophageal sphincter or LES) at the junction of the esophagus and stomach prevents reflux from occurring. Reflux of stomach contents occurs when the LES and diaphragm are unable to provide enough tone or force to squeeze adequately on the esophagus. This may happen in some patients in whom the muscles have weakened over time or in those patients with hiatal hernias. The barrier function in these patients is completely lost, and reflux is present throughout the day.

The majority of patients with GERD, however, have normal LES and diaphragm pressures, yet the sphincter muscles relax frequently throughout the daytime to cause reflux. The relaxation events permit excessive reflux of stomach contents and the patient develops significant symptoms of GERD.

This abnormal event is a neurological reflex, termed transient lower esophageal relaxation (tLESR),[75–79] and is the cause of GERD in over 80% of patients. A tLESR is prompted when there is stretching of the stomach wall, as after a meal. The stretch receptors generate a nerve impulse, which travels upward within the myenteric plexus of the gastroesophageal junction. The myenteric plexus is a network of very small nerves lying between the layers of the stomach and esophagus musculature. The impulses travel through the LES, into the esophagus, and then join the vagus nerve on their way to the brain. When the brain receives these signals, a motor signal is sent to the LES causing prolonged relaxation.

There are hundreds of peer-reviewed scientific publications addressing the importance of tLESR in the development of GERD. Many investigators have collaborated to study the delivery of radio frequency energy for the treatment of GERD.

Investigators at Stanford[80] have recently performed radio frequency ablation of the stomach cardia (Fig. 16.17a,b) in Yucatan mini-pigs to establish the effect on these nerve pathways. These nerve fibers course between the muscle layers of the LES and cardia. These investigators have demonstrated a statistically significant effect of delivering radio frequency energy to the cardia on the parameter of gastric yield pressure. This test is directly related to tLESRs. The stomach is stretched with carbon dioxide gas

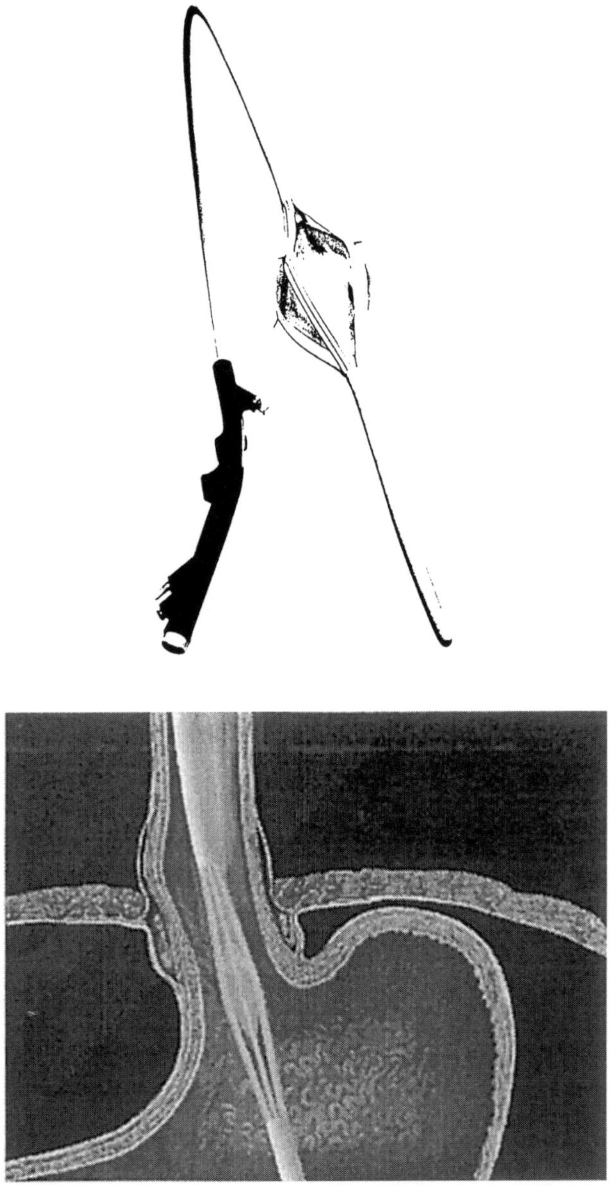

FIGURE 16.17 (top) Catheter used for nerve ablation in the treatment of gastroesophageal reflux disease; (bottom) Catheter positioning within the gastric cardia to deliver radio frequency energy for nerve ablation.

until the LES yields or relaxes in response to pressure. Yield pressures were higher in all animals after treatment, indicating that the nerve reflex arc was modulated to have a higher threshold for stimulation, or a lower frequency of transmission to the brain.

16.2.7.1 Step-by-Step Treatment for Reflux — Research Carried Out at Conway Stuart Medical, Inc.

1. **Position, inflate, deploy, irrigate, treat.** The physician positions the catheter, inflates the balloon, deploys the needles, and begins irrigation. During treatment, radio frequency energy is delivered in a controlled manner to the tissue surrounding the needle electrodes (Fig. 16.18a);
2. **Treatment at multiple levels.** The treatment sequence is repeated to create well-defined coagulative lesions along the length of the lower esophageal sphincter and cardia (Fig. 16.18b);
3. **Resorption and shrinking.** Over the next few weeks, the coagulated tissue resorbs and shrinks, increasing resistance to reflux (Fig. 16.18c).

In an abstract entitled "Augmentation of lower esophageal sphincter pressure and gastric yield pressure after radiofrequency energy delivery to the lower esophageal sphincter muscle: a porcine model," Drs. D.S. Utley, M.A. Vierra, M.S. Kim, and G. Triadafilopoulos of VA Palo Alto Health Care System and Stanford University in Palo Alto, California,[80] report on their investigation of the technique of endoscopic, submucosal radio frequency energy (Rfe) delivery to the lower esophageal sphincter (LES) as a possible alternative treatment of GERD. Utilizing a porcine reflux model, they determined the effects of Rfe on LES pressure (LESP) and gastric yield pressure (GYP). In summary, Rfe is a promising new modality in the endoscopic treatment of GERD (Fig. 6.18).

16.2.8 RF in the Treatment of Solid Organ Tumors

RITA Medical Systems, Inc. has developed a controlled tissue ablation system to treat solid organ tumors using minimally invasive RF ablation technology. This system includes an RF generator and a family of electrodes for the treatment of solid tumors. The controlled application of RF energy through an electrode placed directly in the tumor heats tissue to the required target temperature.

Each electrode consists of a thin hollow stainless steel shaft that acts as a primary electrode and also allows the introduction into the tumor of a curved array of secondary electrodes. The secondary electrodes have temperature sensors mounted on the tips to provide temperature feedback.

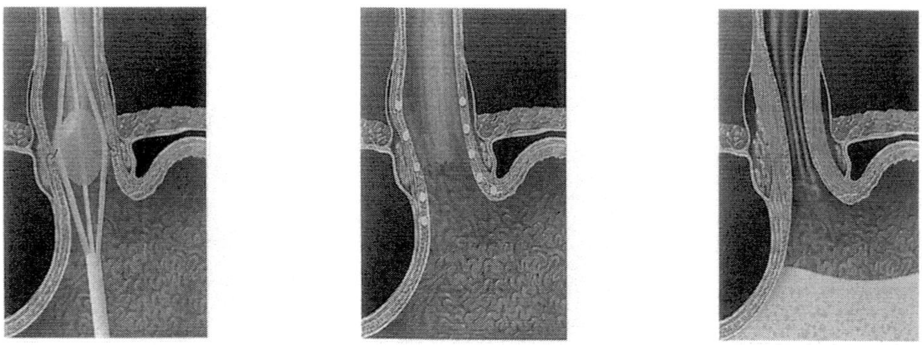

(a) (b) (c)

FIGURE 16.18 Step-by-step treatment for reflux.

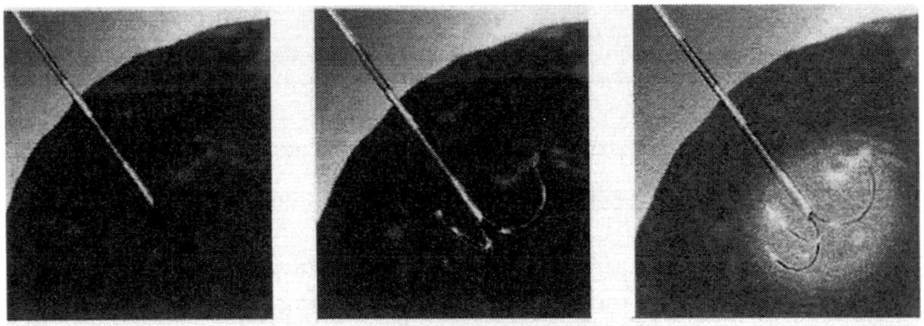

(a) Introduction (b) Deployment (c) Ablation

FIGURE 16.19 RF in the treatment of solid organ tumors. (With permission from RITA Medical Systems Inc.)

The RF generator delivers up to 50 W to destroy or ablate the tumor. The operator sets the desired temperature; the generator automatically adjusts the power to attain the proper temperature and displays delivered power, impedance, and temperature.

The unique features of RF in treating solid tumors are as follows (Fig. 16.19):

- *Minimally invasive* — many procedures can be performed through a laparoscopic or even a percutaneous approach, frequently on an outpatient basis.
- *Creates large volume of ablated tissue* — the current device can ablate a spherical area of around 3 cm, approximating the size and shape of many cancerous lesions.
- *Temperature feedback leads to predictability and controllability* — the system provides temperature feedback at the periphery of the ablation volume to confirm tissue destruction. It also provides impedance feedback, which can be used to guide the application of RF power in order to ablate the tumor.

16.2.9 Application of RF Thermal Arthroscopy[72]

The combined use of RF energy and arthroscopy has been, in recent years, successfully advanced by ORATEC Interventions, Inc. They have accumulated an extensive collection of case reports, and from among these we will discuss one important clinical application. This method will serve as an example of the significance of RF thermal arthroscopy techniques.

In a report entitled, "Arthroscopic shoulder stabilization using suture anchors and capsular shrinkage," Jeffrey S. Abrams, M.D. of Princeton Orthopedic and Rehabilitation Associates, describes the treatment of recurrent shoulder joint instability after traumatic injury. The ligaments that normally anchor the upper arm to the shoulder joint can tear away from their attachments on the joint (glenoid cavity) as a result of trauma. Without these ligamentous attachments (the labrum) the shoulder joint can repeatedly dislocate. Such recurrent instability not only eliminates recreational sports for the patients but can significantly affect their daily life.[73]

The traditional approach to such injury involves orthopedic surgery to reattach the torn labrum and restore tension to the damaged ligaments, thus holding the joint together firmly. Laparoscopic techniques, in general, reduce surgical time, decrease morbidity, and can increase success rates. The arthroscopic technique for shoulder stabilization still involves the use of sutures to anchor the labrum back to the edge of the glenoid. RF electrothermal technology is then used to shrink the tissues of the ligamentous joint capsule and thus increase the tension on these ligaments. Increased tension on the ligaments, in turn, stabilizes the shoulder joint. Such procedures can be performed on an outpatient basis and require minimal postoperative medication.

16.3 Conclusions

In this chapter, we have reviewed a few of the existing applications of RF/microwaves in medicine. We have indicated with some detail the new applications currently under investigation. A more detailed discussion of some of the topics can be found in the book entitled *New Frontiers in Medical Device Technology* edited by Arye Rosen and Harel D. Rosen, published by John Wiley & Sons, 1995, as part of the Wiley Series in Microwave and Optical Engineering/Kai Chang, Series Editor.

Acknowledgments

We wish to recognize the assistance and advice of many who, since the early 1980s, have participated in research in the areas of RF/microwaves in medicine. Some of their research was covered in this chapter: from Jefferson Medical College, Drs. Paul Walinsky and Arnold J. Greenspon; from MMTC, Dr. Fred Sterzer and Mr. Dan Mawhinney; from Temple University, Dr. William Santamore; from the University of Pennsylvania, Dr. Louis Bucky. Gratitude is also due to Mr. John Hendrick of VidaMed, Inc., Mr. Hugh Sharkey of ORATEC, Interventions, Inc., and Mr. Barry Cheskin of RITA Medical Systems, Inc., who were so kind to furnish some of the material; and to Mr. Walter Janton for his technical skills and invaluable support. Finally, recognition is due to Mrs. Daniella Rosen for her contribution to the research on microwave-assisted liposuction, and for revising this manuscript again and again.

References

1. Cosman BJ, Cosman ER: *Guide to Radio Frequency Lesion Generation in Neurosurgery.* Radionics, Inc., 1974.
2. Cosman ER, Cosman BJ: *Methods of making nervous system lesions, Medical Therapy of Movement Disorders Guide to Radio Frequency Lesion Generation in Neurosurgery.* Radionics, Inc., 1974.
3. Alberts WW, Wright EW Jr, Feinstein B, Gleason CA: Sensory responses elicited by subcortical high frequency electrical stimulation in man, *J. Neurosurg.*, 36, 80–82, 1972.
4. Dieckmann G, Gabriel E, Hassler R: Size, form, and structural peculiarities of experimental brain lesions obtained by thermocontrolled radiofrequency, *Confin. Neurol.*, 26, 134–142, 1965.
5. Brodkey JS, Miyazaki Y, Ervin FR, Mark VH: Reversible heat lesions with radiofrequency current: A method of stereotactic localization, *J. Neurosurg.*, 21, 49–53, 1964.
6. Fager CA: Surgical treatment of involuntary movement disorders, *Lahey Clin. Found. Bull.*, 22, 79–83, 1973.
7. Hurt RW, Ballantine HT Jr: Stereotactic anterior cingulate lesions for persistent pain: A report on 68 cases. *Clin. Neurosurg.*, 21, 334–351, 1974.
8. Schwan HP, Electrical properties of tissues and cell suspensions, *Advanced Phys. Med. Biol.*, 5, 147–209, 1957.
9. Stuchly MA, Stuchly SS: Dielectric properties of biological substances-tabulated, *J. Microwave Power*, 15, 19–26, 1980.
10. Foster KR, Schepps JL, Schwan HP: Microwave dielectric relaxation in tissue: a second look, *Biophys. J.*, 29, 271–281, 1980.
11. Thuari M, Steel MC, Sheppard RJ, Grant HE: Dielectric properties of developing rabbit brain at 37 degrees C, *Bioelectromagnetics*, 6, 235–242, 1985.
12. Schwan HP, Foster KR: RF-field interactions with biological systems: electrical properties and biophysical mechanism, *Proc. IEEE*, 68, 104–113, 1980.
13. Pethig R, Kell DB: The passive electrical properties of biological systems: their significance in physiology, biophysics, and biotechnology, *Phys. Med. Biol.*, 32, 933–970, 1987.
14. Duck, FA: *Physical Properties of Tissue: A Comprehensive Reference Book*, Academic Press, New York, 1990.
15. Gabriel C, Gabriel S, Corthout E: The dielectric properties of biological tissues: I. Literature review, *Phys. Med. Biol.*, 41, 2231–2249, 1996.

16. Grant HE, Sheppard RJ, South FP: *Dielectric Behaviour of Biological Molecules in Solution,* Oxford University Press, Oxford, 1978.
17. Foster KR, Schepps JL: Dielectric properties of tumor and normal tissues at radio through microwave frequencies, *J. Microwave Power,* 16, 107–119, 1981.
18. Gabriel S, Lau RW, Gabriel C: The dielectric properties of biological tissues: II. Measurements in the frequency range 10 Hz to 20 Ghz, *Phys. Med. Biol.,* 41, 2251–2269, 1996.
19. Cole KS, Cole RH, Dispersion and absorption in dielectrics: I. Alternating current characteristics, *J. Chem. Phys.,* 9, 341–351, 1941.
20. Debye P, Huckel E: *Phys. Z.,* 24, 185, 1923.
21. Frohlich H: *Theory of Dielectrics,* Oxford University Press, Oxford, England, 1949.
22. Daniel VV: *Dielectric Relaxation.* Academic Press, London, 1967.
23. Sterzer F et al.: RF therapy for malignancy, *IEEE Spectrum,* 17, 12, 32–37, Dec. 1980.
24. Lele, PP: Induction of deep, local hyperthermia by ultrasound and electromagnetic fields: problems and choices. *Radiat. Environ. Biophy.,* 17, 205, 1980.
25. Tofighi M-R: A two-port microstrip test fixture for measurement of complex permittivity of biological tissues at microwave and millimeter wave frequencies, PhD. Dissertation, Department of ECE, Drexel University, Philadelphia, PA.
26. Roberts S, Von Hippel A: A new method for measuring dielectric constant and loss in the range of centimeter waves, *J. Appl. Phys.,* 17, 610–616, 1946.
27. Westphal: Dielectric measuring techniques, in *Dielectric Materials and Applications,* A.R. von Hippel, Ed., Wiley, New York, 63–122, 1954.
28. Burdette EC, Cain FL, Seals J: *In vivo* probe measurement technique for determining dielectric properties at VHF through microwave frequencies, *IEEE Trans. Microwave Theory Tech.,* MTT-28, 414–424, 1980.
29. Stuchly SS, Sibbald CL, Anderson JM: A new aperture admittance model for open-ended waveguides, *IEEE Trans. Microwave Theory Tech.,* MTT-42, 192–198, 1994.
30. Steel MC, Sheppard RJ: The dielectric properties of rabbit tissue, pure water and various liquids suitable for tissue phantoms at 35 Ghz, *Phys. Med. Biol.,* 33, 467–472, 1988.
31. Steel MC, Sheppard RJ, Collin R: Precision waveguide cells for the measurement of complex permittivity of lossy liquids and biological tissue at 35 Ghz, *J. Phys. E. Sci. Instrum.,* 20, 872–877, 1987.
32. Land DV, Campbell AM: A quick accurate method for measuring the microwave dielectric properties of small tissue samples, *Phys. Med. Biol.,* 37, 183–192, 1992.
33. Athey TW, Stuchly MA, Stuchly SS: Measurement of radio frequency permittivity of biological tissues with an open-ended coaxial line: Part I, *IEEE Trans. Microwave Theory Tech.,* MTT-30, 82–86, 1982.
34. Kraszewski A, Stuchly MA, Stuchly SS, Smith M: *In vivo* and *in vitro* dielectric properties of animal tissues at radio frequencies, *Bioelectromagnetics,* 3, 421–432, 1982.
35. Nyshadham A, Sibbald CL, Stuchly SS: Permittivity measurements using open-ended sensors and reference liquid calibration — an uncertainty analysis, *IEEE Trans. Microwave Theory Tech.,* MTT-40, 305–314, 1992.
36. Misra DM, Chabbra M, Epstein BR, Mirotznik M, Foster KR: Noninvasive electrical characterization of materials at microwave frequencies using an open-ended coaxial line: Test of an improved calibration technique, *IEEE Trans. Microwave Theory Tech.,* MTT-38, 8–13, 1990.
37. Gabriel C, Chan TYA, Grant EH: Admittance models for open ended coaxial probes and their place in dielectric spectroscopy, *Phys. Med. Biol.,* 39, 2183–2199, 1994.
38. Bao J, Lu S, Hurt DW: Complex dielectric measurements and analysis of brain tissues in the radio and microwave frequencies, *IEEE Trans. Microwave Theory Tech.,* MTT-45, 1730–1740, 1997.
39. Daryoush A, private communications.
40. King RWP, Smith GS: *Antennas in Matter,* MIT Press, Cambridge, MA, 1981.

41. King RWP, Trembly BS, Strohbehn JW: The electromagnetic field of an insulated antenna in a conducting or dielectric medium, *IEEE Trans. Microwave Theory Tech.*, MTT-31, 7, 574–583, July 1983.
42. Iskander MF, Tumeh AM: Design optimization of interstitial antennas, *IEEE Trans. Biomed. Eng.*, 36, 238–246, Feb. 1989.
43. Tumeh AM, Iskander MF: Performance comparison of available interstitial antennas for microwave hyperthermia, *IEEE Trans. Microwave Theory Tech.*, 37, 7, 1126–1133, July 1989.
44. Debicki PS, Astrahan MA: Calculating input impedance of electrically small insulated antennas for microwave hyperthermia, *IEEE Trans. Microwave Theory Tech.*, 41, 2, 357–360, February 1993.
45. Su DW-F, Wu L-K, Input impedance characteristics of coaxial slot antennas for interstitial microwave hyperthermia, *IEEE Trans. Microwave Theory Tech.*, 47, 3, 302–307, March 1999.
46. Casey JP, Bansal R: The near field of an insulated dipole in a dissipative dielectric medium, *IEEE Trans. Microwave Theory Tech.*, 34, 4, 459–463, April 1986.
47. Greenspon AJ, Walinsky P, Rosen A: Catheter ablation for the treatment of cardiac arrhythmias, in *New Frontiers in Medical Technology*, John Wiley & Sons, Inc., New York, 1995.
48. Rosenbaum RM, Greenspon AJ, Hsu S, Walinsky P, Rosen A: RF and microwave ablation for the treatment of ventricular tachycardia, *IEEE MTT-S Digest*, 1993.
49. Sterzer F: Localized hyperthermia treatment of cancer, *RCA Rev.*, 42, 727, 1981.
50. Hahn GM: *Hyperthermia and Cancer*. Plenum Press, New York, 1982.
51. Storm FK: *Hyperthermia in Cancer Therapy*. GK Hall, Boston, 1983.
52. Walinsky P, Rosen A, Greenspon AJ: Method and apparatus for high frequency catheter ablation. U.S. Pat. 4,641,649.
53. Walinsky P, Rosen A, Martinez-Hernandez A, Smith D, Nardone DO, Brevette B: Microwave balloon angioplasty, *J. Invasive Cardiol.*, 3, 3, May/June 1991.
54. Langberg JJ, Wonnell TL, Chin M, et al.: Catheter ablation of the atrioventricular junction using a helical microwave antenna: A novel means of coupling energy to the endocardium, *PACE*, 14, 2105, 1991.
55. Wonnell TL, Stauffer PR, Langberg JJ: Evaluation of microwave and RF catheter ablation in a myocardial equivalent phantom model, *IEEE Trans. Biomed. Eng.*, 39, 1086, 1992.
56. Rosen A, Walinsky P, Smith D, Kosman Z, Martinez A, Sterzer F, Presser A, Mawhinney D, Chou J-S, Goth P: Studies of microwave thermal balloon angioplasty in rabbits, *IEEE MTT-S Digest*, 1993.
57. Rosen A, Rosen HD: The efficacy of transurethral thermal ablation in the management of benign prostatic hyperplasia, in *New Frontiers in Medical Technology*, John Wiley & Sons, Inc., New York, 1995.
58. Rosen A, Walinsky P: Microwave balloon angioplasty, in *New Frontiers in Medical Technology*, John Wiley & Sons, Inc., New York, 1995.
59. Rosen HD, Rosen A: RF/Microwaves, a hot topic in medicine, *IEEE Potentials*, August/September 1999.
60. Sterzer F: Localized heating of deep-seated tissues using microwave balloon catheters, in *New Frontiers in Medical Technology*, John Wiley & Sons, Inc., New York, 1995.
61. Schmidt-Nowara W et al.: Oral appliances for the treatment of snoring and Obstructive Sleep Apnea, *Sleep*, 18, 6, 501–510, 1995.
62. Simmons FB, Guilleminault C, Miles LE: The palatopharyngoplasty operation for snoring and sleep apnea: An interim report, *Otolaryngol. Head Neck Surg.*, 92, 375–380, 1984; and Conway W, Fujita S, Zorick F, et al.: Uvulopalatopharyngoplasty: One year follow-up, *Chest*, 88, 385–387, 1985.
63. Riley et al. 1995, The description of this approach and the outcomes data are taken from this paper, though the team has published several papers on the method since 1988.
64. Powell NB, Riley RW, Troell RJ, Blumen MB, and Guilleminault C: Radiofrequency volumetric reduction of the tongue — A porcine pilot study for the treatment of obstructive sleep apnea syndrome, laboratory and animal investigations, *Chest*, 111, 1348–1355, 1997.

65. Rosen A, Rosen D, Tuma G, Bucky L,: RF/Microwave aided tumescent liposuction, *IEEE MTT Trans.*, (Nov. 2000), in press.
66. Jemison W et al.: New antenna design for microwave assisted liposuction, to be published; private communications.
67. Hurter W, Reinbold F, Lorenz WJ: A dipole antenna for interstitial microwave hyperthermia, *IEEE Trans. Microwave Theory Tech.*, 34, 4, 459–463, April 1986.
68. Labonte S, Blais A, Legault SR, Ali HO, Roy L: Monopole antennas for microwave catheter ablation, *IEEE Trans. Microwave Theory Tech.*, 44, 10, 1832–1840, Oct. 1996.
69. Santamore W: Private communication.
70. *Health aspects of radio frequency and microwave radiation exposure, Part 1*, 77-EHD-13, National Health and Welfare, Canada, November 1977.
71. Paglione R: Medical applications of microwave energy, *RCA Engineer*, 27, 5, 17–21, Sept./Oct. 1982.
72. Sharkey H: Private communication.
73. Abrams JS: Arthroscopic shoulder stabilization using suture anchors and capsular shrinkage, ORATEC Interventions, Inc., Applications in Electrothermal Arthroscopy, Case Report.
74. Tell RA: Microwave Energy Absorption in Tissue, *Techn. Rep. PB*, Environmental Protection Agency, Feb. 1972.
75. Mittal RK, Holloway RH, Penagini R, Blackshaw LA, Dent D: Transient lower esophageal sphincter relaxation, *Gastroenterology*, 109, 601–610, 1995.
76. Rawahara H, Dent J, Davidson G: Mechanisms responsible for gastroesophageal reflux in children, *Gastroenterology*, 113, 399–408, 1997.
77. Panagini R, Bianchi PA: Effect of morphine on gastroesophageal reflux and transient lower esophageal sphincter relaxation, *Gastroenterology*, 113, 409–414, 1997.
78. Blackshaw LA, Haupt JA, Omari T, Dent J: Vagal and sympathetic influences on the ferret lower esophageal sphincter, *J. Autonomic Nerv. Syst.*, 66, 179–188, 1997.
79. Stakeberg J, Lehman A: Influence of different intragastric stimuli on triggering of transient lower esophageal sphincter relaxation in the dog, *Neurogastroenterology*, 11, 125–132, 1999.
80. Private communications with Edwards, SD.

III

System and Electromagnetic Simulation

17

System Simulation

Joseph Staudinger
Motorola

The concept of system simulation is an exceptionally broad topic. The term itself, *system*, does not have a rigid definition and in practice the term is applied to represent dramatically differing levels of circuit integration, complexity, and interaction. In a simple sense, the term is applied to represent the interconnection and interaction of several electrical circuits. In a broader sense, such as in communication systems, the term may be applied to represent a much higher level of complexity including part of, or the composite mobile radio, base unit, and the transmission medium (channel). Regardless of complexity of the level, the issue of simulation is of critical importance in the area of design and optimization.

As one might expect, the techniques and methods available in the engineering environment to simulate system level performance are quite diverse in technique and complexity [1–3]. Techniques include mathematically simple formula-based methods based upon simplified models of electrical elements. Such methods tend to be useful in the early design phase and are applied with the intent of providing insight into performance level and trade-off issues. While such methods tend to be computationally efficient allowing simulations to be performed rapidly, accuracy is limited in large part due to the use of simplified models representing electrical elements. Other techniques tend to be computationally intensive computer-aided design (CAD) based where waveforms are generated and calculated throughout the system. The waveform level technique is more versatile and can accommodate describing electrical elements to almost any level of detail required, thereby exploring the system design and performance space in fine detail. The models may be simple or complex in mathematical form. In addition, it is possible to use measured component data (e.g., scattering parameters) or results from other simulators (e.g., small- and large-signal circuit simulations using harmonic balance where the active device is represented by a large-signal electrical model). The price for the improvement in accuracy is significantly longer, perhaps much longer simulation times and the requirement to very accurately describe the characteristics of the electrical components.

The intent of this section is to examine fundamental issues relative to simulating and evaluating performance of microwave/radio frequency (RF) related system components. A number of terms describing system level performance characteristics will be examined and defined. In addition, first-order methods for calculating the performance of systems consisting of cascaded electrical circuits will be examined. To begin, consider three parameters of interest in nearly all systems: gain, noise figure (NF), and intermodulation distortion (IMD).

0-8493-1732-0/03/$0.00+$1.50

17.1 Gain

A usual parameter of interest in all systems is the small signal (linear) gain relationship describing signal characteristics at the output port relative to the input port and/or source for a series of cascaded circuits. Numerous definitions of gain have been defined in relationship to voltage, current, and power (e.g., power gain can be defined in terms of transducer, available, maximum stable) [1]. In general for system analysis, the concept of transducer power gain (G_T) is often applied to approximate the small signal gain response of a series of cascaded elements. Transducer power gain (G_T) is defined as the magnitude of the forward scattering parameter (S_{21}) squared (i.e., $G_T = |S_{21}|^2$, the ratio of power delivered to the load to that available from the source). This assumes the source (Γ_S) and load (Γ_L) voltage reflection coefficient are equal to zero, or alternatively defined as a terminating impedance equal to characteristic impedance Z_0 (typically 50 Ω).

Consider several two-port networks cascaded together as illustrated in Fig. 17.1 where the transducer power gain of the ith element is represented as G_{Ti}. The transducer power gain of the cascaded network is:

$$G_{T_T} = G_{T1}(\text{dB}) + G_{T2}(\text{dB}) + G_{T3}(\text{dB}) + \ldots G_{Tn}(\text{dB}) \tag{17.1}$$

The accuracy of Eq. (17.1) relies on the assumption that that i^{th} two-port network is terminated by characteristic impedance Z_0 per the above definition. In practice, the source and load termination provided to the ith element is defined by the $i^{th} - 1$ and $i^{th} + 1$ elements, respectively. Even though in a well-designed subsystem, each two-port network is designed such that the input (S_{11}) and output (S_{22}) scattering parameters are near zero in magnitude, they cannot be exactly zero resulting in impedance mismatch effects. Hence, Eq. (17.1) is approximate and its accuracy dependent on each element satisfying the above criteria. A more thorough analysis accounting for impedance mismatches can be performed at the expense of more complexity. In general this requires a more precise description of each element using perhaps some form of network parameters. For example, the T and scattering parameters (T_T and S_T, respectively) for two networks, A and B, cascaded together are given by [3,7]

$$T_T = \begin{bmatrix} T_{11}^A & T_{12}^A \\ T_{21}^A & T_{22}^A \end{bmatrix} \begin{bmatrix} T_{11}^B & T_{12}^B \\ T_{21}^B & T_{22}^B \end{bmatrix} \tag{17.2}$$

$$S_T = \begin{bmatrix} S_{11}^A & 0 \\ 0 & S_{22}^B \end{bmatrix} + \begin{bmatrix} S_{12}^A & 0 \\ 0 & S_{21}^B \end{bmatrix} \begin{bmatrix} -S_{22}^A & 1 \\ 1 & -S_{11}^B \end{bmatrix}^{-1} \begin{bmatrix} S_{21}^A & 0 \\ 0 & S_{12}^B \end{bmatrix} \tag{17.3}$$

While the above methods allow an exact analysis for cascaded linear circuits, it is often difficult to apply them to practical systems since the network parameters for each of the elements comprising the system are usually not known precisely. For example, in systems consisting of interconnected circuits, board layout effects (e.g., coupling between elements) and interconnecting structures (board traces) must

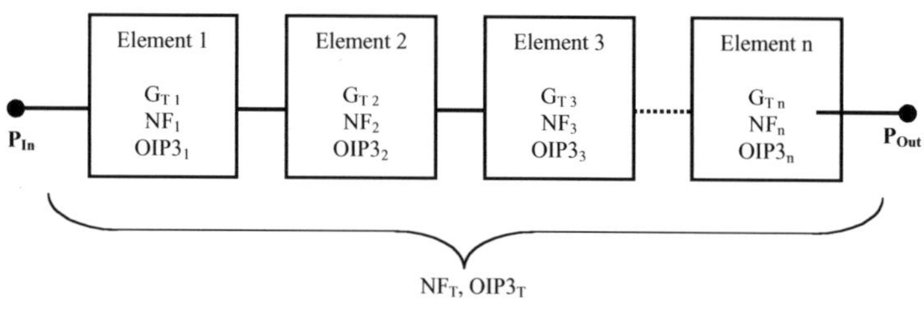

FIGURE 17.1 System formed by cascading three electrical elements.

also be included in applying the network analysis techniques shown in Eqs. (17.2) and (17.3). This, of course, requires accurate knowledge of the electrical nature of these structures, which is often unknown. In critical situations, the network parameters for these elements can be determined by measurement or through the use of electromagnetic simulations assuming the geometric and physical nature of these structures are known.

17.2 Noise

A second parameter of interest important to all systems is noise. In receivers, noise performance is often specified by noise figure, defined as

$$NF(\mathrm{dB}) = \frac{S_i/N_i}{S_o/N_o} \tag{17.4}$$

where S_i/N_i and S_o/N_o are the signal-to-noise ratio at the input and output ports, respectively. Note that NF is always greater than or equal to unity (0 dB). When several circuits are cascaded together as illustrated in Fig. 8.1, the cascaded noise figure (NF_T) is given by

$$NF_T = NF_1 + \frac{NF_2 - 1}{G_1} + \frac{NF_3 - 1}{G_1 G_2} + \frac{NF_4 - 1}{G_1 G_2 G_3} + \ldots \frac{NF_n - 1}{G_1 G_2 G_3 \ldots G_n} \tag{17.5}$$

where G_i and NF_i are the gain and noise figure of the ith element, respectively. Note the importance of the contribution of the first element's noise figure to the total cascaded noise figure. Hence, the noise figure of the low noise amplifier contained in a receiver is a major contributor in setting the noise performance of the receiver.

17.3 Intermodulation Distortion

Intermodulation distortion (IMD) has been a traditional spectral measure (frequency domain) of linearity applied to both receiver and transmitter elements. The basis of IMD is formed around the concept that the input-output signal relationship of an electrical circuit can be expressed in terms of a series expansion taking the form:

$$E_o = a_1 E_{in} + a_2 E_{in}^2 + a_3 E_{in}^3 + a_4 E_{in}^4 + a_5 E_{in}^5 + . \tag{17.6}$$

where E_{in} and E_o are instantaneous signal levels at the input and output ports of the electrical circuit, respectively. If the circuit is exactly linear, all terms in the expansion are zero except for a_1 (i.e., gain). In practice, all circuits exhibit some nonlinear behavior and hence higher order coefficients are nonzero. Of particular interest is the spectral content of the output signal when the circuit is driven by an input consisting of two sinusoids separated slightly in frequency taking the form

$$E_{in}(t) = \cos(\omega_1 \mathrm{t}) + \cos(\omega_2 \mathrm{t}) \tag{17.7}$$

where ω_1 and ω_2 are the angular frequencies of the input stimuli and where ω_2 is slightly greater than ω_1. The output signal will exhibit spectral components at frequencies $m\omega_1 \pm n\omega_2$, m = 0, 1, 2, … and n = 0, 1, 2, … as illustrated in Fig. 17.2. Notice that the third series term (a_3 Ein^3) generates spectral tones at $2\omega_1 - \omega_2$ and $2\omega_2 - \omega_1$. These spectral components are termed third-order intermodulation distortion (*IM3*). It can also be shown that series terms greater than three also produce spectral

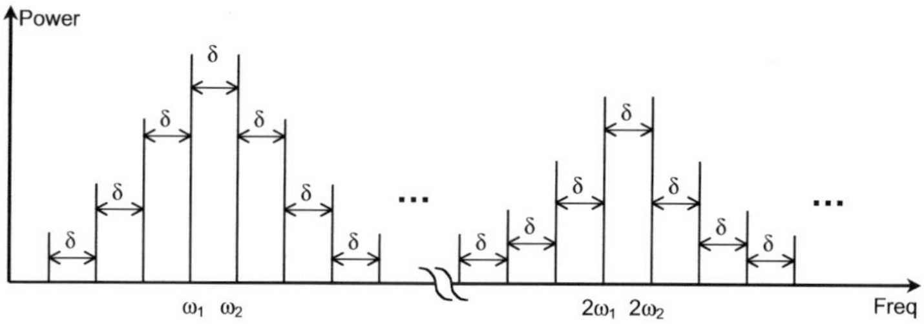

FIGURE 17.2 Resultant spectrum representing the nonlinear amplification of two equal amplitude sinusoidal stimuli at frequencies ω_1 and ω_2 ($\delta = \omega_2 - \omega_1$).

components at these frequencies and hence the total IM3 is the vector sum from Ein^3, Ein^4.... In a similar manner, higher order IMD products exist (e.g., IM5 @ $3\omega_1 - 2\omega_2$ and $3\omega_2 - 2\omega_1$) due to the higher order series terms. Notice, however, that all IMD products are close in frequency to ω_1 and ω_2, fall within the desired frequency band of the circuit, and hence cannot be removed by external filtering. In practice, third-order products are often the highest in magnitude and thus of greatest concern, although in some cases fifth and higher order products may also be of interest. Spectral analysis of the circuit can be greatly simplified if the input signal is assumed small in magnitude such that the dominant contributor to IM3 is from Ein^3.

Intermodulation distortion products (IM3, IM5,...) can be expressed in terms of power, either absolute (dBm) or relative to the carrier (dBc), or by a fictitious intercept point. For certain circuits where the input stimulus is small in magnitude (e.g., low noise amplifier and certain receiver components), an intercept point distortion specification is useful. Consider the nonlinear response of a circuit represented by Eq. (17.6) and driven with an equal amplitude two-tone sinusoidal stimuli as given in Eq. (17.7). Assume further than only a_1 and a_3 in Eq. (17.6) are nonzero. A plot of the circuit's output power spectrum as a function of input power is illustrated in Fig. 17.3. The output spectral tones at ω_1 and ω_2 increase on a 1:1 (dB) basis with input power. The IM3 products ($2\omega_1 - \omega_2$, $2\omega_2 - \omega_1$) increase on a 3:1 (dB) basis with input power (due to E_{in}^3). The intersection of the fundamental tones with the third-order products is defined as the third-order intercept point. Note that the intercept point can be specified relative to input or output power of each tone, IIP_3 and OIP_3, respectively. Given this linear

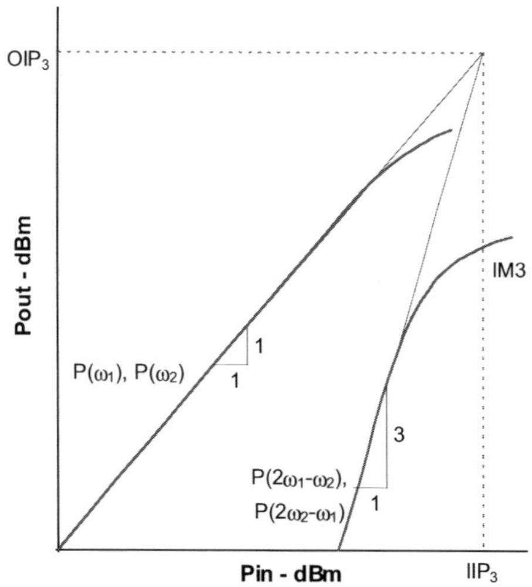

FIGURE 17.3 Expected relationship between fundamental frequency components and third-order intermodulation distortion products neglecting effects due to higher order series coefficients.

relationship, the output intercept point can easily be calculated based on the power of the fundamental and third-order terms present at the output port [2, 6]

$$OIP3(\text{dBm}) = P_{out}(\omega_1) + \frac{P_{out}(\omega_1) - P_{out}(2\omega_1 - \omega_2)}{2} \tag{17.8}$$

where $P_{out}(\omega_1)$ and $P_{out}(2\omega_2 - \omega_1)$ is the power (dBm) in the fundamental and third-order products referenced to the output port. Notice that since the input stimuli is an equal amplitude two-tone sinusoidal stimulus, like order spectra products are assumed equal in magnitude (i.e., $P_{out}(\omega_1) = P_{out}(\omega_2)$, $P_{out}(2\omega_2 - \omega_1) = P_{out}(2\omega_1 - \omega_2)$, ...).

The relationship between input and output intercept points is given by

$$IIP3(\text{dBm}) = OIP3(\text{dBm}) - G(\text{dBm}) \tag{17.9}$$

In a similar manner, the fifth-order output intercept point can be defined as [5]

$$OIP5(\text{dBm}) = P_{out}(\omega_1) + \frac{P_{out}(\omega_1) - P_{out}(3\omega_1 - 2\omega_2)}{4} \tag{17.10}$$

where $P_{out}(3\omega_2 - 2\omega_1)$ is power (dBm) of fifth-order products referenced to the output port.

Similarly,

$$IIP3(\text{dBm}) = OIP3(\text{dBm}) - G(\text{dBm}) \tag{17.11}$$

In system analysis, it is often desirable to consider the linearity performance for a series of cascaded two-port circuits and the contribution of each circuit's nonlinearity to the total. The IM3 distortion of the complete cascaded network can be approximated based on the third-order intercept point of each element. Consider the two-tone sinusoidal stimulus [Eq. (17.7)] applied to the input of the cascaded circuits shown in Fig. 17.1. The magnitude of the fundamental and IM3 products (i.e., ω_1, ω_2, $2\omega_1 - \omega_2$, and $2\omega_2 - \omega$) at the output port of the first element can be calculated based on knowledge of the input power level of tones ω_1 and ω_2, transducer gain, and the third-order intercept point of element one using Eq. (17.8). Next, consider the second element where the input stimulus now consists of spectral components at $2\omega_1 - \omega_2$ and $2\omega_2 - \omega_1$ in addition to those at ω_1 and ω_2. The IM3 spectral products at the second element's output port will be the result of contributions from two sources, (1) those due to intermodulation distortion of ω_1 and ω_2 in element 2, and (2) those due to amplifying spectral products $2\omega_1 - \omega_2$, and $2\omega_2 - \omega$ present at the input port of element 2. The IM3 products due to the former are, again, calculated from Eq. (17.8). The IM3 products at the output of element 2 due to the latter will be the IM3 products at the input amplified by G_2. Hence, the total IM3 spectral products are the vector sum from each. Both a minimum and maximum value is possible depending on the vector relationship between the various signals. A worst case (lowest *OIP*3) results when they combine in phase and are given by [2]

$$\frac{1}{IIP3_T} = \frac{1}{IIP3_1} + \frac{G_1}{IIP3_2} + \frac{G_1 G_2}{IIP3_3} \cdots \tag{17.12}$$

with *IIP*3 expressed in watts.

Alternatively, from [6]

$$OIP3_{T\,\min} = \left(\sum_{i=1}^{n} \frac{1}{OIP3_i\, g_i} \right)^{-1} \tag{17.13}$$

with *OIP*3 expressed in watts and where g_i is the cascaded gain from the output of the ith element to the system output, including impedance mismatch effects. A best case scenario (highest *OIP*3) results when they combine out of phase with the results given by [6]:

$$OIP3_{T\max} = \left(\sum_{i=1}^{n} \frac{1}{OIP3_i^2\, g_i^2} - 2 \sum_{\substack{i=2 \\ i>j}}^{n} \sum_{j=1}^{n} \frac{1}{OIP3_i OIP3_j g_i g_j} \right)^{-1/2} \tag{17.14}$$

Hence, Eqs. (17.13) and (17.14) specify bounds for intercept performance of cascaded networks.

An illustration of the measured spectral content of an amplifier driven with a two-tone sinusoidal stimuli is shown in Fig. 17.4. At low power levels, third- and fifth-order IM products closely follow a 3:1 and 5:1 (dB) relationship with input power. Hence, per the previous discussion, *OIP*3 and *OIP*5 can be calculated based on measurements of the output spectral products at a given input power level. In this example, the spectral content is $P_{out}(3\omega_2 - 2\omega_1) = -87.4$ dBm, $P_{out}(2\omega_2 - \omega_1) = -50.2$ dBm, $P_{out}(\omega_1) = +3.0$ dBm, and G = 10.4 dB for the input level shown. Applying Eqs. (17.8) and (17.10) yield $OIP3 = 29.6$ dBm and $OIP5 = 25.6$ dBm, respectively. The input intercept points are determined from Eqs. (17.9) and (17.11).

Limitations rooted in the approximations in deriving intercept points become more apparent at higher power levels where the relationship between input power and spectral products deviates dramatically from their assumed values. At some increased power level, the effects due to the higher order series coefficients in Eq. (17.6) become significant and cannot be ignored. Hence, for certain circuits, such as power amplifiers, for example, the concept of intercept point is meaningless. A more meaningful measure of nonlinearity is the relative power in the IMD products (dBc) referenced to the fundamental tones, with the reference generally made to output rather than input power.

17.4 System Simulation with Digitally Modulated RF Stimuli

Many modern communications systems, including second- and third-generation cellular, satellite communications, and wireless local area networks (WLAN), to name but a few, utilize some form of digital modulation to encode information onto an RF carrier. As discussed earlier in this chapter, these signals are complex in that the RF carrier's phase, amplitude, or both are modulated in some manner to represent digital information. An extensive examination of the mathematical techniques and available methods to simulate the system response of passing such signals through various RF circuits is well beyond the scope of this section. The reader is referred to [1] for a more detailed discussion of simulation techniques. Nevertheless, some of the fundamental RF-related aspects of system simulation will be examined in the context of a mobile wireless radio. Consider the architecture illustrated in Fig. 17.5 which is intended to represent major elements of wireless radio such as presently utilized in 2G and 3G cellular systems. The radio utilizes frequency division multiplexing whereby a diplexer confines RX and TX signals to the respective receiver and transmitter paths.

To begin, consider the TX path where digital information is first generated by a DSP. These data are modulated onto an RF carrier whereby the information is encoded and modulated onto a carrier conforming to a particular modulation format. From this point, the signal is injected into a mixer where it is raised in frequency to coincide with the TX frequency band. The signal is then amplified by a power amplifier and passed to the antenna via a diplexer.

Simulation of the TX signal path begins by considering the digital information present at the modulator. In a cellular system, this information corresponds to a digital sequence representing voice information and/or data. The precise structure of the sequence (i.e., patterns of zeros and ones) is important in that it is a major contributor in defining the envelope characteristics of the RF signal to be transmitted.

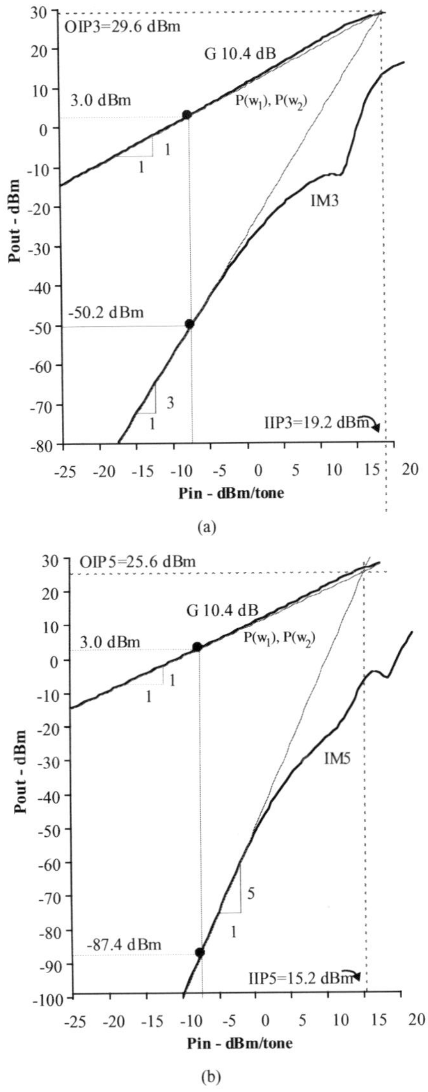

FIGURE 17.4 Typical relationship between fundamental frequency components and third- and fifth-order intermodulation distortion products including effects due to higher order series coefficients. (a) Third-order IMD and (b) fifth-order IMD.

Also note for simulation purposes, the RF stimulus is generally formed by repeating this sequence and modulating it onto an RF carrier. Hence, the resultant RF signal is periodic per the digital bit sequence. The effect of a particular digital bit sequence in defining the RF signal is illustrated by considering the two randomly generated NRZ bit patterns shown in Fig. 17.6. The amplitude modulated RF envelope voltage developed by utilizing these sequences in a $\pi/4$ DQPSK modulator is also shown in Fig. 17.6. While the two envelope signals are nearly identical in their average power levels, they are substantially different, especially in their peak voltage excursions. Hence, the digital bit sequence and the resultant modulated RF waveform can be particularly important when evaluating the performance of nonlinear circuits such as power amplifiers in that the bit sequence can affect spectral distortion. Nevertheless for simulation purposes, it is necessary to choose a suitable sequence to represent the digital information to

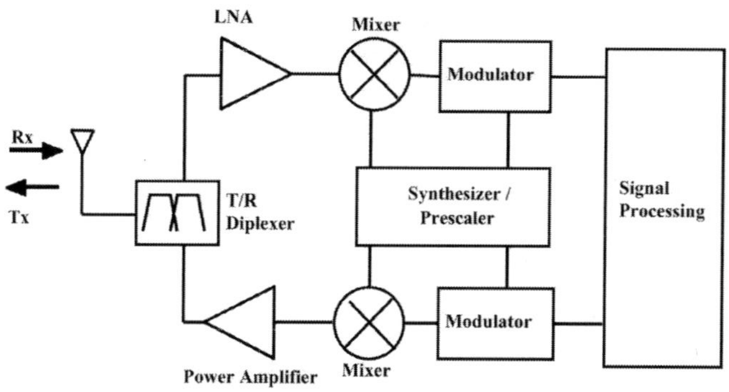

FIGURE 17.5 Block diagram representing major elements in mobile cellular radio.

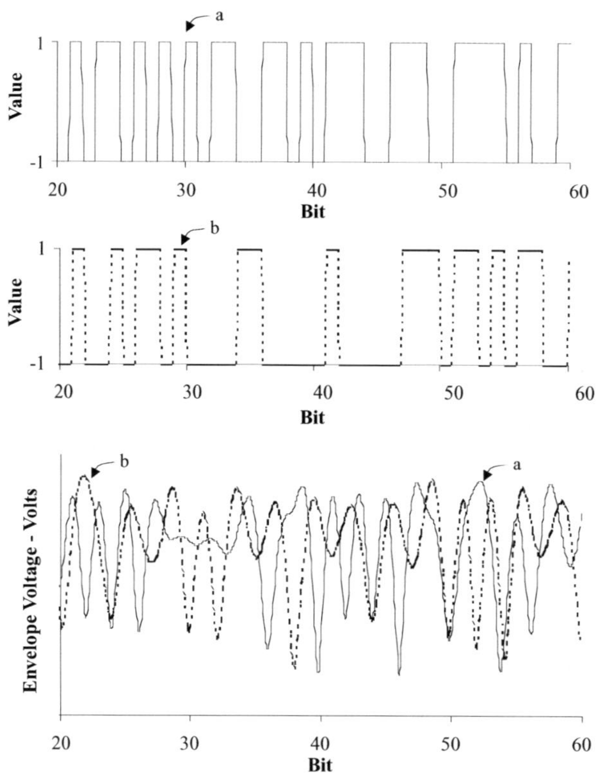

FIGURE 17.6 The RF-modulated envelope formed by considering two randomly generated bit sequences.

be transmitted. In general, either a predefined binary NRZ sequence, a randomly generated one, or a pseudo-noise (PN) sequence is generally chosen. Often the latter is considered a more desirable choice due to its statistical properties. For example, a maximal length PN sequence of length 2^m-1 contains all but one m bit combinations of 1's and 0's. This property is particularly important in that it allows all possible bit patterns (except for the all zeros pattern) to be utilized in generating the RF-modulated waveform. In contrast, a much longer random sequence would be needed with no guarantee of this property. Further, the autocorrelation function of a PN sequence is similar to a random one [1]. A potentially significant disadvantage of applying a random sequence in evaluating nonlinear circuit blocks

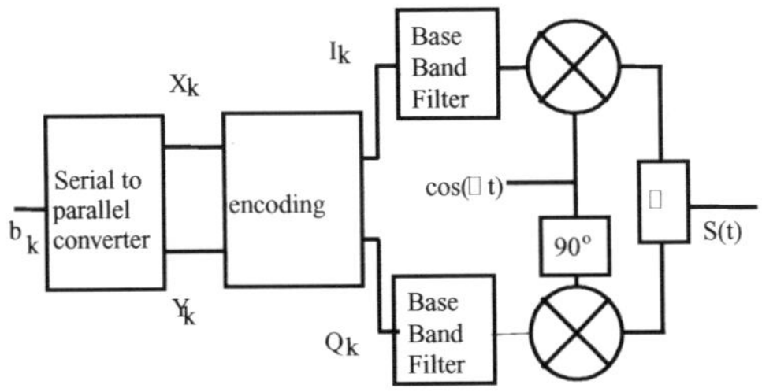

FIGURE 17.7 The process of generating modulated signal $S(t)$ in a QPSK system.

is that the simulation results will change from evaluation to evaluation, as the randomly generated sequence is not identical for each simulation.

Once a sequence is chosen, the performance of the modulator can be evaluated. The modulator can be modeled and simulated at the component level using time-based methods such as SPICE. However, in the early system design phase, such detail and the time required to perform a full circuit level simulation may be unattractive. On the other hand, it may be more appropriate to model the modulator at a higher level (e.g., behavioral model) and only consider selected first-order effects. For example, the simplified diagram in Fig. 17.7 depicts the functionality of a quadrature modulator (in this case to represent a quaternary phase-shift keying [QPSK] modulator). Starting with a data stream, every 2 b are grouped into separate binary streams representing even and odd bits as indicated by X_K and Y_K. These signals (X_K and Y_K) are encoded in some manner (e.g., as relative changes in phase for IS-136 cellular) and are now represented as I_K and Q_K with each symbol extending over 2 b time intervals. These signals now pass though a baseband filter, often implemented as a finite impulse response filter with impulse response h(t). The filtered signal can be calculated based upon the convolution theorem for time sampled signals. These signals are then modulated onto a carrier (IF) with the output modulated signal taking the form

$$S(t) = \sum_n g(t-nT)\cos(\Phi_n)\cos(\omega_c t) - \sum_n g(t-nT)\sin(\Phi_n)\sin(\omega_c t) \tag{17.15}$$

where ω_c is the radian carrier frequency, Φ_n represents phase, $g(t)$ is a pulse shaping factor, and n = 0, 1, 2, … are discrete time samples.

At this point, some first-order effects can be evaluated by considering the mathematical nature of Eq. (17.15). For example, phase imbalance (Φ_{imb}) within the modulator can be modeled as:

$$S(t) = \sum_n g(t-nT)\cos(\Phi_n)\cos(\omega_c t) - \sum_n g(t-nT)\sin(\Phi_n)\sin(\omega_c t + \Phi_{imb}) \tag{17.16}$$

Given a higher level model, the modulated envelope can be simulated using time-based methods to determine $S(t)$.

The power amplifier represents an element in the transmitter chain where linearity is of concern, especially in those RF systems employing modulation methods resulting in a nonconstant amplitude envelope. These cases, which incidentally include a number of cellular and PCS systems, require a linear power amplifier to preserve the amplitude/phase characteristics of the signal. The nonlinear characteristics of the amplifier can be simulated at the circuit level using time- and/or frequency-based methods. However, circuit-based simulations require accurate and detailed knowledge of all circuit components within the amplifier, including an appropriate large signal model for all active devices as well as highly accurate models for all passive structures. Such knowledge is often unavailable at the system level with sufficient detail and accuracy to perform such a simulation. In addition, circuit level nonlinear simulations of the amplifier driven by digitally modulated RF stimulus are generally quite computationally intensive resulting in long simulation times, making this approach even more unattractive.

A more common approach to modeling the nonlinearity of a power amplifier at the system level is through the use of behavioral models [1, 7, 8]. While a number of behavioral models have been proposed with varying levels of complexity and sophistication, all of them rely to some extent on certain approximations regarding the circuit nonlinearity. A common assumption in many of the behavioral models is that the nonlinear circuit/RF-modulated stimulus can be represented in terms of a memoryless and bandpass nonlinearity [1]. Although the active device within the amplifier generally exhibits some memory behavior, and additional memory-like effects can be caused by bias networks, these assumptions are generally not too limiting for power amplifiers utilized in cellular communications systems. Further, when the above-noted assumptions are met or nearly met, the simulation results are very accurate and the needed simulation time is very short.

In general, an input–output complex envelope voltage relationship is assumed with the complex output envelope voltage $v_{out}(t)$ taking the form

$$v_{out}(t) = RE\left\{G(V(t))e^{j\{\Phi(t)+\varphi(V(t))+\omega_c t\}}\right\} \tag{17.17}$$

where $G(V(t))$ and $\varphi(V(t))$ describe the instantaneous input–output envelope voltage gain and phase. Note that functions $G(V)$ and $\varphi(V)$ represent the amplifier's am–am and am–pm response, respectively. The term ω_c represents the carrier frequency.

An inspection of Eq. (17.17) suggests the output envelope voltage can be calculated in a time-based method by selecting time samples with respect to the modulation rate rather than at the RF carrier frequency. This feature is particularly advantageous in digitally modulated systems where the bit rate and modulation bandwidth are small in comparison to the carrier frequency. Significantly long and arbitrary bit sequences can be simulated very quickly since the time steps are at the envelope rate. For example, consider an NRZ bit sequence on the order of several ms which is filtered at baseband and modulated onto an RF carrier with a 1-ns period (i.e., 1 GHz). Time-based simulation at the RF would likely require time samples significantly less than 0.1 ns and the overall number of sample points would easily exceed 10^7. Alternatively, simulating the output envelope voltage by choosing time steps relative to the modulation rate would result in several orders of magnitude fewer time samples.

A particular advantage of the above model is that the entire power amplifier nonlinearity is described in terms of its am–am (gain) and am–pm (insertion phase) response. The am–am response is equivalent to RF gain and the am–pm characteristics are equivalent to the insertion phase — both measured from input to output ports with each expressed as a function of input RF power. For simplicity, both characteristics are often determined using a single tone CW stimulus, although modulated RF signals can be used at the expense of more complexity. The nature of the behavioral model allows it to be characterized based on the results from another simulator (e.g., harmonic balance simulations of the amplifier circuit), an idealized or assumed response represented in terms of mathematical functions describing am–am and am–pm characteristics, or from measurements made on an actual amplifier.

Computation of the output envelope voltage from Eq. (17.17) allows evaluation of many critical system characteristics, including ACPR (Adjacent and/or Alternate Channel Power Ratio), Error Vector Magnitude (EVM), and Bit Error Rate (BER) to name but a few. For example, using Fourier methods, the spectral properties of the output signal expressed in Eq. (17.17) can be calculated. Nonlinear distortion mechanisms are manifested in spectral regrowth or equivalently a widening of the spectral properties. This regrowth is quantified by figure of merit ACPR, which is defined as a ratio of power in the adjacent (or in some cases alternative) transmit channel to that in the main channel. In some systems (e.g., IS-136), the measurement is reference to the receiver (i.e., the transmitted signal must be downconverted in frequency and filtered at baseband). An illustration of spectral distortion for a $\pi/4$ DQPSK stimulus compliant to the IS-136 cellular standard is shown in Fig. 17.8. In this system the channel bandwidth is

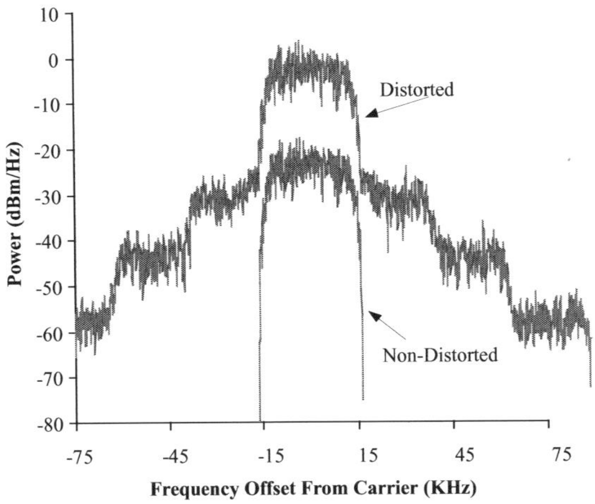

FIGURE 17.8 Nonlinear distortions due to a power amplifier result in a widening of the spectral properties of the RF signal.

30 kHz. Hence, using the specified root raised cosine baseband filter with excess bandwidth ratio $\alpha = 0.35$, a 24.3 kS/s data sequence will result in a non-distorted modulated RF stimulus band limited to 32.805 kHz (i.e., 1.35 $*$ 24.3 kS/s) as illustrated in Fig. 17.8. Nonlinear distortion mechanisms cause regrowth as illustrated. Note the substantial increase in power in adjacent and alternate channels.

Another measure of nonlinear distortions involves examining the envelope characteristics of $S(t)$ in time as measured by the receiver. Consider an input $\pi/4$ DPSK modulated RF signal which is developed from a 256 length randomly selected NRZ symbol sequence and passed through both linear and nonlinear amplifiers. Figure 17.9 illustrates the constellation diagram for both signals as measured at the receiver. Note that both signals represent the I and Q components of the envelope where $S(t)$ has been passed through a root-raised-cosine receiver filter. The constellation plot only shows the values of the I and Q components at the appropriate time sample intervals. The non-distorted signal exhibits expected values of ± 1, $\pm\sqrt{1/2} \pm i\sqrt{1/2}$, and $\pm i$. The distorted signal is in error from these expected values in terms of both amplitude and phase.

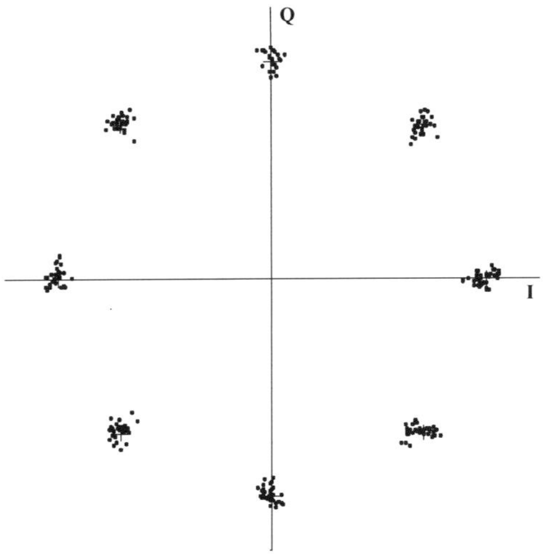

FIGURE 17.9 Constellation plot for both a non-distorted and distorted $\pi/4$ DPSK stimulus.

References

1. M.C. Jeruchim, P. Balaban, and K. Sam Shanmugan, *Simulation of Communication Systems,* Plenum Press, New York, 1992.
2. J. Tsui, *Microwave Receivers and Related Components,* Air Force Avionics Lab., Wright-Patterson AFB, OH, 1983.
3. G. Gonzalez, *Microwave Transistor Amplifiers,* Prentice Hall, Inc., Englewood Cliffs, NJ, 1984.
4. J. Sevice, Nonlinear Analysis Methods for the Simulation of Digital Wireless Communication Systems, *Int. Journal of Microwave and Millimeter-Wave Computer-Aided Eng.,* 6, 1997, 197–216.
5. H. Xiao, Q. Wu, and F. Li, Measure a Power Amplifier's Fifth-Order Intercept Point, *RF Design,* April 1999, 54–56.
6. N. Kanaglekar, R. Mcintosh, and W. Bryant, Analysis of Two-Tone, Third-Order Distortion in Cascaded Two-Ports, *IEEE Transactions on Microwave Theory and Techniques,* April 1988, 701–705.
7. System Theory of Operation, OmniSys Manual, Hewlett-Packard Company.
8. J. Staudinger, Applying the Quadrature Modeling Technique to Wireless Power Amplifiers, *Microwave Journal,* Nov. 1997, 66–86.

18

Numerical Techniques for the Analysis and Design of Radio Frequency and Microwave Structures

Manos M. Tentzeris
Georgia Institute of Technology

Recent advances in wireless and microwave communication systems — higher operation frequency bands, more compact topologies containing monolithic microwave integrated circuits (MMICs) and microelectro-mechanical systems (MEMS) — have increased the necessity of fast and accurate numerical simulation techniques [1–20]. Unlike hybrid microwave integrated circuits at low frequencies, it is extremely difficult and essentially impossible to adjust the circuit and radiation characteristics of communication modules once they are fabricated. The starting point for the development of efficient numerical algorithms is an accurate characterization of the passive and active structures involved in the topologies. Although most commercial computer-aided design (CAD) programs are based on curve-fitting formulas and lookup tables and not on accurate numerical characterization, the latter can be used if it is fast enough. In addition, it can be used to generate lookup tables and to check the accuracy of empirical formulas.

Any numerical method for characterization needs to be as efficient and economical as possible in both central processing unit (CPU) time and temporary storage requirement, although recent rapid advances in computers impose less severe restrictions on the efficiency and economy of the method. Another important aspect in the development of numerical methods has been the versatility of the method. In reality, however, numerical methods are chosen on the basis of trade-offs between accuracy, speed, storage requirements, versatility, etc., and are often structure dependent. Among these techniques, the most popular ones include the Moment Method (MoM), Integral Equation Based Techniques, Mode Matching

0-8493-1732-0/03/$0.00+$1.50

(MM), Finite Difference Time Domain (FDTD), Transmission Line Matrix (TLM) method, and Finite Element Method (FEM).

18.1 Integral Equation Based Techniques

18.1.1 Method of Moments (MoM) — Integral Equation

The term *moment method* was introduced in electromagnetics by Harrington [3] in 1968 to specify a certain general method for reducing linear operator equations to finite matrix solutions [4–7]. MoM computations invariably reduce the physical problem, specified by the Maxwell's equations and the boundary conditions, into integral equations having finite and preferably small domains. In this small domain, the discretization is performed through the expansion of unknowns as a series of basis functions. An example is the magnetic field integral equation (MFIE) for the scattering of a perfectly conducting body illuminated by an incident field H^i [21],

$$\hat{n} \times \mathbf{H}(\mathbf{r}) = 2\hat{n} \times \mathbf{H}^{\mathbf{i}}(\mathbf{r}) + 2\hat{n} \times \int_S \left[\hat{n}' \times \mathbf{H}(\mathbf{r}')\right] \times \nabla' G(r, r') ds' \quad on \;\; S \tag{18.1}$$

where H^i is defined as the field due to the source in the absence of the scattering body S, and

$$G(r, r') = \frac{e^{-jk|r-r'|}}{4\pi|r-r'|} \tag{18.2}$$

where r and r' are the position vectors for the field and source positions, respectively. A continuous integral, such as that in Eq. (18.1) can be written in an abbreviated form as

$$L(f) = g \tag{18.3}$$

where f denotes the unknown, which is H as in Eq. (18.1), and g denotes the given excitation, which is H^i. Also, L is a linear operator. Let f be expanded in a series of functions $f_1, f_2, f_3, \ldots$ in the domain S of L, as

$$f = \sum_n a_n f_n \tag{18.4}$$

where a_n are constants and f_n are called expansion (basis) functions. For exact solutions, the above summation is infinite and f_n forms a complete set of basis functions. For approximate solutions, this solution is truncated to

$$\sum_n a_n L(f_n) = g \tag{18.5}$$

Assume that a suitable inner product $<f, g> = \int f(x)g(x)dx$ has been determined for the problem. Defining a set of weighting (testing) functions, $w_1, w_2, w_3, \ldots$ in the range of L, and taking the inner product of the summation in Eq. (18.5) with each w_m leads to the result

$$\sum_n a_n \langle w_m, L(f_n) \rangle = \langle w_m, g \rangle, \quad m = 1, 2, 3, \ldots \tag{18.6}$$

which can be written in matrix form as

$$[l_{mn}][a_n] = [g_m], \tag{18.7}$$

where

$$[l_{mn}] = \begin{bmatrix} \langle w_1, L(f_1)\rangle & \langle w_1, L(f_2)\rangle & \cdots \\ \langle w_2, L(f_1)\rangle & \langle w_2, L(f_2)\rangle & \cdots \\ \cdots & \cdots & \cdots \end{bmatrix}, \quad [a_n] = \begin{bmatrix} a_1 \\ a_2 \\ \vdots \end{bmatrix}, \quad [g_m] = \begin{bmatrix} \langle w_1, g\rangle \\ \langle w_2, g\rangle \\ \vdots \end{bmatrix}. \tag{18.8}$$

If the matrix [l] is nonsingular, its inverse $[l]^{-1}$ exists and the a_n are given by

$$[a_n] = [l_{mn}]^{-1}[g_m] \tag{18.9}$$

and the unknown f is given from the weighted summation — Eq. (18.4). Assuming that the finite expansion basis is defined by$[\sim f_n] = [f_1, f_2, f_3, \ldots]$, the approximate solution for f is

$$f = [\tilde{f}_n][a_n] = [\tilde{f}_n][l_{mn}]^{-1}[g_m] \tag{18.10}$$

Depending on the choice of f_n and w_n this solution could be exact or approximate [22]. The most important aspect of MoM is the choice of expansion and testing functions. The f_n should be linearly independent and chosen such that a finite-term superposition approximates f quite well. The w_n should also be linearly independent. In addition, this choice is affected by the size of the matrix that has to be inverted (should be minimal), the ease of evaluation of the matrix elements, the accuracy of the desired solution, and the realization of a well-conditioned matrix [l]. The special choice $w_n = f_n$ gives Galerkin's method. The two most popular subsectional bases are the pulse function (step approximation) and the triangle function (piecewise linear approximation), as shown in Fig. 18.1. The numerical Gaussian quadrature rule [7] is used when the integrations involved in the evaluation of $l_{mn} = \langle w_m, L(f_n)\rangle$ are difficult to perform for common w_m's for a specific problem or when a more complex expansion basis is used. An alternative approach makes use of the Dirac delta functions for testing. This technique is called point matching and effectively satisfies Eq. (18.1) at discrete points in the region of interest. When

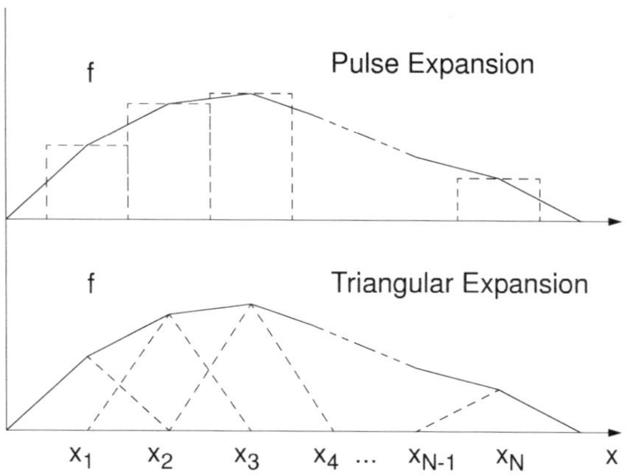

FIGURE 18.1 Moment method expansion in pulse and triangular basis.

the basis functions f_n exist only over subsections of the domain of f, each a_n affects the approximation of f only over a subsection of the region of interest (method of subsections). In this way, the evaluation of l_{mn} is simplified and the form of the matrix [l] is stripped and easier to invert.

The MoM involves setting up and solving dense, full, complex-valued systems. For a one-dimensional (1D) structure (e.g., a wire) of length L wavelengths, the size of the matrix is typically on the order of 10 L; for a three-dimensional (3D) structure with surface area S square wavelengths, the size is typically on the order of 100 S. Consequently, there are applications such as radar scattering from aircraft when the system of equations has order in excess of 1 million. Though MoM has been proved to be a very robust technique, it is plagued by the significant computational burden of having to solve a dense matrix equation of large order. In addition, modeling of a new structure requires the reformulation of the integral equation, a task that may require the very difficult derivation of a geometry-specific Green's function. MoM is used for the solution of various integral equations such as the electric field integral equation (EFIE) and the magnetic field integral equation (MFIE). The integral equation approach has the advantage that it rigorously takes into account the radiation conditions of any open structure and therefore it is not necessary to implement absorbing boundary conditions. The kernel of the integral equation is Green's function that accurately describes all possible wave propagation effects, such as radiation, leakage, anisotropy, and stratification.

The MoM has been used in many scattering problems [23–25], microstrip lines on multilayered dielectrics [26], microstrip antennas [27, 28], integrated waveguides in microwave [29, 30] and optical frequencies [31] even on anisotropical substrates [32]. In addition, results have been derived for the characterization of junctions [33], high-speed interconnects [34], viaholes [35], couplers [36], and infinite aperture arrays [37]. It should be emphasized that the discretization may be nonuniform, something that has been demonstrated in the application of MoM to planar circuits by Eleftheriades et al. [38].

Zhao and Chew [39] introduced a new method to precondition the matrix equation resulting from applying MoM to EFIE, This method leads to dramatic reductions in iteration count and allows for the use of fast solvers such as the low-frequency multilevel fast multipole algorithm (LF-MLFMA) [40]. In addition, Brown et al. [41] suggested one computationally effective way to enable the efficient large-domain MoM solutions to electrically large practical electromagnetic (EM) problems. Johnson and Rahmat-Samii [42] used genetic algorithms to facilitate the MoM modeling of integrated antenna elements and arrays. Recently, the MoM has been used for the accurate modeling and design of smart antennas [43], three-dimensional multilayer RF geometries [44, 45], and electromagnetic band gaps (EBGs) [46].

18.1.2 Spectral Domain Approach

Electromagnetic (EM) analysis in the spectral (Fourier transform) domain is preferable to many spatial domain numerical techniques, especially for planar transmission lines, microstrip antennas, and other planar-layered structures. If applied to an equation describing EM fields in planar media, a Fourier transform reduces a fully coupled, 3D equation to a 1D equation that depends on two parameters (the transform variables), but is uncoupled and can be solved independently at any value of those parameters. After solving the 1D equation, the inverse Fourier transform performs a superposition of the uncoupled 1D solutions to obtain the 3D solution. Thus, the process offers substantial savings because it effectively converts a 3D problem into a series of 1D problems.

Yamashita and Mittra [47] introduced the Fourier domain analysis for the calculation of the phase velocity and the characteristic impedance of a microstrip line. Using the quasi-TEM approximation (negligible longitudinal E- and H-fields), the line capacitance is determined by the assumed charge density through the application of the Fourier domain variational method. Denlinger [48] extended this approach to the full wave analysis of the same line. However, his solution depends strongly on the assumed current strip distributions. Itoh and Mittra [49] introduced a new technique, the Spectral Domain Approach (SDA), that allows for the systematic improvement of the solution accuracy to a desired degree. In SDA, the transformed equations are discretized using MoM, yielding a homogeneous system of equations to determine the propagation constant and the amplitude of current distributions from which the characteristic impedance is derived.

For metallic strip problems, the Fourier transform is performed along the direction parallel to the substrate interface and perpendicular to the strip. The first step is the formulation of the integral equation that correlates the E-field and the current distribution J along the strip and the application of the boundary conditions for E- and H-fields. Then, the Fourier transform is applied over E and J, and the MoM technique produces a system of algebraic equations that can be solved. Different choices of expansion and testing functions have been discussed by Aksun and Mittra [50]. SDA is applicable to most planar transmission lines (microstrips, fin lines, CPWs) [51–56]; microstrip antennas and arrays [57, 58]; interconnects [59]; junctions [60]; dielectric waveguides [61]; resonators of planar configurations [62]; and embedded passives [63] and micromachined devices [64] on single or multilayered [65] dielectric substrates, including conductors with finite conductivity [66]. This method requires significant analytical preprocessing, something that improves its numerical efficiency, but also restricts its applicability especially for structures with finite conductivity strips and arbitrary dielectric discontinuities.

18.1.3 Mode-Matching Technique

This technique is usually applied to the analysis of wave guiding/packaging discontinuities that involve numerous field modes. The fields on both sides of the discontinuity are expanded in summations of the modes in the respective regions with unknown coefficients [67] and appropriate continuity conditions are imposed at the interface to yield a system of equations. As an example, to analyze a waveguide step discontinuity with TE_{n0} excitation, E_y and H_x fields are written as the superposition of the modal functions $f_{an}(x)$ and $f_{bn}(x)$ for $n = 1, 2, \ldots$, respectively, for the left waveguide (waveguide A) and the right waveguide (waveguide B), as it is displayed in Fig. 18.2. Both of these fields should be continuous at the interface $z = 0$. Thus

$$\sum_{n=1}^{\infty}\left(A_n^+ + A_n^-\right)\phi_{an} = \begin{cases} \sum_{n=1}^{\infty}\left(B_n^+ + B_n^-\right)\phi_{bn}, & 0 < x < b \\ 0, & b < x < a \end{cases} \tag{18.11}$$

$$\sum_{n=1}^{\infty}\left(A_n^+ - A_n^-\right)Y_{an}\,\phi_{an} = \begin{cases} \sum_{n=1}^{\infty}\left(B_n^+ - B_n^-\right)Y_{bn}\,\phi_{bn}, & 0 < x < b \\ 0, & b < x < a \end{cases} \tag{18.12}$$

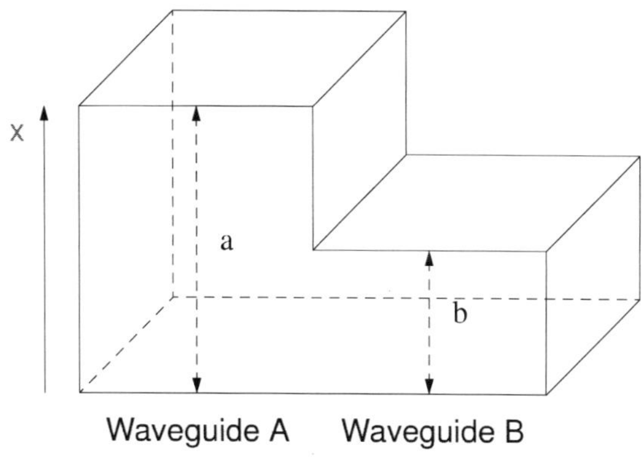

FIGURE 18.2 Rectangular waveguide discontinuity.

where (+) and (–) indicate the modal waves propagating to the positive and negative z direction and Y_{an}, Y_{bn} are the mode impedances. Sampling these equations with f_{bm} (m-mode in waveguide B) and making use of the mode orthogonality in waveguide B,

$$\sum_{n=1}^{\infty} H_{nm}\left(A_n^+ + A_n^-\right) = B_m^+ + B_m^- \tag{18.13}$$

$$\sum_{n=1}^{\infty} H_{nm} Y_{an}\left(A_n^+ - A_n^-\right) = Y_{bm}\left(B_m^+ - B_m^-\right) \tag{18.14}$$

with $H_{nm} = \int_0^b f_{an}(x) f_{bm}(x)dx$. Similarly, sampling Eqs. (18.11) and (18.12) with f_{am} (m-mode in waveguide A) and making use of mode orthogonality in waveguide A, we have

$$A_m^+ + A_m^- = \sum_{n=1}^{\infty} H_{mn}\left(B_n^+ + B_n^-\right) \tag{18.15}$$

$$Y_{am}\left(A_m^+ - A_m^-\right) = \sum_{n=1}^{\infty} H_{mn} Y_{an}\left(B_n^+ - B_n^-\right) \tag{18.16}$$

Assuming that the structure is excited through A_n^+ terms, the calculation of A_n^- (reflected modes in A) and B_m^+ (transmitted modes in B) provide the scattering parameters for the analyzed structure through a procedure that involves matrix inversions.

The foundation of this technique is the expansion of an electromagnetic field in terms of an infinite series of normal modes. Because a computer's capacity for numerical calculation is finite, these summations have to be truncated, something that could lead to incorrect solutions if not performed efficiently.

The main criterion for this truncation is the convergence of the summation. A natural way to check it is to plot the numerical values of some desired parameters vs. the number of retained terms. The truncation is considered sufficient when the change in the parameters is smaller than prespecified criteria. The procedure becomes more complicated where there is a need for the truncation of two or more infinite series (bifurcated waveguide, step junction). The numerical results appear to converge to different values depending on the manner of the truncation, a phenomenon that is called relative convergence. It has been found that relative convergence is related to the violation of field distributions at the edge of a conductor at the boundary [9] and to the ill-conditioned situation of the linear system of the computation process [68]. Thus, either the edge condition or the condition number of the linear system can be used as a criterion to ensure the validity of modal analyses. Another common criterion is to plot the field distributions on both sides of the boundary and observe their matching conditions. The mode-matching method has been applied to analyze various discontinuities in waveguides with rectangular or circular cross sections [69–72], fin lines [73, 74], microstrip lines [75, 76], and coplanar waveguides [77, 78]. In addition, this method is used for closed-region scattering geometries involving a discrete set of modes, such as E-plane filters [79, 80], waveguide impedance transformers [81], power dividers [82], and microstrip filters [83]. Moreover, this technique has been implemented for the solution of eigenvalue problems, such as the resonant frequency of a cavity [84] and the performance of evanescent mode filters [85], because it can efficiently model both evanescent and propagating modes.

18.2 Partial Differential Equation Based Techniques

In contrast to the previous techniques, numerical methods based on the partial differential equation (PDE) solutions of Maxwell's equations yield either sparse matrices (frequency domain, finite-element methods) or no matrices at all (time-domain, finite-difference, or finite-volume methods). In addition, specifying a new geometry is reduced to a problem of mesh generation only. Thus, PDE solvers could provide a framework for a space and time (frequency) microscope permitting the EM designer to visualize with submicron and subpicosecond resolution the dynamics of EM wave phenomena propagating at light speed within proposed geometries.

18.2.1 Finite-Difference Time-Domain (FDTD) Technique

The Finite-Difference Time-Domain (FDTD) technique [10–13] is an explicit solution method for Maxwell's time-dependent curl equations. It is based on volumetric sampling of the electric and magnetic field distribution within and around the structure of interest over a period of time. The sampling is set below the Nyquist limit and typically more than ten samples per wavelength are required. The time step has to satisfy a stability condition. For simulations of open geometries, absorbing boundary conditions (ABC) are employed at the outer grid truncation planes to reduce spurious numerical reflections from the grid termination.

In 1966, Yee [86] suggested the solution of the first-order Maxwell equations in time and space instead of solving the second-order wave equation. In this way, the solution is more robust and more accurate for a wider class of structures. In Yee's discretization cell, E- and H-fields are interlaced by half space and time-gridding steps, as shown in Fig. 18.3. The spatial displacement is very useful in specifying field boundary conditions and singularities and leads to finite-difference expressions for the space derivatives that are central in nature and second-order accurate. The time displacement (leapfrog) is fully explicit, completely avoiding the problems involved with simultaneous equations and matrix inversion. The resulting scheme is nondissipative; numerical wave modes propagating in the mesh do not spuriously decay due to a nonphysical artifact of the time-stepping algorithm. Denoting any function u of space and time evaluated at a discrete point in the grid and at a discrete point in time as $u(i\Delta x, j\Delta y, k\Delta z, l\Delta t) = {}_l u_{i,j,k}$, where Δt is the time step and Δx, Δy, and Δz define the cell size along the x, y and z directions, the first partial space derivative of u in the x direction and the first time derivative of u are approximated with the following central differences, respectively:

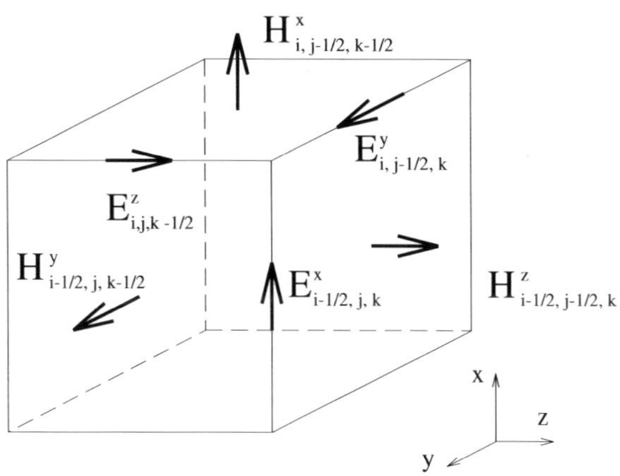

FIGURE 18.3 FDTD Yee cell.

$$\frac{\partial u}{\partial x}(i\Delta x, j\Delta y, k\Delta z, l\Delta t) = \frac{{}_l u_{i+1/2,j,k} - {}_l u_{i-1/2,j,k}}{\Delta x} + O\left[(\Delta x)^2\right]$$
$$\frac{\partial u}{\partial t}(i\Delta x, j\Delta y, k\Delta z, l\Delta t) = \frac{{}_{l+1/2} u_{i,j,k} - {}_{l-1/2} u_{i,j,k}}{\Delta t} + O\left[(\Delta t)^2\right] \tag{18.17}$$

By applying Eq. (18.17), the FDTD equations are derived for all field components. For example,

$$\begin{aligned}{}_{l+0.5}H^x_{i,j-0.5,k-0.5} &= \left(\frac{1-\dfrac{\rho'_{i,j,k}\Delta t}{2\mu_{i,j,k}}}{1+\dfrac{\rho'_{i,j,k}\Delta t}{2\mu_{i,j,k}}}\right){}_{l-0.5}H^x_{i,j-0.5,k-0.5} \\ &+ \left(\frac{\dfrac{\Delta t}{\mu_{i,j,k}}}{1+\dfrac{\rho'_{i,j,k}\Delta t}{2\mu_{i,j,k}}}\right)\left(\frac{{}_lE^y_{i,j-0.5,k} - {}_lE^y_{i,j-0.5,k-1}}{\Delta z} - \frac{{}_lE^z_{i,j,k-0.5} - {}_lE^z_{i,j-1,k-0.5}}{\Delta y}\right)\end{aligned} \tag{18.18}$$

where $\rho'_{i,j,k}$ is the magnetic loss coefficient for the (i, j, k) cell. It can be observed that a new value of a field vector component at any space lattice point depends only on its previous value and the previous values of the components of the other field vectors at adjacent points. Therefore, at any given time step, the value of a field vector component at p different lattice points can be calculated simultaneously if p parallel processors are employed, demonstrating that the FDTD algorithm is highly parallelizable. Holland [87] suggested an exponential time stepping to model the exponential decay of propagating waves in certain highly lossy media that the standard Yee time-stepping algorithm fails to describe. Stability analysis [88] has shown that the upper bound for the FDTD time step for a homogeneous region of space (ϵ_r, μ_r) is given by

$$\Delta t \le \frac{\sqrt{\epsilon_r \mu_r}}{c\sqrt{\dfrac{1}{(\Delta x)^2} + \dfrac{1}{(\Delta y)^2} + \dfrac{1}{(\Delta z)^2}}} \ (\textbf{3D simulations}), \quad \Delta t \le \frac{\sqrt{\epsilon_r \mu_r}}{c\sqrt{\dfrac{1}{(\Delta x)^2} + \dfrac{1}{(\Delta y)^2}}} \ (\textbf{2D simulations})$$

Lower values of upper bounds are used in case a highly lossy material or a variable grid is employed. Discretization with at least 10 to 20 cells per wavelength almost guarantees the FDTD algorithm to have satisfactory dispersion characteristics (phase error smaller than 5°/λ for a time step close to the upper bound value).

The computational domain must be large enough to enclose the structure of interest, and a suitable ABC on the outer perimeter of the domain must be used to simulate its extension to infinity to minimize numerical reflections for a wide range of incidence angles and frequencies. Central differences cannot be implemented at the outermost lattice planes, because by definition no information exists concerning the fields at points one-half space cell outside these planes. The perfectly matched layer (PML) ABC, introduced in 2D by Berenger in 1994 [89] and extended to 3D by Katz et al. [90], provides numerical reflection comparable to the reflection of anechoic chambers with values –40 dB lower than the previous absorbers. A new ABC based on Green's functions that absorbs efficiently propagating and evanescent modes has also been demonstrated [91, 92] for waveguide and RF packaging structures.

An incoming plane wave source [86] is very useful in modeling radar scattering problems, because in most cases of this type the target of interest is in the near field of the radiating antenna, and the incident illumination can be considered to be a plane wave.

The hard source [93] is another common FDTD source implementation. It is set up simply by superimposing a desired time function onto specific electric or magnetic field components in the FDTD space lattice that are regularly updated by the FDTD equations. Collinear arrays of hard source field vector components in 3D can be useful for exciting waveguides and strip lines. In the FDTD simulations of microstrip and stripline structures, the Gaussian pulse (nonzero DC content) is used as the excitation of the microstrip and strip line structures. The Gabor function

$$s(t) = e^{-((t-t_o)/(pw))^2}\sin(\omega t) \tag{18.19}$$

where $pw = \dfrac{2\sqrt{6}}{\pi\left(f_{max} - f_{min}\right)}$, $t_o = 2pw$, $\omega = \pi(f_{\min} + f_{\max})$, is used as the excitation of the waveguide structures, because it has zero DC content. By modifying the parameters *pw* and *w*, the frequency spectrum of the Gabor function can be practically restricted to the interval $[f_{\min}, f_{\max}]$. As a result, the envelope of the Gabor function represents a Gaussian function in both time and frequency domain. Monochromatic simulations are performed through the use of continuous-wave (sinusoidal) excitations.

It is very common, especially for high-speed circuit structures, to use a cell size Δ that is dictated by the very fine dimensions of the circuit and is almost always much finer than that needed to resolve the smallest spectral wavelength propagating in the circuit. As a result, with the time step Δt bound to Δ by numerical stability considerations, FDTD simulations have to run for tens of thousands of time steps to fully evolve the impulse responses needed for calculating impedances, S parameters, or resonant frequencies. One popular way to avoid virtually prohibitive execution time has been to apply contemporary analysis techniques from the discipline of digital signal processing and spectrum estimation. The strategy is to extrapolate the EM field time waveform by 10:1 or more beyond the actual FDTD time window, allowing a very good estimate of the complete system response with 90% or greater reduction in computation time. This extrapolation can be performed using forward–backward predictors [94] or autoregressive (AR) models [95].

The FDTD technique has found numerous applications in modeling microwave devices such as waveguides, resonators, transmissions lines, vias, antennas, and active and passive elements. In 1985, DePourcq [96] used FDTD to analyze various three-dimensional waveguide devices. Navarro et al. [97] investigated rectangular, circular, and T junctions in square coaxial waveguides and narrow-wall, multiple-slot couplers. Wang et al. [98] studied the Q factors of resonators using FDTD. Liang et al. [99] used FDTD to analyze coplanar waveguides and slot lines and Sheen et al. [100] presented FDTD results for various microstrip structures including a rectangular patch antenna, a low-pass filter, and a branch line coupler. Cangellaris and Wright [101] estimated the effect of the staircasing approximation of conductors of arbitrary orientation.

The characterization of interconnect transitions in multichip and microwave circuit modules has also been investigated using FDTD. Lam et al. [102] used a nonuniform mesh to model microstrip-to-via-to-strip line connections. Piket-May et al. [103] studied pulse propagation and cross talk in a computer module with more than ten metal-dielectric-metal layers and numerous vias. Luebbers et al. [104] and Shlager and Smith [105] developed and described in detail efficient 3D, time-domain, near-to-far-field transformations. In 1990, Maloney et al. [106] presented accurate results for the radiation from rotationally symmetric simple antennas such as cylindrical and conical monopoles, whereas Luebbers and coworkers [107, 108] presented mutual coupling and gain computations for a pair of wire dipoles and Tirkas and Balanis [109] modeled 3D horn antennas. Uehara and Kagoshima [110] analyzed microstrip phased-array antennas and Jensen and Rahmat-Samii [111] presented results for the input impedance and gain of monopoles, planar inverted-F antennas (PIFAs), and loop antennas on handheld transceivers.

In addition, Taflove [112] used FDTD to model scattering and compute near and far fields and radar cross section (RCS) for 2D and 3D structures. Britt [113] calculated the RCS of both two- and three-dimensional perfectly conducting and dielectric scatterers. In 2002, Bushyager et al. [114] proposed an adaptive FDTD-based modeling technique that couples Maxwell's, mechanical and solid-state equations for the simulation of MEMS devices with moving parts [115] and of submicron transistors in wireless and microwave transceivers.

Another area of FDTD applications is active and passive device modeling. Two different approaches are used. In the first, analytical device models are coupled directly with FDTD. In the second, lumped-element subgrid models are used with the device behavior determined by other software, something that may be preferable in the modeling of active devices with complicated equivalent circuits. In 1992, Sui et al. [116] reported a two-dimensional FDTD model with lumped circuit elements, including nonlinear devices, such as diodes and transistors, this approach was extended to 3D by Piket-May et al. [103] and Ciampolini et al. [117]. Kuo et al. [118] presented a large-signal analysis of packaged nonlinear active microwave circuits. Alsunaidi et al. [119] developed an active device model that couples the Yee update equations with the solution of the current continuity equation, the energy-conservation equation, and the momentum-conservation equations. Thomas et al. [120] developed an approach for coupling SPICE-lumped elements into the FDTD method.

Dey and Mittra [121, 122] introduced a very simple, stable and accurate contour-path technique to model curved metal surfaces in practical microwave and mixed-signal structures. Maloney and Kesler [123] introduced several novel means to analyze periodic (PBG, EBG) structures using FDTD; and Painter et al. [124] employed FDTD to design, construct, and successfully test the world's smallest microcavity laser based on a 2D photonic bandgap structure. In 2000, Zheng et al. [125] introduced the first 3D alternating-direction implicit (ADI) FDTD algorithm with provable unconditional stability regardless of the size of the time step, something that could be useful for large geometries such as photonic structures [126].

18.2.2 Transmission Line Matrix Method (TLM)

The TLM method [127–129] is similar to FDTD. The main difference is that the EM problem is analyzed through the use of a 3D equivalent network problem [130]. It is a very versatile time-domain technique and discretizes the computational domain using cubic cells with a period Δl. Boundaries corresponding to perfect electric (magnetic) conductors are represented by short-circuited (open-circuited) parallel nodes on the boundary. Variations of dielectric and diamagnetic constants [131] are introduced by adding short-circuited series stubs of length $\Delta l/2$ at the series (H-field) nodes and open-circuited $\Delta l/2$ stubs at the shunt (E-field) nodes. Losses can be introduced by resistively loading the shunt nodes. After the time-domain response is obtained, the frequency response is calculated using the Fourier transform. Due to the introduction of periodic lattice structures, a typical passband–stopband phenomenon appears in the frequency domain data. The frequency range must be below the upper bound of the lowest passband and is determined by the mesh size Δl.

The TLM technique has been used in the analysis of wave guiding structures [132], making use of the Diakoptics theorem [133], and has been extensively compared with the FDTD technique [134]. Effective numerical absorbers [135–137] including the FDTD-popular PML absorber have been derived and implemented in the modeling of radiating structures [138, 139]. In addition, TLM has been employed in the analysis of bondwire packaging [140] and MEMS switches [141]. So and Hoefer [142] coupled TLM with SPICE and performed a coupled field and circuit time-domain simulation. Paul et al. presented ways to model with TLM materials with frequency-dependent [143], anisotropic [144], nonlinear [145] properties. Recently TLM was applied to the calculation of SAR properties of cellular phones [146] and to the modeling of longitudinally periodic waveguides [147].

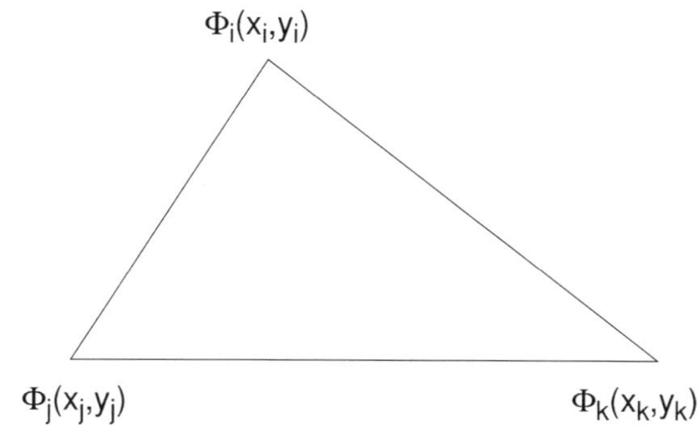

FIGURE 18.4 FEM triangular (2D) element.

18.2.3 Finite Element Method (FEM)

In the finite element method (FEM) [14–20], instead of partial differential equations with boundary conditions, corresponding functionals (e.g., power) are set up and variational expressions are applied to each cell (element) of the area of interest. Most of the time the elements are rectangles or triangles for 2D problems and parallelepiped (bricks) or tetrahedra for 3D problems, which allow for the efficient representation of most arbitrary shapes.

Assume that the 2D (x–y) Laplace equation is to be solved

$$\frac{\partial^2\phi}{\partial x^2}+\frac{\partial^2\phi}{\partial y^2}=0 \tag{18.20}$$

The solution is equivalent to the minimization of the functional

$$I(\phi)=\langle\phi,\nabla^2\phi\rangle=\iint_S \phi\left(\frac{\partial^2\phi}{\partial x^2}+\frac{\partial^2\phi}{\partial y^2}\right)dx\,dy=-\iint_S\left[\left(\frac{\partial\phi}{\partial x}\right)^2+\left(\frac{\partial\phi}{\partial y}\right)^2\right]dx\,dy. \tag{18.21}$$

The last integral is the superposition of each element's contribution. In each two-dimensional element ϕ can be approximated as a polynomial of variables x and y. For example, for a triangular element p (Fig. 18.4),

$$\phi=a+a_x x+a_y y, \tag{18.22}$$

where the constants a, a_x, a_y depend on the ϕ values at the three vertices of the triangle

$$\phi_p=a+a_x x_p+a_y y_p, \tag{18.23}$$

where $p = i, j, k$ are the three triangle vertices. Due to the first derivatives of Eq. (18.22), only a_x, a_y are needed for the calculation of I(ϕ), thus

$$\begin{bmatrix} a_x \\ a_y \end{bmatrix} = A \begin{bmatrix} \phi_i \\ \phi_j \\ \phi_k \end{bmatrix} \tag{18.24}$$

For each (i, j, k) element, I(ϕ) gets the value

$$I_{i,j,k}(\phi) = \left[\phi_i, \phi_j, \phi_k\right] A^T A \begin{bmatrix} \phi_i \\ \phi_j \\ \phi_k \end{bmatrix} \cdot |\Delta S|, \tag{18.25}$$

where A^T is the transpose of A and ΔS is the area of the triangle (element) calculated by

$$\Delta S = \frac{1}{2} \begin{vmatrix} 1 & x_i & y_i \\ 1 & x_j & y_j \\ 1 & x_k & y_k \end{vmatrix}. \tag{18.26}$$

The Rayleigh–Ritz technique is used for the minimization of $I_{i,j,k}$

$$\frac{\partial I_{i,j,k}}{\partial \phi_i} = \frac{\partial I_{i,j,k}}{\partial \phi_j} = \frac{\partial I_{i,j,k}}{\partial \phi_k} = 0 \tag{18.27}$$

As a result, for each (i, j, k) element

$$A^T A \begin{bmatrix} \phi_i \\ \phi_j \\ \phi_k \end{bmatrix} = 0 \tag{18.28}$$

After iterating this procedure to all elements of the computational domain S and using a connection matrix to account for points that are vertices common to more than one element, the following matrix equation is derived

$$\mathbf{B} \begin{bmatrix} \phi_1 \\ \phi_2 \\ \vdots \\ \phi_N \end{bmatrix} = 0 \tag{18.29}$$

After plugging in the known values of the ϕ_i's that are located on the boundaries, the rest of the ϕ_i's are calculated through the inversion of matrix **B**. In this way, the potentials of all interior points can be given by Eq. (18.22).

Various element forms [148–152] have been used to minimize the memory requirements and facilitate the gridding procedure and the modeling of boundary conditions (PECs, dielectric interfaces). The effect of discretization error on the numerical dispersion has been extensively studied and gridding guidelines have been derived by Lee and Cangellaris [153] and Warren and Scott [154].

In addition, the analysis of radiating (antennas) and scattering problems has led to the development of numerical absorbers [155–157] with very low numerical reflection coefficients. Due to the shape of the finite elements, the FEM technique can accurately represent very complex geometries and is one of the most popular techniques for scattering problems [158], discontinuities and transitions [159], packaging [160], interconnects [161], and MMIC modeling [162]. Li et al. [163] optimized patch antennas on ferrite substrates and Zhu and Cangellaris [164] presented a way to model lossy media. Hung and Senturia [165] applied FEM to MEMS including mechanical motion equations, and Ammous et al. [166] modeled the thermal characteristics of power semiconductor devices.

Anderson and Volakis [167] and Zhu and Cangellaris [168] used hierarchical vector finite elements to perform an adaptive multiresolution modeling of antennas and microwave structures.

The boundary element (BE) technique [169, 170] is a combination of the boundary integral equation and a finite-element discretization applied to the boundary. In essence, it is a form of the integral equation–MoM approach discussed previously. The wave equation for the volume is converted to the surface integral equation through Green's identity. The surface integrals are discretized into N elements, and the evaluation is performed for each element after E- and H-fields are approximated by polynomials. Due to the reduction of the number of dimensions, there is a significant reduction in memory and CPU time requirements. The BE technique has been utilized in the analysis of cavities [171] and of planar layered media [172], and in the incorporation of lumped elements in FEM analysis [173].

18.3 Hybrid Techniques

As has become clear from the previous discussion, all numerical techniques have advantages and disadvantages depending on the geometry to be modeled. Integral equation techniques allow for the quick and efficient modeling of radiation phenomena, but derivation of Green's function for complex structures is tedious. The MM method is more appropriate for wave-guiding structures where modes are easily determined. The FDTD (and TLM) technique is quite general and requires no preprocessing, though it must often be run for medium to large execution times. The FEM technique is adaptive due to the shape of the elements, but gridding and functional optimization demands significant computational effort.

Thus, there have been numerous efforts for the development of hybrid simulation approaches that use different techniques for different subgeometries and utilize connection relationships for the areas of numerical interfaces. Jin and Volakis, [174], Gedney et al. [175], and Musolino and Raugi [176] proposed a hybrid FEM/MoM method for the modeling of wave scattering by 3D apertures, wave diffraction in gratings, and nonlinear analysis of microwave devices. Wu and Itoh [177] and Monorchio and Mittra [178] suggested an FEM/FDTD approach for the multifrequency modeling of complex geometries. The MM technique has been coupled with the integral equation [179], spectral domain [180], and FEM [181,182] to analyze complicated passive microwave circuits and wave-guiding problems including inductive loading and wave scattering. Lindenmeier et al. [183, 184] introduced a hybrid TLM/MIE for thin wire modeling and for the estimation of the EM interaction between complex objects (e.g., MMICs) separated by large free-space regions, and Pierantoni et al. [185] analyzed numerical aspects of MoM FDTD and TLM integral equation used in EMC modeling.

18.4 Wavelets: A Memory-Efficient Adaptive Approach?

The term *wavelet* [186–190] has a very broad meaning, ranging from singular integral operators in harmonic approach to sub-band coding algorithms in signal processing, from coherent states in quantum analysis to spline analysis in approximation theory, from multiresolution transform in computer vision to a multilevel approach in the numerical solution of partial differential equations, and so on. Most of the time wavelets could be considered mathematical tools for waveform representations and segmentations, time-frequency analysis, and fast and efficient algorithms for easy

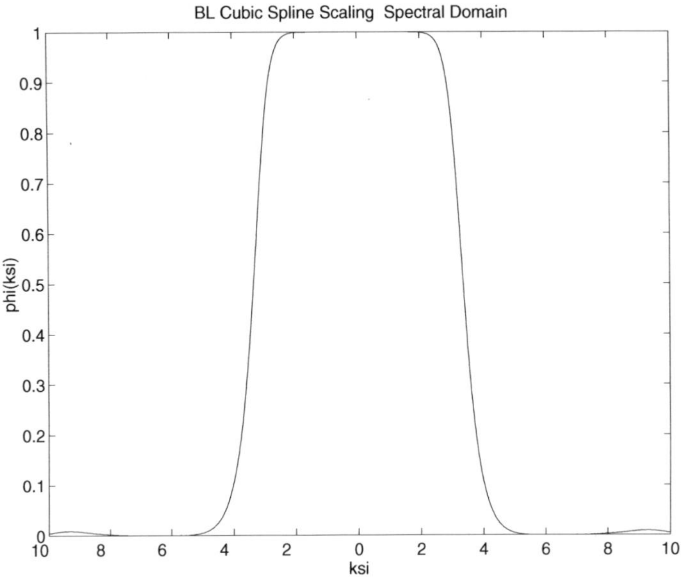

FIGURE 18.5 Battle–Lemarie scaling-spectral (low-pass) domain.

implementation in both time and frequency domains. One of the most important characteristics of expansion to scaling and wavelet functions is the time (domain)-frequency (Fourier-transformed domain) localization.

Another very salient feature of these new expansions is that the entire set of basis functions is generated by the dilation and shifting of a single function, called the mother wavelet. The standard approach in ideal low-pass (scaling, Fig. 18.5) and band-pass (wavelet, Fig. 18.6) filtering for separating an analog signal into different frequency bands emphasizes the importance of time localization. Multiresolution analysis (MRA), introduced by Mallat [191] and Meyer [192], provides a very powerful tool for the construction of wavelets and implementation of the wavelet decomposition and reconstruction algorithms. In the case of cardinal B splines [193], an orthonormalization process is used to produce an orthonormal scaling function and, hence, its corresponding orthonormal wavelet by a suitable modification of the two-scale sequence. The orthonormalization process was introduced by Schweinler and Wigner [194]; and the resulting wavelets are the Battle–Lemarie wavelets, obtained independently by Battle [195] and Lemarie [196] using different methods. The only orthonormal wavelet that is symmetric or antisymmetric and has compact support (to give finite decomposition and reconstruction series) is the Haar [197] wavelet. Nevertheless, these wavelets exhibit poor time-frequency localization. Another set of orthonormal basis is the Daubechies wavelets [198]. At present, MRA has been applied to alleviate the numerical disadvantages mainly of the integral equation (IE) and FDTD methods, though preliminary efforts for FEM are currently under investigation

It is well known that the IE method described in previous sections offers a straightforward and efficient numerical solution when applied to small- to medium-scale problems. Difficulties arise when the complexity of the geometry and subsequently the number of the unknowns increases, resulting in very large matrices. All conventional basis functions traditionally used in MoM generate full moment matrices. The computational problems associated with the storage and manipulation of large, densely populated matrices easily rule out the practicality of the integral equation techniques. The potential application of wavelet theory in the numerical solution of integral equations led to the finding that wavelet expansion of certain types of integral operators generates highly sparse linear systems [199]. This proposition was used [200–202] in the MoM formulation of one-dimensional EM scattering problems and in the analysis of

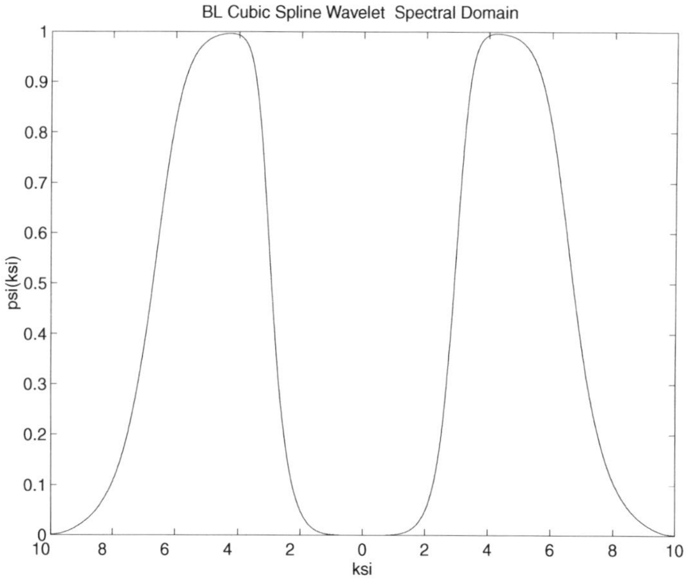

FIGURE 18.6 Battle–Lemarie wavelet-spectral (band-pass) domain.

integrated millimeter and submillimeter waveguides with a Battle–Lemarie orthonormal basis [203]. Sparsity results above 90% allowed for the accurate modeling of structures that could not be analyzed with IE using conventional expansions. Various expansion basis functions have been used [204] leading to significant computational economies in the modeling of 2D and 3D path propagation and scattering problems from regular or rough surfaces [205] and of high-speed interconnects [206].

As far as it concerns FDTD, despite its numerous applications, many practical geometries, especially in microwave and millimeter-wave integrated circuits (MMIC), packaging, interconnects, subnanosecond digital electronic circuits such as multichip modules (MCM), and antennas used in wireless and microwave communication systems, have been left untreated due to their complexity and the inability of the existing techniques to deal with requirements for large size and high resolution. Krumpholz and Katehi [207] have shown that Yee's FDTD scheme can be derived by applying the MoM for the discretization of Maxwell's equations using pulse basis functions for the expansion of the unknown fields.

The use of scaling and wavelet functions as a complete expansion basis of the fields demonstrates that MultiResolution Time Domain (MRTD) [207] schemes are generalizations of Yee's FDTD and can extend this technique's capabilities by improving computational efficiency and substantially reducing computer resources.

In an MRTD scheme, the fields are represented by a twofold expansion in scaling and wavelet functions with respect to time and space. Scaling functions (low pass) guarantee a correct modeling of smoothly varying fields. In regions characterized by strong field variations or field singularities, higher resolution is enhanced by incorporating wavelets in the field expansions. The major advantage of the use of MRTD in the time domain is the capability to develop time and space adaptive grids through the thresholding of the wavelet coefficients for each time step throughout the grid. MRTD schemes based on cubic spline Battle–Lemarie scaling and wavelet functions have been used for the derivation of time/space-adaptive schemes [208, 209] in real time.

Various types of wavelets have been successfully applied to the simulation of nonlinear effects [210], 3D dielectric cavities [211], filters [212], mixers [213], RF packaging structures [214], optical waveguides [215], mine detection [216], and millimeter-wave integrated circuits [217] offering economies in memory

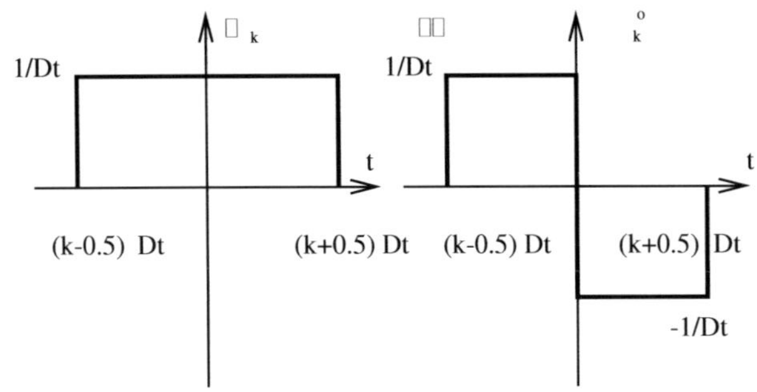

FIGURE 18.7 Haar expansion basis

and execution time of orders of magnitude with respect to FDTD. Dispersion analysis [218] shows the capability of excellent accuracy with up to two points per wavelength (Nyquist limit) for the Battle–Lemarie basis. Nevertheless, the functions of this family do not have compact support; thus, the MRTD schemes have to be truncated with respect to space, something that requires the use of image theory for the modeling of hard boundaries (e.g., PECs).

Therefore, specific problems may require the use of functions with compact support (e.g., Haar, Fig. 18.7), especially for the approximation of time derivatives. In this way, the modeling of boundary conditions is most straightforward, though the computational economies are not as dramatic [219–221]. Sarris and Katehi [222] proposed a way for an improvement of the accuracy of MRTD algorithms modifying the offset of electric and magnetic fields, and Cao et al. [223] presented an anisotropic PML for the modeling of open structures. Goasguen [224] and Bushyager et al. [114] coupled the EM and the semiconductor device simulations using interpolating and Haar wavelets, respectively, and achieving very significant computational economies. In addition, Bushyager et al. [225] proposed a way of modeling multiple PEC's within a single MRTD cell, something that would accelerate EBC and RF packaging structure simulations.

18.5 Conclusions

This chapter has briefly presented various numerical techniques that are commonly used for the analysis and design of RF and microwave circuits. Their fundamental features as well as their advantages and disadvantages have been discussed and representative references have been reported. It must be emphasized that there is no numerical scheme that can achieve optimal performance for all types of structures. Thus, the hybridization of these techniques or the implementation of novel approaches (e.g., wavelets), successfully applied in other research areas, would be better candidates for the generalized, efficient, and accurate modeling of modern complex devices used in telecommunication, radar, and computing applications.

References

1. T. Itoh, *Numerical Techniques for Microwave and Millimeter-Wave Passive Structures,* John Wiley & Sons, New York, 1989.
2. R.C. Booton, *Computational Methods for Electromagnetics and Microwaves,* John Wiley & Sons, New York, 1992.
3. R.F. Harrington, *Field Computation by Moment Methods*, Macmillan, New York, 1968.
4. R.C. Hansen, *Moment Methods in Antennas and Scattering,* Artech House, Norwood, MA, 1990.
5. J.H. Wang, *Generalized Moment Methods in Electromagnetics*, John Wiley & Sons, New York, 1991.
6. R. Bancroft, *Understanding Electromagnetic Scattering Using the Moment Method: A Practical Approach*, Artech House, Norwood, MA, 1996.

7. A.F. Peterson, S.L. Ray, and R. Mittra, *Computational Methods for EM*, IEEE Press, Piscataway, NJ, 1998.
8. C.R. Scott, *The Spectral Domain Method in Electromagnetics*, Artech House, Norwood, MA, 1989.
9. R. Mittra and W.W. Lee, *Analytical Techniques in the Theory of Guided Waves*, Macmillan, New York, 1971.
10. K.S. Kunz and R.J. Luebbers, *The Finite Difference Time-Domain Method for Electromagnetics*, CRC Press, Boca Raton, FL, 1993.
11. A. Taflove and S.C. Hagness, Computational Electrodynamics: *The Finite-Difference Time-Domain Technique*, 2nd ed., Artech House, Norwood, MA, 2000.
12. A. Taflove, *Advances in Computational Electrodynamics*, Artech House, Norwood, MA, 1998.
13. T. Itoh and B. Housmand, *Time-Domain Methods for Microwave Structures: Analysis and Design*, IEEE Press, Piscataway, NJ, 1998.
14. G. Strang and G.J. Fix, *An Analysis of the Finite Element Method*, Prentice Hall, Englewood Cliffs, NJ, 1973.
15. P.P. Silvester and R.L. Ferrari, *Finite Elements for Electrical Engineers*, Cambridge University Press, Cambridge, 1983.
16. J.N. Reddy, *An Introduction to the Finite Element Method*, McGraw-Hill, New York, 1984.
17. J.-C. Sabonnadiere and J.-L. Coulomb, *Finite Element Methods in CAD: Electric and Magnetic Fields*, Springer-Verlag, New York, 1987.
18. O.C. Zienkiewicz and R.L. Taylor, *The Finite Element Method*, McGraw-Hill, London, 1988.
19. T. Itoh, G. Pelosi, and P.P. Silvester, *Finite Element Software for Microwave Engineering*, John Wiley & Sons, New York, 1996.
20. J.L. Volakis, A Chaterjee, and L.C. Kempel, *Finite Elements Method for Electromagnetics*, IEEE Press, Piscataway, NJ, 1998.
21. A.W. Glisson and D.R. Wilton, Simple and efficient methods for problems of radiation and scattering from surfaces, *IEEE Trans. Antennas Propag.*, 28, 593–603, 1980.
22. T.K. Sarkar, A.R. Djordjevic, and E. Arvas, On the choice of expansion and weighting functions in the numerical solution of operator equations, *IEEE Trans. Antennas Propag.*, 33, 988–996, 1985.
23. J.R. Mossig and F.E. Gardiol, Analytical and numerical techniques in the Green's function treatment of microstrip antennas and scatterers, *IEEE Proc.*, Part H, 130, 175–182, 1983.
24. A.J. Poggio and E.K. Miller, Integral equation solutions of three-dimensional scattering problems, in *Computer Techniques for Electromagnetics,* R. Mittra, Ed., Hemisphere, New York, 1987.
25. A.F. Peterson and P.W. Klock, An improved MFIE formulation for TE-wave scattering from lossy, inhomogeneous dielectric cylinders, *IEEE Trans. Antennas Propag.*, 25, 518–524, 1987.
26. F. Ling, D. Jiao, and J.M. Jin, Efficient EM modeling of microstrip structures in multilayer media, *IEEE Trans. Microwave Theory Tech.*, 47, 1810–1818, 1999.
27. D.M. Pozar, Input impedance and mutual coupling of rectangular microstrip antennas, *IEEE Trans. Antennas Propag.*, 30, 1191–1196, 1982.
28. W.-T. Chen and H.-R. Chuang, Numerical computation of the EM coupling between a circular loop antenna and a full-scale human-body model, *IEEE Trans. Microwave Theory Tech.*, 1516–1520, 1998.
29. G. Athanasoulias and N.K. Uzunoglu, An accurate and efficient entire-domain basis Galerkin's method for the integral equation analysis of integrated rectangular dielectric waveguides, *IEEE Trans. Microwave Theory Tech.*, 43, 2794–2804, 1995.
30. M. Swaminathan, T.K. Sarkar, and A.T. Adams, Computation of TM and TE modes in waveguides based on surface integral formulation, *IEEE Trans. Microwave Theory Tech.*, 40, 285–297, 1992.
31. S.J. Polychronopoulos and N.K. Uzunoglu, Propagation and coupling properties of integrated optical waveguides — an integral equation formulation, *IEEE Trans. Microwave Theory Tech.*, 44, 641–650, 1996.

32. P. Cottis and N. Uzunoglu, Integral equation approach for the analysis of anisotropic channel waveguides, *J. Opt. Soc. A. A*, 8, 4, 608–614, 1991.
33. P. Guillot, P. Couffignal, H. Baudrand, and B. Theron, Improvement in calculation of some surface integrals: application to junction characterization in cavity filter design, *IEEE Trans. Microwave Theory Tech.*, 41, 2156–2160, 1993.
34. T.E. van Deventer, L.P.B. Katehi, and A.C. Cangellaris, Analysis of conductor losses in high-speed interconnects, *IEEE Trans. Microwave Theory Tech.*, 42, 78–83, 1994.
35. A.W. Mathis and A.F. Peterson, Efficient electromagnetic analysis of a doubly infinite array of rectangular apertures, *IEEE Trans. Microwave Theory Tech.*, 46–54, 1998.
36. S-G. Hsu and R.-B. Wu, Full-wave characterization of a through hole via in multi-layered packaging, *IEEE Trans. Microwave Theory Tech.*, 1073–1081, 1995.
37. A.M. Rajeek and A. Chakraborty, Analysis of a wide compound slot-coupled parallel waveguide coupler and radiator, *IEEE Trans. Microwave Theory Tech.*, 43, 802–809, 1995.
38. G.V. Eleftheriades, J.R. Mosig, and M. Guglielmi, A fast integral equation technique for shielded planar circuits defined on nonuniform meshes, *IEEE Trans. Microwave Theory Tech.*, 44, 2293–2296, 1996.
39. J.-S. Zhao and W.C. Chew, Integral equation solution of Maxwell's equations from zero frequency to microwave frequencies, *IEEE Trans. Antennas Propag.*, 48, 10, 1635–1645, 2000.
40. W.C. Chew, J.M. Jin, C.C. Lu, E. Michielssen and J.M. Song, Fast solution methods in electromagnetics, *IEEE Trans. Antennas Propag.*, 45, 3, 533–543, 1997.
41. R.A. Brown, B.M. Notaros, B.D. Popovic, Z. Popovic, and J.P. Weem, Efficient large-domain MoM solutions to electrically large practical EM problems, *IEEE Trans. Microwave Theory Tech.*, 49, 1, 151–159, 2001.
42. J.M. Johnson and Y. Rahmat-Samii, Genetic algorithms and method of moments (GA/MOM) for the design of integrated antennas, *IEEE Trans. Antennas Propag.*, 47, 10, 1606–1614, 1999.
43. K.R. Dandekar, H. Ling, and G. Xu, Experimental study of mutual coupling compensation in smart antenna applications, *IEEE Trans. Wireless Communications*, 1, 3, 480–487, 2002.
44. F. Ling, J. Liu, and J.-M. Jin, Efficient electromagnetic modeling of three-dimensional multilayer microstrip antennas and circuits, *IEEE Trans. Microwave Theory Tech.*, 50, 6, 1628–1635, 2002.
45. M.R. Abdul-Gafoor, H.K. Smith, A.A. Kishk, and A.W. Glisson, Simple and efficient full-wave modeling of electromagnetic coupling in realistic RF multilayer PCB layouts, *IEEE Trans. Microwave Theory Tech.*, 50, 6, 1445–1457, 2002.
46. A.S. Barlevy and Y. Rahmat-Samii, Characterization of electromagnetic band-gaps composed of multiple periodic tripods with interconnecting vias: concept, analysis and design, *IEEE Trans. Antennas Propag.*, 49, 3, 343–353, 2001.
47. E. Yamashita and R. Mittra, Variational method for the analysis of microstrip line, *IEEE Trans. Microwave Theory, Tech.*, 16, 251–256, 1968.
48. E.J. Denlinger, A frequency dependent solution for microstrip transmission lines, *IEEE Trans. Microwave Theory, Tech.*, 19, 30–39, 1971.
49. T. Itoh and R. Mittra, Spectral-domain approach for calculating the dispersion characteristics of microstrip lines, *IEEE Trans. Microwave Theory, Tech.*, 21, 496–499, 1973.
50. M.I. Aksun and R. Mittra, Choices of expansion and testing functions for MoM applied to a class of EM problems, *IEEE Trans. Microwave Theory, Tech.*, 41, 503–509, 1993.
51. R.H. Jansen, Unified user-oriented computation of shielded, covered and open planar microwave and millimeter wave transmission-line characteristics, *IEEE J. Microwave Opt. Acoust.*, 3, 14–22, 1979.
52. T. Itoh, Spectral domain immitance approach for dispersion characteristics of generalized printed transmission lines, *IEEE Trans. Microwave Theory Tech.*, 28, 733–736, 1980.

53. J. Sercu, N. Fache, F. Libbrecht, and D. De Zutter, Full-wave space-domain analysis of open microstrip discontinuities including the singular current-edge behavior, *IEEE Trans. Microwave Theory Tech.*, 41, 1581–1588, 1993.
54. F. Olyslager, D. De Zutter, and K. Blomme, Rigorous full-wave analysis of propagation characteristics of general lossless and lossy multiconductor transmission lines in multilayered media, *IEEE Trans. Microwave Theory Tech.*, 41, 79–88, 1993.
55. K.K.M. Cheng and J.K.A. Everard, A new technique for the quasi-TEM analysis of conductor-backed coplanar waveguide structures, *IEEE Trans. Microwave Theory Tech.*, 41, 1589–1592, 1993.
56. N. Gupta and M. Singh, Investigation of periodic structures in a fin line: A space-spectral domain approach, *IEEE Trans. Microwave Theory Tech.*, 43, 2708–2710, 1995.
57. Y. Imaizumi, M. Shinagawa, and H. Ogawa, Electric field distribution measurement of microstrip antennas and arrays using electro-optic sampling, *IEEE Trans. Microwave Theory Tech.*, 43, 2402–2407, 1995.
58. Y.-D. Lin, J.-W. Sheen, and C.-K.C. Tzuang, Analysis and design of feeding structures for microstrip leaky wave antenna, *IEEE Trans. Microwave Theory Tech.*, 44, 1540–1547, 1996.
59. P. Petre and M. Swaminathan, Spectral domain technique using surface wave excitation for the analysis of interconnects, *IEEE Trans. Microwave Theory Tech.*, 42, 1744–1749, 1994.
60. B.L. Ooi, M.S. Leong, P.S. Kooi, and T.S. Yeo, Enhancements of the spectral-domain approach for analysis of microstrip Y-junction, *IEEE Trans. Microwave Theory Tech.*, 45, 1800–1805, 1997.
61. K. Sabetfakhri and L.P.B. Katehi, Analysis of general class of open dielectric waveguides by spectral-domain technique, *Proc. IEEE-MTT Symp.*, 3, 1523–1526, 1993.
62. T. Itoh, Analysis of microstrip resonators, *IEEE Trans. Microwave Theory Tech.*, MTT-22, 946–952, 1974.
63. S.-H. Song, H.-B. Lee, H.-K. Jung, S.-Y. Hahn, K.-S. Lee, C. Cheon, and H.-S. Kim, Spectral domain analysis of the spiral inductor on multilayer substrates, *IEEE Trans. Magnetics*, 33, 2, 1488–1491, 1997.
64. T.M. Weller, K.J. Herrick, and L.P.B. Katehi, Quasi-static design technique for mm-wave micromachined filters with lumped elements and series stubs, *IEEE Trans. Microwave Theory Tech.*, 45, 931–938, 1997.
65. Y.L. Chow, N. Hojjat, S. Safavi-Naeini, and R. Faraji-Dana, Spectral Green's functions for multilayer media in a convenient computational form, *IEEE Proc. Microwave, Antennas Propag.*, 145, 1, 85–91, 1998.
66. M. Farina and T. Rozzi, Spectral domain approach to two-dimensional modeling of open planar structures with thick lossy conductors, *IEEE Proc. Microwave, Antennas Propag.*, 147, 5, 321–324, 2000.
67. Y.C. Shih and K.G. Gray, Convergence of numerical solutions of step-type waveguide discontinuity problems by modal analysis, Proc. *IEEE-MTT Symp.*, 233–235, 1983.
68. M. Leroy, On the convergence of numerical results in modal analysis, *IEEE Trans. Antennas Propag.*, 31, 655–659, 1983.
69. A. Wexler, Solution of waveguide discontinuities by modal analysis, *IEEE Trans. Microwave Theory Tech.*, 15, 508–517, 1967.
70. W. Wessel, T. Sieverding, and F. Arndt, Mode-matching analysis of general waveguide multiport junctions, *Proc. IEEE-MTT Symp.*, 1273–1276, 1999.
71. U. Balaji and R. Vahldieck, Mode-matching analysis of circular-ridged waveguide discontinuities, *IEEE Trans. Microwave Theory Tech.*, 46, 2, 191–195, 1998.
72. K.L. Chan and S.R. Judah, Mode-matching analysis of a waveguide junction formed by a circular and a larger elliptic waveguide, *IEEE Proc. Microwaves, Antennas Propag.*, 145, 1, 123–127, 1998.
73. A.S. Omar and K. Schunemann, Transmission matrix representation of finite discontinuity, *IEEE Trans. Microwave Theory Tech.*, 33, 830–835, 1985.

74. R. Vahldieck and W.J.R. Hoefer, Finline and metal insert filters with improved passband separation and increased stopband attenuation, *IEEE Trans. Microwave Theory Tech.*, 33, 1333–1339, 1985.
75. W. Menzel and I. Wolff, A method for calculating the frequency-dependent properties of microstrip discontinuities, *IEEE Trans. Microwave Theory Tech.*, 25, 107–112, 1977.
76. T.S. Chu, T. Itoh, and Y.-C. Shih, Comparative study of mode-matching formulations for microstrip discontinuity problems, *IEEE Trans. Microwave Theory Tech.*, 33, 1018–1023, 1985.
77. F. Alessandri, G. Baini, M. Mongiardo, and R. Sorrentino, A three-dimensional mode matching technique for the efficient analysis of coplanar MMIC discontinuities with finite metallization thickness, *IEEE Trans. Microwave Theory Tech.*, 41, 1625–1629, 1993.
78. R. Schmidt and P. Russer, Modeling of cascaded coplanar waveguide discontinuities by the mode-matching approach, *IEEE Trans. Microwave Theory Tech.*, 43, 2910–2917, 1995.
79. Y.C. Shih, Design of waveguide E-plane filters with all metal insert, *IEEE Trans. Microwave Theory Tech.*, 32, 695–704, 1984.
80. F. Arndt et al., E-plane integrated filters with improved stopband attenuation, *IEEE Trans. Microwave Theory Tech.*, 32, 1391–1394, 1984.
81. F. Arndt et al., Computer-optimized multisection transforms between rectangular waveguides of adjacent frequency bands, *IEEE Trans. Microwave Theory Tech.*, 32, 1479–1484, 1984.
82. F. Arndt et al., Optimized E-plane T-junction series power divider, *IEEE Trans. Microwave Theory Tech.*, 35, 1052–1059, 1987.
83. R. Mehran, Computer-aided design of microstrip filters considering dispersion, loss and discontinuity effects, *IEEE Trans. Microwave Theory Tech.*, 27, 239–245, 1979.
84. L. Accatino, G. Bertin, and M. Mongiardo, Elliptical cavity resonators for dual-mode narrowband filters, *IEEE Trans. Microwave Theory Tech.*, 45, 2393–2401, 1997.
85. J. Bornemann and F. Arndt, Rigorous design of evanescent-mode E-plane finned waveguide bandpass filters, *Proc. IEEE-MTT Symp.*, 603–606, 1989.
86. K.S. Yee, Numerical solution of initial boundary value problems involving Maxwell's equations in isotropic media, *IEEE Trans. Antennas Propag.*, 14, 302–307, 1966.
87. R. Holland, THREDE: A free-field EMP coupling and scattering code, *IEEE Trans. Nuclear Science*, 24, 2416–2421, 1977.
88. A. Taflove and M.E. Brodwin, Numerical solution of steady-state electromagnetic scattering problems using the time-dependent Maxwell's equations, *IEEE Trans. Microwave Theory Tech.*, 23, 623–630, 1975.
89. J.-P. Berenger, A perfectly matched layer for the absorption of electromagnetic waves, *Computational Phys.*, 114, 185–200, 1994.
90. D.S. Katz, E.T. Thiele, and A. Taflove, Validation and extension to three dimensions of the Berenger PML absorbing boundary conditions for FDTD meshes, *IEEE Microwave Guided Wave Lett.*, 4, 344–346, 1994.
91. E. Tentzeris, M. Krumpholz, N. Dib, J.-G. Yook, and L.P.B. Katehi, FDTD characterization of waveguide probe structures, *IEEE Trans. Microwave Theory Tech.*, 46, 10, 1452–1460, 1998.
92. M. Werthen, M. Rittweger, and I. Wolff, FDTD simulation of waveguide junctions using a new boundary condition for rectangular waveguides, Proc. 24th European Microwave Conf., Cannes, France, 1715–1719, 1994.
93. A. Taflove, Computation of the electromagnetic fields and induced temperatures within a model of the microwave-irradiated human eye, Ph.D. Dissertation, Department of Electrical Engineering, Northwestern University, Evanston, IL, June 1975.
94. J. Chen, C. Wu, T.K. Lo, K.-L. Wu, and J. Litva, Using linear and nonlinear predictors to improve the computational efficiency of the FDTD algorithm, *IEEE Trans. Microwave Theory Tech.*, 42, 1992–1997, 1994.
95. V. Jandhyala, E. Michielssen, and R. Mittra, On the performance of different AR methods in the spectral estimation of FDTD waveforms, *Microwave Optical Technol. Lett.*, 7, 690–692, 1994.

96. M. DePourcq, Field and power density calculations in closed microwave systems by three-dimensional finite differences, *IEE Proc. H: Microwaves, Antennas Propag.*, 132, 360–368, 1985.
97. E.A. Navarro, V. Such, B. Gimeno, and J.L. Cruz, T-junctions in square coaxial waveguide: an FDTD approach, *IEEE Trans. Microwave Theory Tech.*, 42, 347–350, 1994.
98. C. Wang, B.-Q. Gao, and C.-P. Ding, Q factor of a resonator by the FDTD method incorporating perturbation techniques, *Electron. Lett.*, 29, 1866–1867, 1993.
99. G.-C. Liang, Y.-W. Liu, and K.K. Mei, Full-wave analysis of coplanar waveguide and slotline using FDTD, *IEEE Trans. Microwave Theory Tech.*, 37, 1949–1957, 1989.
100. D.M. Sheen, S.M. Ali, M.D. Abouzahra, and J.A. Kong, Application of the three-dimensional FDTD method to the analysis of planar microstrip circuits, *IEEE Trans. Microwave Theory Tech.*, 38, 849–857, 1990.
101. A.C. Cangellaris and D.B. Wright, Analysis of the numerical error caused by the stair-stepped approximation of a conducting boundary in FDTD simulations of electromagnetic phenomena, *IEEE Trans. Antennas Propag.*, 39, 1518–1525, 1991.
102. C.-W. Lam, S.M. Ali, and P. Nuytkens, Three-dimensional modeling of multichip module interconnects, *IEEE Trans. Components, Hybrids Manufacturing Technol.*, 16, 699–704, 1993.
103. M. Piket-May, A. Taflove, and J. Baron, FDTD modeling of digital signal propagation in three-dimensional circuits with passive and active loads, *IEEE Trans. Microwave Theory Tech.*, 42, 1514–1523, 1994.
104. R.J. Luebbers, K.S. Kunz, M. Schneider, and F. Hunsberger, An FDTD near zone to far zone transformation, *IEEE Trans. Antennas Propag.*, 39, 429–433, 1991.
105. K.L. Shlager and G.S. Smith, Comparison of two FDTD near-field to near-field transformations applied to pulsed antenna problems, *Electron. Lett.*, 31, 936–938, 1995.
106. J.G. Maloney, G.S. Smith, and W.R. Scott, Jr., Accurate computation of the radiation from simple antennas using the FDTD method, *IEEE Trans. Antennas Propag.*, 38, 1059–1069, 1990.
107. R. Luebbers and K. Kunz, FDTD calculations of antenna mutual coupling, *IEEE Trans. Electromagn. Compat.*, 34, 357–359, 1992.
108. R.J. Luebbers and J. Beggs, FDTD calculation of wideband antenna gain and efficiency, *IEEE Trans. Antennas Propag.*, 40, 1403–1407, 1992.
109. P.A. Tirkas and C.A. Balanis, FDTD method for antenna radiation, *IEEE Trans. Antenna Propag.*, 40, 334–340, 1992.
110. K. Uehara and K. Kagoshima, Rigorous analysis of microstrip phased array antennas using a new FDTD method, *Electron. Lett.*, 30, 100–101, 1994.
111. M.A. Jensen and Y. Rahmat-Samii, Performance analysis of antennas for hand-held transceivers using FDTD, *IEEE Trans. Antennas Propag.*, 42, 1106–1113, 1994.
112. A Taflove and K.R. Umashankar, Radar cross section of general three-dimensional scatterers, *IEEE Trans. Electromagn. Compat.*, 25, 433–440, 1983.
113. C.L. Britt, Solution of electromagnetic scattering problems using time-domain techniques, *IEEE Trans. Antennas Propag.*, 37, 1181–1192, 1989.
114. N. Bushyager, B. McGarvey, and E.M. Tentzeris, Introduction of an Adaptive Modeling Technique for the simulation of RF structures requiring the coupling of Maxwell's, mechanical and solid-state equations, *Appl. Computational Electromagn. (ACES) Soc. J.*, 17, 1, 104–111, 2002.
115. N. Bushyager, K. Lange, M. Tentzeris, and J. Papapolymerou, Modeling and optimization of RF-MEMS reconfigurable tuners with computationally efficient time-domain techniques, *Proc. 2002 IEEE-IMS Symp.*, Seattle, WA, 2002, II.883–886.
116. W. Sui, D.A. Christensen, and C.H. Durney, Extending the two-dimensional FDTD method to hybrid electromagnetic systems with active and passive lumped elements, *IEEE Trans. Microwave Theory Tech.*, 40, 724–730, 1992.
117. P. Ciampolini, P. Mezzanotte, L. Roselli, and R. Sorrentino, Accurate and efficient circuit simulation with lumped-element FDTD, *IEEE Trans. Microwave Theory Tech.*, 44, 2207–2215, 1996.

118. C.-N. Kuo, B. Housmand, and T. Itoh, Full-wave analysis of packaged microwave circuits with active and nonlinear devices: an FDTD approach, *IEEE Trans. Microwave Theory Tech.*, 45, 819–826, 1997.
119. M.A. Alsunaidi, S.M. Sohel Imtiaz, and S.M. El-Ghazaly, Electromagnetic wave effects on microwave transistors using a full-wave time-domain model, *IEEE Trans. Microwave Theory Tech.*, 44, 799–808, 1996.
120. V.A. Thomas, M.E. Jones, M. Piket-May, A. Taflove, and E. Harrigan, The use of SPICE lumped circuits as sub-grid models for FDTD analysis, *IEEE Microwave Guided Wave Lett.*, 4, 141–143, 1994.
121. S. Dey and R. Mittra, A locally conformal finite-difference time-domain algorithm for modeling three-dimensional perfectly conducting objects, *IEEE Microwave Guided Wave Lett.*, 7, 273–275, 1997.
122. S. Dey and R. Mittra, A modified locally conformal finite-difference time-domain algorithm for modeling three-dimensional perfectly conducting objects, *Microwave Optical Technology Lett.*, 17, 349–352, 1998.
123. J.G. Maloney and M.P. Kesler, Analysis of Periodic Structures, in *Advances in Computational Electrodynamics: The Finite-Difference Time-Domain Method*, A. Taflove, Ed., Artech House, Norwood, MA, 1998, chap. 6.
124. O.R. Painter, K. Lee, A. Scherer, A. Yariv, J.D. O'Brien, P.D. Dapkus, and I. Kim, Two-dimensional photonic band-gap defect mode laser, *Science*, 284, 1819–1821, June 1999.
125. F. Zheng, Z. Chen, and J. Zhang, Towards the development of a three-dimensional stable finite-difference time-domain method, *IEEE Trans. Microwave Theory Tech.*, 48, 9, 1550–1558, 2000.
126. H. Rao, R. Scarmozzino, and R.M. Osgood, Jr., An improved ADI-FDTD method and its application to photonic simulations, *IEEE Photonics Technol. Lett.*, 14, 4, 477–479, 2002.
127. S. Akhtarzad and P.B. Johns, Three-dimensional transmission-line matrix computer analysis of microstrip resonators, *IEEE Trans. Microwave Theory Tech.*, MTT-23, 990–997, 1975.
128. W.J.R. Hoefer, The transmission-line matrix method-theory and applications, *IEEE Trans. Microwave Theory Tech.*, MTT-33, 882–893, 1985.
129. C. Christopoulos, *The Transmission-Line Modeling Method*, IEEE/OUP, 1995.
130. P.B. Johns, A symmetrical condensed node for the TLM-method, *IEEE Trans. Microwave Theory Tech.*, 35, 370–377, 1987.
131. L.R.A.X. de Menezes and W.J.R. Hoefer, Modeling of general (nonlinear) constitutive relationships in SCN TLM, *IEEE Trans. Microwave Theory Tech.*, MTT-44, 854–861, 1996.
132. M. Krumpholz and P. Russer, Discrete time-domain Green's functions for three-dimensional TLM modeling of the radiating boundary conditions, 9th Annual Review of Progress in ACES Digest, 458–466, 1993.
133. M. Righi and W.J.R. Hoefer, Efficient three-dimensional-SCN-TLM diakoptics for waveguide components, *IEEE Trans. Microwave Theory Tech.*, 42, 2381–2384, 1994.
134. M. Krumpholz, C. Huber, and P. Russer, A field theoretical comparison of FDTD and TLM, *IEEE Trans. Microwave Theory Tech.*, 43, 1935–1950, 1995.
135. C. Eswarappa and W.J.R. Hoefer, One-way equation absorbing boundary condition for three-dimensional TLM analysis of planar and quasi-planar structures, *IEEE Trans. Microwave Theory Tech.*, 42, 1669–1677, 1994.
136. J.L. Dubard and D. Pompei, A modified three-dimensional-TLM variable node for the Berenger's perfectly matched layer implementation, 13th Annual Review of Progress in ACES Digest, 661–665, 1997.
137. L. Pierantoni, C. Tomassoni and T. Rozzi, A new termination condition for the application of the TLM method to discontinuity problems in closed homogeneous waveguide, *IEEE Trans. Microwave Theory Tech.*, 50, 11, 2513–2518, 2002.
138. M.I. Sobhy, M.W.R. Ng, R.J. Langley, and J.C. Batchelor, TLM simulation of patch antenna on magnetized ferrite substrate, 16th Annual Review of Progress in ACES Digest, 562–569, 2000.

139. J.-L. Dubard and D. Pompei, Optimization of the PML efficiency in three-dimensional TLM method, *IEEE Trans. Microwave Theory Tech.*, 48, 7, 1081–1088, July 2000.
140. A.P. Duffy, J.L. Herring, T.M. Benson, and C. Christopoulos, Improved wire modeling in TLM, *IEEE Trans. Microwave Theory Tech.*, 42, 1978–1983, 1994.
141. F. Coccetti, L. Vietzorreck, V. Chtchekatourov, and P. Russer, A numerical study of MEMS capacitive switches using TLM, 16th Annual Review of Progress in ACES Digest, 580–587, 2000.
142. P.P.M. So and W.J.R. Hoefer, A TLM-SPICE interconnection framework for coupled field and circuit analysis in time-domain, *IEEE Trans. Microwave Theory Tech.*, 50, 12, 2728–2733, 2002.
143. J. Paul, C. Christopoulos, and D.W.P. Thomas, Generalized material models in TLM – Part I: materials with frequency-dependent properties, *IEEE Trans. Antennas Propag.*, 47, 10, 1528–1534, 1999.
144. J. Paul, C. Christopoulos, and D.W.P. Thomas, Generalized material models in TLM – Part II: materials with anisotropic properties, *IEEE Trans. Antennas Propag.*, 47, 10, 1535–1542, 1999.
145. J. Paul, C. Christopoulos, and D.W.P. Thomas, Generalized material models in TLM – Part III: materials with nonlinear properties, *IEEE Trans. Antennas Propag.*, 50, 7, 997–1004, 2002.
146. H. Dominguez, A. Raizer, and W.P. Carpes, Jr., Electromagnetic fields radiated by a cellular phone in close proximity to metallic walls, *IEEE Trans. Magn.*, 38, 2, 793–796, 2002.
147. M. Walter, O. Pertz, and A. Beyer, A contribution to the modeling of longitudinally periodic waveguides by the help of the TLM method, *IEEE Trans. Microwave Theory Tech.*, 48, 9, 1574–1576, September 2000.
148. D.H. Schaubert, D.R. Wilton, and A.W. Glisson, A tetrahedral method for electromagnetic scattering by arbitrarily shaped inhomogeneous dielectric bodies, *IEEE Trans. Antennas Propag.*, 32, 77–85, 1984.
149. Z.J. Cendes, Vector finite elements for electromagnetic field computation, *IEEE Trans. Magn.*, 27, 3958–3966, 1991.
150. A. Chatterjee, J.M. Jin, and J.L. Volakis, Edge-based finite elements and vector ABC's applied to three-dimensional scattering, *IEEE Trans Antennas Propag.*, 41, 221–226, 1993.
151. A.F. Peterson and D.R. Wilton, Curl-conforming mixed-order edge elements for discretizing the two-dimensional and three-dimensional vector Helmholtz equation, in *Finite Element Software for Microwave Engineering*, T. Itoh, G. Pelosi, and P.P. Silvester, Eds., Wiley, New York, 1996.
152. Z. Pantic-Tanner, J.S. Savage, D.R. Tanner, and A.F. Peterson, Two-dimensional singular vector elements for finite-element analysis, *IEEE Trans. Microwave Theory Tech.*, 46, 178–184, 1998.
153. R. Lee and A.C. Cangellaris, A study of discretization error in the finite element approximation of wave solutions, *IEEE Trans. Antennas Propag.*, 40, 542–549, 1992.
154. G.S. Warren and W.R. Scott, An investigation of numerical dispersion in the vector finite element method, *IEEE Trans. Antennas Propag.*, 42, 1502–1508, 1994.
155. L.W. Pearson, R.A. Whitaker, and L.J. Bahrmasel, An exact radiation boundary condition for the finite-element solution of electromagnetic scattering on an open domain, *IEEE Trans. Magn.*, 25, 3046–3048, 1989.
156. J.P. Webb and V.N. Kanellopoulos, A numerical study of vector absorbing boundary conditions for the finite element solution of Maxwell's equations, *IEEE Microwave Guided Wave Lett.*, 1, 325–327, 1991.
157. W. Sun and C.A. Balanis, Vector one-way absorbing boundary conditions for FEM applications, *IEEE Trans. Antennas Propag.*, 42, 872–878, 1994.
158. J.M. Jin and V.V. Liepa, A note on hybrid finite element for solving scattering problems, *IEEE Trans. Antennas Propag.*, 36, 1486–1490, 1988.
159. J.S. Wang, Analysis of microstrip discontinuities based on edge-element three-dimensional FEM formulation and absorbing boundary conditions, *Proc. IEEE-MTT Symp.*, 2, 745–748, 1993.
160. J.-S. Wang and R. Mittra, Finite element analysis of MMIC structures and electronic packages using absorbing boundary conditions, *IEEE Trans. Microwave Theory Tech.*, 42, 441–449, 1994.

161. J.-G. Yook, N.I. Dib, and L.P.B. Katehi, Characterization of high frequency interconnects using finite difference time-domain and finite element methods, *IEEE Trans. Microwave Theory Tech.*, 42, 1727–1736, 1994.
162. A.C. Polycarpou, M.R. Lyons, and C.A. Balanis, Finite element analysis of MMIC waveguide structures with anisotropic substrates, *IEEE Trans. Microwave Theory Tech.*, 44, 1650–1663, 1996.
163. Z. Li, J.L. Volakis and P.Y. Papalambros, Optimization of patch antennas on ferrite substrate using the finite element methods, *Proc. 1999 IEEE AP-Symp.* II-1026–1029, 1999.
164. Y. Zhu and A.C. Cangellaris, Robust finite-element solution of lossy and unbounded electromagnetic eigenvalue problems, *IEEE Trans. Microwave Theory Tech.*, 50, 10, 2331–2338, 2002.
165. E.S. Hung and S.D. Senturia, Generating efficient dynamical models for MEMS from a few finite-element simulation runs, *J. Microelectromechanical Systems*, 8, 3, 280–289, 1999.
166. A. Ammous, S. Ghedira, B. Allard, H. Morel, and D. Renault, Choosing a thermal model for electrothermal simulation of power semiconductor devices, *IEEE Trans. Power Electron.*, 14, 2, 300–307, 1999.
167. L.S. Andersen and J.L. Volakis, Adaptive multiresolution antenna modeling using hierarchical mixed-order tangential vector finite elements, *IEEE Trans. Antennas Propag.*, 49, 2, 211–222, 2001.
168. Y. Zhu and A.C. Cangellaris, Hierarchical multilevel potential preconditioner for fast finite-element analysis of microwave devices, *IEEE Trans. Microwave Theory Tech.*, 50, 8, 1984–1989, 2002.
169. C.A. Brebbia, *The Boundary Element Method for Engineers*, Pentech Press, London, 1978.
170. S. Kagami and I. Fukai, Application of boundary element method to EM field problems, *IEEE Trans. Microwave Theory Tech.*, 32, 455–461, 1984.
171. H. Cam, S. Toutain, P. Gelin, and G. Landrac, Study of Fabry-Perot cavity in the microwave frequency range by the boundary element method, *IEEE Trans. Microwave Theory Tech.*, 40, 298–304, 1992.
172. T.F. Eibert and V. Hansen, three-dimensional FEM/BEM-hybrid approach based on a general formulation of Huygen's principle for planar layered media, *IEEE Trans. Microwave Theory Tech.*, 45, 1105–1112, 1997.
173. K. Guillouard, M.F. Wong, and V. Fouad Hanna, A new global time domain electromagnetic simulator of microwave circuits including lumped elements based on finite element method, *Proc. IEEE-MTT Symp.*, 1239–1242, 1997.
174. J.M. Jin and J.L. Volakis, A FE-boundary integral formulation for scattering by three-dimensional cavity-blocked apertures, *IEEE Trans. Antennas Propag.*, 39, 97–104, 1991.
175. S.D. Gedney, J.F. Lee, and R. Mittra, A combined FEM/MoM approach to analyze the plane wave diffraction by arbitrary gratings, *IEEE Trans. Antenna Propag.*, 40, 363–370, 1992.
176. A. Musolino and M. Raugi, Dual formulations of a hybrid FEM/MoM method for nonlinear analysis of EM fields, *IEEE Trans. Magn.*, 35, 3, 1380–1383, 1999.
177. R.-B. Wu and T. Itoh, Hybridizing FDTD analysis with unconditionally stable FEM for objects of curved boundary, *Proc. 1995 IEEE-MTT Symp.*, 833–836, 1995.
178. A. Monorchio and R. Mittra, A hybrid FE/FDTD technique for solving complex electromagnetic problems, *IEEE Microwave Guided Wire Lett.*, 8, 93–95, 1998.
179. A.G. Engel and L.P.B. Katehi, Frequency and time domain characterization of microstrip-ridge structures, *IEEE Trans. Microwave Theory Tech.*, 1251–1262, 1993.
180. H. Esteban et al., A new hybrid mode-matching method for the analysis of inductive obstacles and discontinuities, *Proc. 1999 IEEE-AP Symp.*, 966–969, 1999.
181. D.C. Ross, J.L. Volakis, and H. Anastasiu, Hybrid FE-modal analysis of jet engine inlet scattering, *IEEE Trans. Antennas Propag.*, 43, 1995.
182. J. Rubio, J. Arroyo, and J. Zapata, Analysis of passive microwave circuits by using a hybrid two-dimensional and three-dimensional finite-element mode-matching method, *IEEE Trans. Microwave Theory Tech.*, 47, 9, 1746–1749, 1999.

183. S. Lindenmeier, C. Christopoulos, and P. Russer, Thin wire modeling with TLM/MIE method, 16th Annual Review of Progress in ACES Digest, 587–594, 2000.
184. S. Lindenmeier, L. Pierantoni, and P. Russer, Hybrid space discretizing-integral equation methods for numerical modeling of transient interference, *IEEE Trans. Electromagn. Compat.*, 41, 4, 425–430, November 1999.
185. L. Pierantoni, G. Cerri, S. Lindemeier, and P. Russer, Theoretical and numerical aspects of the hybrid MoM-FDTD, TLM-IE and ARB methods for the efficient modeling of EMC problems, *Proc. 29 EuMC*, 2, 313–316, 1999.
186. I. Daubechies, Ten Lectures on Wavelets, SIAM, 1992.
187. G. Kaiser, *A Friendly Guide to Wavelets*, Springer Verlag, Berlin, 1994.
188. C. Burrus, R.A. Gopinath, and H. Guo, *Introduction to Wavelets and Wavelet Transforms: A Primer*, Prentice Hall, Englewood Cliffs, NJ, 1997.
189. J. Goswami and A. Chan, *Fundamentals on Wavelets: Theory, Algorithms and Applications*, John Wiley & Sons, New York, 1999.
190. W. Dahmen, A.J. Kurdila, and P. Oswald, *Multiscale Wavelet Methods for Partial Differential Equations*, Academic Press, New York, 1997.
191. S. Mallat, Multiresolution representation and wavelets, Ph.D. Thesis, University of Pennsylvania, Philadelphia, 1988.
192. Y. Meyer, Ondelettes et Fonctions Splines, Seminaire EDP, Ecole Polytechnique, Paris, December 1986.
193. L.L. Schumaker, *Spline Functions: Basic Theory*, Wiley-Interscience, New York, 1981.
194. H.C. Schweinler and E.P. Wigner, Orthogonalization methods, *J. Math. Phys.*, 11, 1693–1694, 1970.
195. G. Battle, A block spin construction of ondelettes, Part I: Lemarie functions, *Comm. Math. Phys.*, 110, 601–615, 1987.
196. P.G. Lemarie, Ondelettes a localization exponentielles, *J. Math. Pures Appl.*, 67, 227–236, 1988.
197. A. Haar, Zur theorie der orthogonalen funktionsysteme, *Math. Ann.*, 69, 331–371, 1910.
198. I. Daubechies, Orthonormal bases of compactly supported wavelets, *Comm. Pure Appl. Math.*, 41, 909–996, 1988.
199. G. Beylkin, R. Coifman, and V. Rokhlin, Fast wavelet transforms and numerical algorithms I, *Commun. Pure Appl. Math.*, 44, 141–183, 1991.
200. B.Z. Steinberg and Y. Leviatan, On the use of wavelet expansions in the method of moments, *IEEE Trans. Antennas Propag.*, 41, 610–619, 1993.
201. G. Wang, On the utilization of periodic wavelet expansions in the moment methods, *IEEE Trans. Microwave Theory Tech.*, 43, 2495–2498, 1995.
202. J.C. Goswami, A.K. Chan, and C.K. Chui, On solving first-kind integral equations using wavelets on a bounded interval, *IEEE Trans. Antennas Propag.*, 43, 614–622, 1995.
203. K. Sabetfakhri and L.P.B. Katehi, Analysis of integrated millimeter-wave and submillimeter-wave waveguides using orthonormal wavelet expansions, *IEEE Trans. Microwave Theory Tech.*, 42, 2412–2422, 1994.
204. G. Pan, M.V. Toupikov, and B.K. Gilbert, On the use of Coifman intervallic wavelets in the method of moments for the construction of wavelet sparsified matrices, *IEEE Trans. Antennas Propag.*, 47, 7, 1189–1200, 1999.
205. D. Zahn, K. Sarabandi, K.F. Sabet, and J.F. Harvey, Numerical Simulation of scattering from rough surfaces: a wavelet-based approach, *IEEE Trans. Antennas Propag.*, 48, 2, 246–253, 2000.
206. J. Zheng, Z.-F. Li, and X.-N. Qian, An efficient solver for the three-dimensional capacitance of the interconnects in high speed digital circuit by the multiresolution method of moments, *IEEE Trans. Advanced Packaging*, 22, 1, 9–15, 1999.
207. M. Krumpholz and L.P.B. Katehi, MRTD: new time domain schemes based on multiresolution analysis, *IEEE Trans. Microwave Theory Tech.*, 44, 4, 555–561, April 1996.

208. M.M. Tentzeris, J. Harvey, and L.P.B. Katehi, Time adaptive time-domain techniques for the design of microwave circuits, *IEEE Microwave Guided Wave Lett.*, 9, 96–98, 1999.
209. E.M. Tentzeris, A. Cangellaris, L.P.B. Katehi, and J. Harvey, Multiresolution time-domain (MRTD) adaptive schemes using arbitrary resolutions of wavelets, *IEEE Trans. Microwave Theory Tech.*, 50, 2, 501–516, February 2002.
210. M. Krumpholz, H.G. Winful, and L.P.B. Katehi, Nonlinear time-domain modeling by MRTD, *IEEE Trans. Microwave Theory Tech.*, 45, 3, 385–393, May 1997.
211. R.L. Robertson, M. Tentzeris, and L.P.B. Katehi, Modelling of dielectric-loaded cavities using MRTD, *Int. J. Numerical Modeling, Special Issue on Wavelets in Electromagn.*, 11, 55–68, 1998.
212. M. Tentzeris and L.P.B. Katehi, Space adaptive analysis of evanescent waveguide filters, *Proc. IEEE-MTT Symp.*, 481–484, 1998.
213. L. Roselli, M. Tentzeris, and L.P.B. Katehi, Nonlinear circuit characterization using a multiresolution time domain technique, *Proc. IEEE-MTT Symp.*, 1387–1390, 1998.
214. B. McGarvey, D. Staiculescu, E.M. Tentzeris, and J. Laskar, Adaptive modeling of complex packaging geometries using Haar-based MRTD algorithms, Proc. 2000 IEEE European Microwave Conference, Paris, France, October 2000, 413–416 (Vol. I).
215. M. Fujii and W.J.R. Hoefer, A wavelet formulation of the finite-difference method: full-vector analysis of optical waveguide junctions, *IEEE J. Quantum Electron.*, 37, 8, 1015–1029, 2001.
216. T. Dogaru and L. Carin, Time-domain sensing of targets buried under a Gaussian exponential, or fractal rough surface, *IEEE Trans. Geosci. Remote Sensing*, 39, 8, 1807–1819, 2001.
217. Q. Cao and Y. Chen, Scaling-function based multiresolution time domain analysis for planar printed millimeter-wave integrated circuits, *IEE Proc. Microwave, Antennas Propag.*, 148, 3, 179–187, 2001.
218. M. Tentzeris, R. Robertson, J. Harvey, and L.P.B. Katehi, Stability and dispersion analysis of Battle-Lemarie based MRTD schemes, *IEEE Trans. Microwave Theory Tech.*, 47, 7, 1004–1013, 1999.
219. K. Goverdhanam, M. Tentzeris, M. Krumpholz, and L.P.B. Katehi, An FDTD multigrid based on multiresolution analysis, *Proc. IEEE-AP Symp.*, 352–355, 1996.
220. M. Fujii and W.J.R. Hoefer, Multiresolution analysis similar to the FDTD method-derivation and application, *IEEE Trans. Microwave Theory Tech.*, 46, 12, 2463–2475, December 1998.
221. C. Sarris and L.P.B. Katehi, Multiresolution time-domain (MRTD) schemes with space-time Haar wavelets, *Proc. IEEE-MTT Symp.*, 1459–1462, 1999.
222. C.D. Sarris and L.P.B. Katehi, Fundamental gridding-related dispersion effects in multiresolution time-domain schemes, *IEEE Trans. Microwave Theory Tech.*, 49, 12, 2248–2257, 2001.
223. Q. Cao, Y. Chen and R. Mittra, Multiple image technique (MIT) and anisotropic perfectly matched layer (APML) in implementation of MRTD scheme for boundary truncations of microwave structures, *IEEE Trans. Microwave Theory Tech.*, 50, 6, 1578–1589, 2002.
224. S. Goasguen, M.M. Tomeh, and S.M. El-Ghazaly, Electromagnetic and semiconductor device simulation using interpolating wavelets, *IEEE Trans. Microwave Theory Tech.*, 49, 12, 2258–2265, 2001.
225. N. Bushyager, E. Dalton, and M. Tentzeris, Modeling of complex RF structures using computationally optimized time-domain techniques, *Proc. 5th International Workshop on CEM-TD*, Halifax, Canada, June 2003.

IV

Underlying Physics

19

Maxwell's Equations

Nicholas E. Buris
Motorola Labs

Microwaves and radio frequency (RF) comprise a branch of electrical engineering that deals ultimately with special cases of the physics of electrically charged particles and their interactions via electromagnetic waves. The fundamental branch of science describing the physics of electrically charged particles is electromagnetism. Electromagnetism deals with the electromagnetic force and is based on the concept of electric and magnetic vector fields, $\boldsymbol{E}(\boldsymbol{r}, t)$ and $\boldsymbol{H}(\boldsymbol{r}, t)$, respectively. The fields $\boldsymbol{E}(\boldsymbol{r}, t)$ and $\boldsymbol{H}(\boldsymbol{r}, t)$ were first introduced to resolve the issues of the "action at a distance" experienced between charges. Maxwell's equations are four coupled partial differential equations describing the electromagnetic field in terms of its sources, the charges, and their associated currents (charges in motion). Electromagnetic waves are one special solution of Maxwell's equations on which microwave engineering is built. In engineering, of course, depending on the technology of interest, we deal with a full range of special circumstances of electromagnetism. At one end of the spectrum are applications such as solid-state devices where electromagnetics is applied to just a few charges, albeit in a phenomenological sense and in conjunction with quantum mechanics. In this realm the forces on individual charges are important. At the other end we have applications where the wavelength of the electromagnetic waves is much smaller than the dimensions of the problem and electromagnetics is reduced to optics where only simple, plane wave phenomena are at play. In the middle of the spectrum, we deal with structures whose size is comparable to the wavelength and electromagnetics is treated as a rigorous mathematical boundary value problem. The majority of microwave applications is somewhat in the middle of this spectrum with some having connections to either end. When studying time harmonic events in microwaves, the frequency domain version of Maxwell's equations is very convenient. In the frequency domain, we have developed a number of high level descriptions of electromagnetic phenomena and several specialized disciplines such as circuits, filtering, antennas, and others have been created to efficiently address the engineering problems at hand.

0-8493-1732-0/03/$0.00+$1.50

This chapter of the book will describe Maxwell's equations and their solution in order to establish the connection between the various microwave and RF topics and their basic physics, electromagnetism.

19.1 Time Domain Differential Form of Maxwell's Equations

As implied earlier, the fundamental description of the physics involved in the study of microwaves and RF is based on the concept of electric and magnetic vector fields, $\boldsymbol{E}(\boldsymbol{r}, t)$ and $\boldsymbol{H}(\boldsymbol{r}, t)$, respectively. According to the well-known Helmholtz theorem, any vector field can be uniquely specified in terms of its rotation (curl) and divergence components [3]. Maxwell's equations in vacuum essentially define the sources of the curl and divergence of $\boldsymbol{E}$ and $\boldsymbol{H}$. In the International System of units (SI) these equations take the following form:

$$\nabla \times \mathrm{E} = -\mathbf{J}_m - \mu_o \frac{\partial \mathbf{H}}{\partial t} \tag{19.1}$$

$$\nabla \cdot \mathbf{E} = \frac{\rho_e}{\varepsilon_o} \tag{19.2}$$

$$\nabla \times \mathbf{H} = \mathbf{J}_e + \varepsilon_o \frac{\partial \mathbf{E}}{\partial t} \tag{19.3}$$

$$\nabla \cdot \mathbf{H} = \frac{\rho_m}{\mu_o} \tag{19.4}$$

where the constants ε_o and μ_o are the permittivity and permeability of vacuum, ρ_e and $\boldsymbol{J}_e$ are the electric charge and current densities, and ρ_m and $\boldsymbol{J}_m$ are the magnetic charge and current densities.

Equations (19.1) through (19.4) indicate that, in addition to the charges and the currents, time variation in one field serves as a source to the other. In that sense, in microwaves, dealing with high frequency harmonic time variations, the electric and magnetic fields are always coupled and we refer to them combined as the electromagnetic field. It is interesting to note that Maxwell developed his equations by abstraction and generalization from a number of experimental laws that had been discovered previously. Up to Maxwell's time, electromagnetism was a collection of interesting experimental and theoretical laws from Coulomb, Gauss, Faraday, and others. Maxwell, in 1864, combined and extended all these into a remarkably complete system of equations, thus founding the science of electromagnetism. Maxwell's generalizations helped start and propel work in electromagnetic waves, and also facilitated the introduction of the special theory of relativity. Interestingly, Maxwell's equations were not covariant under a Galilean transformation (observer moving with respect to the environment). However, after the postulates of the special theory of relativity, no modification of any kind was needed to Maxwell's equations. The speed of light, derived from the wave solutions to Maxwell's equations, is a constant for all inertial frames of reference.

As mentioned above, the electromagnetic fields are a conceptual contraption, the result of an effort to systematically describe how electrically charged particles move. Maxwell's equations describe the field in terms of its sources but they do not describe how the charges move. The motion of charges is governed by Coulomb's law. The force $\boldsymbol{F}$ on an electric charge, q, moving with velocity v inside an electromagnetic field is

$$\mathbf{F} = q\mathbf{E} + q\mu_o \mathbf{v} \times \mathbf{H}$$

A very important assumption made here is that the charge q is a test charge, that is, small enough that it does not alter the field in which it exists. The problem of the self fields of charges cannot be solved by classical means, but only through quantum electrodynamics [12]. The fields themselves cannot be measured directly. It is through their effects on charge particles that they are experienced. In most microwave applications we do not see Coulomb's force because we seldom deal with just a few particles. Instead, we devised complicated quantities such as voltage, impedance, and others to arrive at efficient engineering designs.

19.2 Some Comments on Maxwell's Equations

Because the divergence of the curl of any vector is identically equal to zero, combining Eqs. (19.2) and (19.3) results in the current continuity equation (charges conservation), i.e.,

$$\nabla \cdot \mathbf{J}_e + \frac{\partial \rho_e}{\partial t} = 0 \tag{19.5a}$$

Similarly,

$$\nabla \cdot \mathbf{J}_m + \frac{\partial \rho_m}{\partial t} = 0 \tag{19.5b}$$

This is a peculiar fact, as the continuity equation is somewhat of a statement on the nature of the field sources, and as such, one would not expect it to be an intrinsic property of Maxwell's equations.

Another peculiar property of Maxwell's equations is that they are covariant under a duality transformation (see part on duality later in this text). Critical quantities quadratic in the fields, remain invariant under such a transformation. Under the proper choice of a duality transformation, Eqs. (19.1) through (19.4) can be made to have only electric, or only magnetic sources. Therefore, the question of whether magnetic sources exist is equivalent to whether the ratio of electric to magnetic sources is the same for all charged particles [6]. Again, this is another statement related to the structure of the field sources and it is rather peculiar that it should be contained in Maxwell's equations. In microwaves, we frequently analyze problems where we consider both electric and magnetic sources to be present. The concept of duality is used extensively to simplify certain problems where apertures are significant parts of the geometry at hand.

19.3 Frequency Domain Differential Form of Maxwell's Equations

It is customary to restrict investigation of Maxwell equations to the case where the time variations are harmonic, adopting the phasor convention $e^{j\omega t}$ where ω represents the angular frequency. According to this convention, $E(r, t) = \mathrm{Re}\{\mathbf{E}(\mathbf{r},\omega)\, e^{j\omega t}\}$. Similar expressions hold for all scalar and vector quantities. We say that $E(r, t)$ is the time domain representation of the field while $\mathbf{E}(\mathbf{r},\omega)$ is the phasor, or frequency domain representation. This convention simplifies the mathematics of the partial differential equations with respect to time, reducing time derivatives to simple multiplication by $j\omega$. It should be noted, however, that the product of two harmonic signals in the time domain does not correspond to the product of their phasors.

To effectively treat complicated materials such as dielectrics, ferrites, and others, the electric and magnetic flux density vector fields are introduced, $\mathbf{D}(\mathbf{r},\omega)$ and $\mathbf{B}(\mathbf{r},\omega)$, respectively. These fields essentially account for complicated material mechanisms such as losses and memory (dispersion), in a phenomenological way. They represent average field quantities when large quantities of particles are present. In fact, in the Gaussian system of units, **E** and **D** have the same units and so do **H** and **B**. To first-order approximation, for example, $\mathbf{D}(\mathbf{r},\omega)$, the electric flux density field in a dielectric equals the externally applied electric field plus the field due to the dipole moment, created by the atoms being "stretched" by the external field. Similar arguments could be made for the magnetic field flux density, although to be correct, quantum mechanical considerations are needed for a more satisfying explanation. Additional discussions on materials will be made in the sections to follow. Maxwell's equations in complicated media and in the frequency domain become

$$\nabla \times \mathbf{E} = -\mathbf{J}_m - j\omega\mathbf{B} \tag{19.6a}$$

$$\nabla \cdot \mathbf{D} = \rho_e \tag{19.6b}$$

$$\nabla \times \mathbf{H} = \mathbf{J}_e + j\omega\mathbf{D} \tag{19.6c}$$

$$\nabla \cdot \mathbf{B} = \rho_m \tag{19.6d}$$

with the constitutive relations

$$\mathbf{D} = \bar{\bar{\varepsilon}} \cdot \mathbf{E} \tag{19.7}$$

and

$$\mathbf{B} = \bar{\bar{\mu}} \cdot \mathbf{H} \tag{19.8}$$

and with the continuity equations

$$\nabla \cdot \mathbf{J}_e + j\omega\rho_e = 0 \tag{19.9}$$

$$\nabla \cdot \mathbf{J}_m + j\omega\rho_m = 0 \tag{19.10}$$

In this formulation, $\bar{\bar{\mu}}$ and $\bar{\bar{\varepsilon}}$ are the generalizations of the parameters ε_o and μ_o and they can be complex, tensorial functions of frequency. There are also some rare materials for which the constitutive relations are even more general. For a simplified, molecular level derivation of ε for dielectrics, see the text by Jackson [5].

To study inhomogeneous materials, we need proper boundary conditions to govern the behavior of the fields across the interface between two media. Consider such an interface as depicted in Fig. 19.1.

Equations (19.1) through (19.4) are associated with the following boundary conditions (derivable from Maxwell's equations themselves).

$$\hat{n} \cdot (\mathbf{D}_2 - \mathbf{D}_1) = \rho_{es} \tag{19.11a}$$

$$\hat{n} \cdot (\mathbf{B}_2 - \mathbf{B}_1) = \rho_{ms} \tag{19.11b}$$

$$-\hat{n} \times (\mathbf{E}_2 - \mathbf{E}_1) = \mathbf{J}_{ms} \tag{19.11c}$$

$$-\hat{n} \times (\mathbf{H}_2 - \mathbf{H}_1) = \mathbf{J}_\epsilon \tag{19.11d}$$

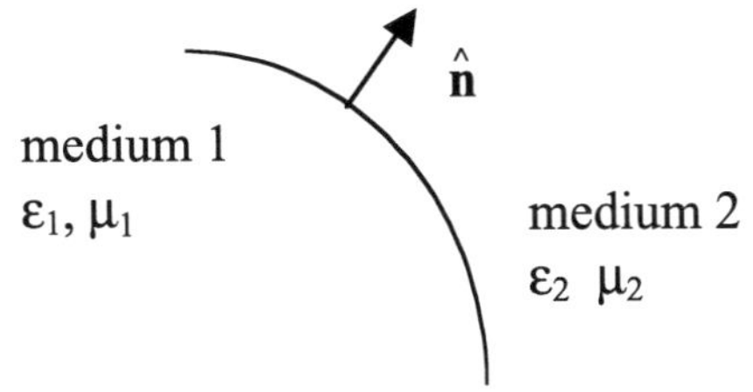

FIGURE 19.1 Geometry for the boundary conditions at the interface between two media.

where $\hat{n}$ is the unit vector normal to the interface pointing from medium 1 to medium 2. The quantities ρ_{es}, $\mathbf{J}_{es}$, ρ_{ms}, and $\mathbf{J}_{ms}$, represent the free electric, magnetic charge, and current surface densities on the interface, respectively. Note that these are free sources, and as such, only exist on conductors and are equal to zero on the interface between two dielectric media. There are bound (polarization) charges on dielectric interfaces, but they are accounted for by Eq. (19.11a) and the constitutive relation in Eq. (19.7).

19.4 General Solution to Maxwell's Equations (the Stratton–Chu Formulation)

One of the most comprehensive solutions to Maxwell's equations in a general homogeneous and isotropic domain is given by Stratton and Chu [1] and is also further discussed by Silver [2]. Consider a volume V with a boundary consisting of a collection of closed surfaces, S_1, S_2, ..., S_n. Next consider the unit vectors, $\hat{n}$, normal to the boundary surface with direction pointing inside the volume of interest as depicted in Fig. 19.2. The volume V is occupied uniformly by a material with dielectric constant ε, and magnetic permeability μ. Inside V there exist electric and magnetic charges and current density distributions, ρ_e, $\mathbf{J}_e$, ρ_m, and $\mathbf{J}_m$, respectively.

The electric and magnetic fields at an arbitrary point, **r**, inside V are then given in terms of the sources and the values of the fields at the boundary by the following equations.

$$\begin{aligned}\mathbf{E}(\mathbf{r}) = &-\int_V \left[j\omega\mu G(\mathbf{r},\mathbf{r}')\mathbf{J}_e(\mathbf{r}') + \mathbf{J}_m(\mathbf{r}') \times \nabla' G - \frac{\rho_e(\mathbf{r}')}{\varepsilon}\nabla' G \right] dV' \\ &- \int_{S_1+S_2+\cdots+S_n} \left[j\omega\mu G(\mathbf{r},\mathbf{r}')(\hat{n}' \times \mathbf{H}(\mathbf{r}')) + (-\hat{n}' \times \mathbf{E}(\mathbf{r}')) \times \nabla' G - (\hat{n}' \cdot \mathbf{E}(\mathbf{r}'))\nabla' G \right] dS'\end{aligned} \tag{19.12}$$

and

$$\begin{aligned}\mathbf{H}(\mathbf{r}) = &-\int_V \left[j\omega\varepsilon G(\mathbf{r},\mathbf{r}')\mathbf{J}_m(\mathbf{r}') - \mathbf{J}_e(\mathbf{r}') \times \nabla' G - \frac{\rho_m(\mathbf{r}')}{\mu}\nabla' G \right] dV' \\ &- \int_{S_1+S_2+\cdots+S_n} \left[j\omega\varepsilon G(\mathbf{r},\mathbf{r}')(-\hat{n}' \times \mathbf{E}(\mathbf{r}')) - (\hat{n}' \times \mathbf{H}(\mathbf{r}')) \times \nabla' G - (\hat{n}' \cdot \mathbf{H}(\mathbf{r}'))\nabla' G \right] dS'\end{aligned} \tag{19.13}$$

where

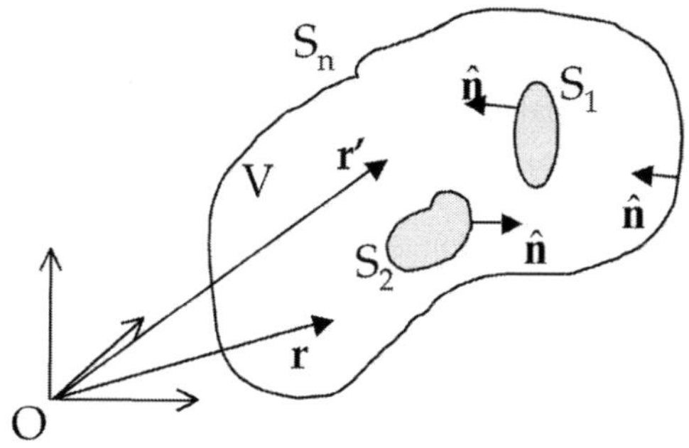

FIGURE 19.2 The volume of interest and its boundary surface consisting of $S_1 + S_2 + \ldots + S_n$.

$$G(\mathbf{r},\mathbf{r}') = \frac{e^{-jk|\mathbf{r}-\mathbf{r}'|}}{4\pi|\mathbf{r}-\mathbf{r}'|} \tag{19.14}$$

is called the free-space Green's function and

$$k = \sqrt{\omega^2 \mu \varepsilon} \tag{19.15}$$

is called the wavenumber with ω representing the angular frequency.

When the surface S_n is at infinity, it can be shown that the fields there satisfy the radiation conditions (**E** and **H** attenuate at least as fast as 1/r, they are perpendicular to each other and to **r**, and their magnitudes are related by the wave impedance, $\sqrt{\mu/\varepsilon}$). Moreover, the continuity equations can be employed to eliminate ρ_e and ρ_m from the field solution. When the volume V is unbounded ($S_n \rightarrow$ infinity) and the charges are substituted by their current density expressions, the solution of Maxwell's equations becomes

$$\mathbf{E}(\mathbf{r}) = -\frac{j}{\omega\varepsilon}\int_V \left[\left(\mathbf{J}_e(\mathbf{r}')\cdot\nabla'\right)\nabla' + k^2\,\mathbf{J}_e(\mathbf{r}') - j\omega\varepsilon\,\mathbf{J}_m(\mathbf{r}')\times\nabla'\right]G(\mathbf{r},\mathbf{r}')dV' \tag{19.16}$$

and

$$\mathbf{H}(\mathbf{r}) = -\frac{j}{\omega\mu}\int_V \left[\left(\mathbf{J}_m(\mathbf{r}')\cdot\nabla'\right)\nabla' + k^2\,\mathbf{J}_m(\mathbf{r}') + j\omega\mu\,\mathbf{J}_e(\mathbf{r}')\times\nabla'\right]G(\mathbf{r},\mathbf{r}')dV'. \tag{19.17}$$

These forms of the fields are used routinely for the numerical solution to Maxwell's equations, particularly using the method of moments.

It should be noted here that in the special case where the boundary surfaces $S_1, S_2, \ldots, S_{n-1}$ are perfect electric conductors, the field solution can be expressed by Eqs. (19.12) and (19.13), or their unbounded counterparts, Eqs. (19.16) and (19.17) provided one recognizes that the free currents on an electric conductor are $\hat{n} \times \mathbf{H}$ and the charges are $\hat{n} \times \mathbf{D}$. From the materials point of view (also discussed later in this chapter) a perfect conductor has unlimited capacity to provide free charges that distribute themselves on its surface in such a way as to effectively eliminate their internal fields regardless of the external field they are in. Therefore, since the fields vanish inside, perfect conductors have no energy in them. They simply shape and guide the energy in the space around them. Thus, a coil has no energy inside its metal; it simply stores energy inside and outside its windings. In this light, electromagnetic design of especially small, densely populated electronic structures is very difficult and suffers from potential ElectroMagnetic Interference (EMI) problems.

19.5 Far Field Approximation

At the limit where the observation point is at infinity ($r \to \infty$, far field) it can be shown that Eqs. (19.16) and (19.17) simplify to

$$\mathbf{E}(\mathbf{r}) = -j\omega\mu \frac{e^{-jkr}}{4\pi r} \int_V \left[\mathbf{J}_e(\mathbf{r}') - (\hat{\mathbf{r}} \cdot \mathbf{J}_e(\mathbf{r}'))\hat{\mathbf{r}} + \sqrt{\frac{\varepsilon}{\mu}} \mathbf{J}_m(\mathbf{r}') \times \hat{\mathbf{r}}' \right] e^{jk\hat{\mathbf{r}}\cdot\mathbf{r}'} dV' \tag{19.18}$$

with

$$\mathbf{H}(\mathbf{r}) = \sqrt{\frac{\varepsilon}{\mu}} \hat{\mathbf{r}} \times \mathbf{E}(\mathbf{r}). \tag{19.19}$$

$\mathbf{E}(\mathbf{r})$ and, consequently, $\mathbf{H}(\mathbf{r})$ in Eqs. (19.18) and (19.19) have zero radial components and can be found often in the following, more practical forms.

$$\mathbf{E}_\theta(\mathbf{r}) = -j\omega\mu \frac{e^{-jkr}}{4\pi r} \int_V \left[\mathbf{J}_\theta(\hat{\mathbf{r}}') + \sqrt{\frac{\varepsilon}{\mu}} \mathbf{J}_{m_\phi}(\mathbf{r}') \right] e^{jk\hat{\mathbf{r}}\cdot\mathbf{r}'} dV' \tag{19.20}$$

and

$$\mathbf{E}_\phi(\mathbf{r}) = -j\omega\mu \frac{e^{-jkr}}{4\pi r} \int_V \left[\mathbf{J}_\phi(\mathbf{r}') - \sqrt{\frac{\varepsilon}{\mu}} \mathbf{J}_{m_\theta}(\mathbf{r}') \right] e^{jk\hat{\mathbf{r}}\cdot\mathbf{r}'} dV' \tag{19.21}$$

or, equivalently,

$$\mathbf{E}_\theta(\mathbf{r}) = -j\omega\mu \frac{e^{-jkr}}{4\pi r} \int_V \left[\mathbf{J}_x \cos\theta\cos\phi + \mathbf{J}_y \cos\theta\sin\phi - \mathbf{J}_z \sin\theta + \sqrt{\frac{\varepsilon}{\mu}} \left(-\mathbf{J}_{m_x} \sin\phi + \mathbf{J}_{m_y} \cos\phi \right) \right] e^{jk\hat{\mathbf{r}}\cdot\mathbf{r}'} dV' \tag{19.22}$$

and

$$\mathbf{E}_\phi(\mathbf{r}) = -j\omega\mu \frac{e^{-jkr}}{4\pi r} \int_V \left[-\mathbf{J}_x \sin\phi + \mathbf{J}_y \cos\phi - \sqrt{\frac{\varepsilon}{\mu}} \left(\mathbf{J}_{m_x} \cos\theta\cos\phi + \mathbf{J}_{m_y} \cos\theta\sin\phi - \mathbf{J}_{m_z} \sin\theta \right) \right] e^{jk\hat{\mathbf{r}}\cdot\mathbf{r}'} dV' \tag{19.23}$$

where it also holds that

$$\mathrm{H}_\phi = \sqrt{\frac{\varepsilon}{\mu}} \mathrm{E}_\theta \tag{19.24}$$

and

$$H_\theta = \sqrt{\frac{\varepsilon}{\mu}} E_\phi . \tag{19.25}$$

These solutions to Maxwell's equations can also be derived via a potential field formulation. The interested reader can refer to the treatment in Harrington [10] and Balanis [7].

19.6 General Theorems in Electromagnetics

19.6.1 Uniqueness of Solution

It can be shown that the field in the volume V, depicted in Fig. 19.2, is uniquely defined by the source distributions in it and the values of the tangential electric field, or tangential magnetic field on all the boundary surfaces, $S_1, S_2, \ldots, S_n$. Uniqueness is also guaranteed when the tangential **E** is specified for part of the boundary while tangential **H** is specified for the remaining part. A detailed discussion of the uniqueness of the solution can be found in Stratton [9].

19.6.2 Duality

As briefly mentioned earlier, Maxwell's equations are covariant under duality transformations. Consider the transformation,

$$\begin{pmatrix} \mathbf{E} \\ \mathbf{H} \end{pmatrix} = \begin{bmatrix} \cos\xi & \sin\xi \\ -\sin\xi & \cos\xi \end{bmatrix} \cdot \begin{pmatrix} \mathbf{E}' \\ \mathbf{H}' \end{pmatrix} \tag{19.26}$$

$$\begin{pmatrix} \mathbf{D} \\ \mathbf{B} \end{pmatrix} = \begin{bmatrix} \cos\xi & \sin\xi \\ -\sin\xi & \cos\xi \end{bmatrix} \cdot \begin{pmatrix} \mathbf{D}' \\ \mathbf{B}' \end{pmatrix} \tag{19.27}$$

$$\begin{pmatrix} \mathbf{J}_e \\ \mathbf{J}_m \end{pmatrix} = \begin{bmatrix} \cos\xi & \sin\xi \\ -\sin\xi & \cos\xi \end{bmatrix} \cdot \begin{pmatrix} \mathbf{J}'_e \\ \mathbf{J}'_m \end{pmatrix} \tag{19.28}$$

and

$$\begin{pmatrix} \rho_e \\ \rho_m \end{pmatrix} = \begin{bmatrix} \cos\xi & \sin\xi \\ -\sin\xi & \cos\xi \end{bmatrix} \cdot \begin{pmatrix} \rho'_e \\ \rho'_m \end{pmatrix} \tag{19.29}$$

It can be easily shown that under this duality transformation, Maxwell's equations are covariant. That is, if the primed fields satisfy Maxwell's equations with the primed sources, the non-primed fields satisfy Maxwell's equations with the non-primed sources. Moreover, critical quantities quadratic in the fields, such as $\mathbf{E} \times \mathbf{H}$ and $\mathbf{E} \cdot \mathbf{D} + \mathbf{B} \cdot \mathbf{H}$ remain invariant under this duality transformation (for real ξ). That is, for example, $\mathbf{E} \times \mathbf{H} = \mathbf{E}' \times \mathbf{H}'$ [6]. As a special case of this property, $\xi = -\pi/2$, we have the following correspondence:

$$\mathbf{E} \leftrightarrow -\mathbf{H}$$

$$\mathbf{H} \leftrightarrow \mathbf{E}$$

$$\mathbf{J}_m \leftrightarrow \mathbf{J}_e$$

$$\rho_m \leftrightarrow \rho_e$$

$$\mu \leftrightarrow \varepsilon$$

$$\varepsilon \leftrightarrow \mu$$

It is worth mentioning here that, since the fields themselves cannot be measured, and since all the important quantities (such as the power and energy terms, $\mathbf{E} \times \mathbf{H}$ and $\mathbf{E} \cdot \mathbf{D} + \mathbf{B} \cdot \mathbf{H}$) remain invariant under duality, it is just a matter of convenience which of several possible dual solutions to an electromagnetic problem we consider. This is precisely the reason why in some antenna problems, for example, one can formulate them with electric currents on the conductors, or with magnetic currents on the apertures and obtain the same detectable (measurable) results.

19.6.3 Lorentz Reciprocity Theorem

Consider the fields $\mathbf{E}_1$ and $\mathbf{H}_1$, generated by the system of sources $\mathbf{J}_{e1}$ and $\mathbf{J}_{m1}$ and the fields $\mathbf{E}_2$ and $\mathbf{H}_2$, generated by the system of sources $\mathbf{J}_{e2}$ and $\mathbf{J}_{m2}$. Then it can be shown that [8]

$$\oiint_S (\mathbf{E}_1 \times \mathbf{H}_2 - \mathbf{E}_2 \times \mathbf{H}_1) \cdot d\mathbf{S} = \int_V (\mathbf{H}_1 \cdot \mathbf{J}_{m2} - \mathbf{E}_1 \cdot \mathbf{J}_{m2} - \mathbf{H}_2 \cdot \mathbf{J}_{m1} + \mathbf{E}_2 \cdot \mathbf{J}_{e1}) dV \tag{19.30}$$

As a special case of this theorem, consider the open volume of interest V and two sources in it, $\mathbf{J}_{e1}$ and $\mathbf{J}_{e2}$. When $\mathbf{J}_{e1}$ is present alone, it generates the field $\mathbf{E}_1$. When $\mathbf{J}_{e2}$ is present alone, it generates the field $\mathbf{E}_2$. The Lorentz reciprocity theorem as stated by Eq. (19.30) implies that

$$\mathbf{E}_1 \cdot \mathbf{J}_{e2} = \mathbf{E}_2 \cdot \mathbf{J}_{e1} \tag{19.31}$$

The reciprocity theorem is used frequently in electromagnetic problem studies to facilitate solution of complicated problems. One well-known application of reciprocity has been in the expressions of the mutual and self-impedance of simple radiators. Reciprocity also manifests itself in some linear circuits so that Eq. (19.13) above holds when the electric fields and the current densities are replaced by voltages and currents, respectively. For details of this property see the text by Valkenburg [13].

19.6.4 Equivalent Principles (Theory of Images)

Equivalent principles are used to describe the electromagnetic field problem in a volume of interest in more than one configuration of sources. While the volume of interest is the same in all the equivalent configurations (materials, geometry, and source distributions), the geometry outside the volume of interest is different for each equivalent configuration. A detailed discussion of field equivalence principles can be found in Collin [8]. Here only the method of images is mentioned. The method of images is based on the uniqueness of the field. Namely, as long as the sources inside the volume of interest V and the boundary conditions on S_1, S_2, … remain the same, the fields inside V are unique regardless of what sources exist outside V.

The method of images is based on this principle. Its applicability is limited to just a few canonical geometries. Consider, for example, a current-carrying wire over an infinite perfect electric conductor, as shown in Fig. 19.3(a). The tangential electric field is zero on the pec ground plane. Next consider the

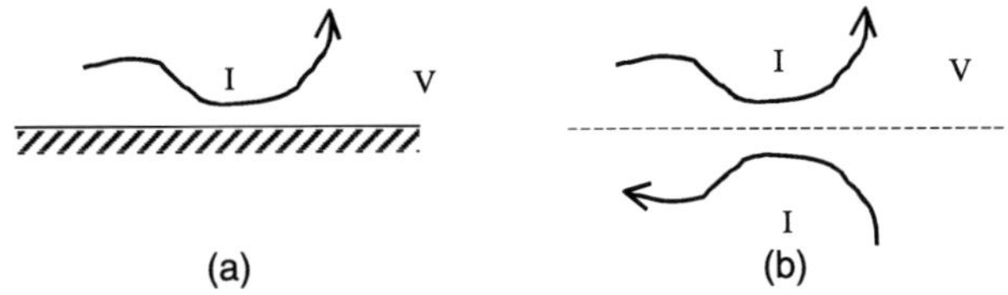

FIGURE 19.3 The fields in V are identical for both configurations, (a) and (b). The fields outside V vanish in (a) while they are certainly nonzero in (b).

configuration (b) where the ground plane has been removed and the volume below it has a second current-carrying wire, shaped in the mirror image of the original that takes an opposite direction for the current. It can be shown with simple symmetry arguments that the tangential electric field on the dashed line (where the ground plane used to be) in configuration (b) vanishes. Therefore, since (a) and (b) are identical within V and they have the same boundary conditions (zero tangential electric field) on the boundary, the fields inside V are the same.

Clearly, it is much easier to numerically evaluate the solution in configuration (b) than that of configuration (a). Images are very often used in electromagnetic problems when appropriate. It is often easy to extend the theory of images to current-carrying wires in the presence of corners of ground planes. Moreover, there have been several studies to extend the theory of images to more complex geometries [11]. However, the complexity of the images quickly escalates diminishing the benefits of removing the ground planes.

19.7 Simple Solution to Maxwell's Equations I (Unbounded Plane Waves)

In electromagnetics, a homogeneous and isotropic region of space with no sources in it is called free space. In free space, $\bar{\bar{\mu}} = \mu_o$, $\bar{\bar{\varepsilon}} = \varepsilon_o$ and Eqs. (19.6a) through (19.6d) decouple, giving rise to the following second-order wave equations:

$$\nabla^2 \mathbf{E} + \omega^2 \mu_o \varepsilon_o \mathbf{E} = 0 \tag{19.32}$$

$$\nabla^2 \mathbf{H} + \omega^2 \mu_o \varepsilon_o \mathbf{H} = 0 \tag{19.33}$$

Applying separation of variables, Eqs. (19.32) and (19.33) are found to have plane wave solutions of the form

$$\mathbf{E} = \mathbf{E}_o^+ e^{-j\mathbf{k}\cdot\mathbf{r}} + \mathbf{E}_o^- e^{+j\mathbf{k}\cdot\mathbf{r}} \tag{19.34}$$

$$\mathbf{H} = \mathbf{H}_o^+ e^{-j\mathbf{k}\cdot\mathbf{r}} + \mathbf{H}_o^- e^{+j\mathbf{k}\cdot\mathbf{r}} \tag{19.35}$$

The electromagnetic wave propagating in the $+\mathbf{k}$ direction ($e^{-j\mathbf{k}\cdot\mathbf{r}}$) travels with the speed of light, $1/(\mu_o\varepsilon_o)^{1/2}$ and has a wave impedance,

$$\left|\frac{\mathbf{E}_o^+}{\mathbf{H}_o^+}\right| = \sqrt{\frac{\mu_o}{\varepsilon_o}} \tag{19.36}$$

The wavenumber, k, obeys the relation

$$k = \frac{\omega}{c} = \omega\sqrt{\mu_o \varepsilon_o} \tag{19.37}$$

where c is the speed of light. It can be shown that the power density carried by the plane wave is along the direction of propagation k and its average value over one period is given by the Poynting vector

$$\mathbf{S}^+ = \frac{1}{2}\mathbf{E}_o^+ \times \mathbf{H}_o^+ \tag{19.38}$$

where the fields **E** and **H** are represented by their amplitude values in the phasor convention. It is worth noting that the Poynting vector itself is not a phasor. Being the product of two harmonic signals, the power density represented by the Poynting vector has a steady (DC) component and a harmonic component of twice the frequency of the electromagnetic field. The DC component of the Poynting vector is exactly given by Eq. (19.38). Similar discussions hold for the wave propagating in the –**k** direction. These wave solutions obey all the usual wave phenomena of reflection and refraction when incident on interfaces between two different media. For detailed discussions on these phenomena and more general cases of wave solutions to Maxwell's equations, the reader is referred to the following sections of this chapter and the text by Collin [8].

19.8 Simple Solution to Maxwell's Equations II (Guided Plane Waves)

It is possible to have electromagnetic waves inside cavities and waveguides. A waveguide is any structure that supports waves traveling in one direction and confined in the transverse plane by its boundaries. A rectangular duct made out of a conductor and a coaxial cable are waveguides with closed boundaries. A twin lead transmission line or a dielectric plated conductor are among the many open boundary waveguides. Waves supported by waveguides are special solutions to Maxwell's equations.

Let us consider a waveguide along the z-axis. We are looking for solutions to Maxwell's equations that represent waves traveling along z. That is, we are looking for solutions in the form:

$$\mathbf{E} = \mathbf{E}(x, y)e^{-j\beta z} = \left(\mathbf{E}_t(x, y) + \hat{\mathbf{z}}\,\mathrm{E}_z(x, y)\right)e^{-j\beta z} \tag{19.39}$$

and

$$\mathbf{H} = \mathbf{H}(x, y)e^{-j\beta z} = \left(\mathbf{H}_t(x, y) + \hat{\mathbf{z}}\,\mathbf{H}_z(x, y)\right)e^{-j\beta z} \tag{19.40}$$

where $\mathbf{E}_t$ and $\mathbf{H}_t$ are vectors in the transverse, (x,y) plane. Solving Eq. (19.6) in a region of space free from any sources and looking specifically for solutions in the form of Eqs. (19.39) and (19.40) reduces the problem to the following wave equations:

$$\nabla_t^2 E_z + \left(\omega^2\mu\varepsilon - \beta^2\right)E_z = 0 \tag{19.41}$$

and

$$\nabla_t^2 H_z + \left(\omega^2\mu\varepsilon - \beta^2\right)H_z = 0 \tag{19.42}$$

The transverse fields are given in terms of the axial components of the waves as

$$\mathbf{E}_t(x,y) = \frac{1}{k^2 - \beta^2}\left(-j\beta\nabla_t E_z(x,y) + j\omega\mu\,\hat{\mathbf{z}} \times \nabla H_z(x,y)\right) \tag{19.43}$$

and

$$\mathbf{H}_t(x,y) = \frac{1}{k^2 - \beta^2}\left(-j\beta\nabla_t H_z(x,y) - j\omega\varepsilon\,\hat{\mathbf{z}} \times \nabla E_z(x,y)\right) \tag{19.44}$$

∇_t in the above equations stands for the transverse del operator, i.e. ($\nabla_t = \hat{\mathbf{x}}\,\partial/\partial x + \hat{\mathbf{y}}\,\partial/\partial y$).

The wavenumber of the guided wave β is, in general, different from the wavenumber in the medium ($k^2 = \omega^2\mu\varepsilon$). The wavenumber is determined when the boundary condition specific to the waveguide are applied. There are some special cases of these guided wave solutions that are examined at a later section in this chapter in much more detail for practical microwave waveguides.

TEM Modes

$$E_z = H_z = 0$$

$$\nabla_t \times \mathbf{E}_t = 0 \quad \beta^2 = k^2 = \omega^2\mu\varepsilon \quad \text{and} \quad \mathbf{H} = \sqrt{\frac{\varepsilon}{\mu}}\left(\hat{\mathbf{z}} \times \mathbf{E}_t\right)$$

TM Modes

$$E_z \neq 0; \quad H_z = 0$$

$$\nabla_t^2 E_z + \left(\omega^2\mu\varepsilon - \beta^2\right)E_z = 0 \quad \mathbf{E}_t(x,y) = \frac{-j\beta}{k^2 - \beta^2}\nabla_t E_z(x,y) \quad \text{and} \quad \mathbf{H}_t = -\frac{\omega\varepsilon}{\beta}\mathbf{E}_t \times \hat{\mathbf{z}}$$

TE Modes

$$H_z \neq 0; \quad E_z = 0$$

$$\nabla_t^2 H_z + \left(\omega^2\mu\varepsilon - \beta^2\right)H_z = 0 \quad \mathbf{H}_t(x,y) = \frac{-j\beta}{k^2 - \beta^2}\nabla_t H_z(x,y) \quad \text{and} \quad \mathbf{E}_t = \frac{\omega\mu}{\beta}\mathbf{H}_t \times \hat{\mathbf{z}}$$

References

1. J.A. Stratton and L.J. Chu, *Phys. Res.*, 56, 99, 1939.
2. S. Silver, Ed., *Microwave Antenna Theory and Design*, McGraw-Hill, New York, 1949, chap. 3.
3. G. Arfken, *Mathematical Methods for Physicists*, 2nd ed., Academic Press, New York, 1970, 66–70.
4. W. Pauli, *Theory of Relativity*, Dover, New York, 1981.
5. J.D. Jackson, *Classical Electrodynamics*, 2nd ed., John Wiley & Sons, New York, 1975, 226–235.
6. J.D. Jackson, *Classical Electrodynamics*, 2nd ed., John Wiley & Sons, New York, 1975, 251–260.
7. C. Balanis, *Advanced Engineering Electromagnetics*, John Wiley & Sons, New York, 1989.
8. R.E. Collin, *Field Theory of Guided Waves*, 2nd ed., IEEE Press, Piscataway, NJ, 1991.
9. J.A. Stratton, *Electromagnetic Theory*, McGraw-Hill, New York, 1941.
10. R.F. Harrington, *Time-Harmonic Electromagnetic Fields*, McGraw-Hill, New York, 1961.

11. I.V. Lindel, Image Theory for the Soft and Hard Surface, *IEEE Trans. Antennas Propag.*, 43, 1, January 1995.
12. R.P. Feynman, R.B. Leighton, and M. Sands, Electromagnetic Mass, in *The Feynman Lectures on Physics*, Addison Wesley, New York, 1964.
13. M.E. Valkenburg, *Network Analysis*, 3rd ed., Prentice Hall, Englewood Cliffs, NJ, 1974, 255–259.

20

Wave Propagation in Free Space

Matthew N.O. Sadiku
Prairie View A&M University

The concept of propagation refers to the various ways by which an electromagnetic (EM) wave travels from the transmitting antenna to the receiving antenna. Propagation of EM wave may also be regarded as a means of transferring energy or information from one point (a transmitter) to another (a receiver). The transmission of analog or digital information from one point to another is the largest application of microwave frequencies. Therefore, understanding the principles of wave propagation is of practical interest to microwave engineers. Engineers cannot completely apply formulas or models for microwave system design without an adequate knowledge of the propagation issue.

Wave propagation at microwave frequencies has a number of advantages [Veley, 1987]. First, microwaves can accommodate very wide bandwidths without causing interference problems because microwave frequencies are so high. Consequently, a huge amount of information can be handled by a single microwave carrier. Second, microwaves propagate along a straight line like light rays and are not bent by the ionosphere as are lower frequency signals. This straight-line propagation makes communication satellites possible. In essence, a communication satellite is a microwave relay station that is used in linking two or more grounded-based transmitters and receivers. Third, it is feasible to design highly directive antenna systems of a reasonable size at microwave frequencies. Fourth, compared with low-frequency electromagnetic waves, microwave energy is more easily controlled, concentrated, and directed. This makes it useful for cooking, drying, and physical diathermy. Moreover, the microwave spectrum provides more communication channels than the radio and TV bands. With the ever-increasing demand for channel allocation, microwave communication has become more common.

EM wave propagation is achieved through guided structures such as transmission lines and waveguides or through space. In this chapter, our major focus is on EM wave propagation in free space and the power resident in the wave.

EM wave propagation can be described by two complimentary models. The physicist attempts a theoretical model based on universal laws, which extends the field of application more widely than currently known. The engineer prefers an empirical model based on measurements, which can be used immediately. This chapter presents the complimentary standpoints by discussing theoretical factors affecting wave propagation and the semiempirical rules allowing handy engineering calculations. First, we consider wave propagation in idealistic simple media, with no obstacles. We later consider the more

0-8493-1732-0/03/$0.00+$1.50

realistic case of wave propagation around Earth, as influenced by its curvature and by atmospheric conditions.

20.1 Wave Equation

The conventional propagation models, on which the basic calculation of microwave links is based, result directly from Maxwell's equations [Sadiku, 2001]:

$$\nabla \cdot \mathbf{D} = \rho_v \tag{20.1}$$

$$\nabla \cdot \mathbf{B} = 0 \tag{20.2}$$

$$\nabla \times \mathbf{E} = -\frac{\partial \mathbf{B}}{\partial t} \tag{20.3}$$

$$\nabla \times \mathbf{H} = \mathbf{J} + \frac{\partial \mathbf{D}}{\partial t} \tag{20.4}$$

In these equations, **E** is electric field strength in volts per meter, **H** is magnetic field strength in amperes per meter, **D** is electric flux density in coulombs per square meter, **B** is magnetic flux density in webers per square meter, **J** is conduction current density in amperes per square meter, and ρ_v is electric charge density in coulombs per cubic meter. These equations go hand in hand with the constitutive equations for the medium:

$$\mathbf{D} = \epsilon \mathbf{E} \tag{20.5}$$

$$\mathbf{B} = \mu \mathbf{H} \tag{20.6}$$

$$\mathbf{J} = \sigma \mathbf{E} \tag{20.7}$$

where $\epsilon = \epsilon_o \epsilon_r$, $\mu = \mu_o \mu_r$, and σ are the permittivity, the permeability, and the conductivity of the medium, respectively.

Consider the general case of a lossy medium that is charge-free ($\rho_v = 0$). Assuming time-harmonic fields and suppressing the time factor $e^{j\omega t}$, Eqs. (20.1) to (20.7) can be manipulated to yield Helmholtz's wave equations

$$\nabla^2 \mathbf{E} - \gamma^2 \mathbf{E} = 0 \tag{20.8}$$

$$\nabla^2 \mathbf{H} - \gamma^2 \mathbf{H} = 0 \tag{20.9}$$

where $\gamma = \alpha + j\beta$ is the *propagation constant,* α is the *attenuation constant* in nepers per meter or decibels per meter, and β is the *phase constant* in radians per meter. Constants α and β are given by

$$\alpha = \omega \sqrt{\frac{\mu \epsilon}{2} \left[\sqrt{1 + \left(\frac{\sigma}{\omega \epsilon} \right)^2} - 1 \right]} \tag{20.10}$$

$$\beta = \omega\sqrt{\frac{\mu\epsilon}{2}\left[\sqrt{1+\left(\frac{\sigma}{\omega\epsilon}\right)^2}+1\right]} \tag{20.11}$$

where $\omega = 2\pi f$ is the frequency of the wave. The wavelength λ and wave velocity u are given in terms of β as

$$\lambda = \frac{2\pi}{\beta} \tag{20.12}$$

$$u = \frac{\omega}{\beta} = f\lambda \tag{20.13}$$

Without loss of generality, if we assume that wave propagates in the z direction and the wave is polarized in the x direction, solving the wave Eqs. (20.8) and (20.9) results in

$$\mathbf{E} = E_o e^{-\alpha z}\cos(\omega t - \beta z)\mathbf{a}_x \tag{20.14}$$

$$\mathbf{H} = \frac{E_o}{|\eta|} e^{-\alpha z}\cos(\omega t - \beta z - \theta_\eta)\mathbf{a}_y \tag{20.15}$$

where $\eta = |\eta|\underline{/\theta_\eta}$ is the *intrinsic impedance* of the medium and is given by

$$|\eta| = \frac{\sqrt{\frac{\mu}{\epsilon}}}{\sqrt[4]{\left[1+\left(\frac{\sigma}{\omega\epsilon}\right)^2\right]}}, \quad \tan 2\theta_\eta = \frac{\sigma}{\omega\epsilon}, \quad 0 \le \theta_\eta \le 45° \tag{20.16}$$

Equations (20.14) and (20.15) show that as the EM wave propagates in the medium, its amplitude is attenuated according to $e^{-\alpha z}$, as illustrated in Fig. 20.1. The distance δ through which the wave amplitude is reduced by a factor of e^{-1} (about 37%) is called the *skin depth* or *penetration depth* of the medium, i.e.,

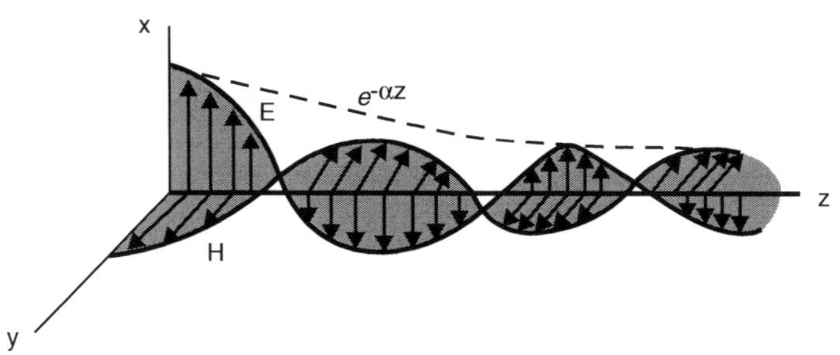

FIGURE 20.1 The magnetic and electric field components of a plane wave in a lossy medium.

$$\delta = \frac{1}{\alpha} \tag{20.17}$$

The power density of the EM wave is obtained from the Poynting vector

$$\mathbf{P} = \mathbf{E} \times \mathbf{H} \tag{20.18}$$

with the time-average value of

$$\begin{aligned} P_{\text{ave}} &= \text{Re}\left(\mathbf{E} \times \mathbf{H}^*\right) \\ &= \frac{E_o^2}{2|\eta|} e^{-2\alpha z} \cos\theta_\eta \mathbf{a}_z \end{aligned} \tag{20.19}$$

It should be noted from Eqs. (20.14) and (20.15) that **E** and **H** are everywhere perpendicular to each other and also to the direction of wave propagation. Thus, the wave described by Eqs. (20.14) and (20.15) is said to be *plane polarized,* implying that the electric field is always parallel to the same plane (the *xz* plane in this case) and is perpendicular to the direction of propagation. Also, as mentioned earlier, the wave decays as it travels in the *z* direction because of loss. This loss is expressed in the *complex relative permittivity* of the medium.

$$\epsilon_c = \epsilon_r' - \epsilon_r'' = \epsilon_r\left(1 - \frac{\sigma}{\omega\epsilon}\right) \tag{20.20}$$

and measured by the *loss tangent,* defined by

$$\tan\delta = \frac{\epsilon_r''}{\epsilon_r'} = \frac{\sigma}{\omega\epsilon} \tag{20.21}$$

The imaginary part $\epsilon_r'' = \sigma/\omega\epsilon_o$ corresponds to the losses in the medium. The refractive index of the medium n is given by

$$n = \sqrt{\epsilon_r} \tag{20.22}$$

Having considered the general case of wave propagation through a lossy medium, we now consider wave propagation in other types of media. A medium is said to be a good conductor if the loss tangent is large ($\sigma \gg \omega\epsilon$) or a lossless or good dielectric if the loss tangent is very small ($\sigma \ll \omega\epsilon$). Thus, the characteristics of wave propagation through other types of media can be obtained as special cases of wave propagation in a lossy medium as follows:

1. Good conductors: $\sigma \gg \omega\epsilon$, $\epsilon = \epsilon_o$, $\mu = \mu_o\mu_r$
2. Good dielectrics: $\sigma \ll \omega\epsilon$, $\epsilon = \epsilon_o$, $\mu = \mu_o\mu_r$
3. Free space: $\sigma = 0$, $\epsilon = \epsilon_o$, $\mu = \mu_o$

where $\epsilon_o = 8.854 \times 10^{-11}$ F/m is the free-space permittivity, and $\mu_o = 4\pi \times 10^{-7}$ H/m is the free-space permeability. The conditions for each medium type are merely substituted in Eqs. (20.10) to (20.21) to obtain the wave properties for that medium.

The classical model of wave propagation presented in this chapter helps us understand some basic concepts of EM wave propagation and the various parameters that play a part in determining the progress of the wave from the transmitter to the receiver. We will apply the ideas to the particular case of wave propagation in free space or the atmosphere in Section 20.3, Propagation in the Atmosphere. Before then, we digress a little and consider the important issue of wave polarization.

20.2 Wave Polarization

The concept of polarization is an important property of an EM wave that has been developed to described the various types of electric field variation and orientation. It is therefore a common practice to describe an EM wave by its polarization. The polarization of an EM wave depends on the transmitting antenna or source. It is determined by the direction of the electric field. It is regarded as the locus of the tip of the electric field (in a plane perpendicular to the direction of propagation) at a given point in space as a function of time. For this reason, there are four types of polarization: linear or plane, circular, elliptic, and random.

In linear or plane polarized waves, the orientation of the field is constant in space and time. For a plane traveling in the $+z$ direction, the electric field may be written as

$$\mathbf{E}(z,t) = E_x(z,t)\mathbf{a}_x + E_y(z,t)\mathbf{a}_y \tag{20.23}$$

where

$$E_x = \mathrm{Re}\left[E_{ox}e^{j(\omega t - kz + \phi_x)}\right] = E_{ox}\cos(\omega t - kz + \phi_x) \tag{20.24a}$$

$$E_y = \mathrm{Re}\left[E_{oy}e^{j(\omega t - kz + \phi_x)}\right] = E_{oy}\cos(\omega t - kz + \phi_y) \tag{20.24b}$$

For linear polarization, the phase difference between the x and y components must be

$$\Delta\phi = \phi_y - \phi_x = n\pi, \qquad n = 0, 1, 2, \ldots \tag{20.25}$$

This allows the two components to maintain the same ratio at all times, which implies that the electric field always lies along a straight line in a constant $-z$ plane. In other words, if we observe the wave in the direction of propagation (z in this case), we will notice that the tip of the electric field follows a line. Hence, we have the term *linear polarization.* Linearly polarized plane waves can be generated by simple antennas (such as dipole antennas) or lasers.

Circular polarized waves are characterized by an electric field with constant magnitude and orientation rotating in a plane transverse to the direction of propagation. Circular polarization takes place when the x and y components are the same in magnitude ($E_x = E_y$) and the phase difference between them is an odd multiple of $\pi/2$, i.e.,

$$\Delta\phi = \phi_y - \phi_x = \pm\left(\frac{1}{2} + 2n\right)\pi, \qquad n = 0, 1, 2, \ldots \tag{20.26}$$

The two components of the electric field rotate around the axis of propagation as a function of time and space. Circularly polarized waves can be generated by a helically wound wire antenna or by two linear sources that are oriented perpendicular to each other and fed with currents that are out of phase by 90°.

Linear and circular polarizations are special cases of the more general case of the elliptic polarization. An elliptically polarized wave is one in which the tip of the field traces an elliptic locus in a fixed transverse plane as the field changes with time. Elliptic polarization is achieved when the x and y components are not equal in magnitude ($E_x \neq E_y$) and the phase difference between them is an odd multiple of $\pi/2$, i.e.,

$$\Delta\phi = \phi_y - \phi_x = \pm\left(\frac{1}{2} + 2n\right)\pi, \quad n = 0, 1, 2, \ldots \tag{20.27}$$

This allows the tip of the electric field to trace an ellipse in the $x - y$ plane.

The polarized waves described so far are illustrated in Fig. 20.2. They are deterministic meaning that the field is a predictable function of space and time. If the field is completely random, the wave is said to be randomly polarized. Typical examples of such waves are radiation from the sun and radio stars.

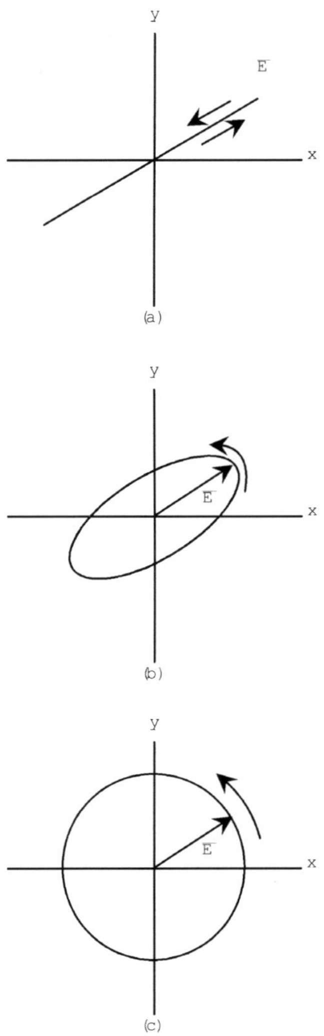

FIGURE 20.2 Wave polarizations: (a) linear, (b) elliptic, (c) circular.

20.3 Propagation in the Atmosphere

Wave propagation hardly occurs under the idealized conditions assumed in previously. For most communication links, the previous analysis must be modified to account for the presence of the earth, the ionosphere, and atmospheric precipitates such as fog, raindrops, snow, and hail. This is done in this section.

The major regions of the earth's atmosphere that are of importance in radio wave propagation are the troposphere and the ionosphere. At radar frequencies (approximately 100 MHz to 300 GHz), the troposphere is by far the most important. It is the lower atmosphere comprised of a nonionized region extending from Earth's surface up to about 15 km. The ionosphere is Earth's upper atmosphere in the altitude region from 50 km to one Earth radius (6370 km). Sufficient ionization exists in this region to influence wave propagation.

Wave propagation over the surface of Earth may assume any of the following three principal modes:

- Surface wave propagation along the surface of Earth;
- Space wave propagation through the lower atmosphere;
- Sky wave propagation by reflection from the upper atmosphere.

These modes are portrayed in Fig. 20.3. The sky wave is directed toward the ionosphere, which bends the propagation path back toward the earth under certain conditions in a limited frequency range (0 to 50 MHz approximately). This is highly dependent on the condition of the ionosphere (its level of ionization) and the signal frequency. The surface (or ground) wave takes effect at the low-frequency end of the spectrum (2 to 5 MHz approximately) and is directed along the surface over which the wave is propagated. Since the propagation of the ground wave depends on the conductivity of the earth's surface, the wave is attenuated more than if it were propagation through free space. The space wave consists of the direct wave and the reflected wave. The direct wave travels from the transmitter to the receiver in

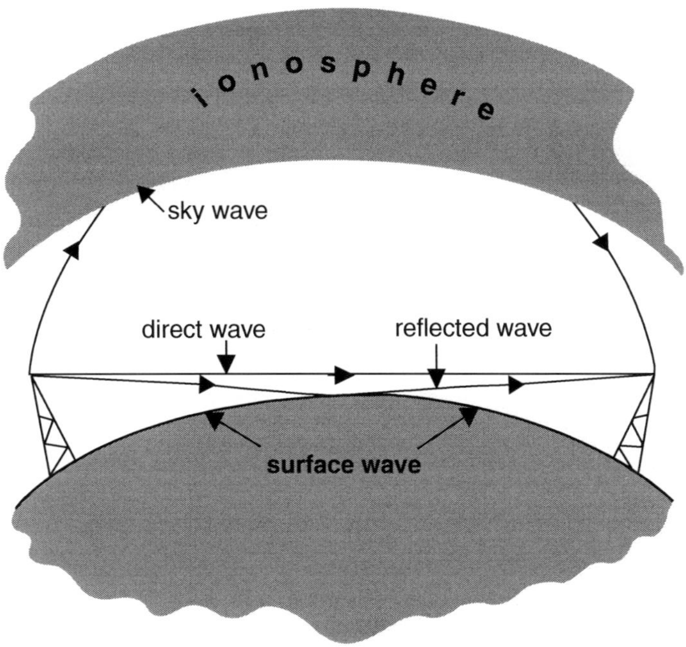

FIGURE 20.3 Modes of wave propagation.

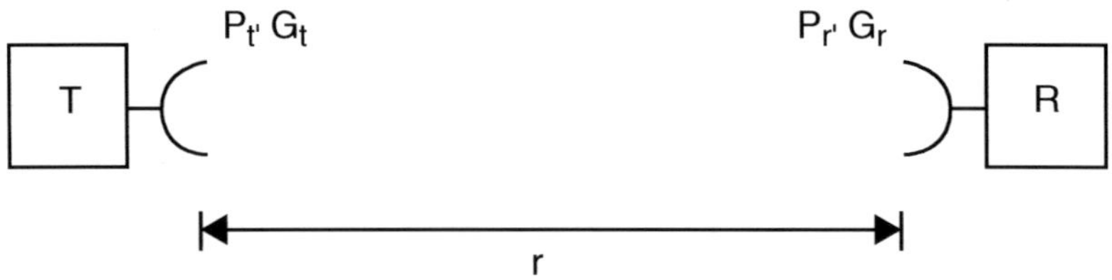

FIGURE 20.4 Transmitting and receiving antennas in free space.

nearly a straight path while the reflected wave is due to ground reflection. The space wave obeys the optical laws in that direct and reflected wave components contribute to the total wave component. Although the sky and surface waves are important in many applications, we will only consider the space wave in this chapter.

Figure 20.4 depicts the EM energy transmission between two antennas in space. As a wave radiates from the transmitting antenna and propagates in space, its power density decreases, as expressed ideally in Eq. (20.29). Assuming that the antennas are in a lossless medium or free space, the power received by the receiving antenna is given by the *Friis transmission equation* (Liu and Fang, 1988):

$$P_r = G_r G_t \left(\frac{\lambda}{4\pi r} \right)^2 P_t \tag{20.28}$$

where the subscripts t and r, respectively, refer to transmitting and receiving antennas. In Eq. (20.28), P = power in watts, G = antenna gain (dimensionless), r = distance between the antennas in meters, and λ = wavelength in meters. The Friis equation relates the power received by one antenna to the power transmitted by the other, provided that the two antennae are separated by $r > 2D^2/\lambda$, and D is the largest dimension of either antenna. Thus, the Friis equation applies only when the two antennas are in the far field of each other. It shows that the received power decays at a rate of 20 dB/decade with distance. In case the propagation path is not in free space, a correction factor F is included to account for the effect of the medium. This factor, known as the *propagation factor,* is simply the ratio of the electric field intensity E_m in the medium to the electric field intensity E_o in free space, i.e.,

$$F = \frac{E_m}{E_o} \tag{20.29}$$

The magnitude of F is always less than unity since E_m is always less than E_o. Thus, for a lossy medium, Eq. (20.28) becomes

$$P_r = G_r G_t \left(\frac{\lambda}{4\pi r} \right)^2 P_t |F|^2 \tag{20.30}$$

For practical reasons, Eqs. (20.28) or (20.29) are commonly expressed in logarithmic form. If the all the terms are expressed in decibels (dB), Eq. (20.30) can be written in logarithmic form as

$$P_r = P_t + G_r + G_t - L_o - L_n \tag{20.31}$$

where P = power in dB referred to 1 W (or simply dBW), G = gain in dB, L_o = free-space loss in dB, and L_m = loss in dB due to the medium. The free-space loss is obtained from standard nomograph or directly from

$$L_o = 20\log\left(\frac{4\pi r}{\lambda}\right) \tag{20.32}$$

while the loss due to the medium is given by

$$L_m = -20\log|F| \tag{20.33}$$

Our major concern in the rest of this section is to determine L_o and L_m for two important cases of space propagation that differ considerably from the free-space conditions.

20.3.1 Effect on Earth

The phenomenon of multipath propagation causes significant departures from free-space conditions. The term *multipath* denotes the possibility of an EM wave propagating along various paths from the transmitter to the receiver. In multipath propagation of an EM wave over Earth's surface, two such paths exist: a direct path and a path via reflection and diffractions from the interface between the atmosphere and Earth. A simplified geometry of the multipath situation is shown in Fig. 20.5. The reflected and diffracted component is commonly separated into two parts: one *specular* (or coherent) and the other *diffuse* (or incoherent), that can be separately analyzed. The specular component is well defined in terms of its amplitude, phase, and incident direction. Its main characteristic is its conformance to Snell's law for reflection, which requires that the angles of incidence and reflection be equal and coplanar. It is a plane wave, and as such, is uniquely specified by its direction. The diffuse component, however, arises out of the random nature of the scattering surface, and as such, is nondeterministic. It is not a plane wave and does not obey Snell's law for reflection. It does not come from a given direction but from a continuum.

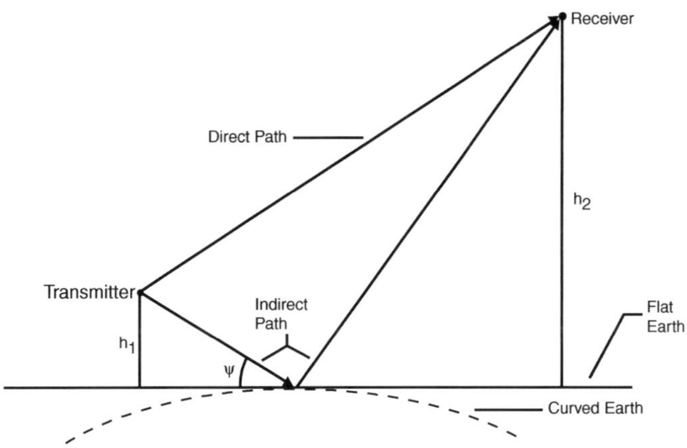

FIGURE 20.5 Multipath geometry.

The loss factor F that accounts for the departures from free-space conditions is given by

$$F = 1 + \Gamma\, \rho_s\, D\, S(\theta)\, e^{-j\Delta} \tag{20.34}$$

where Γ = Fresnel reflection coefficient
ρ_s = roughness coefficient
D = divergence factor
$S(\theta)$ = shadowing function
Δ = phase angle corresponding to path difference

The Fresnel reflection coefficient Γ accounts for the electrical properties of Earth's surface. Since Earth is a lossy medium, the value of the reflection coefficient depends on the complex relative permittivity ϵ_c of the surface, the grazing angle ψ, and the wave polarization. It is given by

$$\Gamma = \frac{\sin\psi - z}{\sin\psi + z} \tag{20.35}$$

where

$$z = \sqrt{\epsilon_c - \cos^2\psi} \quad \text{for horizontal polarization} \tag{20.36}$$

$$z = \frac{\sqrt{\epsilon_c - \cos^2\psi}}{\epsilon_c} \quad \text{for vertical polarization} \tag{20.37}$$

$$\epsilon_c = \epsilon_r - j\frac{\sigma}{\omega\epsilon_o} = \epsilon_r - j60\sigma\lambda \tag{20.38}$$

ϵ_r and σ are the dielectric constant and conductivity of the surface; ω and λ are the frequency and wavelength of the incident wave; and ψ is the grazing angle. It is apparent that $0 < |\Gamma| < 1$.

To account for the spreading (or divergence) of the reflected rays due to Earth's curvature, we introduce the divergence factor D. The curvature has a tendency to spread out the reflected energy more than a corresponding flat surface. The divergence factor is defined as the ratio of the reflected field from a curved surface to the reflected field from a flat surface [Kerr, 1951]. Using the geometry of Fig. 20.6, D is given by

$$D \simeq \left(1 + \frac{2G_1G_2}{a_e G \sin\psi}\right)^{-1/2} \tag{20.39}$$

where $G = G_1 + G_2$ is the total ground range and a_e = 6370 km is the effective earth radius. Given the transmitter height h_1, the receiver height h_2, and the total ground range G, we can determine G_1, G_2, and ψ. If we define

$$p = \frac{2}{\sqrt{3}}\left[a_e\left(h_1 + h_2\right) + \frac{G^2}{4}\right]^{1/2} \tag{20.40}$$

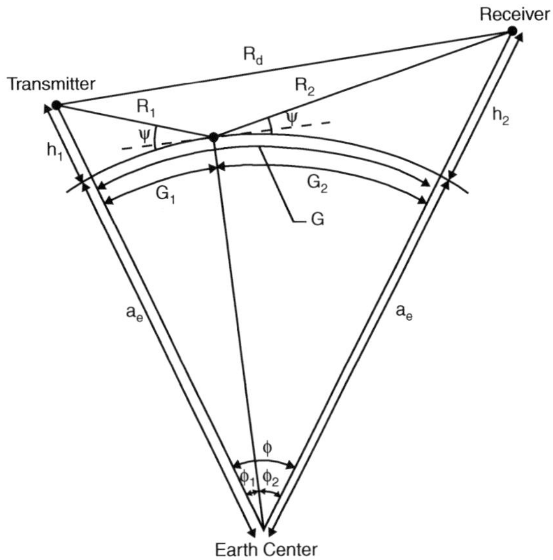

FIGURE 20.6 Geometry of spherical Earth reflection.

$$\alpha = \cos^{-1}\left[\frac{2a_e(h_1 - h_2)G}{p^3}\right] \tag{20.41}$$

and assume $h_1 \le h_2$, $G_1 \le G_2$, using small angle approximation yields

$$G_1 = \frac{G}{2} + p\cos\left(\frac{\pi + \alpha}{3}\right) \tag{20.42}$$

$$G_2 = G - G_1 \tag{20.43}$$

$$\phi_i = \frac{G_i}{a_e}, \quad i = 1, 2 \tag{20.44}$$

$$R_i = \left[h_i^2 + 4a_e(a_e + h_i)\sin^2(\phi_i/2)\right]^{1/2} \quad i = 1, \tag{20.45}$$

The grazing angle is given by

$$\psi = \sin^{-1}\left[\frac{2a_e h_1 + h_1^2 - R_1^2}{2a_e R_1}\right] \tag{20.46}$$

or

$$\psi = \sin^{-1}\left[\frac{2a_e h_1 + h_1^2 + R_1^2}{2(a_e + h_1)R_1}\right] - \phi_1 \tag{20.47}$$

Although D varies from 0 to 1, in practice D is a significant factor at low grazing angle ψ (less than 0.1%).

The phase angle corresponding to the path difference between direct and reflected waves is given by

$$\Delta = \frac{2\pi}{\lambda}\left(R_1 + R_2 - R_d\right) \tag{20.48}$$

The roughness coefficient ρ_s takes care of the fact that the earth's surface is not sufficiently smooth to produce specular (mirror-like) reflection except at a very low grazing angle. Earth's surface has a height distribution that is random in nature. The randomness arises out of the hills, structures, vegetation, and ocean waves. It is found that the distribution of the different heights on Earth's surface is usually the Gaussian or normal distribution of probability theory. If σ_h is the standard deviation of the normal distribution of heights, we define the roughness parameters as

$$g = \frac{\sigma_h \sin\psi}{\lambda} \tag{20.49}$$

If $g < 1/8$, specular reflection is dominant; if $g > 1/8$, diffuse scattering results. This criterion, known as the *Rayleigh criterion,* should only be used as a guideline since the dividing line between a specular and a diffuse reflection or between a smooth and a rough surface is not well defined. The roughness is taken into account by the roughness coefficient ($0 < \rho_s < 1$), which is the ratio of the field strength after reflection with roughness taken into account to that which would be received if the surface were smooth. The roughness coefficient is given by

$$\rho_s = \exp\left[-2(2\pi g)^2\right] \tag{20.50}$$

The shadowing function $S(\theta)$ is important at a low grazing angle. It considers the effect of geometric shadowing — the fact that the incident wave cannot illuminate parts of the earth's surface shadowed by higher parts. In a geometric approach, where diffraction and multiple scattering effects are neglected, the reflecting surface will consist of well-defined zones of illumination and shadow. As there will be no field on a shadowed portion of the surface, the analysis should include only the illuminated portions of the surface. The phenomenon of shadowing of a stationary surface was first investigated by Beckman in 1965 and subsequently refined by Smith [1967] and others. A pictorial representation of rough surfaces illuminated at the angle of incidence θ ($= 90° - \psi$) is shown in Fig. 20.7. It is evident from the figure that the shadowing function $S(\theta)$ equals unity when $\theta = 0$ and zero when $\theta = \pi/2$. According to Smith (1967),

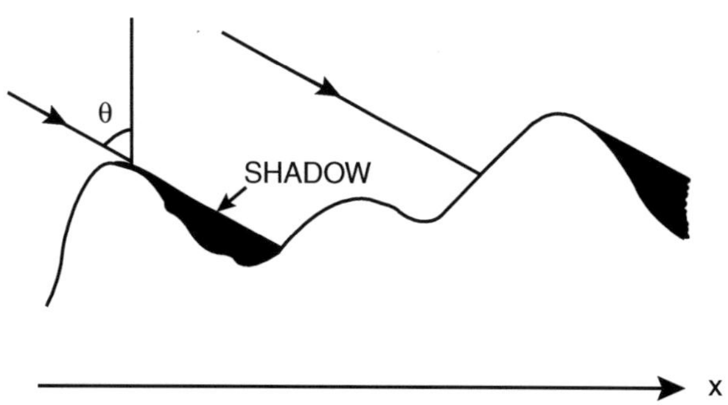

FIGURE 20.7 Rough surface illuminated at an angle of incidence θ.

$$S(\theta) \simeq \frac{\left[1 - \frac{1}{2}\,\mathrm{erfc}(a)\right]}{1 + 2B} \tag{20.51}$$

where erfc(x) is the complementary error function,

$$\mathrm{erfc}(x) = 1 - \mathrm{erf}(x) = \frac{2}{\sqrt{\pi}} \int_x^{\infty} e^{-t^2}\, dt \tag{20.52}$$

and

$$B = \frac{1}{4a}\left[\frac{1}{\sqrt{\pi}} e^{a^2} - a\,\mathrm{erfc}(a)\right] \tag{20.53}$$

$$a = \frac{\cot\theta}{2s} \tag{20.54}$$

$$s = \frac{\sigma_h}{\sigma_l} = \text{rms surface slope} \tag{20.55}$$

In Eq. (20.55), σ_h is the rms roughness height and σ_l is the correlation length. Alternative models for $S(\theta)$ are available in the literature. Using Eqs. (20.35) to (20.55), the loss factor in Eq. (20.34) can be calculated. Thus

$$L_o = 20\,\log\left(\frac{4\pi R_d}{\lambda}\right) \tag{20.56}$$

$$L_m = -20\,\log\left(1 + \Gamma\,\rho_s\,D\,S(\theta)\,e^{-j\Delta}\right) \tag{20.57}$$

20.3.2 Effect of Atmospheric Hydrometeors

The effect of atmospheric hydrometeors on satellite–Earth propagation is of major concern at microwave frequencies. The problem of scattering of electromagnetic waves by atmospheric hydrometeors has attracted much interest since the late 1940s. The main hydrometeors that exist for long duration, and have the greatest interaction with microwaves are rain, snow, and dust particles. At frequencies above 10 GHz, rain has been recognized as the most fundamental obstacle in Earth–space path. Rain has been known to cause attenuation, phase difference, and depolarization of radio waves. For analog signals, the effect of rain is more significant above 10 GHz while for digital signals, rain effects can be significant down to 3 GHz. Attenuation of microwaves due to precipitation becomes severe owing to increased scattering and beam energy absorption by raindrops thus impairing terrestrial as well as Earth–satellite communication links. Cross-polarization distortion due to rain has also been engaging the attention of researchers. This is particularly of interest when frequency reuse employing signals with orthogonal polarizations are used for doubling the capacity of a communication system. Thorough reviews on the interaction of microwaves with hydrometeors have been done by Oguchi [1983].

The loss due to rain-filled medium is given by

$$L_m = \gamma(R)\, \ell_e(R)\, p(R) \tag{20.58}$$

where γ = attenuation per unit length at rain rate R
ℓ_e = equivalent path length at rain rate R
$p(R)$ = probability in percentage of rainfall rate R

The attenuation is a function of the cumulative rain-rate distribution, drop-size distribution, refractive index of water, temperature, and other variables. A rigorous calculation of $\gamma(R)$ using various numerical modeling tools and incorporating raindrop size distribution, velocity of raindrops, and the refractive index of water can be found in Sadiku [2001]. For practical engineering purposes, what is needed is a simple formula relating attenuation to rain parameters. Such is found in the aR^b empirical relationship, which has been employed to calculate rain attenuation directly Collin, [1985], i.e.,

$$\gamma(R) = aR^b \quad \text{dB/km} \tag{20.59}$$

where R is the rain rate and a and b are constants. At 0° C, the values of a and b are related to frequency f in gigahertz as follows:

$$a = G_a\, f^{E_a} \tag{20.60}$$

where

$$G_a = 6.39 \times 10^{-5}, \quad E_a = 2.03, \quad \text{for } f < 2.9 \text{ GHz}$$

$$G_a = 4.21 \times 10^{-5}, \quad E_a = 2.42, \quad \text{for } 2.9 \text{ GHz} \le f \le 54 \text{ GHz}$$

$$G_a = 4.09 \times 10^{-2}, \quad E_a = 0.699, \quad \text{for } 54 \text{ GHz} \le f < 100 \text{ GHz}$$

$$G_a = 3.38, \quad E_a = -0.151, \quad \text{for } 180 \text{ GHz} < f$$

and

$$b = G_b\, f^{E_b} \tag{20.61}$$

where

$$G_b = 0.851, \quad E_b = 0.158, \quad \text{for } f < 8.5 \text{ GHz}$$

$$G_b = 1.41, \quad E_b = -0.0779, \quad \text{for } 8.5 \text{ GHz} \le f < 25 \text{ GHz}$$

$$G_b = 2.63, \quad E_b = -0.272, \quad \text{for } 25 \text{ GHz} \le f < 164 \text{ GHz}$$

$$G_b = 0.616, \quad E_b = 0.0126, \quad \text{for } 164 \text{ GHz} \le f.$$

The effective length $\ell_e(R)$ through the medium is needed since rain intensity is not uniform over the path. Its actual value depends on the particular area of interest and therefore has a number of representations [Liu and Fang, 1988]. Based on data collected in western Europe and eastern North America, the effective path length has been approximated as [Hyde, 1984]

TABLE 20.1 Composition of Dry Atmosphere from Sea Level to about 90 km [Livingston, 1970]

Constituent	Percent by Volume	Percent by Weight
Nitrogen	78.088	75.527
Oxygen	20.949	23.143
Argon	0.93	1.282
Carbon dioxide	0.03	0.0456
Neon	1.8×10^{-3}	1.25×10^{-3}
Helium	5.24×10^{-4}	7.24×10^{-5}
Methane	1.4×10^{-4}	7.75×10^{-5}
Krypton	1.14×10^{-4}	3.30×10^{-4}
Nitrous oxide	5×10^{-5}	7.6×10^{-5}
Xenon	8.6×10^{-6}	3.90×10^{-5}
Hydrogen	5×10^{-5}	3.48×10^{-6}

$$\ell_e(R) = \left[0.00741R^{0.766} + (0.232 - 0.00018R)\sin\theta\right] \tag{20.62}$$

where θ is the elevation angle.

The cumulative probability in percentage of rainfall rate R is given by [Hyde, 1984].

$$p(R) = \frac{M}{87.66}\left[0.03\beta e^{-0.03R} + 0.2(1-\beta)\left(e^{-0.258R} + 1.86e^{-1.63R}\right)\right] \tag{20.63}$$

where M is mean the annual rainfall accumulation in mm and β is the Rice–Holmberg thunderstorm ratio.

The effect of other hydrometeors such as vapor, fog, hail, snow, and ice is governed by fundamental principles similar to the effect of rain [Collin, 1985]. However, their effects are at least an order of magnitude less than the effect of rain in most cases.

20.3.3 Other Effects

Besides hydrometeors, the atmosphere has the composition given in Table 20.1. While attenuation of EM waves by hydrometeors may result from both absorption and scattering, gases act only as absorbers. Although some of these gases do not absorb microwaves, some possess permanent electric and/or magnetic dipole moment and play some part in microwave absorption. For example, nitrogen molecules do not possess permanent electric or magnetic dipole moment and therefore play no part in microwave absorption. Oxygen has a small magnetic moment that enables it to display weak absorption lines in the centimeter- and millimeter-wave regions. Water vapor is a molecular gas with a permanent electric dipole moment. It is more responsive to excitation by an EM field than is oxygen.

Other mechanisms that can affect EM wave propagation in free space, not discussed in this chapter, include clouds, dust, and the ionosphere. The effect of the ionosphere is discussed in detail in standard texts.

References

Collin, R. E., 1985. *Antennas and Radiowave Propagation,* McGraw-Hill, New York, 339–456.

Freeman, R. L., 1994. *Reference Manual for Telecommunications Engineering,* 2nd ed., John Wiley & Sons, New York, 711–768.

Hyde, G., 1984. Microwave propagation, in R. C. Johnson and H. Jasik (Eds.), *Antenna Engineering Handbook,* 2nd ed., McGraw-Hill, New York, 45.1–45.17.

Kerr, D. E., 1951. *Propagation of Short Radio Waves,* McGraw-Hill, New York, (Republished by Peter Peregrinus, London, U.K., 1987), 396–444.

Liu, C. H. and D. J. Fang, 1988. Propagation, in Y. T. Lo and S. W. Lee, *Antenna Handbook: Theory, Applications, and Design,* Van Nostrand Reinhold, New York, 29.1–29.56.

Livingston, D. C., 1970. *The Physics of Microwave Propagation,* Prentice Hall, Englewood Cliffs, NJ, 11.

Oguchi, T., 1983. Electromagnetic Wave Propagation and Scattering in Rain and Other Hydrometeors, *Proc. IEEE,* 71, 1029–1078.

Sadiku, M. N. O., 2001. *Numerical Techniques in Electromagnetics,* 2nd ed., CRC Press, Boca Raton, FL, 95–105.

Sadiku, M. N. O., 2001. *Elements of Electromagnetics,* 3rd ed., Oxford University Press, New York, 410–472.

Smith, B. G., 1967. Geometrical shadowing of a random rough surface, *IEEE Trans. Antennas Propag.,* 15, 668–671.

Veley, V. F., *Modern Microwave Technology,* Prentice Hall, Englewood Cliffs, NJ, 1987, 23–33.

Further Information

The subject of wave propagation could easily fill many chapters, and here it has only been possible to point out some of the main points of concern to microwave systems engineer. There are several sources of information dealing with the theory and practice of wave propagation in space. Some of these are in the reference section. Journals such as *Radio Science, IEE Proceedings Part H, IEEE Transactions on Antenna and Propagation* are devoted to EM wave propagation. *Radio Science* is available at American Geophysical Union, 2000 Florida Avenue, NW, Washington DC 20009; *IEE Proceedings Part H* at IEE Publishing Department, Michael Faraday House, 6 Hills Way, Stevenage, Herts SG1 2AY, U.K.; and *IEEE Transactions on Antenna and Propagation* at IEEE, 445 Hoes Lane, P. O. Box 1331, Piscataway, NJ 08855-1331.

21

Guided Wave Propagation and Transmission Lines

W.R. Deal
Malibu Networks, Inc.

Vesna Radisic
Northrop Grumman

Y. Qian
Microsemi Integrated Products

T. Itoh
University of California

At higher frequencies where wavelength becomes small with respect to feature size, it is often necessary to consider an electronic signal as an electromagnetic wave and the structure where this signal exists as a waveguide. A variety of different concepts can be used to examine this wave behavior. The most simplistic view is transmission line theory, where propagation is considered in a simplistic one-dimensional (1D) manner and the cross-sectional variation of the guided wave is entirely represented in terms of distributed transmission parameters in an equivalent circuit. This is the starting point for transmission line theory that is commonly used to design microwave circuits. In other guided wave structures, such as enclosed waveguides, it is more appropriate to examine the concepts of wave propagation from the perspective of Maxwell's equations, the solutions of which will explicitly demonstrate the cross-sectional dependence of the guided wave structure.

Most practical wave guiding structures rely on single-mode propagation, which is restricted to a single direction. This allows the propagating wave to be categorized according to its polarization properties. A convenient method is classifying the modes as transverse electric and magnetic (TEM), traverse electric (TE), or transverse magnetic (TM). TEM modes have both the electric and magnetic field transverse in the direction of propagation. Only the magnetic field transverses in the direction of propagation in TM modes, and only the electric field transverses in the direction of propagation in TE modes.

In this chapter, we first briefly examine the telegrapher's equation, which is the starting point for transmission line theory. The simple transmission line model accurately describes a number of guided wave structures and is the starting point for transmission line theory. In the next section, enclosed waveguides including rectangular and circular waveguides will be discussed. Relevant concepts such as cutoff frequency and modes will be given. In the final section, four common planar guided wave structures will be discussed. These inexpensive and compact structures are the foundation for the modern commercial RF front end.

0-8493-1732-0/03/$0.00+$1.50

21.1 TEM Transmission Lines, Telegrapher's Equations, and Transmission Line Theory

In this section, the concept of guided waves in simple TEM-guiding structures will be explored in terms of the simple model provided by telegrapher's equations, also referred to as transmission line equations. Telegrapher's equations demonstrate guided wave properties in terms of lumped equivalent circuit parameters available for many types of simple two-conductor transmission lines, and are valid for all types of TEM waveguides if their corresponding equivalent circuit parameters are known. These parameters must be found from Maxwell's equations in their fundamental form. Finally, properties and parameters for several types of two-wire TEM transmission line structures are introduced.

A transmission line or waveguide is used to transmit power and information from one point to another in an efficient manner. Three common types of transmission lines that support TEM guided waves are shown in Figure 21.1a to c, including the parallel-plate transmission line, two-wire line, and coaxial transmission line. The parallel-plate transmission line consists of a dielectric slab sandwiched between two parallel conducting plates of width w. More practical, commonly used variations of this structure at microwave and millimeter-wave frequencies include microstrip and stripline, which will be briefly discussed in the final section of this chapter. A two-wire transmission line, consisting of two parallel conducting lines separated by a distance d is shown in Fig. 21.1b. This is commonly used for power distribution at low frequencies. Finally, the coaxial transmission line consists of two concentric conductors separated by a dielectric layer. This structure is well shielded and commonly used at high frequencies well into the microwave range.

The telegrapher's equations form a simple and intuitive starting point for the physics of guided wave propagation in these structures. An equivalent circuit model is shown in Fig. 21.2 for a two-conductor transmission line of differential length Δz in terms of the following four parameters:

R, resistance per unit length of both conductors (Ω/m);
L, inductance per unit length of both conductors (H/m);
G, conductance per unit length (S/m);
C, capacitance per unit length of both conductors (F/m).

These parameters represent physical quantities for each of the relevant transmission lines. For each of the structures shown in Fig. 21.1a to c, R represents conductor losses, L represents inductance, G represents dielectric losses, and C represents the capacitance between the two lines.

Returning to Fig. 21.2, the quantities $v(z,t)$ and $v(z + \Delta z,t)$ represent change in voltage along the differential length of transmission line, while $i(z,t)$ and $i(z + \Delta z,t)$ represent the change in current. Writing Kirchoff's voltage law and current laws for the structure, dividing by Δz, and applying the fundamental theorem of calculus as $\Delta z \to 0$, two coupled differential equations known as the telegrapher's equations are obtained:

$$-\frac{\partial v(z,t)}{\partial z} = Ri(z,t) + L\frac{\partial i(z,t)}{\partial t} \tag{21.1}$$

$$-\frac{\partial i(z,t)}{\partial z} = Gi(z,t) + D\frac{\partial v(z,t)}{\partial t} \tag{21.2}$$

However, typically we are interested in signals with harmonic time dependence ($e^{j\omega t}$). In this case, the time harmonic forms of the telegrapher's equations are given by

$$-\frac{dV(z)}{dz} = (R + j\omega L)\,I(z) \tag{21.3}$$

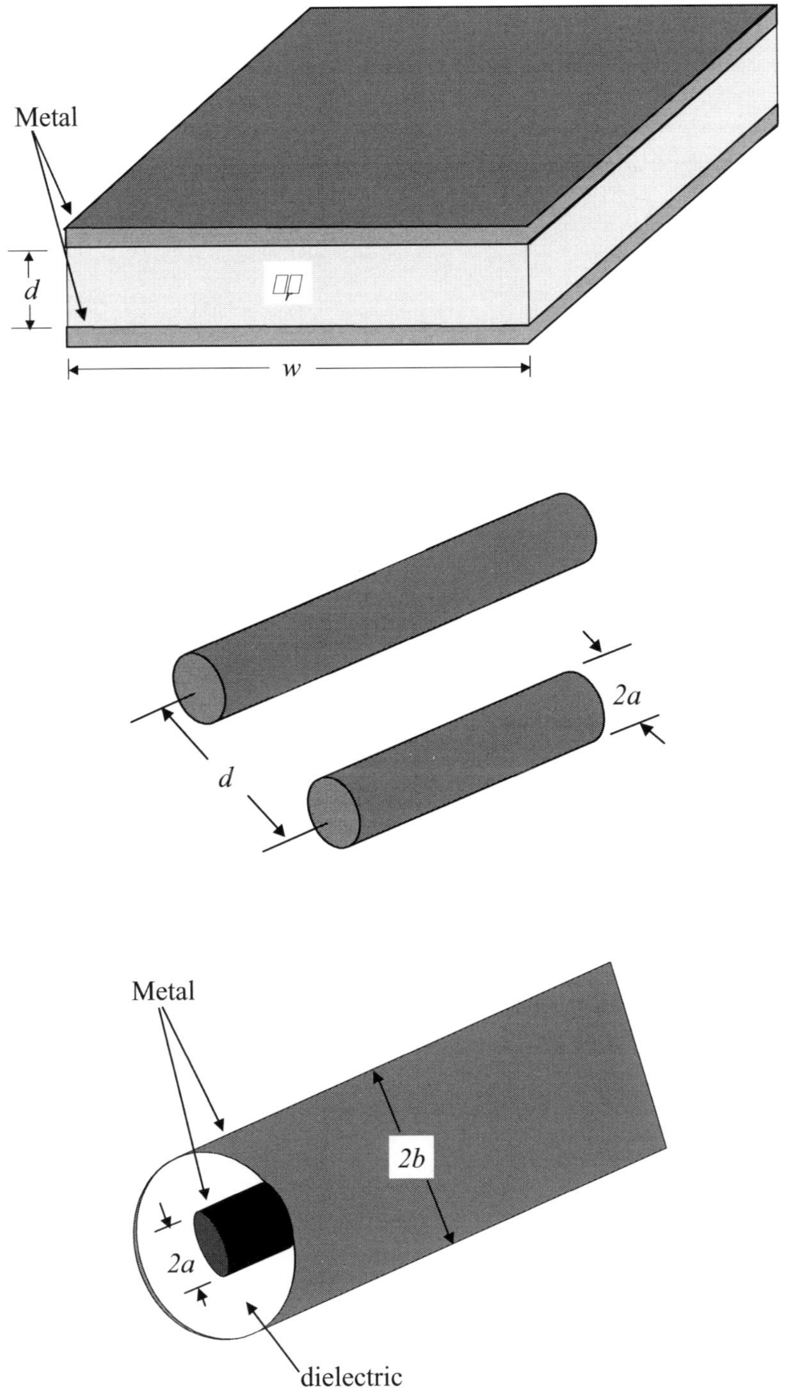

FIGURE 21.1 Three simple TEM-type transmission line geometries including (a) parallel-plate transmission line, (b) two-wire line, and (c) coaxial line.

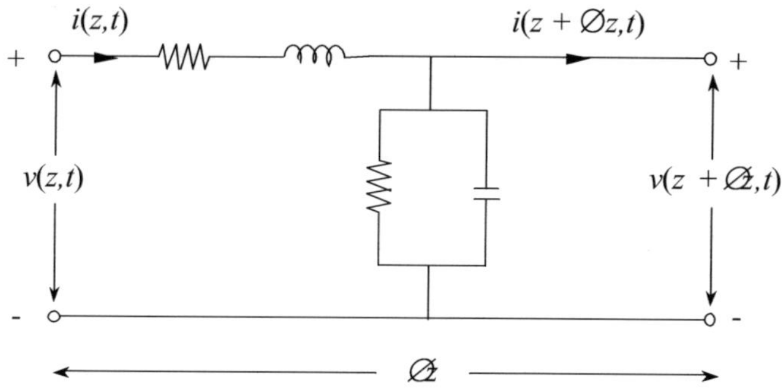

FIGURE 21.2 Distributed equivalent circuit model for a transmission line.

$$-\frac{dI(z)}{dz} = (G + j\omega C)V(z) \quad (21.4)$$

The constant γ is defined as the propagation constant with real and imaginary parts, α and β, corresponding to the attenuation constant (Np/m) and phase constant (rad/m) in the following manner

$$\gamma = \alpha + j\beta = \sqrt{(R + j\omega L)(G + j\omega C)} \quad (21.5)$$

This may then be substituted into the telegrapher's equations, which may then be solved for $V(z)$ and $I(z)$ to yield the following 1D wave equations:

$$\frac{d^2V(z)}{dz^2} - \gamma^2 V(z) = 0 \quad (21.6)$$

$$\frac{d^2I(z)}{dz^2} - \gamma^2 I(z) = 0 \quad (21.7)$$

The form of this equation is the well-known wave equation. This indicates that the transmission line will support a guided electromagnetic wave traveling in the z direction. The telegrapher's equations use a physical equivalent circuit and basic circuit theory to demonstrate the wave behavior of an electromagnetic signal on a transmission line. Alternatively, the same result can be obtained by starting directly with Maxwell's equations in their fundamental form, which may be used to derive the wave equation for a propagating electromagnetic wave. In this case, the solution of the wave equation will be governed by the boundary conditions. Similarly, the parameters R, L, G, and C are determined by the geometry of the transmission line structures.

Returning to the telegrapher's equations, several important facts may be noted. First, the characteristic impedance of the transmission line may be found by taking the ratio of the forward traveling voltage and current wave amplitudes, and is given in terms of the equivalent circuit parameters as

$$Z_0 = \sqrt{\frac{R + j\omega L}{G + j\omega C}} \quad (21.8)$$

In the case of a lossless transmission line, this reduces to $Z_o = \sqrt{L/C}$. The phase velocity, also known as the propagation velocity, is the velocity of the wave as it moves along the waveguide. It is defined as

$$v_p = \frac{\omega}{\beta} \tag{21.9}$$

In the lossless case, this reduces to:

$$v_p = \frac{1}{\sqrt{LC}} = \frac{1}{\sqrt{\mu\varepsilon}} \tag{21.10}$$

This shows that the velocity of the signal is directly related to the medium. In the case of an air-filled, purely TEM mode, the wave will propagate at the familiar value $c = 3 \times 10^8$ m/s. Additionally, it provides the relationship between L, C, and the medium in which the wave is guided. Therefore, if the properties of the medium are known, it is only necessary to determine either L or C. Once C is known, G may be determined by the following relationship:

$$\frac{G}{C} = \frac{\sigma}{\varepsilon} \tag{21.11}$$

Note that σ is the conductivity of the medium, not of the metal conductors. The final parameter, the series resistance R, is determined by the power loss in the conductors. Simple approximations for the transmission line parameters R, L, G, and C for the three types of transmission lines shown in Figs. 21.1a to c are well known and are shown in Table 21.1. Note that μ, ε, and σ relate to the medium separating the conductors, and σ_c refers to the conductor. Once the equivalent circuit parameters are determined, the characteristic impedance and propagation constant of the transmission line may be determined. Note that R_s represents the surface resistance of the conductors, given as

$$R_s = \sqrt{\frac{\pi f \mu_c}{\sigma_c}} \tag{21.12}$$

TABLE 21.1 Transmission Line Parameters for Parallel-Plate, Two-Wire Line and Coaxial Transmission Lines

	Parallel-Plate Waveguide	Two-Wire Line	Coaxial Line
R (Ω/m)	$\frac{2}{w}R_s$	$\frac{R_s}{\pi a}$	$\frac{R_s}{2\pi}\left(\frac{1}{a}+\frac{1}{b}\right)$
L (H/m)	$\mu\frac{d}{w}$	$\frac{\mu}{\pi}\cosh^{-1}\left(\frac{D}{2a}\right)$	$\frac{\mu}{2\pi}\ln\left(\frac{b}{a}\right)$
G (S/m)	$\sigma\frac{w}{d}$	$\frac{\pi\sigma}{\cosh^{-1}(D/2a)}$	$\frac{2\pi\sigma}{\ln(b/a)}$
C (F/m)	$\varepsilon\frac{w}{d}$	$\frac{\pi\varepsilon}{\cosh^{-1}(D/2a)}$	$\frac{2\pi\varepsilon}{\ln(b/a)}$

21.2 Guided Wave Solution from Maxwell's Equations, Rectangular Waveguide, and Circular Waveguide

A waveguide is any structure that guides an electromagnetic wave. In the preceding section, several simple TEM transmission structures were discussed. While these structures do support a guided wave, the term waveguide more commonly refers to a closed metallic structure with a fixed cross section within which a guided wave propagates, as shown for the arbitrary cross section in Fig. 21.3. The guide is filled with a material of permittivity ε and permeability μ, and is defined by its metallic wall parallel to the z-axis. These structures demonstrate lower losses than the simple transmission line structures of the first section, and are used to transport power in the microwave and millimeter-wave frequency range. Ohmic losses are low and the waveguide is capable of carrying large power levels. Disadvantages are bulk, weight, and limited bandwidth, which cause planar transmission lines to be used wherever possible in modern communications circuits. However, a wide variety of components are available in this technology, including high performance filters, couplers, isolators, attenuators, and detectors.

Inside this type of enclosed waveguide, an infinite number of distinct solutions exist, each of which is referred to as a *waveguide mode.* At a given operating frequency, the cross section of the waveguide and the type of material in the waveguide determine the characteristics of these modes. These modes are usually classified by the longitudinal components of the electric and magnetic fields, E_z and H_z, where propagation is in the z direction. The most common classifications are TE, TM, EH, and HE modes. The basic characteristics are described in the next two paragraphs. The TEM modes that were discussed in the previous section do not propagate in this type of metallic enclosed waveguide. This is because a TEM mode requires two conductors to propagate, where a conventional enclosed waveguide has only a single enclosing conductor.

The two most common waveguide modes are the TE and TM modes. TE modes have no component of E in the z direction, which means that E is completely transverse to the direction of propagation. Similarly, TM modes have no component of H in the z direction.

EH and HE modes are hybrid modes that may be present under certain conditions, such as a waveguide partially filled with dielectric. In this case, pure TE and TM are unable to satisfy all of the necessary boundary conditions and a more complex type of modal solution is required. With both EH and HE, neither E nor H are zero in the direction of propagation. In EH modes, the characteristics of the transverse fields are controlled more by H_z than by E_z. HE modes are controlled more by E_z than by H_z. These types

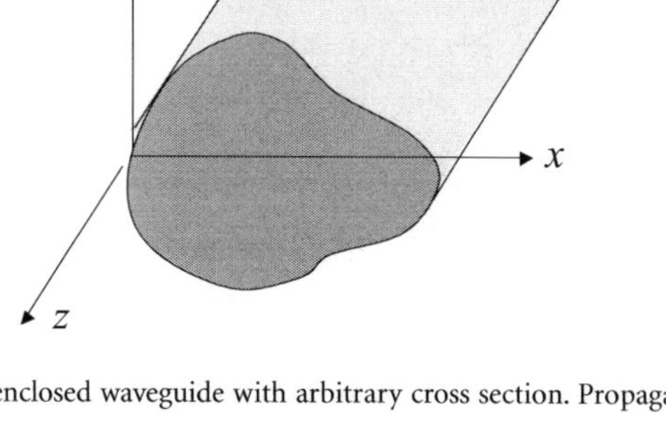

FIGURE 21.3 Geometry of enclosed waveguide with arbitrary cross section. Propagation is in the z direction.

of hybrid modes may also be referred to as LSE (Longitudinal Section Electric) and LSM (Longitudinal Section Magnetic). It should be noted that most commonly used waveguides are homogeneous, being entirely filled with material of a single permittivity (which may of course be air) and these types of modes will not be present.

Inside a homogeneous waveguide, E_z and H_z satisfy the scalar wave equation inside the waveguide:

$$\left(\frac{\partial^2}{\partial x^2}+\frac{\partial^2}{\partial y^2}\right)E_z+h^2E_z=0 \tag{21.13}$$

$$\left(\frac{\partial^2}{\partial x^2}+\frac{\partial^2}{\partial y^2}\right)H_z+h^2H_z=0 \tag{21.14}$$

Note that h is given as:

$$h^2=\omega^2\mu\varepsilon+\gamma^2=k^2+\gamma^2 \tag{21.15}$$

The wavenumber k is for the material filling the waveguide. For several simple homogenous waveguides with commonly used waveguide geometries, applying boundary equations on the walls of the waveguide may be used to solve these equations to obtain closed form solutions. The resulting modal solution will possess distinct eigenvalues determined by the cross section of the waveguide. One important result obtained from this procedure is that waveguide modes, unlike the fundamental TEM mode that propagates in two-wire structures at any frequency, will have a distinct cutoff frequency. It may be shown that the propagation constant varies with frequency as

$$\gamma=\alpha+j\beta=h\sqrt{1-\left(\frac{f}{f_c}\right)^2} \tag{21.16}$$

where the cutoff frequency, f_c is given by:

$$f_c=\frac{h}{2\pi\sqrt{\mu\varepsilon}} \tag{21.17}$$

By inspection of Eq. (21.16), and recalling the $\exp(j\omega t-\gamma z)$ dependence of the wave propagating in the $+z$ direction (for propagation in the $-z$ direction, replace z with $-z$), the physical significance of the cutoff frequency is clear. For a given mode, when $f > f_c$, the propagation constant γ is imaginary and the wave is propagating. Alternatively, when $f < f_c$, the propagation constant γ is real and the wave decays exponentially. In this case, modes operated below the cutoff frequency attenuate rapidly and are therefore referred to as evanescent modes. In practice, a given waveguide geometry is seldom operated at a frequency where more than one mode will propagate. This fixes the bandwidth of the waveguide to operate at some point above the cutoff frequency of the fundamental mode and below the cutoff frequency of the second-order mode, although in some rare instances higher order modes may be used for specialized applications.

The guided wavelength is also a function of the cross-section geometry of the waveguide structure. The guided wavelength is given as:

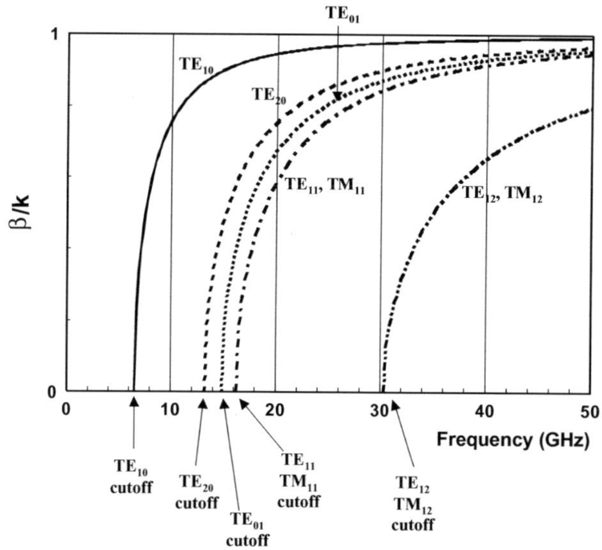

FIGURE 21.4 β/k diagram for WR-90 waveguide illustrating the concept of higher mode propagation and cutoff frequency.

$$\lambda_g = \frac{\lambda_0}{\sqrt{1-\left(\frac{f_c}{f}\right)^2}} \tag{21.18}$$

Note that λ_0 is the wavelength of a plane wave propagating in an infinite medium of the same material as the waveguide. Two important facts may be noted about this expression. First, at frequencies well above the cutoff frequency, $\lambda_g \approx \lambda$. Second, as $f \to f_c$, $\lambda \to \infty$, further illustrating that the mode does not propagate. This is another reason that the operating frequency is always chosen above the cutoff frequency. This concept is graphically depicted in Fig. 21.4, a β/k diagram for a standard WR-90 waveguide. At the cutoff frequency, the phase constant goes to zero, indicating that the wave does not propagate. At high frequencies, β approaches the phase constant in an infinite region of the same medium. Therefore, β/k approaches one.

The wave impedance of the waveguide is given by the ratio of the magnitudes of the transverse electric and magnetic field components, which will be constant across the cross section of the waveguide. For a given mode, the wave impedance for the TE and TM modes are given as:

$$Z_{TE} = \frac{E_T}{H_T} = \frac{j\omega\mu}{\mu} \tag{21.19}$$

$$Z_{TM} = \frac{E_T}{H_T} = \frac{\gamma}{j\omega\varepsilon} \tag{21.20}$$

E_T and H_T represent the transverse electric and magnetic fields. Note that at frequencies well above cutoff, the wave impedance for both the TE and TM modes approaches $\sqrt{\mu/\varepsilon}$, the characteristic impedance of

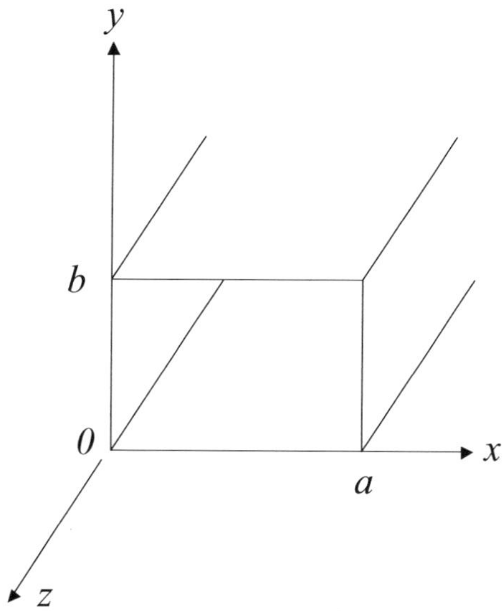

FIGURE 21.5 Geometry of a rectangular waveguide.

a plane wave propagating in an infinite medium of the same material as the waveguide. Further, as $f \to f_c$, then $Z_{TE} \to \infty$ and $Z_{TM} \to 0$, again demonstrating the necessity of choosing an operating point well above cutoff.

A variety of geometries are used for waveguides, the most common being the rectangular waveguide, which is used in the microwave and well into the millimeter-wave frequency regime. Shown in Fig. 21.5, it is a rectangular metallic guide of width a and height b. Rectangular waveguide propagate both TE and TM modes. For conciseness, the field components of the TE_{mn} and TM_{mn} modes are presented in Table 21.2. From the basic form of the equations, we see that the effect of the rectangular cross section is a standing wave dependence determined by the dimensions of the cross section, a and b. Further, h (and therefore the propagation constant γ) are determined by a and b. The dimensions of the waveguide are chosen so that only a single mode propagates at the desired frequency, with all other modes cut off. By convention, $a > b$ and a ratio of $a/b = 2.1$ is typical for commercial waveguide types.

The dominant mode in rectangular waveguide is the TE_{10} mode, which has a cutoff frequency of:

$$f_{c_{10}} = \frac{1}{2a\sqrt{\mu\varepsilon}} = \frac{c}{2a} \tag{21.21}$$

The concept of cutoff frequency is further illustrated in Fig. 21.4, a β/k diagram for a lossless WR-90 waveguide (note that in the lossless case, the propagation constant will be equal to $j\beta$). It is apparent that higher order modes may propagate as the operating frequency increases. At the cutoff frequency, β is zero because the guided wavelength is infinity. At high frequencies, the ratio β/k approaches one.

A number of variations of the rectangular waveguide are available, including single and double-ridged waveguides, which are desirable because of increased bandwidth. However, closed solutions for the fields in these structures do not exist and numerical techniques must be used to solve for the field distributions, as well as essential design information such as guided wavelength and characteristic impedance. Additionally, losses are typically higher than standard waveguides.

The circular waveguide is also used in some applications, although not nearly as often as rectangular geometry guides. Closed form solutions for the fields in a circular geometry, perfectly conducting

TABLE 21.2 Field Components for Rectangular Waveguide

	TE	TM
E_z	0	$E_0 \sin\left(\frac{m\pi x}{a}\right)\sin\left(\frac{n\pi y}{b}\right)e^{-\gamma_{mn}z}$
H_z	$H_0 \cos\left(\frac{m\pi x}{a}\right)\cos\left(\frac{n\pi y}{b}\right)e^{-\gamma_{mn}z}$	0
E_x	$H_0 \frac{j\omega\mu n\pi}{h_{mn}^2 b}\cos\left(\frac{m\pi x}{a}\right)\sin\left(\frac{n\pi y}{b}\right)e^{-\gamma_{mn}z}$	$-E_0 \frac{\gamma_{mn} m\pi}{h_{mn}^2 a}\cos\left(\frac{m\pi x}{a}\right)\sin\left(\frac{n\pi y}{b}\right)e^{-\gamma_{mn}z}$
H_x	$H_0 \frac{\gamma_{mn} m\pi}{h_{mn}^2 a}\sin\left(\frac{m\pi x}{a}\right)\cos\left(\frac{n\pi y}{b}\right)e^{-\gamma_{mn}z}$	$H_0 \frac{j\omega\varepsilon n\pi}{h_{mn}^2 b}\sin\left(\frac{m\pi x}{a}\right)\cos\left(\frac{n\pi y}{b}\right)e^{-\gamma_{mn}z}$
E_y	$-H_0 \frac{j\omega\mu m\pi}{h_{mn}^2 a}\sin\left(\frac{m\pi x}{a}\right)\cos\left(\frac{n\pi y}{b}\right)e^{-\gamma_{mn}z}$	$-E_0 \frac{\gamma_{mn} n\pi}{h_{mn}^2 b}\sin\left(\frac{m\pi x}{a}\right)\cos\left(\frac{n\pi y}{b}\right)e^{-\gamma_{mn}z}$
H_y	$H_0 \frac{\gamma_{mn} n\pi}{h_{mn}^2 b}\cos\left(\frac{m\pi x}{a}\right)\sin\left(\frac{n\pi y}{b}\right)e^{-\gamma_{mn}z}$	$-E_0 \frac{j\omega\varepsilon m\pi}{h_{mn}^2 a}\cos\left(\frac{m\pi x}{a}\right)\sin\left(\frac{n\pi y}{b}\right)e^{-\gamma_{mn}z}$
h_{mn}	$\sqrt{\left(\frac{m\pi x}{a}\right)^2+\left(\frac{n\pi y}{b}\right)^2} = 2\pi f_c\sqrt{\mu\varepsilon}$	$\sqrt{\left(\frac{m\pi x}{a}\right)^2+\left(\frac{n\pi y}{b}\right)^2} = 2\pi f_c\sqrt{\mu\varepsilon}$

waveguide with an inside diameter of $2a$ are given in Table 21.3. Note that these equations use a standard cylindrical coordinate system with ρ the radial distance from the z-axis, and ϕ is the angular distance measured from the y-axis. The axis of the waveguide is aligned along the z-axis. For both the TE_{mn} and TM_{mn} modes, any integer value of $n \geq 0$ is allowed, and $J_n(x)$ and $J'_n(x)$ are Bessel functions of order n and its first derivative. As with the rectangular waveguide, only certain values of h are allowed. For the TE_{mn} modes, the allowed values of the modal eigenvalues must satisfy the roots of $J'_n(h_{mn}a) = 0$, where m signifies the root number and may range from one to infinity with $m = 1$ the smallest root. Similarly,

TABLE 21.3 Field Components for Circular Waveguide

	TE	TM
E_z	0	$E_0 J_n(h_{nm}\rho)\cos(n\phi)e^{-\gamma_{nm}z}$
H_z	$H_0 J_n(h_{nm}\rho)\cos(n\phi)e^{-\gamma_{nm}z}$	0
E_ρ	$H_0 \frac{j\omega\mu n}{h_{nm}^2 \rho} J_n(h_{nm}\rho)\sin(n\phi)e^{-\gamma_{nm}z}$	$-E_0 \frac{\gamma_{nm}}{h_{nm}} J'_n(h_{nm}\rho)\cos(n\phi)e^{-\gamma_{nm}z}$
H_ρ	$-H_0 \frac{\gamma_{nm}}{h_{nm}} J'_n(h_{nm}\rho)\cos(n\phi)e^{-\gamma_{nm}z}$	$-E_0 \frac{j\omega\varepsilon n}{h_{nm}^2 \rho} J'_n(h_{nm}\rho)\sin(n\phi)e^{-\gamma_{nm}z}$
E_φ	$-H_0 \frac{j\omega\mu}{h_{nm}} J'_n(h_{nm}\rho)\cos(n\phi)e^{-\gamma_{nm}z}$	$E_0 \frac{\gamma_{nm}}{h_{nm}^2 \rho} J'_n(h_{nm}\rho)\sin(n\phi)e^{-\gamma_{nm}z}$
H_φ	$H_0 \frac{\gamma_{nm}}{h_{nm}^2} J_n(h_{nm}\rho)\sin(n\phi)e^{-\gamma_{nm}z}$	$-E_0 \frac{j\omega\varepsilon}{h_{nm}} J'_n(h_{nm}\rho)\cos(n\phi)e^{-\gamma_{nm}z}$

TABLE 21.4 Cutoff Frequencies for Several Lower Order Waveguide Modes for Circular Waveguide

$f_c/f_{c_{10}}$	Modes
1.0	TE_{11}
1.307	TM_{01}
1.66	TE_{21}
2.083	TE_{01}, TM_{11}
2.283	TE_{31}
2.791	TE_{21}
2.89	TE_{41}
3.0	TE_{12}

Note: Frequencies have been normalized to the cutoff frequency of the TE_{10} mode.

for the TM_{nm} modes, the values of the modal eigenvalues are the solutions of $J_n\ (h_{mn}a) = 0$. The dominant mode in the circular waveguide is the TE_{11} mode, with a cutoff frequency given by:

$$f_{c_{11}} = \frac{0.293}{a\sqrt{\mu\varepsilon}} \tag{21.22}$$

The cutoff frequencies for several of the lowest order modes are given in Table 21.4, referenced to the cutoff frequency of the dominant mode.

21.3 Planar Guiding Structures

Planar guiding structures are composed of a comparatively thin dielectric substrate with metallization on one or both planes. By controlling the dimensions of the metallization, a variety of passive components, transmission lines, and matching circuits can be constructed using photolithography and photoetching. Further, active devices are readily integrated into planar guiding structures. This provides a low-cost and compact way of realizing complicated microwave and millimeter-wave circuits. Microwave integrated circuits (MICs) and monolithic microwave integrated circuits (MMICs) based on this concept are commonly available.

A variety of planar transmission lines have been demonstrated, including microstrip, coplanar waveguide (CPW), slot line, and coplanar strip line. The cross section of each of these planar transmission lines is shown in Figs. 21.6a to d. Once the dielectric substrate is chosen, characteristics of these transmission lines are controlled by the width of the conductors and/or gaps on the top planes of the geometry. Of these, the microstrip is by far the most commonly used planar transmission line. CPW is also often used, with slot lines and coplanar strip lines being the least common at microwave frequencies, for a variety of reasons that will briefly be discussed later. In this section, we will describe the basic properties of planar transmission lines. Because of its prevalence, the microstrip will be described in detail and closed form expressions for the design of the microstrip will be given.

21.3.1 Microstrip

As seen in Fig. 21.6a, the simplest form of microstrip consists of a single conductor on a grounded dielectric slab. Microstrip is the most common type of planar transmission line used in microwave and millimeter-wave circuits, with a great deal of design data freely available. A broad range of passive components may be designed with the microstrip, including filters, resonators, diplexers, distribution

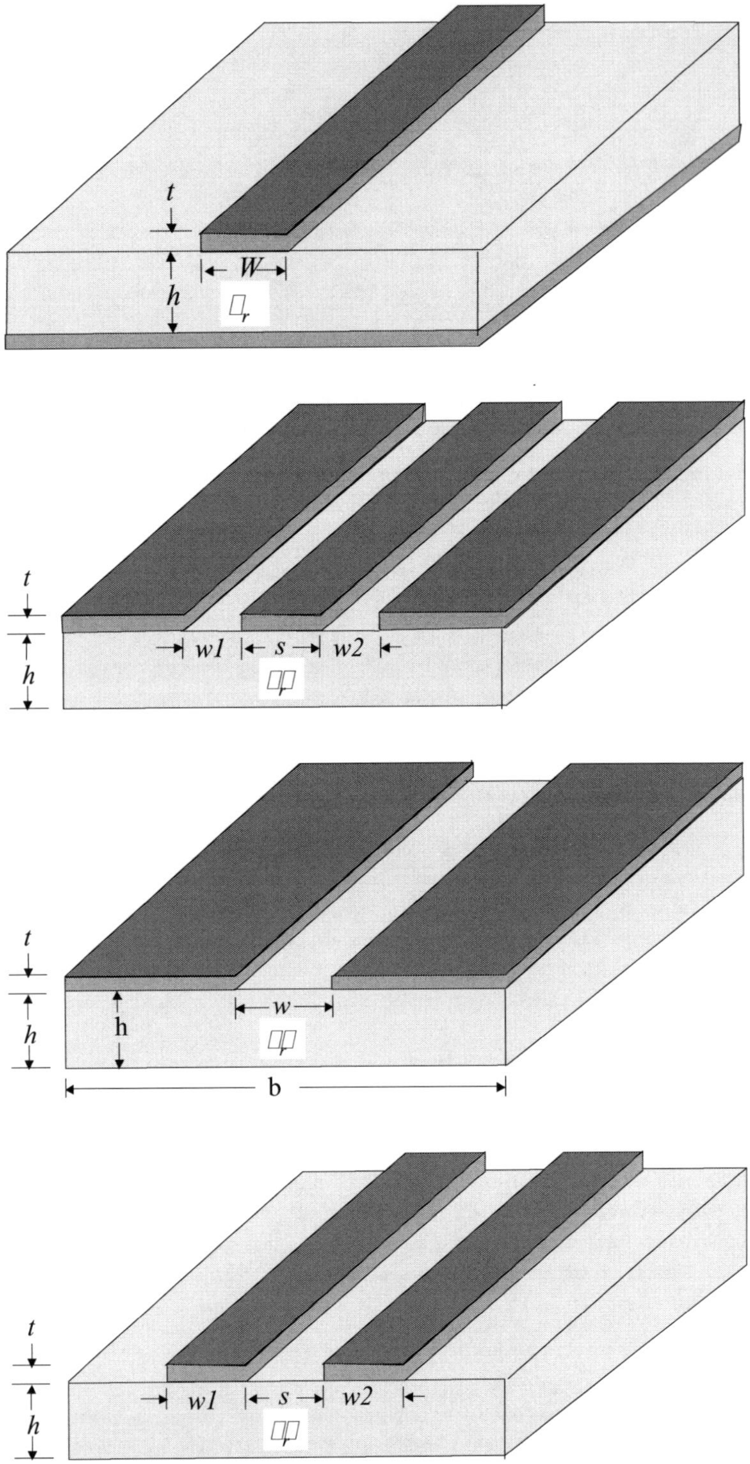

FIGURE 21.6 Cross section of four of the most popular types of planar guiding structures, including (a) microstrip, (b) coplanar waveguide, (c) slot line, and (d) coplanar strip line.

networks, and matching components. Additionally, three terminal active components can be integrated by using vias to ground. However, this may introduce considerable inductances at high frequencies.

The fundamental mode of propagation for this type of planar waveguide is often referred to as quasi-TEM, because of its close resemblance to pure TEM modes. In fact, noting that the majority of the power is confined in the region bounded by the width of the microstrip, the basic characteristics of microstrip are quite similar to the parallel-strip transmission line of Fig. 21.1a. Because of the presence of the air–dielectric interface, it is not a true TEM mode. The use of the dielectric between the ground and top conductor confines the majority of the fields in this region, but some energy may radiate from the structures. Using a high permittivity substrate and shielding the structure helps to minimize this factor. Microstrip is capable of carrying moderate power levels (a 50-Ω microstrip line on 25 mil alumina can handle several kW of power), is broadband, and enables realization of a variety of circuit topologies, both active and passive.

To design the basic microstrip line, it is necessary to be able to determine characteristic impedance and effective permittivity, preferably as a function of frequency. A wide variety of approximations have been presented in the literature, with most techniques using a quasi-static approximation for the characteristic impedance Z_0 at low frequencies, and a dispersion model for the characteristic impedance as a function of frequency $Z_0(f)$ in terms of Z_0. One fairly accurate and simple model commonly used to obtain Z_0 and the effective permittivity ε_{re}, neglecting the effect of conductor thickness is given as[1]:

$$Z_0 = \frac{\eta}{2\pi\sqrt{\varepsilon_{re}}} \ln\left(\frac{8h}{W} + 0.25\frac{W}{h}\right) \quad \text{for} \left(\frac{W}{h} \le 1\right) \tag{21.23}$$

$$Z_0 = \frac{\eta}{\sqrt{\varepsilon_{re}}} \left\{\frac{W}{h} + 1.393 + 0.667 \ln\left(\frac{W}{h} + 1.444\right)\right\}^{-1} \quad \text{for} \left(\frac{W}{h} \ge 1\right) \tag{21.24}$$

Note that η is 120π-Ω, by definition. The effective permittivity is given as:

$$\varepsilon_{re} = \frac{\varepsilon_r + 1}{2} + \frac{\varepsilon_r - 1}{2} F(W/h) \tag{21.25}$$

$$F(W/h) = (1 + 12h/W)^{-1/2} + 0.04(1 - W/h)^2 \quad \text{for} \left(\frac{W}{h} \le 1\right)$$

$$F(W/h) = (1 + 12h/W)^{-1/2} \quad \text{for} \left(\frac{W}{h} \ge 1\right)$$

With these equations, one can determine the characteristic impedance in terms of the geometry. For a desired characteristic impedance, the line width can be determined from:

$$W/h = \frac{8\exp(A)}{\exp(2A) - 2} \quad \text{for } A > 1.52 \tag{21.26}$$

$$W/h = \frac{2}{\pi}\left\{B - 1 - \ln(2B - 1) + \frac{\varepsilon_r - 1}{2\varepsilon_r}\left[\ln(B - 1) + 0.39 - \frac{0.61}{\varepsilon_r}\right]\right\} \quad \text{for } A > 1.52 \tag{21.27}$$

where

$$A = \frac{Z_0}{60}\left\{\frac{\varepsilon_r + 1}{2}\right\}^{1/2} + \frac{\varepsilon_r - 1}{\varepsilon_r + 1}\left\{0.23 + \frac{0.11}{\varepsilon_r}\right\}$$

$$B = \frac{60\pi^2}{Z_0\sqrt{\varepsilon_r}}$$

Once Z_0 and ε_{re} have been determined, effects of dispersion may also be determined using expressions from Hammerstad and Jensen[2] for $Z_0(f)$ and Kobayashi[3] for $\varepsilon_{re}(f)$. To illustrate the effects of dispersion, the characteristic impedance and effective permittivity of several microstrip lines on various substrates are plotted in Figs. 21.7a and b using the formulas from the previously mentioned papers. The substrates

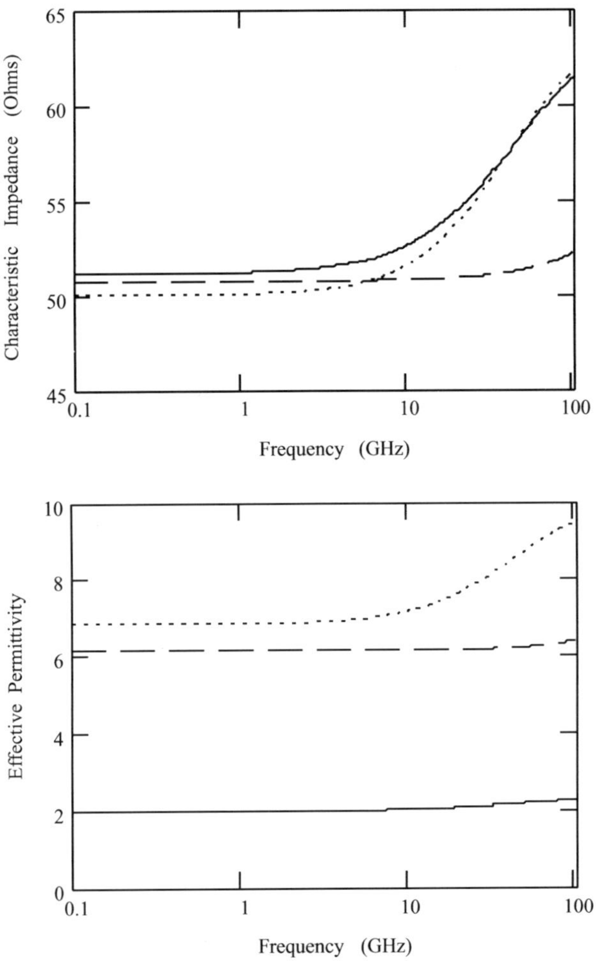

FIGURE 21.7 Dispersion characteristics of 50-Ω line on three substrates (solid line is $\varepsilon_r = 2.33$, $h = 31$ mils, $W = 90$ mils; dotted line is $\varepsilon_r = 10.2$, $h = 25$ mils, $W = 23$ mils; and the dashed line is $\varepsilon_r = 9$, h = 2.464 mils, W = 2.5 mils). Shown in (a), the impedance changes significantly at high frequencies for the thicker substrates as does the effective permittivity shown in (b).

indicated by the solid (ε_r = 2.33, h = 31 mils, W = 90 mils) and dashed (ε_r = 10.2, h = 25 mils, W = 23 mils) lines in these figures are typical for those that might be used in a hybrid circuit at microwave frequencies. We can see in Fig. 21.7a that the characteristic impedance is fairly flat until X-band, above which it may be necessary to consider the effects of dispersion for accurate design. The third line in the figure is an alumina substrate (ε_r = 9, h = 2.464 mils, W = 2.5 mils) on a thin substrate. The characteristic impedance is flat until about 70 GHz, indicating that this thin substrate is useful at a higher frequency operation. The effective permittivity as a function of frequency is shown in Fig. 21.7a. Frequency variation for this parameter is more dramatic. However, it must be remembered that guided wavelength is inversely proportional to the square root of the effective permittivity. Therefore, variation in electrical length will be less pronounced than the plot suggests.

In addition to dispersion, higher frequency operation is complicated by a number of issues, including decreased Q-factor, radiation losses, surface wave losses, and higher order mode propagation. The designer must be aware of the limitations of both the substrate on which he is designing and the characteristic impedance of the lines he is working with. In terms of the substrate, a considerable amount of energy can couple between the desired quasi-TEM mode of the microstrip and the lowest order surface wave mode of the substrate. In terms of the substrate thickness and permittivity, an approximation for determining the frequency where this coupling becomes significant is given by the following expression:[4]

$$f_T = \frac{150}{\pi h}\sqrt{\frac{2}{\varepsilon_r - 1}\arctan(\varepsilon_r)} \tag{21.28}$$

Note that f_T is in gigahertz and h is in millimeters. In addition to the quasi-TEM mode, microstrip will propagate undesired higher order TE and TM-type modes with cutoff frequency roughly determined by the cross section of the microstrip. The excitation of the first mode is approximately given by the following expression:[4]

$$f_c = \frac{300}{\sqrt{\varepsilon_r}\,(2W + 0.8h)} \tag{21.29}$$

Again, note that f_c is in gigahertz, and h and W are both in millimeters. This expression is useful in determining the lowest impedance that may be reliably used for a given substrate and operating frequency. As a rule of thumb, the maximum operating frequency should be chosen somewhat lower. A good choice for maximum frequency may be 90% of this value or lower.

A variety of techniques have also been developed to minimize or characterize the effects of discontinuities in microstrip circuits, a variety of which are shown in Figs. 21.8a and b including a microstrip bend and a T-junction. Another common effect is the fringing capacitance found at impedance steps or open-circuited microstrip stubs.

The microstrip bend allows flexibility in microstrip circuit layouts and may be at an arbitrary angle with different line widths at either end. However, by far the most common is the 90° bend with equal widths at either end, shown on the left of Fig. 21.8a. Due to the geometry of the bend, excess capacitance is formed causing a discontinuity. A variety of techniques have been used to reduce the discontinuity by eliminating a sufficient amount of capacitance, including the mitered bend shown on the right. Note that another way of reducing this effect is to use a curved microstrip line with sufficiently large radius to minimize the effect. A second type of discontinuity commonly encountered by necessity in layouts is the T-junction, shown in Fig. 21.8b, which is formed at a junction of two lines. As with the bend, excess capacitance is formed, degrading performance. The mitered T-junction below is used to reduce this problem. Again, a variety of other simple techniques have also been developed.

Fringing capacitance will be present with microstrip open-circuited stubs and at impedance steps. With the open-circuited stub, this causes the electrical length of the structure to be somewhat longer.

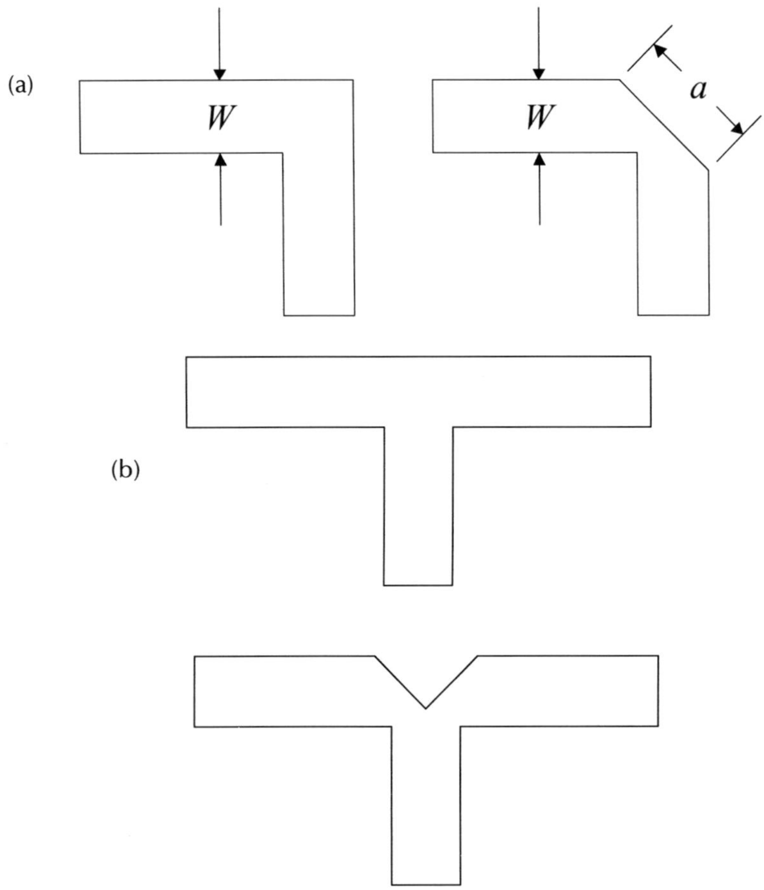

FIGURE 21.8 Two common microstrip discontinuities encountered in layout, including (a) the microstrip bend and (b) the T-junction.

For an impedance step, the lower impedance line will also appear to be electrically longer. The simplest way of compensating for this problem is by modeling the capacitance and effective length of the fringing fields. Again, a variety of simple models have been developed to perform this task, most based on quasi-static approximations. A commonly used expression for the length extension of an open end based on empirical data is given by the following expression.[5]

$$\frac{\Delta l_{oc}}{h} = 0.412\frac{\varepsilon_{re}+0.3}{\varepsilon_{re}-0.258}\left[\frac{W/h+0.264}{W/h+0.8}\right] \tag{21.30}$$

This expression is reported to yield relatively accurate results for substrates with permittivity in the range of 2 to 50, but is not as accurate for wide microstrip lines. For the impedance step, a first-order approximation for determining the excess length of the impedance step is to multiply the open-end extension, $\Delta l_{oc}/h$ by an appropriate factor to obtain a useful value (i.e., $\Delta l_{step}/h \approx \Delta l_{oc}(w_1/w_2 - 1)/h$).

Because of the prevalence of microstrip, modern microwave CAD tools typically have extensive libraries for microstrip components, including discontinuity effects.

21.3.2 Coplanar Waveguide (CPW)

Coplanar waveguide (CPW), shown in Fig. 21.6b, consists of a signal line and two ground planes on a dielectric slab with metallization on one side. For a given substrate, characteristic impedance is determined by the signal line width s, and the two gaps w_1 and w_2. This structure often demonstrates better dispersion characteristics than microstrip. Additionally, three terminal devices are easily integrated into this uniplanar transmission line that requires no vias for grounding. For this reason, parasitics are lower than microstrip making CPW a good choice for a high-frequency operation where this is a primary design concern.

The three-conductor line shown in Fig. 21.6b supports two fundamental modes, including the desired CPW mode and an undesired coupled slot line mode if the two ground planes separating the signal line are not kept at the same potential. For this reason, wires or metal strips referred to as *air bridges* are placed at discontinuities where mode conversion may occur.

Packaging may be a problem for this type of structure, because the bottom plane of the dielectric may come in close proximity with other materials, causing perturbations of the transmission line characteristics. In practice, this is remedied by using *grounded* or *conductor-backed* CPW (CB-CPW) where a ground plane is placed on the backside for electrical isolation. At high frequencies, this may present a problem with additional losses through coupling to the parallel-plate waveguide mode. These losses can be minimized using vias in the region around the transmission line to suppress this problem.

Although CPW was first proposed by Wen[6] in 1969, acceptance of CPW has been much slower than microstrip. For this reason, simple and reliable models for CPW are not as readily available as for microstrip. A compilation of some of the more useful data can be found in Reference 6.

21.3.3 Slot Line and Coplanar Strip Line

Two other types of planar transmission lines are slot line and coplanar strip line (CPS). These structures are used less often than either microstrip or CPW, but do find some applications. Both of these structures consist of a dielectric slab with metallization on one side. Slot line has a slot of width w etched into the ground plane. CPS consists of two metal strips of width w_1 and w_2 separated by a distance s on the dielectric slab. Due to their geometry, both of these structures are balanced transmission line structures, and are useful in balanced circuits such as mixers and modulators. Only limited design information is available for these types of transmission lines.

The slot line mode is non-TEM and is almost entirely TE. However, no cutoff frequency exists as with the waveguide TE modes discussed previously in this section. Microwave circuits designed solely in slot line are seldom used. However, slot line is sometimes used in conjunction with other transmission line types such as microstrip or CPW for increased versatility. Examples of these include filters, hybrids, and resonators. Additionally, slot line is sometimes used in planar antennas, such as the slot antenna or some kinds of multilayer patch antennas.

The CPS transmission line has two conductors on the top plane of the circuit, allowing series or shunt elements to be readily integrated into CPS circuits. CPS is often used in electro-optic circuits such as optic traveling wave modulators, as well as in high-speed digital circuits. Due to its balanced nature, CPS also makes an ideal feed for printed dipoles. Difficulties (or benefits, depending on the application) with CPS include high characteristic impedances.

References

1. E. Hammerstad, Equations for microstrip circuit design, *Proc. European Microwave Conf.*, 1975, 268–272.
2. E. Hammerstad and O. Jensen, Accurate models for microstrip computer-aided design, *IEEE MTT-S Int. Microwave Symp. Dig.*, 1980, 407–409.
3. M. Kobayashi, A dispersion formula satisfying recent requirements in microstrip CAD, *IEEE Trans.*, MTT-36, August 1988, 1246–1250.

4. G.D. Vendelin, Limitations on stripline Q, *Microwave J.*, 13, May 1970, 63–69.
5. R. Garg and I.J. Bahl, Microstrip discontinuities, *Int. J. Electron.*, 45, July 1978, 81–87.
6. C.P. Wen, Coplanar wave guide: A surface strip transmission line suitable for non-reciprocal gyromagnetic device applications, *IEEE Trans.*, MTT-23, 1975, 541–548.
7. K.C. Gupta, R. Garg, I. Bahl, and R. Bhartia, *Microstrip Lines and Slot lines*, Artech House, Inc., Norwood, MA, 1996.

22

Effects of Multipath Fading in Wireless Communication Systems

Wayne E. Stark
University of Michigan

The performance of a wireless communication system is heavily dependent on the channel over which the transmitted signal propagates. Typically in a wireless communication system the channel consists of multiple paths between the transmitter and receiver with different attenuation and delay. The paths have different attenuation and delays because of the different distances between transmitter and receiver along different paths. For certain transmitted signals the entire transmitted signal may experience a deep fade (large attenuation) due to destructive multipath cancellation. The multipath signals may also add constructively giving a larger amplitude. In addition to the multipath effect of the channel on the transmitted signal there are other effects on the transmitted signal due to the channel. One of these is distance related and is called the propagation loss. The larger the distance is between the transmitter and receiver the smaller the received power. Another effect is known as shadowing. Shadowing occurs due to buildings and other obstacles obstructing the line-of-sight path between the transmitter and receiver. This causes the received signal amplitude to vary as the receiver moves out from behind buildings or moves behind buildings.

In this chapter we examine models of fading channels and methods of mitigating the degradation in performance due to fading. We first discuss in detail models for the multipath fading effects of wireless channels. We then briefly discuss the models for propagation loss as a function of distance and shadowing. Next we show an example of how diversity in receiving over multiple, independent faded paths can significantly improve performance. We conclude by discussing the fundamental limits on reliable communication in the presence of fading.

0-8493-1732-0/03/$0.00+$1.50

22.1 Multipath Fading

In this section we discuss the effects of multiple paths between the transmitter and receiver. These effects depend not only on the delays and amplitudes of the paths but also on the transmitted signal. We give examples of frequency selective fading and time selective fading.

Consider a sinusoidal signal transmitted over a multipath channel. If there are two paths between the transmitter and receiver with the same delay and amplitude but opposite phase (180° phase shift), the channel will cause the received signal amplitude to be zero. This can be viewed as destructive interference. However, if there is no phase shift, the received signal amplitude will be twice as large as the signal amplitude on each of the individual paths. This is constructive interference.

In a digital communication system data are modulated onto a carrier. The data modulation causes variations in the amplitude and phase of the carrier. These variations occur at a rate proportional to the bandwidth of the modulating signal. As an example, in a cellular system the data rates are on the order of 25 kb/s and carrier frequency is about 900 MHz. So the amplitude and phase of the carrier is changing at a rate on the order of 25 kHz. Equivalently, the envelope and phase of the carrier might change significantly every 1/25 KHz = 0.04 ms = 40 μs. If this signal is transmitted over a channel with two paths with differential delay of 1 μs, the modulation part of the signal would not differ significantly. However, if this signal was received on two paths with a differential delay of 40 μs, then there would be a significant difference in the modulated part of the signal. If the data rate of the signal was increased to 250 kb/s, then the modulated signal would change significantly in a 4 μs time frame and thus the effect of multipath would be different.

Thus the type of fading depends on various parameters of the channel and the transmitted signal. Fading can be considered as a filtering operation on the transmitted signal. The filter characteristics are time varying due to the motion of the transmitter/receiver. The faster the motion is the faster the change in the filter characteristics operation. Fading channels are typically characterized in the following ways.

1. *Frequency Selective Fading:* If the transfer function of the filter has significant variations within the frequency band of the transmitted signal, the fading is called frequency selective.
2. *Time Selective Fading:* If the fading changes relatively quickly (compared to the duration of a data bit), the fading is said to be time selective.

If the channel is both time and frequency selective, it is said to be doubly selective.

To illustrate these types of fading we consider some special cases. Consider a simple model for fading where there are a finite number, k, of paths from the transmitter to the receiver. The transmitted signal is denoted by $s(t)$. The signal can be represented as a baseband signal modulated onto a carrier as

$$s(t) = \mathrm{Re}\left[s_0(t)\exp\{j2\pi f_c t\}\right]$$

where f_c is the carrier frequency and $s_0(t)$ is the baseband signal or the envelope of the signal $s(t)$. The paths between the transmitter and receiver have delays τ_k and amplitudes α_k. The received signal can thus be expressed as

$$r(t) = \mathrm{Re}\left[\sum_k \alpha_k s_0(t-\tau_k)\exp\{j2\pi f_c(t-\tau_k) + j\phi_k\}\right.$$

where ϕ_k is a phase term added due the kth path that might be due to a reflection off an object. The baseband received signal is given by

$$r_0(t) = \sum_k \alpha_k s_0(t-\tau_k)\exp\{j\phi_k - j2\pi f_c \tau_k\}$$

To understand the effects of multipath we will consider a couple different examples.

22.1.1 Frequency/Time Nonselective (Flat) Fading

First, we consider a frequency and time nonselective fading model. In this case the multipath components are assumed to have independent phases. If we let W denote the bandwidth of the transmitted signal, the envelope of the signal does not change significantly in time smaller than $1/W$. Thus if the maximum delay satisfies $\tau_{max} \ll 1/W$, that is

$$\frac{1}{f_c} \ll \tau_k \ll T = W^{-1}$$

then $s_0(t - \tau_k) \approx s_0(t)$. In this case

$$r_0(t) = s_0(t)\left(\sum_k \alpha_k \exp\{j\theta_k\}\right)$$

$$= Xs_0(t)$$

where $\theta_k = \phi_k - 2\pi f_c \tau_k$. The factor $X = \Sigma_k \alpha_k \exp\{j\theta_k\}$ by which the signal is attenuated/phase shifted is usually modeled by a complex Gaussian distributed random variable. The magnitude of X is a Rayleigh distributed random variable. The phase of X is uniformly distributed. The fading occurs because the random phases sometimes add destructively and sometimes add constructively. Thus for narrow enough signal bandwidths ($\tau_k \ll W^{-1}$) the multipath results in an amplitude attenuation by a Rayleigh distributed random variable. It is important to note that the transmitted signal in this example has not been distorted. The only effect on the transmitted signal is an amplitude and phase change. This will not be true in a frequency selective channel.

Usually the path lengths change with time due to motion of the transmitter or receiver. Here we have assumed that the motion is slow enough relative to the symbol duration so that $\alpha_k(t)$ and $\phi_k(t)$ are constants. In this model the transmitted signal is simply attenuated by a slowly varying random variable. This is called a flat fading model, or frequency and time nonselective fading.

22.1.2 Frequency Selective/Time Nonselective Fading

Now consider the case where the bandwidth of the modulating signal $s_0(t)$ is W and the delays satisfy

$$\tau_k \gg T = W^{-1}$$

In this case we say the channel exhibits frequency selective fading. For example, consider a discrete multipath model. That is,

$$r_0(t) = \alpha_1 e^{j\theta_1} s_0(t - \tau_1) + \cdots \alpha_M s_0(t - \tau_M) e^{j\theta_M}$$

The impulse response of this channel is

$$h(t) = \sum_{k=1}^{M} \alpha_k e^{j\theta_k} \delta(t - \tau_k)$$

and the transfer function is

$$H(f)=\sum_{k=1}^{M}\alpha_k \exp\{j\theta_k - j2\pi f\tau_k\}$$

More specifically, assume that $M = 2$ and the receiver is synchronized to the first path (so that we assume $\tau_1 = \phi_1 = \theta_1 = 0$). Then

$$H(f)=1+\alpha_2 \exp\{j\theta_k - j2\pi f\tau_k\}$$

At frequencies where $2\pi f\tau_2 = \theta_2 + 2n\pi$ or $f = (\theta_2 + 2n\pi)/2\pi\tau_2$ the transfer function will be $H(f) = 1 + \alpha_2$. If $\alpha_2 > 0$, the amplitude of the received signal will be larger because of the second path. This is called constructive interference. At frequencies where $2\pi f\tau_2 = \theta_2 + (2n + 1)\pi$ or $f = (\theta_2 + (2n + 1)\pi)/2\pi\tau_2$, the transfer will be $H(f) = 1 - \alpha_2$. Again, for $\alpha_2 > 0$ the amplitude of the received signal will be smaller due to the second path. This is called destructive interference. The frequency range between successive nulls (destructive interference) is $1/\tau$. Thus if $\tau \gg \frac{1}{W}, \frac{1}{\tau} \ll W$ there will be multiple nulls in the spectrum of the received signal. In Fig. 22.1 we show the transfer function of a multipath channel with two equal strength paths with differential delay of 1 μs. In Fig. 22.2 we show the transfer function of a channel with eight equal strength paths with delays from 0 to 7 μs. The frequency selectivity of the channel is seen in the fact that the transfer function varies as a function of frequency. Narrowband systems have the potential for the whole signal band to experience a deep fade, while in wideband systems the transfer function of the channel varies within the band of the transmitted signal and thus the channel causes distortion of the transmitted signal.

22.1.3 Time Selectivity

The dual concept to frequency selectivity is time selectivity. In this case the path strength is changing as a function of time (e.g., due to vehicle motion) and the envelope of the received signal (as the vehicle moves) undergoes time-dependent fading. A model for this would be that of a time-varying impulse response (without frequency selectivity):

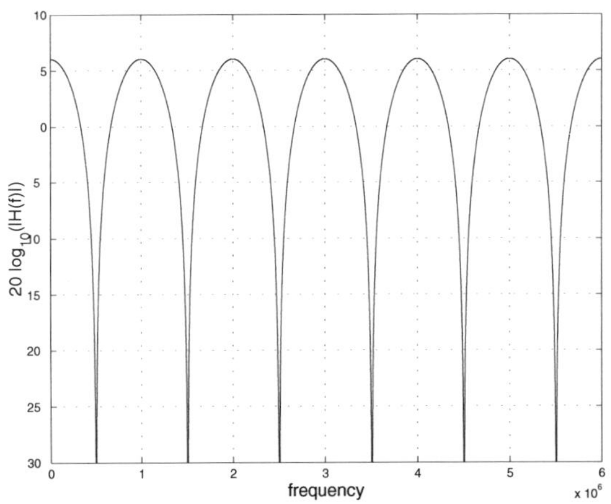

FIGURE 22.1 Transfer function of multipath channel with eight equal strength paths.

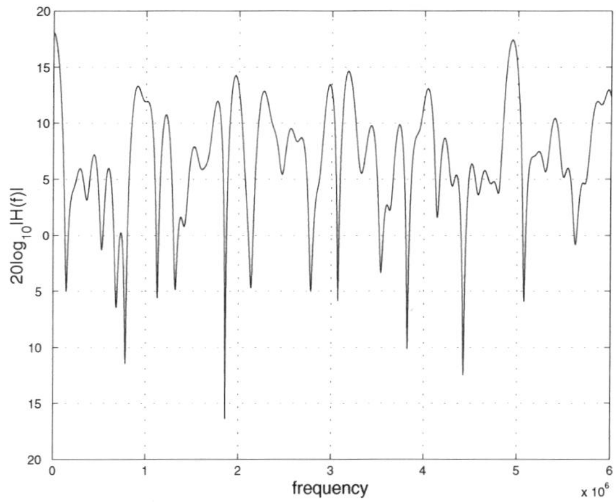

FIGURE 22.2 Transfer function of multipath channel with two equal strength paths and relative delay of 1 μs.

$$h(t;t-\beta)=\alpha_k(t)e^{j\theta(t)}\,\delta(t-\beta-\tau(t))$$

where $\tau(t)$ is the time-varying delay between the transmitter and receiver. The output $r_0(t)$ of the channel is related to the input $s_0(t)$ via

$$\begin{aligned}r_0(t)&=\int_{-\infty}^{\infty}h(t;t-\beta)s_0(\beta)\,d\beta\\&=\int_{-\infty}^{\infty}\alpha_k(t)e^{j\theta(t)}\,\delta(t-\beta-\tau(t))s_0(\beta)\,d\beta\\&=\alpha_k(t)e^{j\theta(t)}s_0(t-\tau(t))\end{aligned}$$

Because the impulse response is time varying, the fading at different time instances is correlated if the time instances are very close and uncorrelated if they are very far apart. Consider the simple case of a sinusoidal at frequency f_c as the signal transmitted. In this case, the baseband component of the transmitted signal $s_0(t)$ is a constant DC term. However, the output of the channel is given by

$$r_0(t)=\alpha_k(t)e^{j\theta(t)}s_0$$

The frequency content of the baseband representation of the received signal is no longer just a DC component, but has components at other frequencies due to the time-varying nature of α and θ. If we consider just a single direct path between the transmitter and receiver and assume that the receiver is moving away from the transmitter, then because of the motion there will be a Doppler shift in the received spectrum. That is, the received frequency will be shifted down in frequency. Similarly if the receiver is moving toward the transmitter, there will be a shift up in the frequency of the received signal. Because there can be paths between the transmitter and receiver that are direct and paths that are reflected, some paths will be shifted up in frequency and some paths will be shifted down in frequency. The overall received signal will be spread out in the frequency domain due to these different frequency shifts on different paths. The spread in the spectrum of the transmitted signal is known as the Doppler spread. If the data duration is much shorter than the time variation of the fading process, the fading can be considered a constant or a slowly changing random process. In Fig. 22.3 we plot the fading amplitude for a single path as a function of time for a vehicle traveling at 10 mi/h. In Fig. 22.4 a similar plot is done for a vehicle at 30 mi/h. It is clear that the faster a vehicle is moving the more quickly the fading amplitude varies. Fading amplitude variations with time can be compensated for by power control at low vehicle velocities. At high velocities the changes in amplitude can be averaged out by proper use of error control coding.

22.2 General Model

In this section we describe a general mode for fading channels and discuss the relevant parameters that characterize a fading channel. The most widely used general model for fading channels is the wide-sense stationary, uncorrelated scattering (WSSUS) fading model. In this model the received signal is modeled as a time-varying filter operation on the transmitted signal. That is

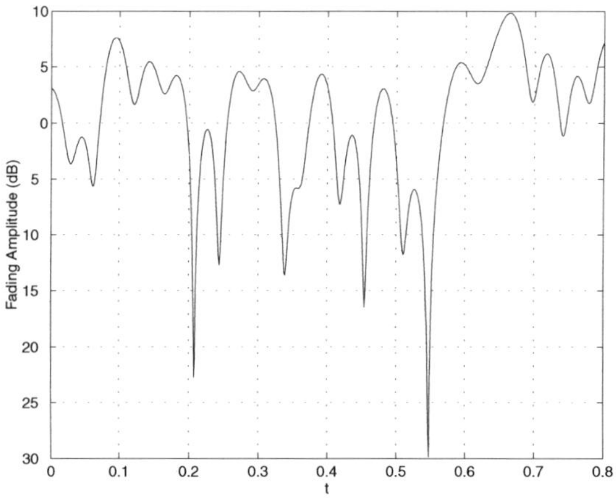

FIGURE 22.3 Received signal strength as a function of time for vehicle velocity 10 mi/h.

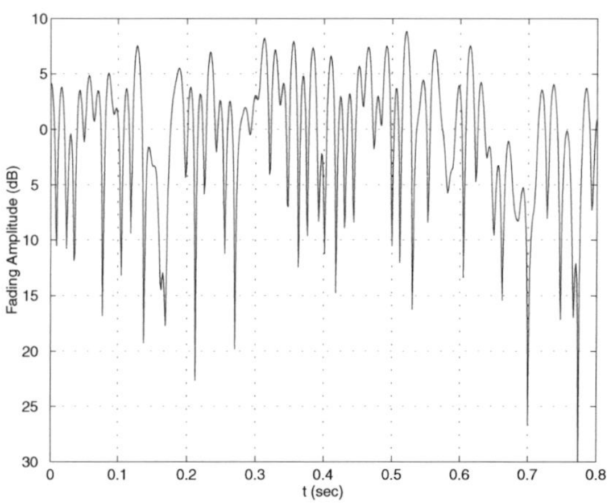

FIGURE 22.4 Received signal strength as a function of time for vehicle velocity 30 mi/h.

$$r_0(t) = \int_{-\infty}^{\infty} h(t;t-\alpha)s_0(\alpha)d\alpha$$

where $h(t; t - \tau)$ is the response due to an impulse at time τ and is modeled as a zero mean complex Gaussian random process. Note that it depends not only on the time difference between the output and the input, but also on the time directly. The first variable in h accounts for the time-varying nature of the channel while the second variable accounts for the delay between the input and output. This is the result of the assumption that there are a large number of (possibly time-varying) paths at a given delay with independent phases. If there is no direct (unfaded) path, then the impulse response will have zero mean. In this case the channel is known as a Rayleigh faded channel. If there is a (strong) direct path between the transmitter and receiver, then the filter $h(t, \tau)$ will have nonzero mean. This case is called a Rican faded channel. In the following we will assume the mean of the channel is zero.

The assumption for WSSUS is that the impulse response $h(t, \tau)$ is uncorrelated for different delays and the correlation at different times depends only on the time difference. Mathematically we write the correlation of the impulse response at different delays and times as an expectation:

$$E\left[h(t;\tau_1)h^*(t+\Delta t;\tau_2)\right] = \phi(\tau_1;\Delta t)\delta(\tau_2-\tau_1)$$

where $E[h(t;\tau_1)h^*(t+\Delta t;\tau_2)]$ denotes the expected (average) value of the impulse response at two different delays and times. The function $\phi(\tau;\Delta t)$ is the intensity delay profile and $\delta(\tau)$ is the usual Dirac delta function.

For a wide-sense stationary uncorrelated scattering (WSSUS) model the correlation between the responses at two different times depends only on the difference between times. This is indicated by the Dirac delta function. Also, the response at two different delays is uncorrelated. The amount of power received at a given delay τ is $\gamma(\tau; 0)$. This is called the intensity delay profile or the delay power spectrum. The mean excess delay μ_m is defined to be the average excess delay above the delay of the first path

$$\mu = \frac{\int_{\tau_{\min}}^{\tau_{\max}} \tau\phi(\tau;0)d\tau}{\int_{\tau_{\min}}^{\tau_{\max}} \phi(\tau;0)d\tau} - \tau_{\min}$$

The rms delay spread is defined as

$$s = \left[\frac{\int_{\tau_{\min}}^{\tau_{\max}} (\tau-\mu-\tau_{\min})^2\phi(\tau;0)d\tau}{\int_{\tau_{\min}}^{\tau_{\max}} \phi(\tau;0)d\tau}\right]^{1/2}$$

The largest value $\tau_{\max}$ of τ such that $\phi(\tau; 0)$ is nonzero is called the multipath spread of the channel. The importance of the rms delay spread is that it is a good indicator of the performance of a communication system with frequency selective fading. The larger the rms delay spread the more inter-symbol interference. In the general model the delays cause distortion in the received signal.

Now consider the frequency domain representation of the channel response. The time-varying function of the channel $H(f; t)$ is given by the Fourier transform of the impulse response with respect to the delay variable. That is,

$$H(f;t) = \int_{-\infty}^{\infty} h(t;\tau) e^{-j2\pi f \tau} d\tau$$

Since $h(t; \tau)$ is assumed to be a complex Gaussian random variable, $H(f; t)$ is also a complex Gaussian random process. The correlation $\Phi(f_1, f_2; \Delta t)$ between the transfer function at two different frequencies and two different times as defined as

$$\Phi(f_1, f_2; \Delta t) = E\left[H(f_1;t) H^*(f_2; t+\Delta t)\right]$$

$$= \int_{-\infty}^{\infty} \phi(\tau; \Delta t) e^{-j2\pi(f_2 - f_1)\tau} d\tau$$

Thus the correlation between two frequencies for the WSSUS model (and at two times) depends only on the frequency difference. If we let $\Delta t = 0$, then we obtain

$$\Phi(\Delta f; 0) = \int_{-\infty}^{\infty} \phi(\tau; 0) e^{-j2\pi(\Delta f)\tau} d\tau$$

As the frequency separation becomes larger the correlation in the response between those two frequencies generally decreases. The smallest frequency separation, B_c such that the correlation of the response at two frequencies separated by B_c is zero, is called the coherence bandwidth of the channel. It is related to the delay spread by

$$B_c \approx \frac{1}{\tau_{\max}}$$

The rms delay spread and coherence bandwidth are important measures for narrowband channels. The performance of an equalizer for narrowband channels often does not depend on the exact delay power profile, but simply on the rms delay spread.

Now consider the time-varying nature of the channel. In particular, consider $\Phi(\Delta f; \Delta t)$, which is the correlation between the responses of the channel at two frequencies separated by Δf and at times separated by Δt. For $\Delta f = 0$, $\Phi(0; \Delta t)$ measures the correlation between two responses (at the same frequency) but separated in time by Δt. The Fourier transform gives the Doppler power spectral density

$$S(\lambda) = \int_{-\infty}^{\infty} \phi(0; \gamma) e^{-j2\pi\lambda\gamma} d\gamma$$

The Doppler power spectral density gives the distribution of received power as a function of frequency shift. Since there are many paths coming from different directions and the receiver is moving, these paths will experience different frequency shifts. Consider a situation where a vehicle is moving toward a base station with velocity v.

22.2.1 Example

If we assume that there are many multipath components that arrive with an angle uniformly distributed over $[0, 2\pi]$, then the Doppler spectral density is given by

$$S(\lambda) = \frac{1}{2\pi f_m}\left[1 - \left(\lambda / f_m\right)^2\right]^{-1/2}, \quad 0 \le |\lambda| \le f_m$$

where $f_m = \nu f_c/c$, f_c is the center frequency and c is the speed of light (3×10^8 *m/s*). For example, a vehicle moving at 100 m/s with 1 GHz center frequency has a maximum Doppler shift of 33.3 Hz. A vehicle moving at 30 m/s would have a maximum Doppler shift of 10 Hz. Thus most of the power is either at the carrier frequency plus 10 Hz or at the carrier frequency minus 10 Hz. The corresponding autocorrelation function is the inverse Fourier transform and is given by

$$\begin{aligned}\Phi(0,\gamma) &= \int_{-\infty}^{\infty} S(\lambda)e^{j2\pi\lambda\gamma}d\lambda \\ &= J_0\left(2\pi f_m \gamma\right)\end{aligned}$$

The channel correlation and Doppler spread are illustrated in Figs. 22.5 and 22.6 for vehicle velocities of 10 km/h and 100 km/h. From these figures it is clear that a lower vehicle velocity implies a small spread in the spectrum of the received signal and a larger correlation between the fading at different times. It is often useful for the receiver in a digital communication system to estimate the fading level. The faster the fading level changes the harder it is to estimate. The product of maximum Doppler spread f_m times the data symbol duration T is a useful tool for determining the difficulty in estimating the channel response. For $f_m T$ products much smaller than 1, the channel is easy to estimate while for $f_m T$ much larger than 1, the channel is hard to estimate. Channel estimation can improve the performance of coded systems as shown in the last section of this chapter. The availability of channel information is sometimes called "side information."

If the channel is not time varying (i.e., time invariant), then the responses at two different times are perfectly correlated so that $\Phi(0; \Delta t) = 1$. This implies that $S(\lambda) = \delta(f)$.

The largest value of λ for which $S(\lambda)$ is nonzero is called the Doppler spread of the channel. It is related to the coherence time T_c, the largest time difference for which the responses are correlated by

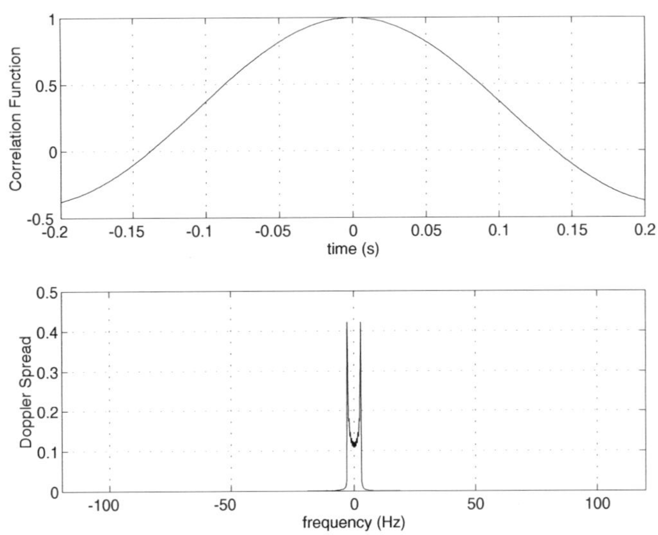

FIGURE 22.5 Channel correlation function and Doppler spread for $f_c = 1$ GHz, $\nu = 100$ km/h.

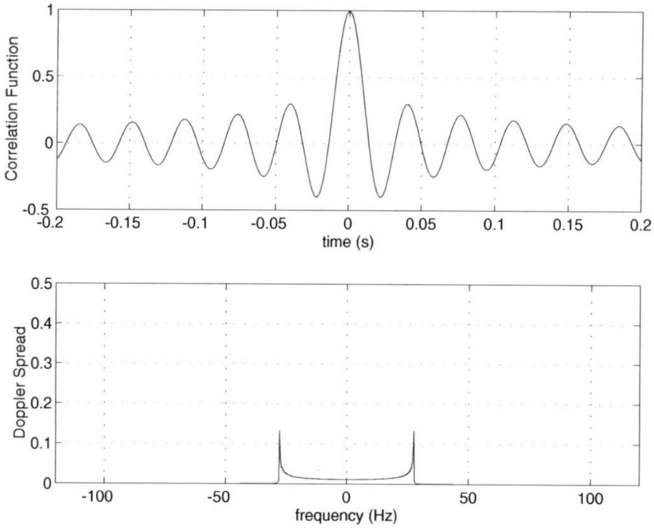

FIGURE 22.6 Channel correlation function and Doppler spread for f_c = 1 GHz, ν = 10 km/h.

$$B_d = \frac{1}{T_c}$$

22.3 GSM Model

The GSM (Global System for Mobile Communications) model was developed in order to compare different coding and modulation techniques. The GSM model is a special case of the general WSSUS model described in the previous section. The model consists of N_p paths, each time varying with different power levels. In Fig. 22.7 one example of the delay power profile for a GSM model of an urban environment is shown. In the model each path's time variation is modeled according to a Doppler spread for a

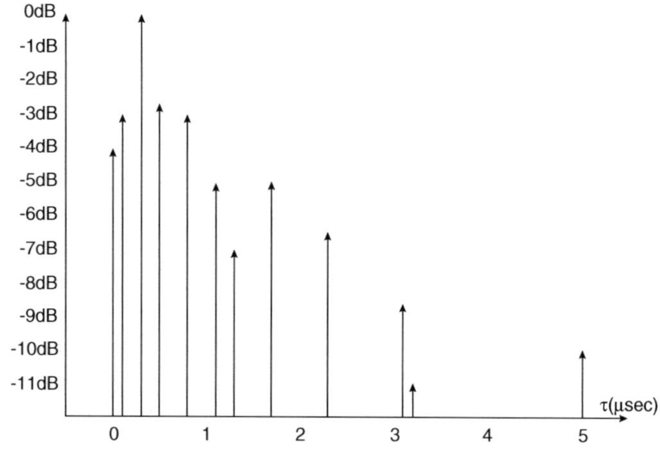

FIGURE 22.7 Power delay profile for the GSM model for typical urban channel.

TABLE 22.1 Parameters for Power Delay Profile GSM Model of Urban Area

Path	Delay (μs)	Average Power (dB)
1	0.0	–4.0
2	0.1	–3.0
3	0.3	0.0
4	0.5	–2.6
5	0.8	–3.0
6	1.1	–5.0
7	1.3	–7.0
8	1.7	–5.0
9	2.3	–6.5
10	3.1	–8.6
11	3.2	–11.0
12	5.0	–10.0

uniform angle of arrival spread for the multipath. Thus the vehicle velocity determines the time selectivity for each path. The power delay profile shown below determines the frequency selectivity of the channel.

In Table 22.1 the parameters for the GSM model are given. The usefulness of this model is that it gives communication engineers a common channel to compare the performance of different designs.

22.4 Propagation Loss

The fading discussed above is referred to as short-term fading as opposed to long-term fading. Long-term fading refers to shadowing of the receiver from the transmitter due to terrain and buildings. The time scale for long-term fading is much longer (on the order of seconds or minutes) than the time scale for short-term fading. It is generally modeled as lognormal. That is, the received power (in dB) has a normal (or Gaussian) distribution.

In this section we discuss the received power as a function of distance from the receiver. Suppose we have a transmitter and receiver separated by a distance d. The transmitter and receiver have antennas with gain G_t and G_r, respectively. If the transmitted power is P_l, the received power is

$$P_r = P_t G_r G_t \left(\frac{\lambda}{4\pi d} \right)^2$$

where $\lambda = c/f$ is the wavelength of the signal. The above equation holds in free space without any reflections or multipath of any sort.

Now consider the case where there is an additional path due to a single reflection from the ground. The multipath has a different phase from the direct path. If we assume the reflection from the ground causes a 180° phase change, then for large distances relative to the heights of the antennas the relation between the transmitted power and the received power changes to

$$P_r = P_t G_r G_t \frac{h_1^2 h_2^2}{d^4}$$

where h_1 and h_2 are the heights of the transmitting and receiving antenna. Thus the relation of received power to distance becomes an inverse fourth power law, or equivalently, the power decreases 40 dB per decade of distance. Experimental evidence for a wireless channel shows that the decrease in power with distance is 20 dB per decade near the base station, but as the receiver moves away, the rate of decrease increases. There are other models based on experimental measurements in different cities that give more complicated expressions for the path loss as a function of distance, antenna height, and carrier frequency. See [1] for further details.

22.5 Shadowing

In addition to the multipath effect on the channel and the propagation loss there is an effect due to shadowing. If a power measurement at a fixed distance from the transmitter was made, there would be local variations due to constructive and destructive interference (discussed previously). At a fixed distance from the transmitter we would also have fluctuations in the received power because of the location of the receiver relative to various obstacles (e.g., buildings). If we measured the power over many locations separated by a distance of a wavelength or more from a given point, we would see that this average would vary depending on the location of measurement. Measurements with an obstacle blocking the direct line-of-sight path would have much smaller averages than measurements without the obstacle. These fluctuations due to obstacles are called shadowing. The fluctuation in amplitude changes much slower than that due to multipath fading. Multipath fading changes as the receiver moves about a wavelength (30 cm for a carrier frequency of 1 GHz) in distance while shadowing causes fluctuations as the receiver moves about 10 m or more in distance.

The model for these fluctuations is typically that of a lognormal distributed random variable for the received power. Equivalently, the power received expressed in dB is a Gaussian distributed random variable with the mean being the value determined by the propagation loss. The variance is dependent on the type of structures where the vehicle is located and varies from about 3 to 6 dB. The fluctuations, however, are correlated. If $\nu(d)$ is a Gaussian random process modeling the shadowing process (in dB) at some location, then the model for the correlation between the shadowing at distance d_1 and the shadowing at distance d_2 is

$$E\left[\nu(d_1)\nu(d_2)\right] = \sigma^2 \exp\left\{-\left|d_1 - d_2\right|/d_0\right\}$$

where d_0 is a parameter that determines how fast the correlation decays with distance. If the velocity is known, then the correlation with time can be determined from the correlation in space. A typical value for d_0 is 10 m. Because shadowing is relatively slow, it can be compensated for by power control algorithms.

22.6 Performance with (Time and Frequency) Nonselective Fading

In this section we derive the performance of different modulation techniques with nonselective fading. We will ignore the propagation loss and shadowing effect and concentrate on the effects due only to multipath fading. First the error probability conditioned on a particular fading level is determined. Then the conditional error probability is averaged with respect to the distribution of the fading level.

22.6.1 Coherent Reception, Binary Phase-Shift Keying (BPSK)

First consider a modulator transmitting a BPSK signal received with a faded amplitude. The transmitted signal is

$$s(t) = \sqrt{2P}b(t)\cos\left(2\pi f_c t\right)$$

where $b(t)$ is a data bit signal consisting of a sequence of rectangular pulses of amplitude +1 or −1. The received signal is

$$r(t) = R\sqrt{2P}b(t)\cos(2\pi f_c t + \phi) + n(t)$$

where $n(t)$ is additive white Gaussian noise with two-side power spectral density $N_0/2$. Assuming the receiver can accurately estimate the phase, the demodulator (matched filter) output at time kT is

$$z_k = R\sqrt{E}b_{k-1} + \eta_k$$

where $E = PT$ is the transmitted energy, b_{k-1} is the data bit transmitted during the time interval $[(k-1)T, kT]$, and η_k is a Gaussian random variable with mean 0 and variance $N_0/2$. The random variable R represents the attenuation due to fading ($R = |X|$) or fading level and has probability density

$$p_R(r) = \begin{cases} 0, & r < 0 \\ \dfrac{r}{\sigma^2} e^{-r^2/2\sigma^2} & r \geq 0 \end{cases}$$

The density function determines the probability that the fading is between any two levels as

$$P\{a < R \leq b\} = \int_a^b p_R(r)\,dr$$

The error probability for a given fading level R is

$$P_e(R) = Q\left(\sqrt{\frac{2ER^2}{N_0}}\right)$$

The unconditional error probability is the average of the conditional error probability for a given fade level with respect to the density of the fading level.

$$\begin{aligned} P_e &= \int_{r=0}^{\infty} p_R(r) Q\left(\sqrt{\frac{2Er^2}{N_0}}\right) dr \\ &= \frac{1}{2} - \frac{1}{2}\sqrt{\frac{\bar{E}/N_0}{1 + \bar{E}/N_0}} \end{aligned}$$

The error probability is shown in Fig. 22.8 for the case of no fading (additive white Gaussian noise) and Rayleigh fading. For the additive white Gaussian noise channel the error probability decreases exponentially with signal-to-noise ratio, E/N_0. However, with fading the decrease in error probability is much slower. In fact, for large E/N_0 the error probability is

$$P_e \simeq \frac{1}{4E/N_0}$$

Thus for high E/N_0, the error probability decreases inverse linearly with signal-to-noise ratio.

To achieve an error probability of 10^{-5} requires a signal-to-noise ratio of 44.0 dB, whereas in additive white Gaussian noise the required signal-to-noise ratio for the same error probability is 9.6 dB. Thus fading causes a loss in signal-to-noise ratio of 34.4 dB. This loss in performance is at the same average

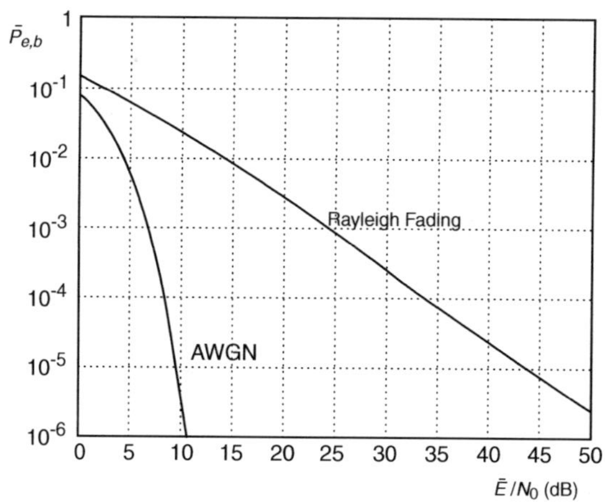

FIGURE 22.8 Bit error probability for BPSK with Rayleigh fading.

received power. The cause of this loss is the fact that the signal amplitude sometimes is very small and causes the error probability to be close to one half. Of course, sometimes the signal amplitude is large and results in a very small error probability (say 0). However, when we average the error probability, the result is going to be much larger than the error probability at the average signal-to-noise ratio because of the highly nonlinear nature of the error probability as a function of signal amplitude without fading.

While the specific error probabilities change when the modulation changes, the general nature of the error probabilities remain the same. That is, without fading the error probability decreases exponentially with signal-to-noise ratio, while with fading the error probability decreases inverse linearly with signal-to-noise ratio. This typically causes a loss in performance of between 30 and 40 dB and forces a designer to consider mitigation techniques as will be discussed subsequently.

22.6.2 BPSK with Diversity

To overcome this loss in performance (without just increasing power) a number of techniques are applied. Many of the techniques attempt to receive the same information with independent fading statistics. This is generally called diversity. The diversity could be the form of L different antennas suitably separated so that the fading on different paths from the transmitter is independent. The diversity could be in the form of transmitting the same data L times suitably separated in time so that the fading is independent.

In any case, consider a system with L-independent paths. The receiver demodulates each path coherently. Assume that the receiver also knows exactly the faded amplitude on each path. The decision statistics are then given by

$$z_l = r_l\sqrt{E}b + \eta_l, \quad l = 1, 2, \ldots, L$$

where r_l are Rayleigh η_l is Gaussian, and b represents the data bit transmitted, which is either +1 or −1. The optimal method to combine the demodulator outputs can be derived as follows. Let $p_1(z_1, \ldots, z_L|r_1, \ldots, r_L)$ be the conditional density function of $z_1, \ldots, z_L$ given the transmitted bit is +1 and the fading amplitude is $r_1, \ldots, r_L$. The unconditional density is

$$p_1(z_1, \ldots, z_L, r_1, \ldots, r_L) = p_1(z_1, \ldots, z_L|r_1, \ldots, r_L)\,p(r_1, \ldots, r_L)$$

The conditional density of z_1 given $b = 1$ and r_1 is Gaussian with mean $r_l\sqrt{E}$ and variance $N_0/2$. The joint distribution of $z_1, \ldots, z_L$ is the product of the marginal density functions. The optimal combining rule is derived from the ratio

$$\Lambda = \frac{p_1(z_1, \ldots, z_L, r_1, \ldots, r_L)}{p_{-1}(z_1, \ldots, z_L, r_1, \ldots, r_L)}$$

$$= \exp\left\{\frac{4}{N_0}\sum_{l=1}^{L} z_l r_l \sqrt{E}\right\}$$

The optimum decision rule is to compare Λ with 1 to make a decision. Thus the optimal rule is

$$\sum_{l=1}^{L} r_l z_l \underset{b=-1}{\overset{b=+1}{\gtrless}} 0$$

The error probability with diversity L can be determined using the same technique as used without diversity. The expression for error probability is

$$P_e(L) = P_e(1) - \frac{1}{2}\sum_{k=1}^{L-1}\frac{(2k)!}{k!k!}\left(1-2P_e(1)\right)\left(P_e(1)\right)^k\left(1-P_e(1)\right)^k$$

The error probability as a function of the signal-to-noise ratio is shown in Fig. 22.9. The signal-to-noise ratio in this case is defined as $E_b/N_0 = E * L/N_0$ where E is the energy transmitted per transmitting antenna or time diversity. Thus we assume that LE_b is the energy needed to have the signal received with L-independent fading amplitudes. If we had L-receiving antennas, then the performance as a function of the transmitting energy would be L times better. In any case, we plot the error probability as a function of the total received energy. In the case of diversity transmission, the energy transmitted per bit E_b is LE. For a fixed E_b, as L increases, each transmission contains less and less energy, but there are more transmissions over independent faded paths. In the limit, as L becomes large using the weak law of large numbers, it can be shown that

$$\lim_{L\to\infty} P_e(L) = Q\left(\sqrt{\frac{2E_b}{N_0}}\right)$$

For a large signal-to-noise ratio the error probability with diversity L is decreasing as $1/(E_b/N_0)^L$. While these curves show it is possible to get back to the performance with additive white Gaussian noise by using sufficient resources (diversity), it is possible to do even better with the right coding. In Fig. 22.10 we show the performance of a rate 1/2 constraint length 7 convolutional code on a Rayleigh faded channel (independent fading on each bit) where the receiver knows the fading level (side information) for each bit and can appropriately weight the metric in the decoder. Notice that the required E_b/N_0 for 10^{-5} bit error probability is about 7.5 dB, which is less than that required for uncoded BPSK without fading. The gain compared to uncoded performance is more than 36 dB.

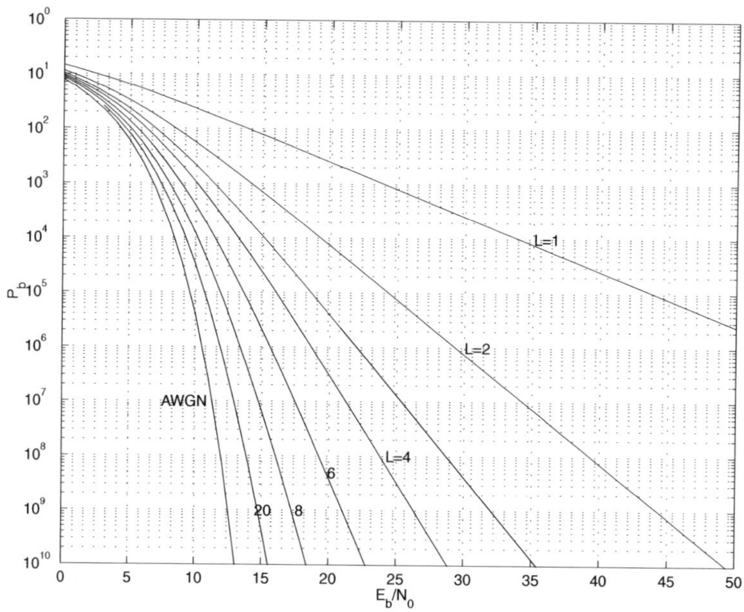

FIGURE 22.9 Error probability for BPSK (coherent demodulation) with and without Rayleigh fading.

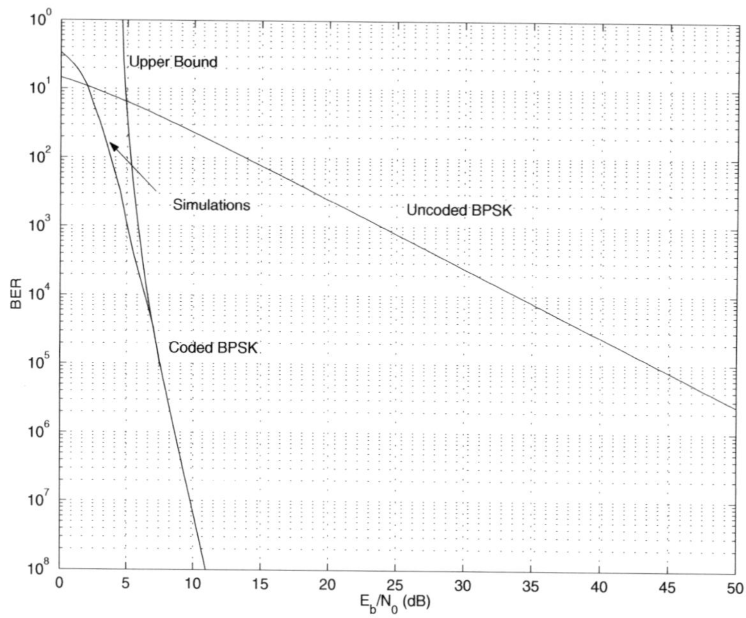

FIGURE 22.10 Error probability for BPSK (coherent demodulation) with Rayleigh fading and convolutional coding.

22.6.3 Fundamental Limits

The fundamental limits on performance can be determined for a variety of circumstances. Here we assume that the transmitter has no knowledge of the fading amplitude and assume the modulation in binary phase-shift keying. When the receiver knows exactly the amplitude (and phase) of the fading process, we say that side information is available. The maximum rate of transmission (in bits/symbol) is called the capacity of the channel C. If an error control code of rate r information bits/channel use is used, then reliable (arbitrarily small error probability) is possible provided the rate is less than the capacity. For the case of side information available this condition is

$$r < C = 1 - \int_{r=0}^{\infty}\int_{y=-\infty}^{\infty} f(r)g(y)\log_2\left(1+e^{-2y\beta}\right)dydr$$

where $f(r) = 2r\exp\{-r^2\}$, $\beta = \sqrt{2\bar{E}/N_0}$ and

$$g(y) = \frac{1}{\sqrt{2\pi}}\exp\left\{-\left(y-\sqrt{2\bar{E}r^2/N_0}\right)^2/2\right\}$$

If the receiver does not know the fading amplitude (but still does coherent demodulation), then we say no side information is available. The rate at which reliable communication is possible in this case satisfies

$$r < C = 1 - \int_{r=0}^{\infty}\int_{y=-\infty}^{\infty} p(y|1)\log_2\left(1+\frac{p(y|0)}{p(y|1)}\right)dy$$

where

$$p(y|0) = \int_0^{\infty} f(r)\frac{1}{\sqrt{2\pi N_0}}e^{-\left(y-\sqrt{\bar{E}}r\right)^2/N_0}dr$$

and

$$p(y|1) = \int_0^{\infty} f(r)\frac{1}{\sqrt{2\pi N_0}}e^{-\left(y+\sqrt{\bar{E}}r\right)^2/N_0}dr$$

If the receiver makes a hard decision about each modulated symbol and the receiver knows the fading amplitude, then the capacity is

$$C = \int_0^{\infty} f(r)\left[1+p(r)\log_2(p(r))+(1-p(r))\log_2(1-p(r))\right]dr$$

where $p(r) = Q\left(\sqrt{\frac{2\bar{E}r^2}{N_0}}\right)$. For a receiver that does not know the fading amplitude and makes hard decisions on each coded bit, the capacity is given by

$$C = 1+\bar{p}\log_2\left(\bar{p}\right)+\left(1-\bar{p}\right)\log_2\left(1-\bar{p}\right)$$

where

$$\bar{p} = \frac{1}{2} - \frac{1}{2}\sqrt{\frac{\bar{E}/N_0}{\bar{E}/N_0 + 1}}$$

Finally if the transmitter is not restricted to binary phase-shift keying but can use any type of modulation, then the capacity when the receiver knows the fading level is

$$C = \int_0^\infty f(r)\frac{1}{2}\log_2\left(1 + 2\bar{E}r^2/N_0\right)dr$$

In Fig. 22.11 we show the minimum signal-to-noise ratio $E_b/N_0 = E/N_0/C$ per information bit required for arbitrarily reliable communication as a function of the code rate ($r = C$) being used. In Fig. 22.11 the top curve (a) is the minimum signal-to-noise ratio necessary for reliable communication with hard decisions and no side information. The second curve (b) is the case of hard decisions with side information. The third curve (c) is the case of soft decisions with side information and binary modulation (BPSK). The bottom curve (d) is the case of unrestricted modulation and side information available at the receiver. There is about a 2-dB gap between hard decisions and soft decisions when side information is available. There is an extra 1-dB degradation in hard decisions if the receiver does not know the amplitude. A roughly similar degradation in performance is also true for soft decisions with and without side information. The model shown here assumes that the fading is constant over one symbol duration, but independent from one symbol to the next. However, for the case of the receiver knowing the fading level (side information) the capacity actually does not depend on the time selectivity as long as the fading is constant for at least one symbol duration. When there is no side information, the capacity gets larger when there is less selectivity. In this case the receiver can better estimate the channel. In fact, as the channel coherence time becomes large, the capacity without side information approaches the capacity with side information.

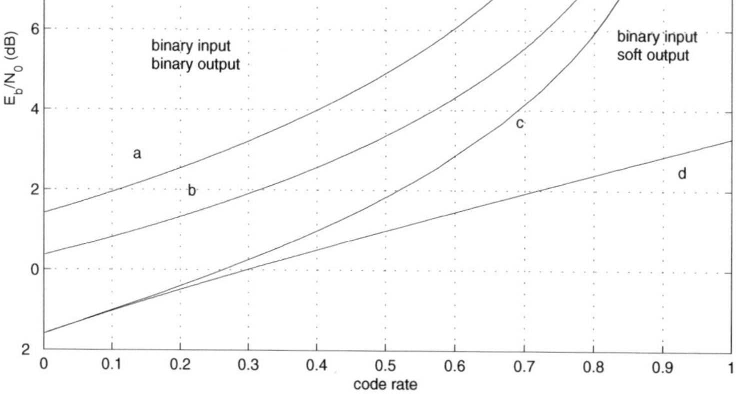

FIGURE 22.11 Capacity of Rayleigh faded channel with coherent detection.

As can be seen in Fig. 22.11 it is extremely important that some form of encoding be used with fading. The required signal-to-noise ratio for small error probabilities has decreased from on the order of 45 dB for an uncoded system to a mere 2 dB for a coded system. When a repetition code is used, the error probability can be made to decrease exponentially with signal-to-noise ratio provided that we use a large number of antennas or repeat the same symbol a large number of times, which results in a small rate of transmission (information bits/modulated symbol). However, with error control coding such as a convolutional code and independent fading we can greatly improve the performance. The minimum required signal-to-noise ratio is no different than an unfaded channel when very low rate coding is used. For rate 1/2 coding, the loss in performance is less than 2 dB compared to an unfaded channel.

In conclusion, multipath fading causes the signal amplitude to vary and the performance of typical modulation techniques to degrade by tens of dB. However, with the right amount of error control coding, the required signal-to-noise ratio can be decreased to less than 2 dB of the required signal-to-noise ratio for an additive white Gaussian channel when the code rate is 0.5.

Reference

1. Pahlavan, K. and Levesque, A. H., *Wireless Information Networks,* John Wiley & Sons, New York, 1995.

23

Electromagnetic Interference (EMI)

Alfy Riddle
Macallan Consulting

23.1 Fundamentals of EMI

Electromagnetic interference (EMI) is a potential hazard to all wireless and wired products. Most EMI concerns are due to one piece of equipment unintentionally affecting another piece of equipment, but EMI problems can arise within an instrument as well. Often the term electromagnetic compatibility (EMC) is used to denote the study of EMI effects. The following sections on generation of EMI, shielding of EMI, and probing for EMI will be helpful in both internal product EMI reduction and external product EMI compliance.

EMI compliance is regulated in the U.S. through the Federal Communications Commission (FCC). Specifically, Parts 15 and 18 of the Code of Federal Regulations (CFR) govern radiation standards and standards for industrial, scientific, and medical equipment. In Europe, Publication 22 from the Comite International Special des Perturbations Radioelectriques (CISPR) governs equipment radiation. Although the primary concern is compliance with radiation standards, conduction of unwanted signals onto power lines causes radiation from the long power lines, so conducted EMI specifications are also included in FCC Part 15 and CISPR 22 [1, 2].

Figure 23.1 shows the allowed conducted EMI. FCC and CISPR specifications do not set any limits above 30 MHz. All of the conducted measurements are to be done with a line impedance stabilization network (LISN) connected in the line. The LISN converts current-based EMI to a measurable voltage. The LISN uses series inductors of 50 μH to build up a voltage from line current interference, and 0.1 μF capacitors couple the noise voltage to 50-Ω resistors for measurement [1]. Capacitors of 1 μF also bridge the output so the inductors see an AC short. The measurements in Fig. 23.1 are reported in dBμV, which is dB with respect to 1 μV. Both FCC and CISPR measurements specify an RF bandwidth of at least 100 kHz. The CISPR limitations given in Fig. 23.1 denote that a quasi-peak (QP) detector should be used. The QP detector is more indicative of human responses to interference. CISPR specifications for an averaging detector are 10 dB below that of the QP detector. FCC specifications require a QP detector.

0-8493-1732-0/03/$0.00+$1.50

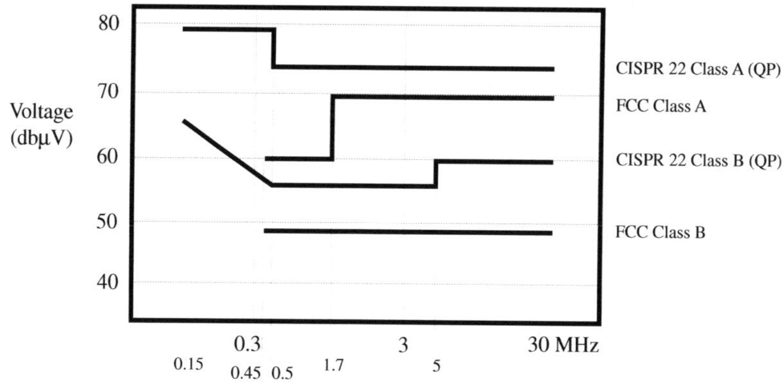

FIGURE 23.1 Conducted EMI specifications, measured with LISN.

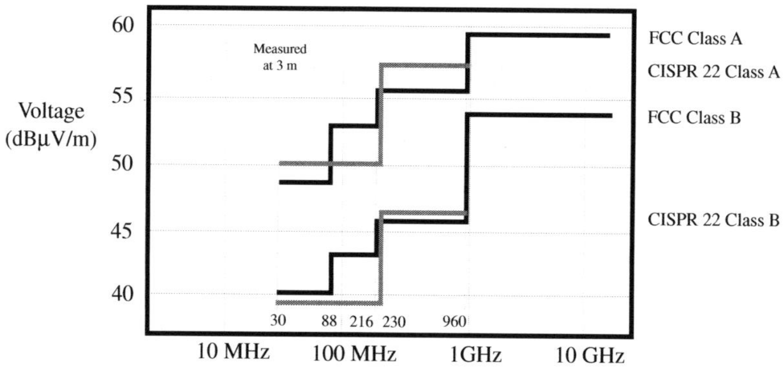

FIGURE 23.2 Radiated EMI specifications referred to 3 m.

Both FCC and CISPR limitations have two classes. Class A is basically for industrial use, while Class B is for residential use.

Both CISPR and FCC radiated EMI specifications begin at 30 MHz. Fig. 23.2 shows the radiation limits for CISPR and FCC Classes A and B [1, 2]. Because these measurements are made with an antenna, they are specified as a field strength in dB referenced to 1 μV/m. The measurement distances for radiation limits vary in the specifications, but all of the limits shown in Fig. 23.2 are referred to 3 m. Other distances can be derived by reducing the limits by 20 dB for every factor of ten increase in distance.

23.2 Generation of EMI

Almost any component can generate EMI. Oscillators, digital switching circuits, switching regulators, and fiber-optic transmitters can radiate through PCB traces, inductors, gaps in metal boxes, ground loops, and gaps in ground planes [1].

23.2.1 Switching Regulators

Because switching regulators have a fundamental frequency below 30 MHz and switch large currents at high speed, they can contribute to both conducted and radiated emissions. Very careful filtering is required so switching regulators do not contaminate ground planes with noise. The input filters on switching

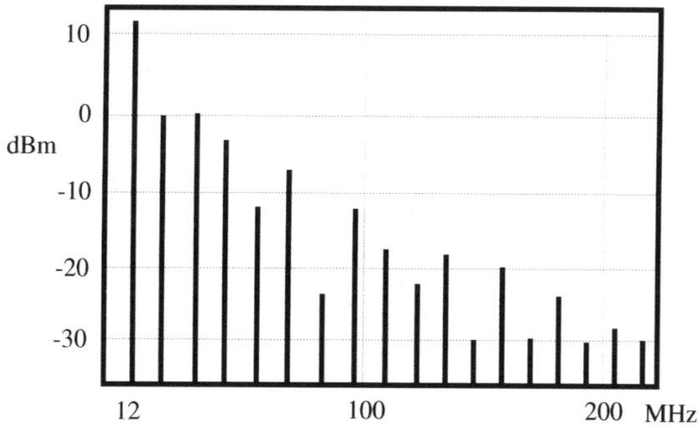

FIGURE 23.3 Harmonic clock spectra.

regulators are just as important as the output filters because the noise can travel to other power supplies and out to the line [3].

23.2.2 Digital Switching

Digital networks have several characteristics that increase EMI. Digital networks tend to have many lines switching simultaneously. The currents on the lines add in phase and increase radiation. Also, as CMOS digital circuits increase in clock frequency they require more current to drive their loads. Increasing both the current and the frequency creates more di/dt noise through the inductive connection from ICs to ground [4]. The fast switching of digital waveforms produces harmonics decades beyond the oscillator fundamental frequency. Fig. 23.3 shows the spectra from a typical 12 MHz crystal square-wave clock oscillator. Note that even though the output appears as a square wave, both even and odd harmonics are present. The harmonics of this oscillator fall off at roughly 20 dB/decade. As will be seen in the next section, most sources of coupling increase at about 20 dB/decade, which causes not only a relatively flat coupling spectrum, but also a significant EMI at 10 or even 100 times the oscillator frequency. While this 12-MHz clock oscillator is at a very low frequency, similar phenomena happens with the laser drivers for high-speed fiber-optic networks that operate with clocks of 2.5 GHz and higher.

23.2.3 Coupling

Any inductor is a potential source of coupling and radiation. Ribbon cables act like very long coupling loops and can spread signals or power supply noise all over an instrument and out to the outside world. Fundamentally, lengths of wire radiate an electric field and loops of wire radiate a magnetic field. The electric field E_{Far}, far from a short radiator, is given by Eq. (23.1) [1]. Eq. (23.1) is in volts per meter where I is the element current, l is the element length, λ is the radiation wavelength, and r is the distance from the radiating element to the measured field. Because the field strength is inversely proportional to frequency, EMI coupling tends to increase with frequency. The far magnetic field, H_{Far}, due to a current loop is given in A/m by Eq. (23.2). In Eq. (23.2) a is the radius of the loop, c is the speed of light, I is the current in the loop, and r is the distance from the loop to the measured radiation.

$$E_{Far} = 377 \text{ I } l/(2\lambda \text{r}) \tag{23.1}$$

$$H_{Far} = \omega^2 \ a^2 \ \mu \text{ I}/(1508 \text{ c } r) \tag{23.2}$$

23.2.4 Cabling

Cables form a source of radiation and susceptibility. In general it is the signal on the shield of a coaxial cable or the common-mode signal on twisted pairs that generates most of the radiation [1]. However, even high-quality coaxial cables will leak a finite amount of signal through their shields. Multiple braids, solid outer conductors, and even solder-filled braid coaxial cables are used to increase shielding. Twisted pair cables rely on twisting to cancel far field radiation from the differential mode. The effectiveness of the twisting reduces as frequency increases [1].

23.3 Shielding

Shielding is the basic tool for EMC work. Shielding keeps unwanted signals out and potential EMI sources contained. For the most part, a heavy metal box with no seams or apertures is the most effective shield. While thin aluminum enclosures with copper tape over the seams appear to enclose the RF currents, in fact they are a poor substitute for a heavy gauge cast box with an EMI gasket. It has been said that if spectrum analyzer manufacturers could make their instruments any lighter, primarily by leaving out some of the expensive casting, they would. The basic equation for shielding is given in Eq. (23.3) [5].

$$S = A + R + B \tag{23.3}$$

In Eq. (23.3), A is the shield absorption in dB, R is the shield reflection in dB, and B is a correction factor for multiple reflections within the shield [5]. Shield effectiveness depends on the nature of the field. Purely electric fields are well isolated by thin conductive layers, while purely magnetic fields are barely attenuated by such layers. Magnetic fields require thick layers of high permeability material for effective shielding at low frequencies. Plane waves contain a fairly high impedance mix of electric and magnetic fields that are both reflected and absorbed by thin metal layers provided the frequency is high enough. One of the subtle points in EMI shielding is that any slot can destroy shielding effectiveness. It is the length of a slot in comparison to a wavelength that determines how easily a wave can pass through the slot [5].

23.4 Measurement of EMI

EMI compliance must be verified. Unfortunately, EMI measurements are time consuming and tedious, are plagued by local interference sources, and often have frustrating variability. The FCC requires measurements to be verified at an Open Area Test Site (OATS) [2]. Many manufactures use a local site or a shielded transverse electric and magnetic (TEM) cell to estimate FCC compliance during product development. With care, OATS and TEM cell measurements can be correlated [6]. In any case, even careful EMI measurements can wander by several dB so most manufacturers design for a healthy margin in their products.

23.4.1 Open Area Test Site (OATS)

A sketch of an OATS site is given in Fig. 23.4a. OATS testing involves setting the antenna at specified distances from the device under test (DUT). FCC and CISPR regulations use distances of 3, 10, and 30 m depending on the verification class [1, 2]. The antenna height must also be varied to account for ground reflections. Finally, the DUT must be rotated about all axes and the antenna must be utilized to test the DUT under radiation by both horizontally and vertically polarized fields.

23.4.2 TEM Cell

TEM cells are a convenient and relatively low-cost method for making accurate and well-isolated EMI measurements [2, 7]. A sketch of one configuration of TEM cell is shown in Fig. 23.4b [2]. The TEM

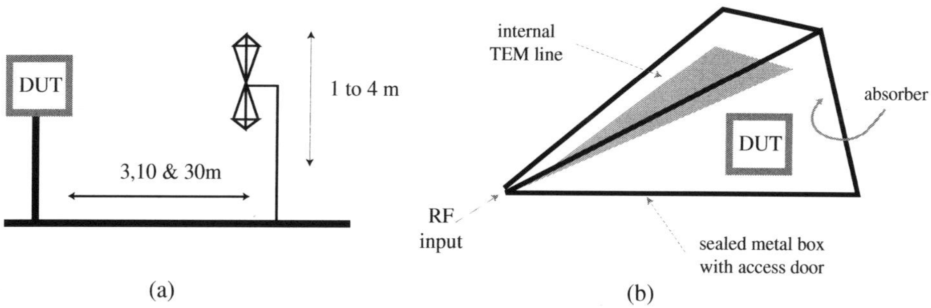

FIGURE 23.4 EMI measurement methods: (a) OATS; and (b) TEM cell.

cell uses a transmission line in a box to create a TEM field for testing devices. The box is driven from a narrow end and usually expands into an area where the DUT can be placed. The box terminates in a resistor surrounded by an RF absorber. For EMI measurements the DUT must be placed away from the box walls and rotated about each axis so that all possible radiated waves can be measured.

23.4.3 Probes

EMI probes can be a very effective way of solving EMI problems. Articles have been written on building probes and commercial probes are available that provide a flat frequency response [8]. These probes can be used to "sniff" around a device until signals with the same spectral spacing as the EMI problem can be found. For clocks and switching power supplies, a close examination of the spectral spacing will indicate the fundamental frequency, which can be traced to a component on a schematic. In the time domain the suspected EMI source can be used to trigger an oscilloscope with the probed signal as the oscilloscope input. If the trace is stable, then the suspected EMI source has been found. Electric field probes are based on Eq. (23.1) and can be as simple as a wire extending from a connector. Magnetic field probes can be as simple as a loop of wire completing the path from a connector's center pin to its flange. A rectangular loop of length l with the near side a distance a from a current source I, and having the far side a distance b from the current source will yield the voltage given in Eq. (23.4). With all small probes it is useful to have at least a 6 dB pad after the probe to establish a load and minimize reflections.

$$V = j\,\omega\,l\,\mu\,I\,\ln(b/a)/(2\,\pi) \tag{23.4}$$

23.5 Summary

EMI problems create an inexhaustible supply of work for those in the field. The EMC field has well documented requirements and a long history of measurement. Many excellent sources of information exist for those working in this area [1, 2, 5, 9].

References

1. Paul, C.R., *Introduction to Electromagnetic Compatibility*, John Wiley & Sons, New York, 1992.
2. Morgan, D., *A Handbook for EMC Testing and Measurement*, Peter Peregrinus Ltd., London, 1994.
3. Lee, F.C. and Yu, Y., Input-Filter Design for Switching Regulators, *IEEE Trans. Aerosp. Electron. Syst.*, 627–634, September 1979.
4. Dolle, M., Analysis of Simultaneous Switching Noise, *IEEE ISCAS*, 904–907, 1995.
5. Ott, H.W., *Noise Reduction Techniques in Electronic Systems*, John Wiley & Sons, New York, 1976.

6. Wilson, P., On Correlating TEM Cell and OATS Emission Measurements, *IEEE Trans. EMC*, 1–16, February 1995.
7. Konigstein, D. and Hansen, D., A New Family of TEM-Cells with Enlarged bandwidth and Optimized Working Volume, *Proc. 7th Int'l. Zurich Symp. on EMC*, 127–132, March 1987.
8. Johnson, F., Simple "Homemade" Sensors Solve Tough EMI Problems, *Electron. Design*, 109–114, November 8, 1999.
9. Kodali, V.P. and Kanda, M., *EMC/EMI Selected Readings*, IEEE Press, Piscataway, NJ, 1996.

Index

A

B

D

F

I

J

K

L

N

O

P

Q

R

S

T

U

X

Y

Z